Advances in Hydraulic Engineering

Advances in Hydraulic Engineering

Edited by
Lilly Martin

WILLFORD PRESS

www.willfordpress.com

Published by Willford Press,
118-35 Queens Blvd., Suite 400,
Forest Hills, NY 11375, USA

ISBN: 978-1-68285-806-6

Cataloging-in-Publication Data

Advances in hydraulic engineering / edited by Lilly Martin.
    p. cm.
Includes bibliographical references and index.
ISBN 978-1-68285-806-6
1. Hydraulic engineering. 2. Hydraulics. 3. Engineering. I. Martin, Lilly.
TC145 .A28 2020
627--dc23

For information on all Willford Press publications
visit our website at www.willfordpress.com

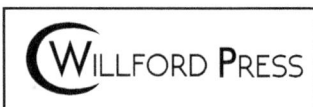

WILLFORD PRESS

Contents

Permissions

List of Contributors

Index

Preface

This book has been a concerted effort by a group of academicians, researchers and scientists, who have contributed their research works for the realization of the book. This book has materialized in the wake of emerging advancements and innovations in this field. Therefore, the need of the hour was to compile all the required researches and disseminate the knowledge to a broad spectrum of people comprising of students, researchers and specialists of the field.

A sub-discipline of civil engineering that is concerned with the flow and conveyance of fluids like water and sewage is known as hydraulic engineering. The force driving the movement of these fluids is the force of gravity. The principles of physical modeling, open channel hydraulics, mechanics of sediment transportation, fluid mechanics, hydrology, etc. are integral to the field of hydraulic engineering. This area of study is vital to the designing of dams, canals, bridges, channels and levees. It is also useful in the construction of hydraulic structures for sewage collection networks, water distribution networks, storm water management, sediment transport, etc. Developing strategies for the control, storage, transport, collection, regulation and use of water is an important dimension of hydraulic engineering. This book includes some of the vital pieces of work being conducted across the world, on various topics related to hydraulic engineering. It strives to provide a fair idea about this discipline and to help develop a better understanding of the latest advances within this field. It aims to serve as a resource guide for students and experts alike and contribute to the growth of hydraulic engineering.

At the end of the preface, I would like to thank the authors for their brilliant chapters and the publisher for guiding us all-through the making of the book till its final stage. Also, I would like to thank my family for providing the support and encouragement throughout my academic career and research projects.

Editor

Sensitivity analysis of fluid flow in a confined aquifer using numerical simulation

Hadi Ghaebi[1,*], Mehdi Bahadorinejad[2], Mohammad Hassan Saidi[2]

[1]Department of Mechanical Engineering, Faculty of Engineering, University of Mohaghegh Ardabili.
[2]Center of Excellence in Energy Conversion (CEEC), School of Mechanical Engineering, Sharif University of Technology, Tehran, Iran.

ARTICLE INFO

Keywords:
Aquifer
Pressure distribution
Numerical simulation
Parametric study

ABSTRACT

Aquifers are underground porous domains containing groundwater. Confined aquifers are surrounded by the impermeable layers. They are saturated by pressurized water and are suitable for energy storage purposes. They have low thermal conductivity and large storage volume. In design of aquifer thermal energy storage (ATES) an applicable model is necessary to predict the aquifer behavior. In this research, by developing a three dimensional finite volume model via FLUENT software, the effects of operative parameters on pressure distribution are investigated. In the ATES, heat transfer is performed by both convection and conduction phenomena. The convective heat transfer in the ATES is occurred because of pressure gradient and hence, recognition of effects of operative parameters on pressure distribution is essential. These effective parameters are some geological parameters such as groundwater natural flow, porosity and permeability, injection and withdrawal rates from wells, number and arrangement (being linear, triangular or rectangular) of wells.

---

1. Introduction

Geothermal systems using heat stored in the underground have been used for cooling and heating of buildings in several countries such as the United States (Meyer and Todd. 1973; Molz et al. 1978; Papadopulos and Larson. 1978; Parr et al. 1983), European countries (Andersson. 2003; Sanner. 2003; Paksoy et al 2000; Preene and Powrie. 2009), and other countries (Umemiya and Satoh. 1990; Gao et al. 2009; Lee. 2008; Fan et al. 2007) since 1970s. Recently, thermal energy storage (TES) systems have become more popular in the world due to the problem caused by depletion of fossil fuels and increase of global warming (Rosen. 1999). There are two main TES systems: A closed system (e.g., borehole thermal energy storage: BTES), and an open system (e.g., aquifer thermal energy storage: ATES). Due to direct usage of groundwater with relatively high volumetric heat storage capacity, the ATES system has higher system performance than the BTES system and any other systems using low temperature geothermal heat. In the ATES system, the contamination and depletion of groundwater can be minimal, since the water circulated from underground to a heat exchanger is immediately re-injected though the injection well into the aquifer (Gao et al. 2009).

The numerical modeling is a powerful tool for flow simulation in porous media such as an aquifer (Shamsai and Vosoughifar. 2004). In the past years, several research works have been performed about the numerical flow simulation in the aquifers. They have also applied some numerical codes such as TRUST (Narasimhan. 1984), TRUMP (Narasimhan, 1973), PORFLOW (Runchal. 1984), UNSAT (Fayer and Jones. 1990), SUTRA (Souza. 1987), MODFLOW (Jobson and Harbaugh. 1999) to model the aquifer system. These codes are generally based on finite difference discretization.

Dong et al. (2012) used MODFLOW code to optimize the rate of pumped water in an aquifer. The weakness of application of this code in defining the geometry of the aquifer that is it has developed for porous media; on the other hand, considering facing the heterogeneous

rocks through increasing the depth and crossing from the alluvium environment, the modelling has to consider its complexities. Krčmář and Sracek (2014) as well, used this code to model the underground water in a mine. Álvarez et al. (2015) modeled groundwater flow in an open pit located in limestones using the same code.

As it is obvious, most of the above mentioned studies have not performed a remarkable research about the parametric study on the pressure distribution inside the aquifer. In the present study a comprehensive investigation is performed on effective parameters that influence pressure distribution inside the aquifer would be used for thermal energy storage (TES) by numerical simulation. Also special consideration is given on the number and arrangement of injection/withdrawal wells.

2. Materials and methods
2.1. Governing equations

The system which is considered in this study comprise of a single phase water flow inside the saturated aquifer that confined by bedrock both above and below. In such a system the water flows only inside the aquifer.

Aquifer is a porous media that its porosity is equal to the volume of voids space to the total volume of the media and is expressed in terms of fraction or percent:

$$\emptyset = \frac{V_V}{V_O} \tag{1}$$

where $\emptyset$, V_V and V_0 are porosity, void and bulk volume of the aquifer, respectively. It is obvious that in saturated aquifers porosity is the ratio of water volume to the total volume. The porosity of the rocks varies from 0 to 45 %. For the TES, 20-30 % porosity is suitable (Tsang. 1980). By considering the porosity definition, the density of aquifer is defined as (Schaetzle. 1980):

Corresponding author Email: hghaebi@uma.ac.ir

$$\rho = \rho_{Water}\emptyset + \rho_{Rock}(1-\emptyset) \tag{2}$$

Water flow depends on the pressure distribution and physical properties of porous media. In general, flow is proportional to pressure gradient, namely head gradient and area (Strack. 1989):

$$Q \propto A\frac{dh}{dL} \tag{3}$$

This flow equation is known as Darcy's law and its proportionality is called permeability or hydraulic conductivity and shown with K:

$$Q = -KA\frac{dh}{dL} \tag{4}$$

Another form of the Darcy's law is defined for the Darcy flux (or the Darcy velocity or specific velocity) which is the discharge rate per unit cross-sectional area. It is hard to define the velocity inside the aquifer because of the existence of pores with different cross sections. The velocity inside of the aquifer must be a rough average number as the cross section is ever not homogeneous at all. As a result, velocity is rarely used in geological evaluations. A velocity defined by dividing the flow rate (Q) by aquifer cross-sectional area (A) is known as the specific velocity, V_S (Strack. 1989):

$$\vec{V}_S = \frac{Q}{A} = -K\frac{dh}{dL} \tag{5}$$

Natural flow in an aquifer is subjected to the equations of flow. The natural flow in an aquifer can be stated as:

$$Q/A = K\vec{\nabla}h \tag{6}$$

The groundwater flow has a 3D pattern. The specific velocity in a Cartesian system can be expressed as:

$$\vec{V}_{S,x} = -K_x\frac{\partial h}{\partial x} \tag{7}$$

$$\vec{V}_{S,y} = -K_y\frac{\partial h}{\partial y} \tag{8}$$

$$\vec{V}_{S,z} = -K_z\frac{\partial h}{\partial z} \tag{9}$$

In general, K_x, K_y and K_z are not the same. In such a case, the medium is called anisotropic. In this study, it is assumed that the aquifer is isotropic.
Generally, charge and discharge rates are performed as a constant value. Therefore, the flow is steady and the continuum equation in porous media satisfies the condition (Strack. 1989):

$$[\vec{\nabla}.(\rho\vec{V}_S)]dV = S \tag{10}$$

where S is related to source term. In this study it is supposed that the porosity and density are constant, hence Eq.10 is converted to:

$$\vec{\nabla}.\vec{V}_S = \frac{S(x,y,z)}{\rho dV} \tag{11}$$

By considering the Darcy's equation, Eq.12 can be rewritten as follows:

$$\nabla^2 h = \frac{S(x,y,z)}{K\rho dV} \tag{12}$$

2.2. Numerical modeling

In this study, FLUENT® software which is a commercial finite volume program was used for simulating the ATES system. The meshed view of the considered domain has been shown in The Figure 1. In this meshing an unstructured 3D mesh with 259346 cells was constructed. The chosen element was Tet/Hybrid and the type was TGrid. For investigation of mesh size independency, so that the unique solution would be obtained, several mesh sizes were examined. Finally, the mesh sizes selected were 1m, 1m and 0.5m in x, y and z directions, respectively. The dependency of solution to the mesh sizes less that these values, was less than 1 percent.

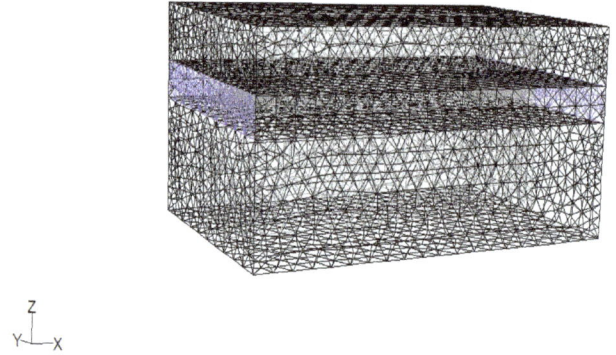

Fig. 1. Meshed view in the FLUENT.

2.2.1. Boundary conditions

According to the requirement, a value for flow rate of injection/withdrawal wells is considered. By considering groundwater velocity, a pressure gradient in the direction of x is added that obtained from the Darcy's equation as below:

$$H_{imp} = V_{gw}\frac{X_{Length}}{K} \tag{13}$$

The head in x=0 and $x = X_{Length}$ were as follows:

$$h_{x=0} = H_{imp} \quad , \quad h_{x=X_{Length}} = 0 \tag{14}$$

The initial head is equal to the model elevation. Boundary condition on the lateral and the lower and upper sides is no flow.

3. Results and discussion

The objective of this research is the investigation of different parameters on pressure distribution inside of the aquifer that will be used for the ATES. As mentioned above, in the ATES the water is withdrawn from one or more wells and after performing heat transfer is justified into the aquifer with the same rate through other well or wells. The physical properties of the aquifer and the specifications of the injection and withdraw wells are listed in Table 1. The length, width and height of the aquifer are 100m, 50m and 6m in x, y and z directions, respectively.

In the following discussions, although the pressure distribution is calculated in 3-dimension, because of symmetry of domain, the distribution is brought only in the xy plane and z=3m.

3.1. The effect of groundwater natural flow on the pressure distribution inside the aquifer

The pressure distribution inside the aquifer is shown in Fig.2 (a, b and c) when the natural flow is 30, 50 and 100 m/year, respectively. Since the natural flow is considered in x-direction, it is influenced by the pressure distribution in the boundaries and perpendicular faces on x-direction. By increasing the natural flow, the pressure in boundaries is increased. Consequently, the pressure distribution in whole of the aquifer is also increasing.

Table 1. Aquifer and well specifications in the base case.

Property	Value
Permeability	$0.0017 \times 10^{-3} \dfrac{m}{s}$
Porosity	0.4
Groundwater natural flow	$10 \dfrac{m}{s}$
Water density	$1000 \dfrac{kg}{m^3}$
Rock density	$1800 \dfrac{Kg}{m^3}$
Injection/withdrawal rate	$740 \dfrac{kg}{s}$
Number of injection/withdrawal well/wells	1
Dimension of injection pump	$x = 10 \ m$ $y = 25 \ m$ $z = 3m$
Dimension of withdrawal pump	$x = 90 \ m$ $y = 25 \ m$ $z = 3m$

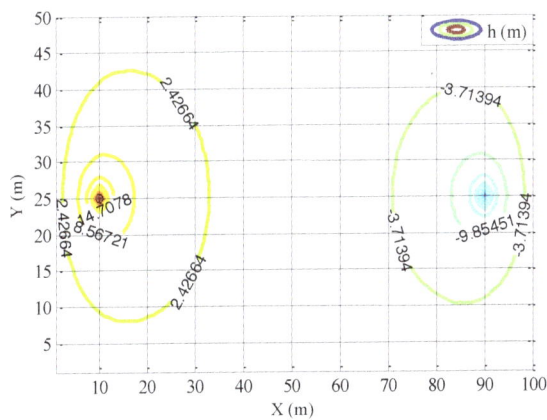

a) Pressure distribution in xy plane and z=3m and k=0.0017 ،
$\emptyset = 0.4$ ،$V_{GW} = 30 \frac{m}{year}$ ،$Q = 0.74 \frac{m^3}{s}$

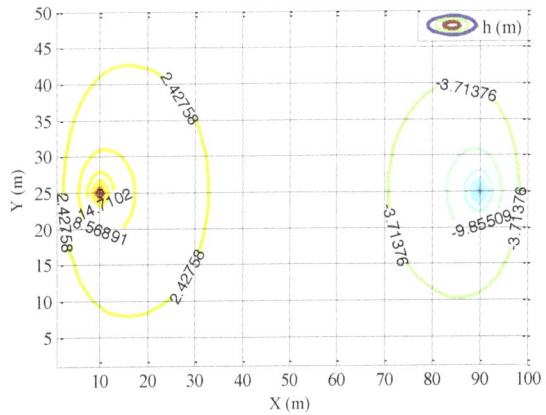

b) Pressure distribution in xy plane and z=3m and
$k=0.0017$ ،$\emptyset = 0.4$ ،$V_{GW} = 50 \frac{m}{year}$ ،$Q = 0.74 \frac{m^3}{s}$

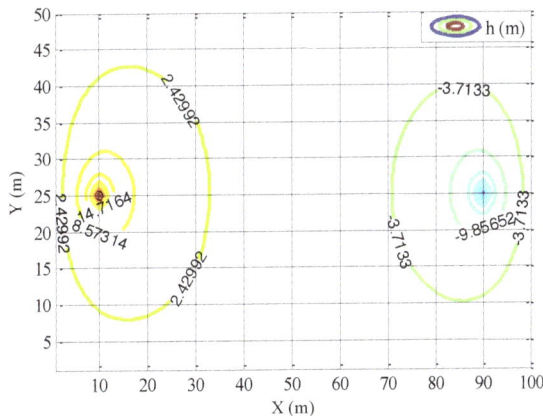

c) Pressure distribution in xy plane and z=3m and k=0.0017 ،$\emptyset = 0.4$ ،$V_{GW} = 100 \frac{m}{year}$ ،$Q = 0.74 \frac{m^3}{s}$

Fig. 2. Variation of pressure distribution with groundwater natural flow.

3.2. The effect of porosity on the pressure distribution inside the aquifer

Fig.3 shows the pressure distribution inside the aquifer with respect to porosity variation. As it is seen, by decreasing the porosity, the pressure distribution decreases. This happens since the increment in porosity tends to decrease aquifer density as stated in Eq. 2 and then the pressure is increasing.

3.3. The effect of permeability on the pressure distribution inside the aquifer

Hydraulic conductivity (permeability) is the ability of porous media for water transition. It should be noted that the hydraulic conductivity implies hydraulic resistance of the aquifer. Fig.4 (a, b and c) shows the variation of pressure distribution with respect to permeability. As permeability increases, the pressure distribution in whole of the aquifer is decreasing. It happens as suit of decrement of water penetration. This phenomenon is also justified mathematically by surveying Eq. 6, the pressure distribution varies with K inversely.

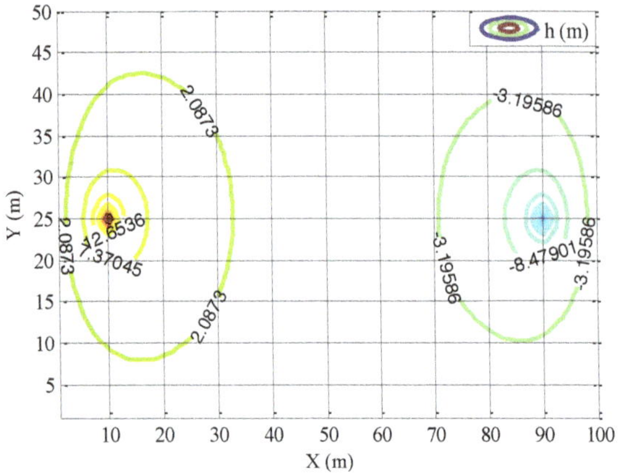

a) Pressure distribution in xy plane and z=3m and $k=0.0017, V_{GW} = 10\frac{m}{year}, Q = 0.74 \frac{m^3}{s}, \emptyset = 0.1$

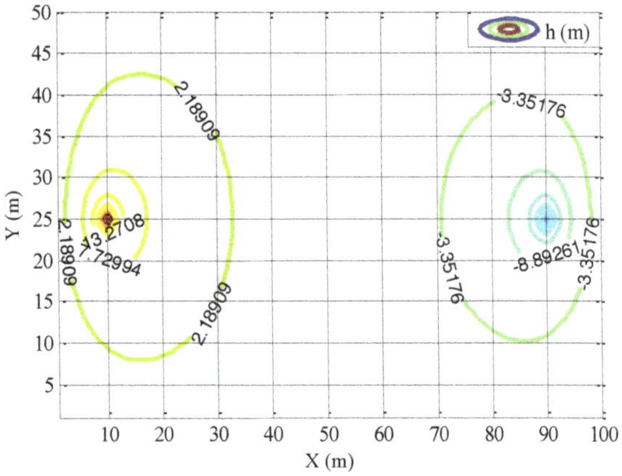

b) Pressure distribution in xy plane and z=3m and $k=0.0017, V_{GW} = 10\frac{m}{year}, Q = 0.74 \frac{m^3}{s}, \emptyset = 0.2$

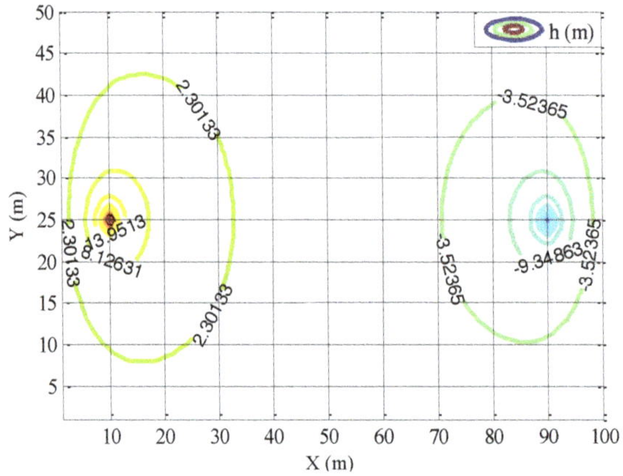

c) Pressure distribution in xy plane and z=3m and $k=0.0017, V_{GW} = 10\frac{m}{year}, Q = 0.74 \frac{m^3}{s}, \emptyset = 0.3$

Fig. 3. Variation of pressure distribution with porosity.

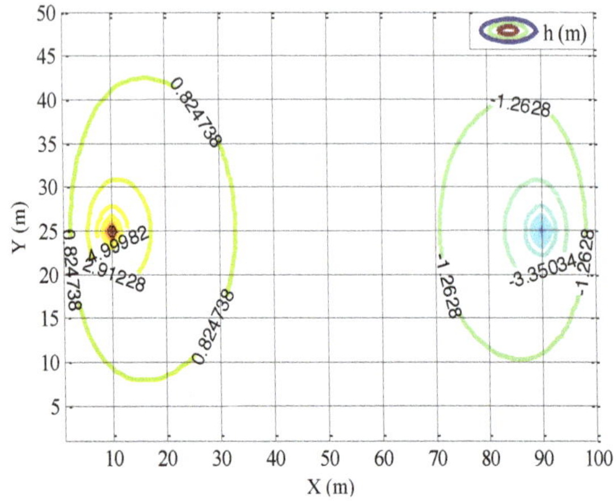

a) Pressure distribution in xy plane and z=3m and $V_{GW} = 10\frac{m}{year}, Q = 0.74 \frac{m^3}{s}, \emptyset = 0.1, k=0.0051$

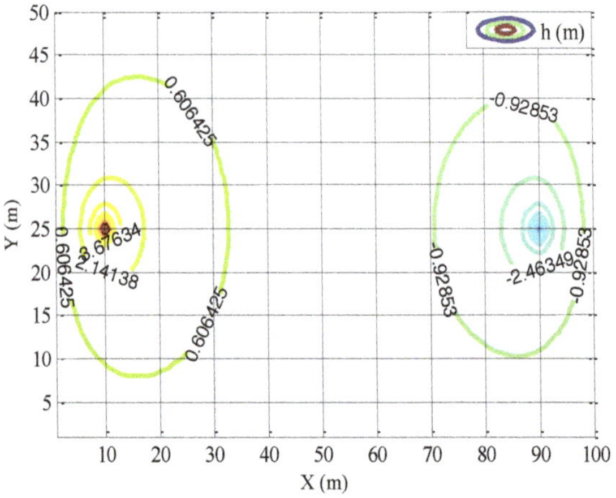

b) Pressure distribution in xy plane and z=3m and $V_{GW} = 10\frac{m}{year}, Q = 0.74 \frac{m^3}{s}, \emptyset = 0.1, k=0.0068$

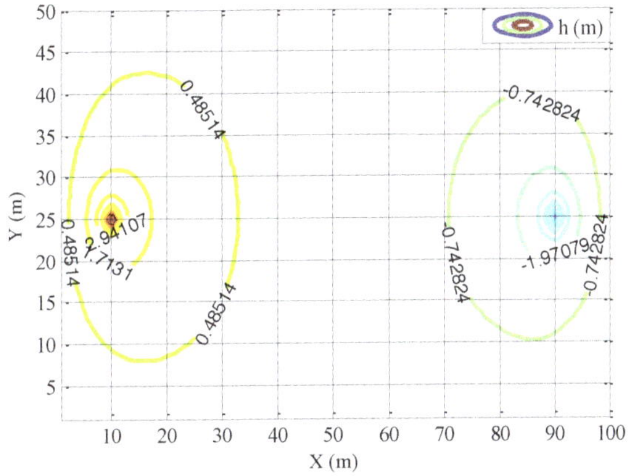

c) Pressure distribution in xy plane and z=3m and $V_{GW} = 10 \frac{m}{year}$ ، $Q = 0.74 \frac{m^3}{s}$ ، $\emptyset = 0.1$ ، k=0.0085

Fig. 4. Variation of pressure distribution with permeability.

3.4. The effect of injection/withdrawal rate on the pressure distribution inside the aquifer

Fig. 5 (a, b and c) shows the effects of increment of injection/withdrawal rates on the pressure distribution. The injection/withdrawal rate is the source term in the pressure distribution equation. As shown in Eq.6, by increasing the flow rate, the pressure distribution is also increasing. When flow rate increases, the pressure in injection/withdrawal pump locations increases and then this effect is diffused all over the aquifer. It should be noted that in this research, the injection rate is the same as the withdrawal rate.

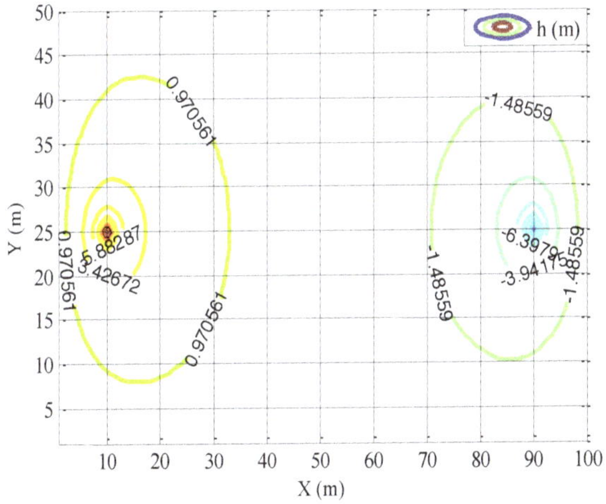

a) Pressure distribution in xy plane and z=3m and k=0.0017 ، $\emptyset = 0.4$ ، $V_{GW} = 10 \frac{m}{year}$ ، $Q = 0.296 \frac{m^3}{s}$

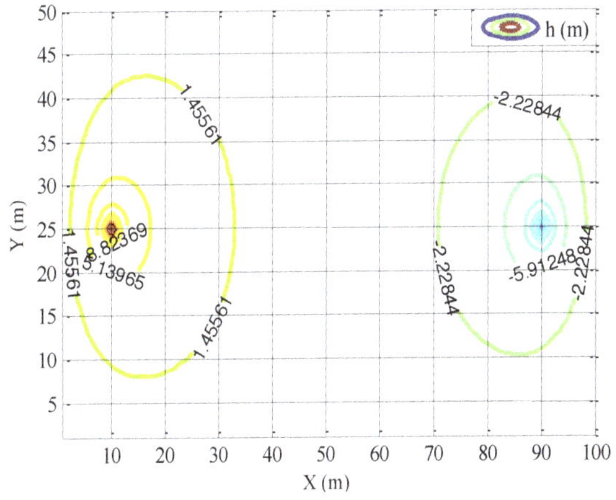

b) Pressure distribution in xy plane and z=3m and k=0.0017 ، $\emptyset = 0.4$ ، $V_{GW} = 10 \frac{m}{year}$ ، $Q = 0.444 \frac{m^3}{s}$

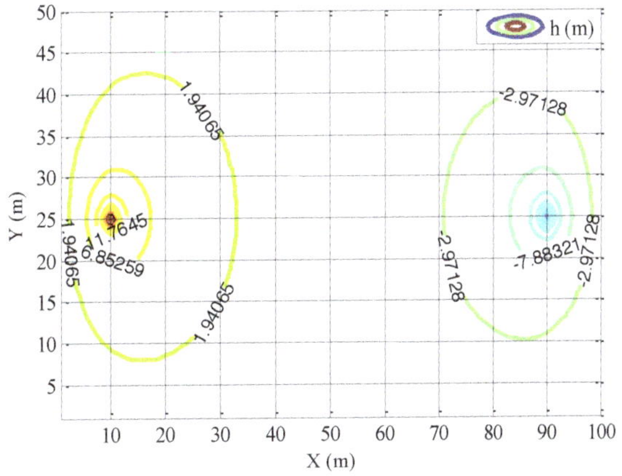

c) Pressure distribution in xy plane and z=3m and k=0.0017 ، $\emptyset = 0.4$ ، $V_{GW} = 10 \frac{m}{year}$ ، $Q = 0.592 \frac{m^3}{s}$

Fig. 5. Variation of pressure distribution with injection/withdrawal flow rate.

3.5. The effect of number and arrangement of wells on the pressure distribution inside the aquifer

In the previous sections one injection/withdrawal well is used. The concession of this research in investigation of number and arrangement of wells (linear, triangular and rectangular) on the pressure distribution inside the aquifer is illustrated as follows:

3.5.1. Three wells array

Fig. 6 indicates the dimensions and arrangements of three wells for injection and three wells for withdrawal). The arrangements can be linear or triangular as shown in Figs. 6a and 6b, respectively.

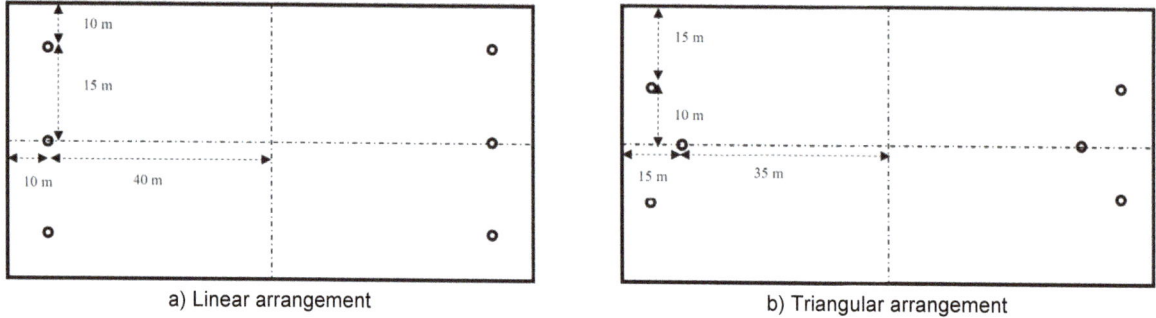

a) Linear arrangement

b) Triangular arrangement

Fig. 6. Three wells application

The pressure distribution in the aquifer is shown in Fig.7(a and b) in three wells application. As it is seen, the pressure value in triangular array is more than linear. It happens because triangular arrangement of pumps tends to increase the pressure distribution.

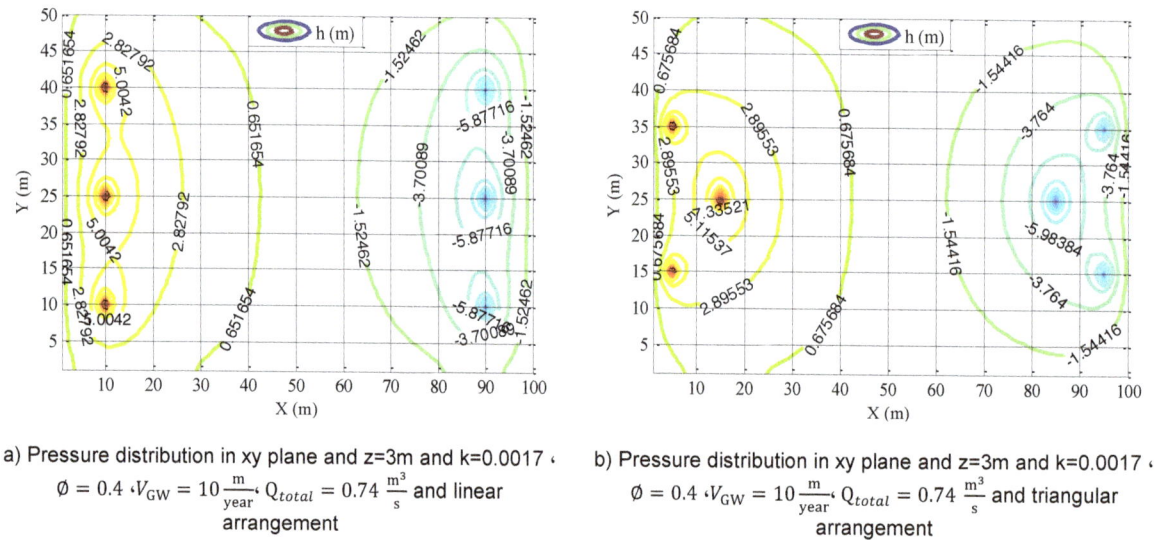

a) Pressure distribution in xy plane and z=3m and k=0.0017 ‹ $\emptyset = 0.4$ ‹$V_{GW} = 10\frac{m}{year}$‹ $Q_{total} = 0.74 \frac{m^3}{s}$ and linear arrangement

b) Pressure distribution in xy plane and z=3m and k=0.0017 ‹ $\emptyset = 0.4$ ‹$V_{GW} = 10\frac{m}{year}$‹ $Q_{total} = 0.74 \frac{m^3}{s}$ and triangular arrangement

Fig. 7. Pressure distribution in three wells application.

3.5.2. Five wells array

Fig.8 shows the arrangement of five wells application. In this case the array can be linear, triangular and rectangular as shown in Fig.

Fig.9 shows pressure distribution in five wells application. As it is seen, the pressure quantity in triangular arrangement is higher.

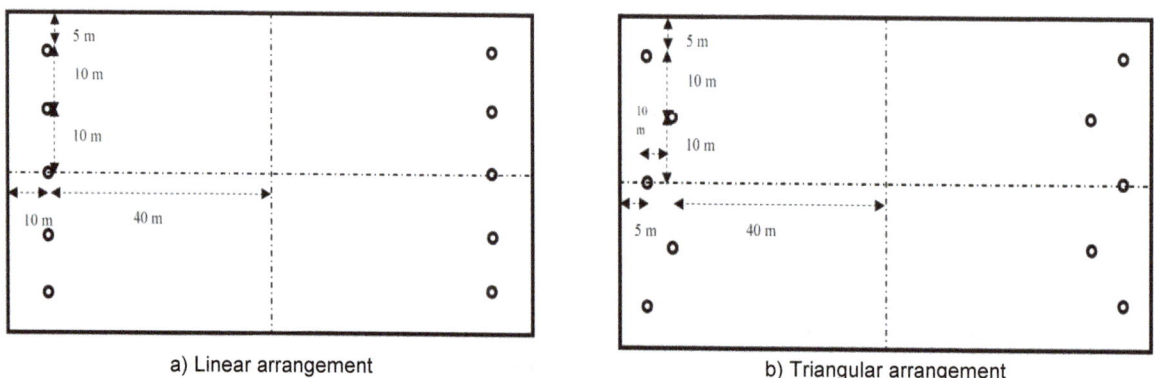

a) Linear arrangement

b) Triangular arrangement

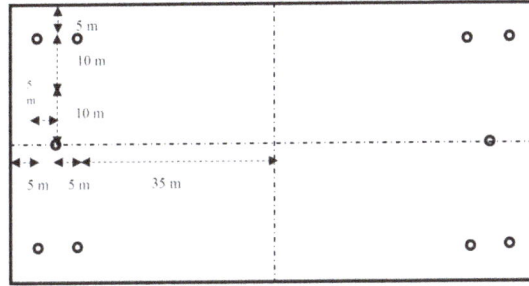

c) Rectangular arrangement

Fig. 8. Five wells application.

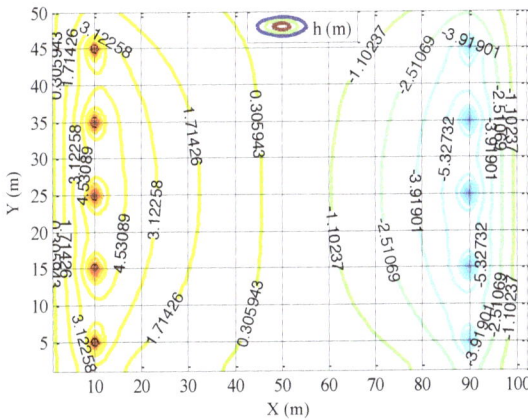

a) Pressure distribution in xy plane and z=3m and k=0.0017 ‹ $\emptyset = 0.4$ ‹ $V_{GW} = 10\frac{m}{year}$ ‹ $Q_{total} = 0.74\frac{m^3}{s}$ and linear arrangement

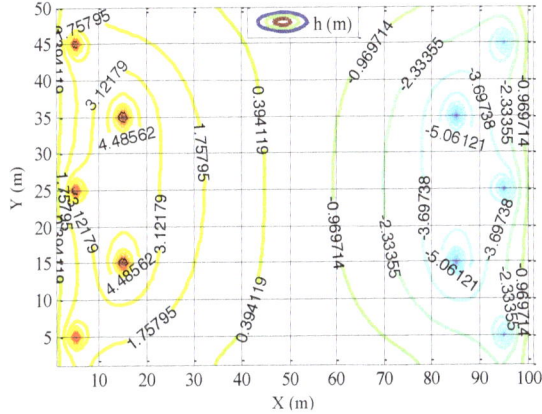

b) Pressure distribution in xy plane and z=3m and k=0.0017 ‹ $\emptyset = 0.4$ ‹ $V_{GW} = 10\frac{m}{year}$ ‹ $Q_{total} = 0.74\frac{m^3}{s}$ and triangular arrangement

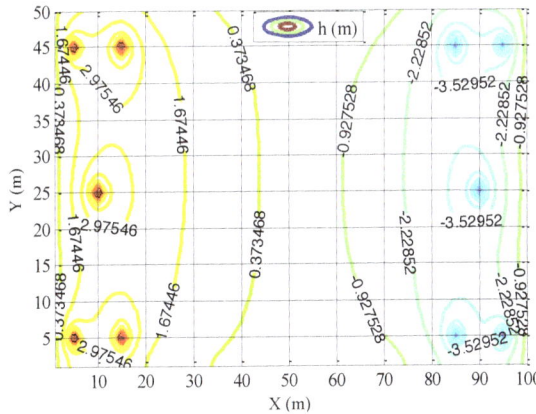

b) Pressure distribution in xy plane and z=3m and k=0.0017 ‹ $\emptyset = 0.4$ ‹ $V_{GW} = 10\frac{m}{year}$ ‹ $Q_{total} = 0.74\frac{m^3}{s}$ and rectangular arrangement

Fig. 9. Pressure distribution in five wells application.

4. Conclusions

In this study sensitivity analysis of the fluid flow in an aquifer respect to some operational and physical parameters was carried out. The aquifer was employed for seasonal thermal energy storage. In the considered system, the flow is withdrawn and after heat transfer reinjected to aquifer through an injection well. The main conclusions of this research are as follows:

- By increasing the natural flow, the distribution in whole of the aquifer is increasing.
- Increment of the porosity tends to increase the aquifer pressure.
- The pressure distribution varies with permeability inversely.
- When the flow rate increases, the pressure increases all over the aquifer.
- The pressure distribution in triangular arrangement of wells is more than other arrangements.

References

Andersson O., Hellstrom G., Nordell B., Heating and cooling with UTES in Sweden-current situation and potential market development, International Proceedings of the 9th international conference on thermal energy storage, Warsaw, Poland, 1 (2003) 359-366.

Dickinson J.S., Buik N., Matthews M.C., Snijders A., Aquifer thermal energy: theoretical and operational analysis, Geotechnique 59 (2009) 249-260.

Dong D., Sun W., Xi S., Optimization of mine drainage capacity using FEFLOW for the No. 14 Coal Seam of China's Linnancang Coal Mine, Mine Water and the Environment 31 (2012) 353-360.

Fan R., Jiang Y., Yao Y., Shiming D., Ma Z., A study on the performance of a geothermal heat exchanger under coupled heat conduction and groundwater advection, Energy 32 (2007) 2199-2209.

Fayer M., Jones T., UNSAT-H version 2.0: Unsaturated soil water and heat flow model, PNL-6779, Pacific Northwest Laboratory, Richland, Washington, 1990.

Fernández-Álvarez J. P., Álvarez-Álvarez L., & Díaz-Noriega R., Groundwater Numerical Simulation in an Open Pit Mine in a Limestone Formation Using MODFLOW, Mine Water and the Environment 34 (2015) 1-11.

Gao Q., Li M., Yu M., Spitler J.D., Yan Y.Y., Review of development from GSHP to UTES in China and other countries, Renewable Sustainable Energy Reviews 13 (2009) 1383-1394.

Jobson H.E., Harbaugh A.W., Modifications to the diffusion analogy syrface-water flow model(daflow) for coupling to the modular finite-difference groundwater flow model(mudflow), U.S. Geological Survey Open-File Report (1999) 99-217.

Krčmář D., Sracek O., MODFLOW-USG: the New Possibilities in Mine Hydrogeology Modelling (or what is Not Written in the Manuals), Mine Water and the Environment 33 (2014) 376-383.

Lee K.S., Performance of open borehole thermal energy storage system under cyclic flow regime, Journal of Geoscience 2008 (12) 169-175.

Meyer C.F., Todd D.K., Heat storage wells, Water Well Journal 10 (1973) 35-41.

Molz F.J., Warman J.C., Jones T.E., Aquifer storage of heated water: Part 1: A field experiment, Ground Water 16 (1978) 234-241.

Narasimhan T.N., TRUST: A computer program for transient and steady state fluid flow in multi-dimensional variably saturated deformable media under isotherm conditions, Lawrence Berkeley Laboratory Memorandum, 1984.

Novo V.A., Bayon R.J., Castro-Fresno D., Rodriguez-Hernandez R., Review of seasonal heat storage in large basins: water tanks and gravel water pits, Applied Energy 2010 (87) 390-397.

Papadopulos S.S., Larson S.P., Aquifer storage of heated water: Part 2: Numerical simulation of field results, Ground Water 16 (1978) 242-248.

Parr D.A., Molz F.J., Melville J.G., Field determination of aquifer thermal energy storage parameters, Ground Water 21 (1983) 22-35.

Paksoy H.O., Andersson O., Abaci S., Evliya H., Turgut B., Heating and cooling of a hospital using solar energy coupled with seasonal thermal energy storage in an aquifer, Renewable Energy 19 (2000) 117-122.

Preene M., Powrie W., Ground energy systems: delivering the potential, Energy 2 (2009) 77-84.

Rosen M.A., Second-law analysis of aquifer thermal energy storage systems, Energy 24 (1999) 167-182.

Runchal A.K., PORFLOW: A software tool for multiphase fluid flow, heat and mass transport in fractured porous media, user's manual version 2.50, Analytical & Computational Research, Inc. Los Angeles, CA 900771984.

Sanner B., Karytsas C., Mendrinos D., Rybach L., Current status of ground source heat pumps and underground thermal energy storage in Europe, Geothermics 32 (2003) 579-588.

Schaetzle W.J., Thermal Energy Storage in Aquifers, Design and Applications, Pergamon Press, 1980.

Schauer D.A., FED: A computer program to generate geometric input for the heat transfer code TRUMP, Report UCRL-50816, 1973.

Shamsai A., Vosoughifar H.R., Finite volume discretization of flow in porous media by MATLAB system, Scientia Iranica 11 (2004) 146-153.

Souza W.R., Documentation of a graphical display program for the saturated- unsaturated transport (SUTRA) finite element simulation model, U.S. Geological Survey Water- Resource Investigation Report 874245, 1987.

Strack O.D.L., Groundwater Mechanics, prentice hall, 1989.

Tsang C.F., Aquifer Thermal Energy Storage A Survey, LBL Report, 1980.

Umemiya H., Satoh Y., A cogeneration system for a heavy-snow fall zone based on aquifer thermal energy storage, Japanese Society of Mechanical Engineering 33 (1990) 757-765.

2

Hydraulic analysis of compound open channel

Abbas Parsaie [*], **Amir Hamzeh Haghiabi**

Department of water Engineering, Lorestan University, Khorramabad, Iran.

ARTICLE INFO	ABSTRACT
Keywords: Stage-discharge relationship Flow discharge Velocity distribution Energy and momentum coefficients	Distribution of velocity of flow in compound open channel due to interaction of floodplains and main channel is strongly non-uniform. Defining the distribution of flow velocity is an important factor in calculation of sediment transport and estimation of flow discharge. One of the correction factors in calculation of flow discharge and shear stress are momentum(β) and energy (α) coefficients. In this study, the effect of β and α coefficients on Froude number and specific energy are assessed. Stage-discharge relationship in compound open channel was assessed using some empirical formula including Single-Channel Method (SCM), Divided-Channel Method (DCM), and modified divided-channel method (MDCM) and compared with together. When the discharge only flows in main channel all the empirical has a same result whereas by increasing the discharge and covering the floodplains by flow the results of them are different. The highest value of outcome of empirical formula is related to the SCM. Results indicated that considering the energy and momentum coefficients have significant effect on distribution of Froude number and specific energy.

1. Introduction

Estimation of flow discharge in the rivers is one of the main parameters in flood management projects (Parsaie et al. 2015a). Prediction of flow discharge in rivers have welcomed to insurance companies. Prediction of flow discharge leads to evaluate the flow depth in streams, using those parameters is necessary to evaluate the risk of insurance of projects located on floodplains (Parsaie and

Haghibia. 2014a). Therefore, they have attempted to develop a damage function proportion of flow discharge and flow depth. To estimate the flow discharge in streams the compound open channel as accurate idea has been proposed (Whyte. 2012). Fig. 1 shows different types of cross section of compound channel. As shown in Fig. 1, the compound open channels cross section includes a main channel and floodplains.

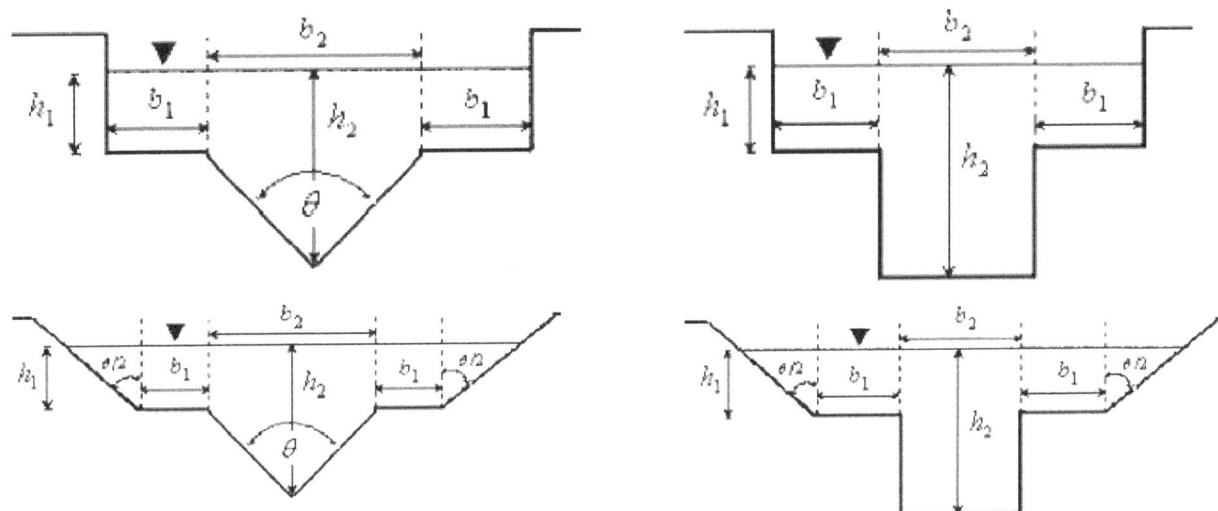

Fig. 1. Compound open channel cross sections.

where, b mc is the main channel width, n mc: main channel roughness, h mc: main channel depth, b fpl: floodplain width, n fpr: floodplain roughness and h fpr: floodplain depth. Several ways as

analytical approaches such as Single-Channel Method (SCM), Divided-Channel Method (DCM), modified divided-channel method (MDCM) have been proposed to estimate the discharge of flow in

Corresponding author Email: Abbas_Parsaie@yahoo.com

compound open channel (Al-Khatib et al. 2013; Al-Khatib et al. 2012; Bousmar and Zech 1999; Huthoff et al. 2008; Khatua et al. 2012; Liao and Knight 2007; Mohanty and Khatua. 2014; Myers. 1987; Naot et al. 1993; Prinos et al. 1985; Seckin. 2004; Seckin et al. 2009; Stephenson and Kolovopoulos. 1990; Tang et al. 1999; Unal et al. 2010). Recently, intelligence techniques such as artificial neural network (Moasheri et al. 2013; Parsaie and Haghiabi. 2015a; Parsaie and Haghiabi. 2014a; Parsaie and Haghiabi. 2014b; Parsaie and Haghiabi 2015b; Parsaie and Haghiabi 2015c; Parsaie and Haghiabi. 2015d; Parsaie et al. 2015b), Genetic Programing (GP) (Azamathulla and Zahiri 2012; Najafzadeh and Zahiri. 2015; Zahiri and Azamathulla. 2014) and Multilayer perceptron neural network (MLP) have been proposed to calculate or predict the flow discharge in compound open channel (Azamathulla and Zahiri. 2012; Najafzadeh and Zahiri. 2015; Sahu et al. 2011; Zahiri and Azamathulla. 2014). Due to interaction of floodplains and main channel the distribution of flow velocity is highly non-uniform. The aim of this study is defining the hydraulic properties of compound open channel such as distribution of energy coefficient, momentum coefficient, is calculated and their effects on the specific energy, Froude number and calculating the head discharge curve by analytical approaches are evaluated.

2. Material and methods

The Bernoulli equation is the basic formula for open channel hydraulic studies. Distribution of hydraulic component such as energy coefficient, momentum coefficient is calculated by equation (1 and 2) which derives from Bernoulli equation and also the specific energy equation by considering the energy coefficient is given in equation (3).

$$\alpha = \frac{\int V^3 dA}{V_m^3 \int dA} \approx \frac{\sum_{i=1}^{N}\left(V_i^3 A_i\right).\left(\sum_{i=1}^{N} A_i\right)^3}{\left(\sum_{i=1}^{N} V_i A_i\right)^3}$$

(1)

$$V_m = \frac{\sum_{i=1}^{N} V_i A_i}{\sum_{i=1}^{N} A_i}$$

$$\beta = \frac{\int V^2 dA}{V_m^2 \int dA}$$

(2)

$$E = y + \frac{\alpha Q^2}{2gA^2}$$

(3)

where, i subscript is related to subsections, V is the flow velocity and A is the area of the subsections respectively. It is notable that the energy and momentum coefficients as seem in equations (1 and 2) are functions of flow discharge, flow depth and cross section area. Calculating the flow discharge by classical formula such as manning formula cases of appear incredible error in compare to measured data so researchers try to improve the accuracy of computation by modifying the manning formula. In the fallow, two main of analytical approaches which was used by investigators in many compound open channel studies are presented. To understand more detail about the hydraulic component of compound open channel an example of hydraulic channel is considered. This example was designed based on the literature section which stated that the roughness of floodplains and main channel usually is different. Table 1 presents the value of

cross section of compound channel which considered in example. In other to the aim of this research is defining the effect of compound cross section of canal and different roughness between the floodplains with main channel on the hydraulic components such as head discharge curve, specific energy distribution and so on.

Table 1. Summary of geometry properties of example compound open channel.

b mc	n mc	h mc	b fpl	n fpl	h fpr	n fpr
0.25	0.012	0.15	1	0.022	1	0.022

2.1. Single channel method (SCM)

The compound open channel cross section has been considered as a unique cross section in the Single channel method (SCM) and there is not any difference between the normal and compound channel. The main point in the SCM is calculating the equivalent roughness for compound channel by prevalent methods such as Horton and Einstein formula (equation 4) and then discharge is calculated by equation (5). The weakness of the SCM is related to calculate the transport capacity, because when the water level increases and flow covers the floodplains especially in the lowest depth of flow at the floodplains, wet perimeter is more increases in compare to the wet area so transport capacity is calculated small and at the end the flow discharge which are calculated by SCM are smaller than the actual values. Increasing the flow depth on the floodplains the accuracy of the SCM may be improved.

$$n_e = \frac{\left[\sum_{i=1}^{N}\left(P_i n_i^{\frac{3}{2}}\right)\right]^{\frac{2}{3}}}{P^{\frac{2}{3}}}$$

(4)

$$Q = \frac{1}{n_e} A R^{\frac{2}{3}} S^{\frac{1}{2}}$$

(5)

2.2. Divided channel method (DCM)

The Divided Channel method (DCM) divides the compound channel to some sub sections. The DCM is based on the uniform velocity in the subsections. In this method the compound channel section is divided to the main channel and floodplains and total discharge is calculated by adding the sub sections discharge. The discharge in subsections calculates by equation (6). as shown in this equation the classical equation such as manning is used for calculating the discharge and i subscription is related to the discharge in each sub sections. The separation line between the main channel and floodplains as shown in the Fig. 1 may be considered as vertical, diagonal or horizontal. Some modification has been conducted on the divided channel method and in this regards the divided channel method with horizontal division lines which are excluded within the calculation of wetted perimeter (DCM (h-e)) can be mentioned.

$$Q_t = \left(\sum_{i=1}^{N} \frac{A_i R_i^{\frac{2}{3}}}{n_i}\right) S_0^{\frac{1}{2}}$$

(6)

3. Results and discussion

The momentum and energy coefficients were calculated from equations (1 and 2) and distribution of them gives in Fig. 2.
As shown in
Fig. 2 the momentum and energy coefficients are constant versus the flow depth until the discharge flows only in the main channel. By increasing the flow depth, the value of the momentum and energy coefficients are increased rapidly and the rate of energy coefficient is much more than the momentum coefficient.

This change in distribution is related to intense non-uniformity in distribution of velocity and entering the properties of floodplains such as area and flow depth in computation process of momentum and energy coefficients. as shown in Fig. 1 and equations (1 and 2), when the flow depth is increased especially when the flow depth increased a modicum value then the main channel, the discharge flows in floodplains and the area of flow increased suddenly, so the value of momentum and energy coefficients increased rapidly. As shown in

Fig. 2 by increasing the flow depth the distribution of momentum and energy coefficients decreases after it's suddenly increase. The reason of this reduction is regulation of velocity distribution and reduction the effect of floodplains on the hydraulic characteristic of compound open channel. The results of discharge calculation in compound open channel by analytical approaches are given in the Fig. 3.

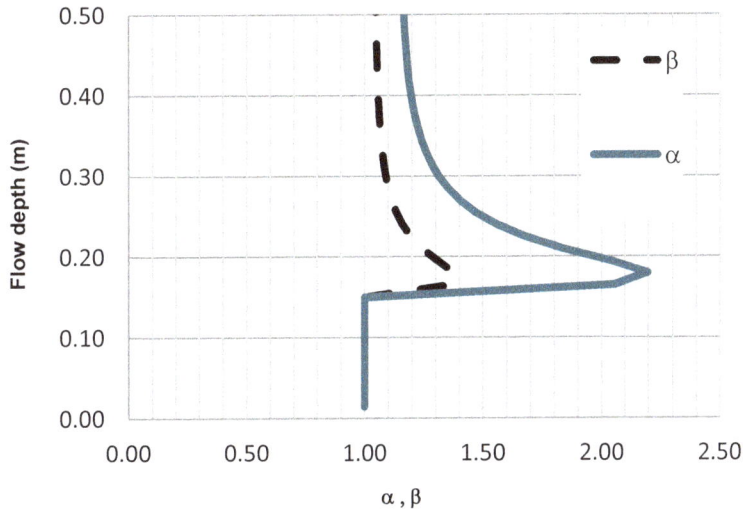

Fig. 2. Distribution of energy and momentum coefficients versus the flow depth.

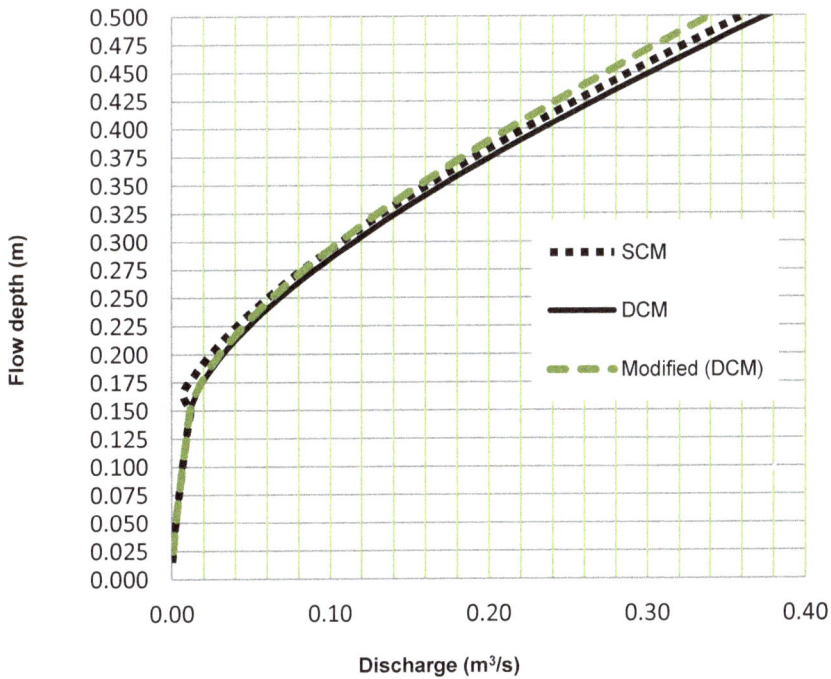

Fig. 3. Result of analytical approaches for calculation of discharge versus the flow depth.

As shown in Fig. 3, the results of the all analytical approaches are similar when the discharge flows in the main channel only. by increasing the flow depth event more than the main channel depth as shown in the Fig. 3 there is no change in the rate of results of the DCM and MDCM methods but the SCM method when the flow depth a lite increased more than main channel, the SCM method results shows a reduction in flow discharge calculation. As shown in the Fig.

3 the results of SCM is more the other analytical approaches. Figs. 4 and 5 show the distribution of Froude number versus the flow depth.

Fig. 4 shows the distribution of Froude number versus of flow depth by considering the energy coefficient in Froude number whereas the Fig. 5 shows the Froude number distribution versus the flow depth without considering the energy coefficient. As seems in both Figs. 4 and 5 by increasing the discharge of flow the flow depth is increased at the constant Froude number. Considering the energy coefficient to calculate the Froude number cases of appears more Froude number in constant flow depth.

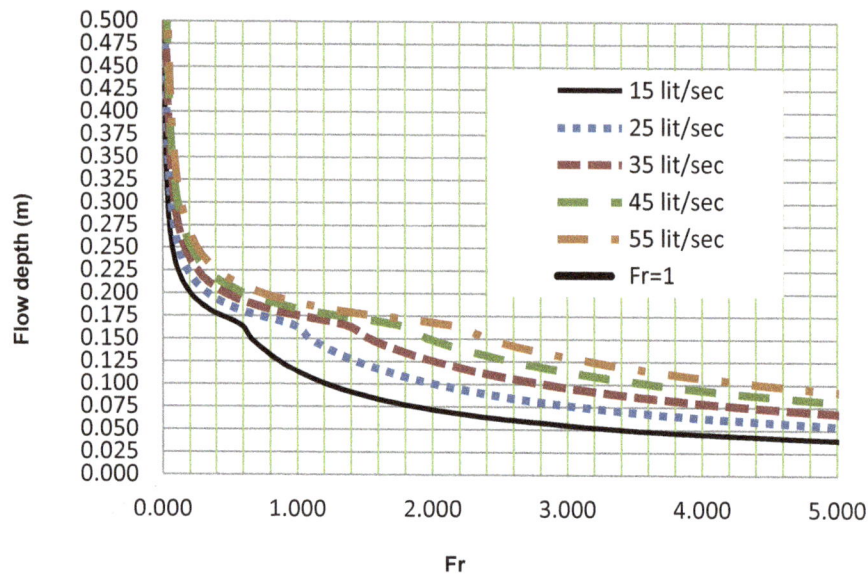

Fig. 4. Distribution of Froude number versus flow depth with energy coefficient.

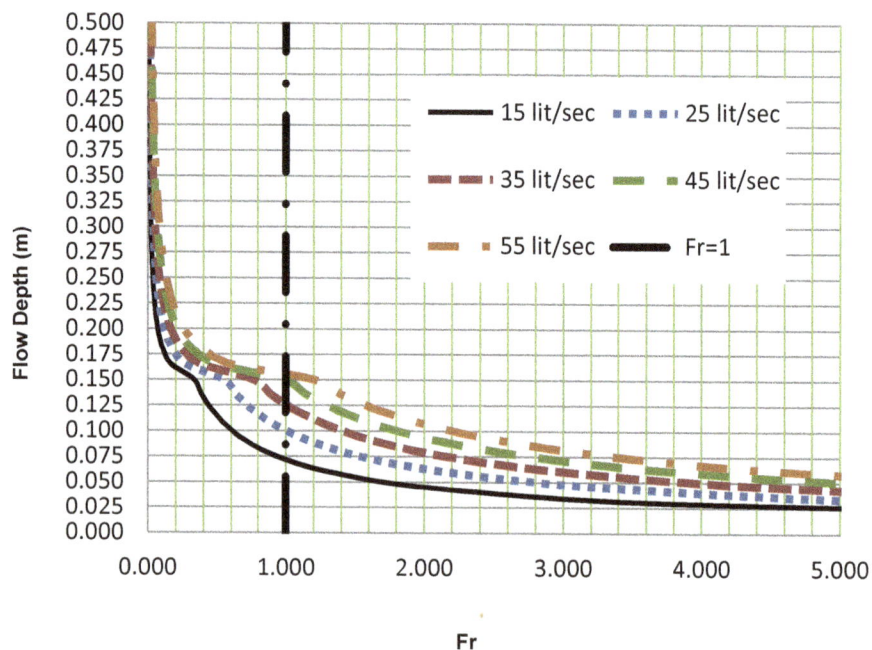

Fig. 5. Distribution of Froude number versus flow depth without energy coefficient.

Both of Figs. 6 and 7 show the distribution of specific energy versus the flow depth. Specific energy increases by increasing the discharge of flow.

Fig. 6 is showing the specific energy by considering the energy coefficient whereas the Fig. 7 is showing the specific energy without the energy coefficient. The energy coefficient is more effective on the specific energy curve especially when the discharge coefficient is increased.

4. Conclusion

In this paper a theoretical hydraulic analysis conducted on the hydraulic components of compound open channel. The result of this study showed that the compound section for open channel cases to non-uniformity in velocity distribution especially when the flow depth is increased in values a lite more than the main channel depth so the energy and momentum coefficients is more variation especially at flow depth values lite more than the main channel depth. Calculation of flow discharge in compound open channel by modified divided-channel method (MDCM) shows the more values in compere to other analytical approaches. Considering the energy coefficient is more effect on the distribution of specific energy versus the flow depth especially is more effective by increasing the discharge.

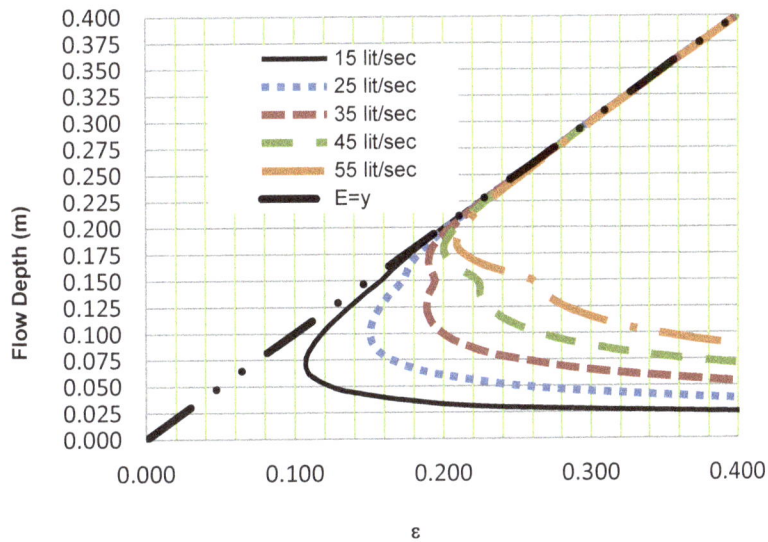

Fig. 6. Distribution of Froude number versus flow depth with energy coefficient.

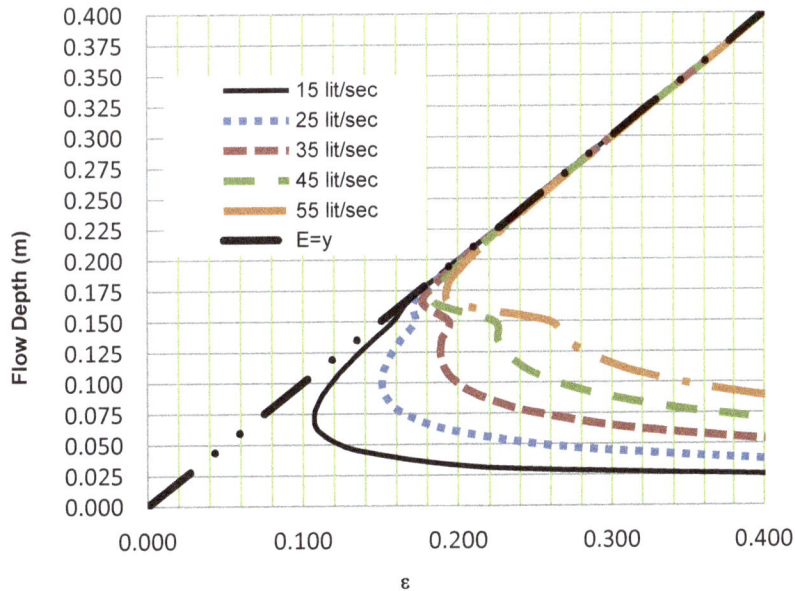

Fig. 7. Distribution of Froude number versus flow depth without energy coefficient.

References

Al-Khatib I., Hassan H., Abaza K., Application and Validation of Regression Analysis in the Prediction of Discharge in Asymmetric Compound Channels, Journal of Irrigation and Drainage Engineering 139 (2013) 542-550.

Al-Khatib I.A., Dweik A.A., Gogus M., Evaluation of separate channel methods for discharge computation in asymmetric compound channels, Flow Measurement and Instrumentation 24 (2012) 19-25.

Azamathulla H.M., Zahiri A., Flow discharge prediction in compound channels using linear genetic programming, Journal of Hydrology 454–455 (2012) 203-207.

Bousmar D., Zech Y., Momentum Transfer for Practical Flow Computation in Compound Channels, Journal of Hydraulic Engineering 125 (1999) 696-706.

Huthoff F., Roos P., Augustijn D., Hulscher S., Interacting Divided Channel Method for Compound Channel Flow, Journal of Hydraulic Engineering 134 (2008) 1158-1165.

Khatua K., Patra K., Mohanty P., Stage-Discharge Prediction for Straight and Smooth Compound Channels with Wide Floodplains Journal of Hydraulic Engineering 138 (2012) 93-99.

Liao H., Knight D., Analytic Stage-Discharge Formulas for Flow in Straight Prismatic Channels, Journal of Hydraulic Engineering 133 (2007) 1111-1122.

Moasheri S.A., Goshki A.S., Parsaie A., "SAR" Qualities parameter persistence by a compound method of geostatic and artificial neural network (Case study of Jiroft plain), International Journal of Agriculture and Crop Sciences 6 (2013) 157-166.

Mohanty P.K., Khatua K.K., Estimation of discharge and its distribution in compound channels Journal of Hydrodynamics, Ser B 26 (2014) 144-154.

Myers W., Velocity and Discharge in Compound Channels Journal of Hydraulic Engineering 113 (1987) 753-766.

Najafzadeh M., Zahiri A., Neuro-Fuzzy GMDH-Based Evolutionary Algorithms to Predict Flow Discharge in Straight Compound Channels, Journal of Hydrologic Engineering 20. (2015).

Naot D., Nezu I., Nakagawa H., Calculation of Compound-Open-Channel Flow Journal of Hydraulic Engineering 119 (1993) 1418-1426.

Parsaie A., Haghiabi A., The Effect of Predicting Discharge Coefficient by Neural Network on Increasing the Numerical Modeling Accuracy of Flow Over Side Weir, Water Resources Management 29 (2015a) 973-985.

Parsaie A., Haghiabi A.H., Evaluation of Selected Formulas and Neural Network Model for Predicting the Longitudinal Dispersion Coefficient in River, Journal of Environmental Treatment Techniques 2 (2014a) 176-183.

Parsaie A., Haghiabi A.H., Predicting the side weir discharge coefficient using the optimized neural network by genetic algorithm Scientific Journal of Pure and Applied Sciences 3 (2014b) 103-112.

Parsaie A., Haghiabi A.H., Calculating the Longitudinal Dispersion Coefficient in River, Case Study: Severn River, UK, International Journal of Scientific Research in Environmental Sciences 3 (2015b)199-207.

Parsaie A., Haghiabi A.H., Computational Modeling of Pollution Transmission in Rivers, Applied Water Science (2015c)1-10.

Parsaie A., Haghiabi A.H., Predicting the longitudinal dispersion coefficient by radial basis function neural network, Modeling Earth Systems and Environment 1 (2015d) 1-8.

Parsaie A., Haghiabi A.H., Moradinejad A., CFD modeling of flow pattern in spillway's approach channel, Sustainable Water Resources Management 1 (2015a) 245-251.

Parsaie A., Yonesi H.A., Najafian S., Predictive modeling of discharge in compound open channel by support vector machine technique, Modeling Earth Systems and Environment 1 (2015b) 1-6.

Prinos P., Townsend R., Tavoularis S., Structure of Turbulence in Compound Channel Flows, Journal of Hydraulic Engineering 111 (1985) 1246-1261.

Sahu M., Khatua K., Mahapatra S., A neural network approach for prediction of discharge in straight compound open channel flow, Flow Measurement and Instrumentation 22 (2011) 438-446.

Seckin G., A comparison of one-dimensional methods for estimating discharge capacity of straight compound channels, Canadian Journal of Civil Engineering 31 (2004) 619-631.

Seckin G., Mamak M., Atabay S., Omran M., Discharge estimation in compound channels with fixed and mobile bed Sadhana 34 (2009) 923-945.

Stephenson D., Kolovopoulos P., Effects of Momentum Transfer in Compound Channels Journal of Hydraulic Engineering 116 (1990) 1512-1522.

Tang X., Knight D.W., Samuels P.G., Variable parameter Muskingum-Cunge method for flood routing in a compound channel, Journal of Hydraulic Research 37 (1999) 591-614.

Unal B., Mamak M., Seckin G., Cobaner M., Comparison of an ANN approach with 1-D and 2-D methods for estimating discharge capacity of straight compound channels Advances in Engineering Software 41 (2010) 120-129.

Whyte D., River Flow: New and Selected Poems (Revised). Many Rivers Press (2012).

Zahiri A., Azamathulla H.M., Comparison between linear genetic programming and M5 tree models to predict flow discharge in compound channels, Neural Comput & Applic 24 (2014) 413-420.

Williams S.C., Beresford J., The effect of anaerobic zone mixing on the performance of a three-stage bardenpho plant, Water Science and Technology 38 (1998) 55-61.

Yamaguchi T., Ishida M., Suzuki T., Biodegradation of hydrocarbon by Protothecazopfii in rotating biological contactors, Process Biochemistry 35 (1999) 403-409.

3

Head-discharge relationship of flumes: Energy loss versus boundary layer

Matthieu Dufresne[1,2,*], José Vazquez[1,2]

[1]National School for Water and Environmental Engineering of Strasbourg (ENGEES), Strasbourg cedex France.
[2]ICube (University of Strasbourg, CNRS, INSA of Strasbourg, ENGEES), Mechanics Department, Fluid Mechanics Team.

ARTICLE INFO	ABSTRACT
Keywords: Venturi flume Boundary layer Energy loss	Long-throated flumes are measurement structures often used in water and wastewater systems to determine the flow discharge. The head-discharge relationship of long-throated flumes is traditionally determined following the critical flow theory and the boundary layer concept. After a review of the traditional approach and an analysis of the approximate assumptions of the boundary layer approach, this study revisits the energy loss approach as an alternative to the questionable boundary layer concept for the determination of the discharge in long-throated flumes. Computational fluid dynamics (CFD) is used for determining the kinetic energy correction coefficient and the piezometric energy correction coefficient along the throat of the flume (especially in the critical section); CFD is also used for locating the critical section and determining the energy loss between the measurement section and the critical section. A new method based on the kinetic energy correction coefficient, the piezometric energy correction coefficient and the energy loss between the measurement section and the control section is proposed. A step-by-step procedure is given for the head-discharge calculation. It appears that the proposed alternative is a simple and promising method to accurately determine the discharge coefficient.

1. Introduction

The discharge Q in long-throated rectangular flumes can be evaluated following the traditional critical flow theory based on a hydrostatic pressure distribution and a uniform velocity distribution (Bos 1977, ISO 2013), which leads to Eq. (1). Here, h is the water depth in the approach channel; Bt, the breadth of the throat; g, the gravitational acceleration; CV = (H/h)3/2, a correction coefficient taking into account the specific energy head H in the approach channel and CD, a discharge coefficient.

$$Q = C_D C_V \left(\frac{2}{3}\right)^{3/2} \sqrt{g} B_t h^{3/2} \tag{1}$$

The discharge coefficient CD can be evaluated using the boundary layer concept (Ackers et al. 1978): it consists in calculating the notional displacement thickness of the boundary layer in the control section in order to determine the effective width and water depth. Nevertheless, this approach presents a number of deficiencies (Yeung 2007). First, the boundary layer is assumed to originate at the leading edge of the throat whereas the flume presents a converging zone that probably influences the development of the boundary layer (results on boundary layer have been obtained on a flat plate). Second, the experimental results about the boundary layer have been obtained for a flow presenting a constant free stream velocity (Harrison 1967) whereas the mean velocity is not constant between the beginning of the throat and the control section. Third, the evaluation of the transition Reynolds between laminar and turbulent boundary layer may lead to significant errors in the evaluation of the displacement thickness. Fourth, the flow is said to become critical at the end f the thoroat whereas some studies show that it may become critical near the beginning (Yeung 2007) or near the middle of the throat (Dabrowski and Polak 2012).

An alternative to this questionable boundary layer concept is the more physical energy loss approach (Yeung 2007). This approach was followed amongst others by Hager (1985) assuming hydrostatic pressure distribution and uniform velocity distribution in the control section. Nevertheless, these assumptions are also questionable in Venturi flumes.

The objective of this study is to revisit this alternative without any assumption about the pressure and the velocity distributions and to determine whether this method is simple and viable for practical applications. This research is motivated by the desire to improve discharge determination with flumes; our belief is that it involves the development of a physically based method.

Generalized critical flow theory for discharge calculation

Without any assumption about the velocity and the pressure distributions, the specific energy head can be written as Eq. (2) (Jaeger 1956, Castro-Orgaz and Chanson 2009), where Ke and α are defined by Eq. (3) and Eq. (4) respectively. Here, U is the mean velocity; z, the vertical coordinate (the origin is located at the invert of the channel); P, the pressure; ρ, the water density; $\vec{V}$, the velocity (three components) and A, the surface area of the flow cross-section.

$$H = K_e h + \alpha \frac{U^2}{2g} \tag{2}$$

$$K_e = \frac{1}{Q} \iint_A \left(\frac{z + \frac{P}{\rho g}}{h}\right) \left|\vec{V}.\vec{dA}\right| \tag{3}$$

*Corresponding author E-mail: matthieu.dufresne@engees.unistra.fr

$$\alpha = \frac{1}{Q} \iint_A \left(\frac{V}{U}\right)^2 \left|\vec{V} . \overrightarrow{dA}\right| \tag{4}$$

The piezometric energy correction coefficient Ke conveys the gap between the pressure distribution and an idealized hydrostatic one. It is equal to 1 when the distribution is hydrostatic; higher (respectively lower) than 1 when the pressure is higher (respectively lower) than the hydrostatic one. The kinetic energy correction coefficient α quantizes the difference between the actual velocity distribution and an idealized uniform one. Clemmens et al. (2001) proposed an expression for the evaluation of α in the control section but this expression is limited to wide channels ($h/Bt < 0.33$).

Critical flow occurs when the specific energy head H has its minimum value for a given discharge Q (Hager 1999), that means $dH/dh = 0$, which leads to Eq. (5) for the critical depth.

$$h_c = \left(\frac{\alpha}{K_e}\right)^{\frac{1}{3}} \left(\frac{Q^2}{gB_t^2}\right)^{\frac{1}{3}} \tag{5}$$

Once the critical depth is known, Eq. (2) can then be used to evaluate the critical specific energy head. The link with the specific energy head H in the measurement section in the approach channel can be made with the energy loss j between the measurement section and the control section. Finally, the discharge equation can be written as Eq. (1) where CD, whose expression is given below, is a correction coefficient taking into account the non-hydrosticity of the pressure and the non-uniformity of the velocity in the control section, and also the energy loss between the measurement section and the control section.

$$C_D = \frac{1}{K_e \sqrt{\alpha}} \left(\frac{H-j}{H}\right)^{\frac{3}{2}} \tag{6}$$

2. Methods

Since experimental studies generally only investigate global variables such as the discharge and the water level in the approach channel, Computational Fluid Dynamics (CFD) has been here used to generate data. The geometry experimentally investigated by Yeung (2007) was chosen as the test case since it follows the requirements of ISO 4359. The throat of the flume is 300 mm long and 101 mm wide; the breadth of the approach channel is 203 mm. Discharges from 2 L/s to 14 L/s were investigated, corresponding to the experimental range studied by Yeung (2007).

Numerical simulations were performed with the computational fluid dynamics finite volume code ANSYS-FLUENT (ANSYS 2010). The three dimensional Reynolds-averaged Navier-Stokes (RANS) equations were used. In order to reproduce the non-uniformity of the water level distribution, the two-phase Volume of Fluid model was chosen. Since this may have a significant influence on the velocity and pressure distributions, two turbulence models were used and compared for the minimum and the maximum discharge: the simple and isotropic k-ε turbulence model and the anisotropic Reynolds Stress Model (RSM) able to reproduce complex velocity distributions such as secondary currents of the second kind driven by turbulence anisotropy. This comparison showed a difference lower than the numerical uncertainty presented below, which means that the influence of the anisotropy of the turbulence is not significant on the investigated variables. For this reason, the simple k-ε model was chosen for all the simulations.

A grid sensitivity analysis was performed in order to evaluate the Grid Convergence Index (GCI) as an estimator of the numerical uncertainty (Roache 1994). This analysis leads to the following conclusions: the precision of the numerical model is about 0.1 mm for the water depths, the specific energy heads and the energy losses; the precision on the correction coefficients α and Ke is less than 0.005; finally, the precision on the location of the control section is about 2 mm.

3. Results
3.1. Distribution of the correction coefficients along the flume

Fig. 1 and Fig. 2 show important changes of the correction coefficients α and Ke along the flume. Indeed, the kinetic energy correction coefficient α is abruptly increasing a few centimeters upstream of the converging zone with a maximum value of approximately 1.16 at the entrance of the converging zone for all the discharges. This behavior can be simply explained by the contraction of the main stream in the middle zone of the channel, which leads to an increase of the heterogeneity of the velocity distribution. In the converging zone, α is rapidly decreasing to values around 1.01 and 1.02 in the throat of the flume, whatever the discharge is. In the diverging zone, the velocity distribution becomes more heterogeneous, leading to an increase of α.

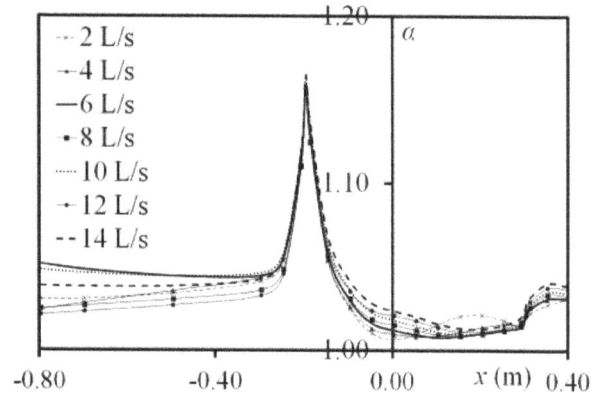

Fig. 1. Distribution of the kinetic energy correction coefficient α along the flume – x = 0 corresponds to the beginning of the throat.

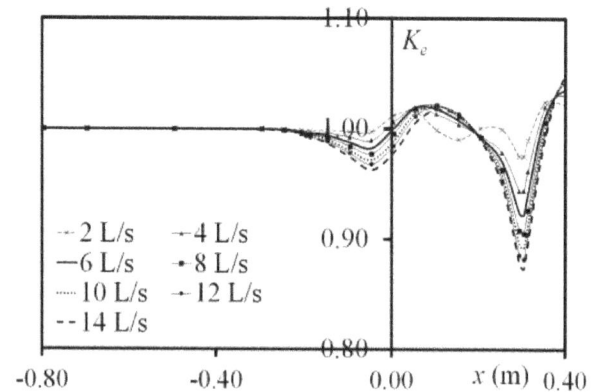

Fig. 2. Distribution of the piezometric energy correction coefficient Ke along the flume.

Concerning the piezometric energy correction coefficient Ke, it is equal to 1 in the approach channel, which can be simply explained by the almost horizontal water level in this region. At the beginning of the converging zone, the water level is decreasing; the streamlines become curved with a center of curvature lying below the free surface, which leads to a decrease of the pressure (Ke < 1). At the end of the entrance, the center of curvature alternatively lies above and below the free surface, which leads to values of Ke respectively higher and lower than 1 (between 0.88 and 1.02 for 14 L/s). Contrarily to the basic assumption of the traditional approach, Ke is not equal to 1 in the throat, even if the length of the throat is long.

3.2. Description of the control section

Using the values of α and Ke given by Computational Fluid Dynamics, the position of the control section can be determined by comparison between the water level and Eq. (5) at each abscissa of the flume. With the exception of the discharge 2 L/s for which the control section is situated near the downstream section of the throat (at 80% of the length Lt), the control section is located between 25% and 40% of the length of the throat, which proves that the assumption of the traditional approach is not correct (see Fig. 3). It should be noticed that the validity of the numerical model is questionable for the discharge 2 L/s, as highlighted by the comparison between the experimental results of Yeung (2007) and the numerical results in Fig. 1 (difference of 2% in the discharge coefficient); it can be explained by the fact that the turbulence model is not completely suitable for low Reynolds numbers (approximately 104 in the throat for a discharge of 2 L/s).

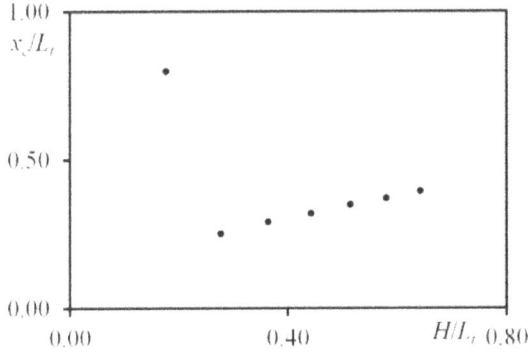

Fig. 3. Position of the critical section xc/Lt as a function of H/Lt – xc measured from the beginning of the throat; Lt is the length of the throat.

The values of the correction coefficients α and Ke in the control section are nearly constant for the whole simulations: 1.01 and 1.02 respectively. This highlights a constant hydraulic behavior for the control section, whatever the discharge is. This observation is very promising for the future generalization of the method proposed in this paper.

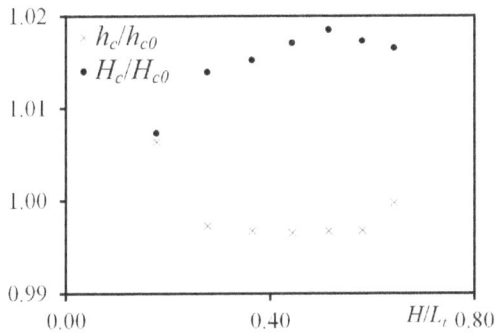

Fig. 4. Ratios hc/hc0 and Hc/Hc0 as a function of H/Lt.

The propagation of these values in Eq. (6) shows that the actual critical depth hc is not very different from the traditional critical depth hc0, as highlighted in Figure 4. Indeed, the correction factor $(α/Ke)^{1/3}$ is very close to 1. On the contrary, the propagation in Eq. (2) highlights that the actual specific energy head Hc is significantly higher than the one calculated following the traditional approach, namely Hc0 (see also Fig. 4). Energy loss between the measurement section and the control section.

The total loss between the measurement section and the control section ranges between 1.0 mm for 4 L/s and 1.9 mm for 14 L/s, which is not negligible if a high accuracy on the discharge is needed. In order to investigate this point, the energy loss has been divided into two parts: the head loss between the measurement section and the upstream section of the converging zone jm_cv, and the head loss between the upstream section of the converging zone and the control section jcv_c. The first one represents between 15% and 20% of the total loss whereas the second one represents up to 85%.

It has been verified that the energy loss between the measurement section and the upstream section of the converging zone jm_cv can be evaluated with a classical friction loss formula such as the Colebrook's formula (assuming that the water depth is almost constant). For the energy loss between the upstream section of the converging zone and the control section jcv_c, such a model is more complicated to use because of the non-homogeneity of the correction coefficients along the flume. It is here proposed to evaluate the friction

loss using a local loss formulation, as written in Eq. (7). Here Uc is the mean velocity in the control section and K a loss coefficient to be calibrated.

$$j_{cv\_c} = K \frac{U_c^{\,2}}{2g} \tag{7}$$

The analysis of the numerical results shows that the value K = 0.025 is adapted between 4 L/s and 14 L/s. The generalization of this approach will need an investigation of K as a function of the geometry of the converging section.

3.3 Procedure of the head-discharge calculation
3.3.1 Description of the method

Even if it is obvious that further investigations are needed to characterize the correction coefficients in the critical section and the energy loss coefficient for other geometric configurations, the previous analysis has shown the capability of the proposed method to determine the head – discharge relationship of long-throated rectangular flumes Venturi flumes. The determination can be done using the following procedure.

- Select a series of values of critical depths hc.
- Calculate the discharge Q and the critical energy head Hc using respectively Eq. (5) and Eq. (2). Consider α = 1.01 and Ke = 1.02.
- Calculate the critical velocity Uc = Q/(Bt×hc).
- Calculate the energy loss between the entrance of the flume and the control section jcv_c using Eq. (7) with K = 0.025.
- Iterative procedure: consider h = Hc as the initial point.
- Calculate the energy loss between the measurement section and the upstream section of the converging zone jm_cv using the Colebrook's equation.
- Calculate the total energy loss j = jm_cv + jcv_c and then the energy head H = Hc + j in the measurement section.
- Modify h until the energy head H in the measurement section is equal to the previous calculation (end of the iterative procedure). In the measurement section, consider Ke = 1 and α = 1.05. One should precise that the influence of α in the measurement section is not significant between 1.00 and 1.10.
- Calculate CV = (H/h)3/2 and CD using Eq. (6).

3.3.2. Verification of the method with the numerical results

The application of the proposed method is illustrated and compared with the experimental/numerical points and the traditional approach in Figure 5. The comparison with experimental highlights that the energy loss based approach leads to accurate results. Moreover, the flatter line of the proposed method seems to be more adapted to the experimental points than the traditional approach. It is also obvious in this figure that the correction coefficients α and Ke have a significant influence on the discharge determination (see the comparison between "α = 1.00, Ke = 1.00" and "α = 1.01, Ke = 1.02").

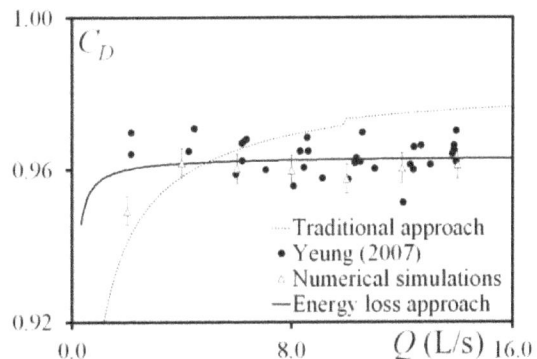

Fig. 5. Experimental results (Yeung 2007) versus numerical results; traditional approach (boundary layer) versus energy loss approach.

4. Conclusions and perspectives

This study has shown a number of deficiencies of the traditional approach for the determination of the head – discharge relationship of flumes. First, the control section is not located at the downstream end of the throat but in the first half of the throat. Second, the pressure distribution in the control section cannot be considered as hydrostatic, even for long-throated flumes. Third, the velocity distribution is not uniform in the control section.

Based on these observations, a new method has been proposed for the calculation of the discharge coefficient as an alternative to the boundary layer concept: this method is based on the utilization of the kinetic energy correction coefficient α and the piezometric energy correction coefficient K_e, and also the calculation of the energy loss along the flume. The comparison with experimental data has shown that the proposed alternative is a simple and promising method to accurately determine the discharge coefficient. Further work is needed to investigate the values of the correction coefficients and the loss coefficient as a function of the geometric and hydraulic characteristics of the flume.

References

Ackers P., White W.R., Perkins J.A., Harrison, A.J.M., Weirs and flumes for flow measurement, Wiley, New-York, (1978).

ANSYS, FLUENT user's guide, ANSYS Ltd, (2000).

Bos M.G., The use of long-throated flumes to measure flows in irrigation and drainage canals, Agricultural Water Management 1 (1977) 111-126.

Castro-Orgaz O., Chanson H., Bernoulli theorem, minimum specific energy, and water wave celerity in open-channel flow, Journal of Irrigation and Drainage Engineering 135(6) (2009) 773-778.

Clemmens A.J., Wahl T.L., Bos M.G., Replogle J.A., Water measurement with flumes and weirs, International Institute for Land Reclamation and Improvement, Wageningen, The Netherlands, (2001).

Dabrowski W., Polak U., Improvements in flow rate measurement by flumes, Journal of Hydraulic Engineering 138(8) (2012) 757-763.

Hager W.H., Critical flow condition in open-channel hydraulics, Acta Mechanica 54 (1985) 157-179.

Hager W.H., Wastewater hydraulics – Theory and practice, Springer, (1999).

Harrison A.J.M., Boundary layer displacement thickness on flat plates, Journal of the Hydraulics Division 93(HY4) (1967) 79-91.

Jaeger C., Engineering fluids mechanics, Blackie and Son, (1956).

ISO, Liquid flow measurement in open channels – Rectangular, trapezoidal and U-shaped flumes, International Standardization Organization, ISO 4359:2012, (2012).

Roache P.J., Perspective: a method for uniform reporting of grid refinement studies, Journal of Fluids Engineering 116 (1994) 405-413.

Yeung H., An examination of BS3680 4C (ISO/DIS 4369) on the measurement of liquid flow in open channels-flumes, Flow Measurement and Instrumentation 18 (2007) 175-182.

4

Assessment of some famous empirical equation and artificial intelligent model (MLP, ANFIS) to predicting the side weir discharge coefficient

Abbas Parsaie[*], Amir Hamzeh Haghiabi

Water Engineering Department, Lorestan University Khoram Abad, Lorestan Province, Iran.

ARTICLE INFO

Keywords:
Side weir
Discharge coefficient
AI model (ANN, ANFIS)
Empirical formula

ABSTRACT

Allocation and removing of excess water from the irrigation and drainage network is one of the most important activities in the management of these networks. Side weir is one of the most common structures for this purpose. Study on the flow Hydraulic characteristics of this structure included two parts, defining the water surface profiles and estimating the discharge coefficient. To estimate the discharge coefficient, many ways as experimental formulas and artificial intelligent models are propose. The empirical formula for simplifying in developing process that assume by the authors, contained significant error so using the AI models are inevitable. In this paper, some of the famous empirical formula and AI models such as Multilayer neural network (MLP) and Adaptive Neuro fuzzy inference system (ANFIS) are assessing with a laboratory experiment. Among the experimental formula, Borghei formula is most accurate (R^2=0.83) and the performance of the AI model in Training and testing stage is more suitable (R^2=0.96).

1. Introduction

Today, due to the shortage of water resources for agriculture, the modern agriculture has increased interest. One of the most important problems in the modern agriculture is the managing the water. One of the most powerful tools to do this much better is modeling of water structure that used common in the irrigation and drainage network (Misra. 200).

Modeling and Predicting the hydraulic phenomenon is an important part of hydraulic engineering activities that leads to do much better managing of the Hydraulic structures.

Estimating the flow rate through the Hydraulic structure is one of the most important issues in Hydraulic engineering (Hager. 1987)

Side weir is an overflow and set into the side of the channel. This structure has been used to Allocation and removing excess flow therefore it used in the most water engineering project such as irrigation network, land drainage, urban sewage systems and sometimes used as an overflow dam (Ghodsian. 2004).

When the water level reaches above the side weir crest its divert a certain amount of discharge (May et al. 2003). like all typical weir, side weir can be sharp, broad crested with various geometric and also the flow in the main channel can be sub-critical and supercritical (Subramanya et al. 1972).

Many studies were due on defining the hydraulic behavior of this type of weir, which are often experimental method (Borghei et al. 2011; Emiroglu et al. 2011; Kumar et al. 2012). The flow over a side weir is a typical case of spatially varied flow (SVF) with decreasing discharge (AL-TAEE. 2012). De Marchi with assuming constant energy, obtained equation of the (SVF)and to calculate the outflow discharge form the side, he solved analytically the (SVF). Finally, he proposed an equation for side weir discharge coefficient(CDSW) that today is known as the De Marchi coefficient (Hadadi et al. 2012).

More recently, Laboratory Studies conducted on the various types of this structure to improve the performance of them (Emiroglu et al 2011; Emiroglu et al 2012; Kaya. 2010). An experimental and theoretical investigations have been done on the flow over labyrinth,

Oblique, semi-elliptical, Curved Plan-form and trapezoidal sharp and broad-crested side weir in rectangular channels in the under subcritical and supercritical flow conditions. the results show that CDSW is related to the Froude number at the upstream of the weir, ratios of weir height to depth of flow, weir length to width of main channel and length of broad-crested weir to width of main channel. The end of all this researches show the effect of each geometric shape of the Side weir on the rising the performances of this structure and for each specific geometric shape of the Side weir experimental formulas are proposed (Emiroglu et al. 2011; Kumar et al. 2012; Cosar et al. 2004, Honar et al. 2007; Kumar et al. 2013; Kumar et al. 2013). The simple Rectangular side weir used very much in water engineering projects because facility in the construction and operation (AL-TAEE. 2012; Rahimi. 2012).

The governing equation on side weir hydraulic characteristics is spatially varied flow (SVF). Computer simulating of flow over side weir included two parts. One- calculating the water surface profile therefore, to this purpose SVF equation must be solved with a suitable numerical and two- estimating the CDSW with empirical formula or Artificial intelligent techniques (Rahimi. 2012) .

In the field of studying the flow characteristics of side weir, predicting the CDSW is more important so, more research conducted on determining this factor in Table (1), some of the empirical equation that proposed for the Sharp crested rectangular side weir was collected (Borghei et al. 2011; Hadadi et al. 2012; Bilhan et al. 2011; Kaya et al. 2011; Parsaie et al. 2013).

To predicting the CDSW, the investigations based on the artificial intelligence techniques have been done also. In Artificial Intelligence Studying a network is develop Instead of a relationship that results from the linear or nonlinear regression. The ANN (Artificial Neural Network), ANFIS (Adaptive Neuro Fuzzy Inference System) and SVM (Support Vector Machine) models in the field of artificial intelligence techniques was used to calculate the CDSW. Based on the reported, accuracy of all these methods is much more than empirical equation (Bilhan et al. 2011; Parsaie et al. 2013). The conclusion that can be drive from a review on literature is that study on hydraulic

[*]Corresponding author E-mail: Abbas_Parsaie@yahoo.com

characteristic of side weir started with Laboratory investigation and to computer simulating of flow over side weir, addition to solve the SVF for determining the water surface profile, the CDSW must predict by a suitable method. The goal of this research is assessing the empirical formula and performance of AI model to calculating and predicting the CDSW in a real laboratory investigation.

Table 1. Some famous empirical formulas to calculation of CDSW

row	Author	Equation
1	Nandesamoorthy et al.	$C_d = 0.432\left(\dfrac{2-Fr_1^2}{1+2Fr_1^2}\right)^{0.5}$
2	Subramanya et al	$C_d = 0.864\left(\dfrac{1-Fr_1^2}{2+Fr_1^2}\right)^{0.5}$
3	Yu-Tech	$C_d = 0.623 - 0.222Fr_1$
4	Ranga Raju et al	$C_d = 0.81 - 0.6Fr_1$
5	Hager	$C_d = 0.485\left(\dfrac{2-Fr_1^2}{2+3Fr_1^2}\right)^{0.5}$
6	Cheong	$C_d = 0.45 - 0.221Fr_1$
7	Singh et al	$C_d = 0.33 - 0.18Fr_1 + 0.49\left(\dfrac{P}{h_1}\right)$
8	Jalili et al	$C_d = 0.71 - 0.41Fr_1 + 0.22\left(\dfrac{P}{h_1}\right)$
9	Borghei	$C_d = 0.7 - 0.48Fr_1 + 0.3\left(\dfrac{P}{h_1}\right) + 0.06\left(\dfrac{L}{h_1}\right)$

2. Methodology

First dimensionless factors influence in the CD_{SW} was extracted with using dimensional analysis technique then to assessing the empirical equation, some data range published in Articles was collect in table 2. The number of these data is about 140. In the Following the MLP and ANFIS models were developed. Training and testing process of these models has done with the same data. All the process of the assessing and preparing the empirical formula and AI model was programing in Matlab software.

Table 2. Range of data collected.

	F1	p/h1	L/b	L/h1	Cd
min	0.09	0.34	0.30	0.35	0.28
max	0.83	0.91	3.00	10.71	1.75
mean	0.41	0.77	1.60	3.99	0.57
stdv	0.20	0.14	1.11	3.10	0.26

2.1. Dimensional analysis

Referring to Fig. 1 the CD_{SW} can be written as a function of width of channel (b), flow depth in the main channel(h1), mean velocity of flow at upstream end of side weir (V1),length of side weir (L), acceleration due to gravity (g), slope of mainchannel bed (So), deviation angle of flow (ψ),

$$C_d = f\left(v_1, L, b, h_1, P, \psi, s_0\right) \quad (1)$$

Using the Buckingham π theorem, nondimensional equations infunctional forms can be obtained as below:

$$C_d = f_1\left(Fr_1 = \frac{v_1}{\sqrt{gh_1}}, \frac{L}{b}, \frac{L}{h_1}, \frac{P}{h_1}, \psi, s_0\right)$$

$$\sin(\psi) = \sqrt{1 - \left(\frac{V_1}{V_s}\right)^2} \quad (2)$$

where Fr is Froude number. El-Khashab also mentioned that the dimensionless length of the side weir (L/b) includes the effect of the deviation angle on the discharge coefficient. Therefore, the deviation angle ψ is not existed in the CD_{SW} equations in the literature. Thus, the CD_{SW} depends on the following dimensionless parameters (Emiroglu et al. 2011; Kaya et al. 2011).

$$C_d = f_2\left(Fr_1, \frac{L}{b}, \frac{L}{h_1}, \frac{P}{h_1}\right) \quad (3)$$

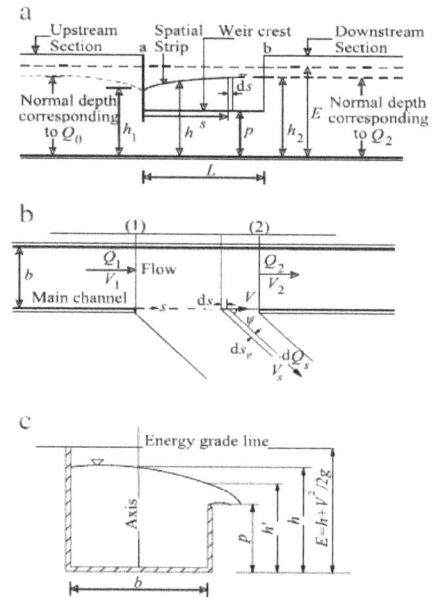

Fig. 1. Definition sketch of subcritical flow over a rectangular side weir.

2.2. Artificial neural network (ANN)

The ANN is a nonlinear mathematical model that is able to simulate arbitrarily complex nonlinear processes that relate the inputs and outputs of any system. In many complex mathematical problems that leads to solve complex nonlinear equations, A Multilayer Perceptron network with definition of Appropriate functions, weights and bias can used. Due to the nature of the problem, different Activity functions in neurons can be used. An ANN has one or more hidden layers. Fig. 2 demonstrates a three-layer neural network consisting of inputs layer, hidden layer (layers) and outputs layer. As shown in the fig.2 the w_i is the weights and b_i is the bias for each neuron. Initial assigned weight values will progressively corrected during a training process that compares predicted outputs to known outputs. Such networks are often trained using back propagation algorithm. In the current study, the ANN was trained using Levenberg–Marquardt technique because this technique is more powerful and faster than the conventional gradient descent technique (Aleksander et al. 1995).

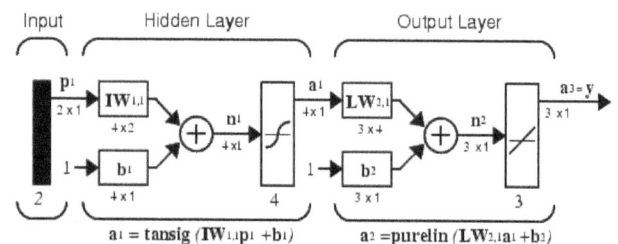

Fig. 2. A three-layer ANN architecture.

2.3. Adaptive neuro fuzzy inference system (ANFIS)

Adaptive Neuro fuzzy Inference Systems (ANFIS) is a powerful tool to modeling of complex system based on input and output data. ANFIS are realized by an appropriate combination of neural and fuzzy systems. This combination enables to use both the numeric power of intelligent systems.

In fuzzy systems, different fuzzification and defuzzificationn strategies with different rule was considering for inputs parameter. For doing the effect of fuzzy logic on the inputs data, selecting of membership function can be selected. In this stage maybe considered a Gaussian function for each of inputs variables. Figure (3.a) shows a fuzzy reasoning process. For simplicity illustrating, a fuzzy system with two inputs variable and one output was choiced. Suppose that the rule base containing two fuzzy if-then rules.

$$Rule\,1 : if\ \ x\ is\ A_1\,and\ y\ is\ B_1 then f_1 = p_1 x + q_1 y + r_1$$

$$Rule\,2 : if\ \ x\ is\ A_2\,and\ y\ is\ B_2 then f_2 = p_2 x + q_2 y + r_2$$

where A_1; A_2 and B_1; B_2 are the MFs for inputs x and y; respectively; p_1; q_1; r_1 and p_2; q_2; r_2 are the parameters of the output function. ANFIS architecture is presented in Fig. 3(b), in the first layer, all the inputs variable gave the grade membership with membership function. and in layer 2, all the membership grade will be multiplying together and in the layer 3, all the grade of member will be normalized and in the layer 4 contribution of all the rule will be compute. In addition, in the last layer output variable will be compute as weighted average of grade membership.

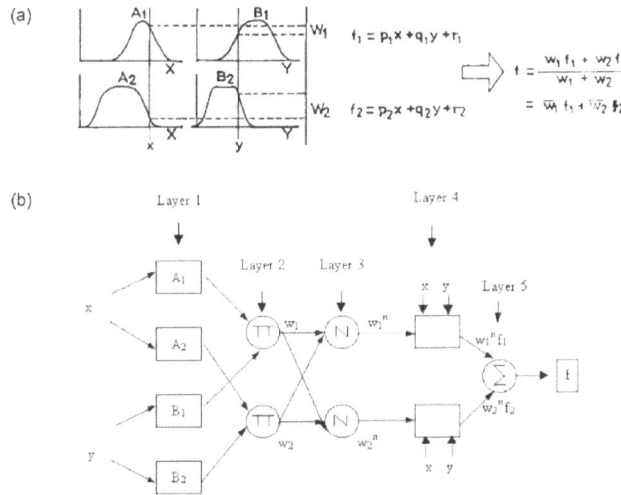

Fig. 3. (a) Fuzzy inference system, (b) Equivalent ANFIS architecture.

3. Results and discussion

Firstly, evaluating the empirical formulas was does with the collected data set. The error indices (eq.11) were calculated and result of them and given in the tables (3). As given in the Table (3), the Borghei equation is most accurate among the empirical equation. To much more accuracy of the predicting the CD_{sw}, the AI models (MLP and ANFIS) was developed. Developing the AI models (MLP and ANFIS) does with the same data collected. The performance of the AI models in each stage of preparing such as training, validation and testing was calculated and shows in the Fig. (6 to 10). For developing the MLP model, 70% of data was used to training, 15 % used for validation and 15%used for testing of models. For developing the ANFIS model, the 80% of data was used for training 25%used for testing of models. The three-layer has been considered for MLP model structure. The first layer consider as inputs and second as hidden layer and third as outputs. In the hidden layer as shows in the Fig 4, five neurons with tansig transfer function was considered. The levenberge _Marquat technique was used to training this model. In the ANFIS model as shows in the Fig. 5, the four-membership function is considered for each inputs parameters. A hybrid technique is implementing to training the ANFIS model. To assessing the accuracy of the empirical formulas and AI models in the real engineering problem. These models were used to calculate the CD_{sw} in the experiment laboratory study that did with the Emiroglu et al (2011). The experiment did in a flume with 12m in long, 0.5 m in width, 0.7 m deep and longitudinal slope was equal to 0.001. The result of the Borgei equation and AI model was calculates for two runs of the experiments and give in the Table (4).

The result of this study showing that using the suitable AI model is much more accuracy in the prediction of the CD_{sw}. This increasing in accuracy in prediction, caused to better operation of this structure.

$$RSME = \sqrt{\frac{1}{N}\sum_{i=1}^{N}(C_d Obs - C_d est)}$$

$$MAE = \frac{1}{N}\sum_{i=1}^{N} abs(C_d Obs - C_d est) \qquad (11)$$

$$APE = \frac{1}{N}\sum_{i=1}^{N} abs\left(\frac{C_d Obs - C_d est}{C_d Obs}\right)$$

Table 3. The RMSE, MAE, AP and R^2 statistics of the empirical equations

Equation	R^2	RSME	MAE	APE
Nandesamoorthy	0.29	0.315	0.205	32.53
Subramanya	0.34	0.315	0.207	33.88
Yu-Tech	0.277	0.285	0.192	32.23
Ranga Raju	0.277	0.336	0.237	41.72
Hager	0.17	0.312	0.19	25.53
Cheong	0.34	0.325	0.194	25.89
Singh	0	0.273	0.209	40.86
Jalili	0.337	0.372	0.235	32.41
Borghei	0.834	0.288	0.247	51.79

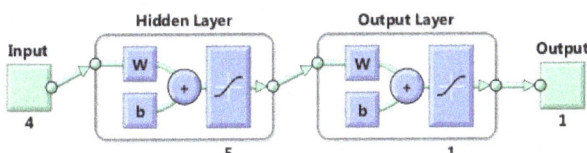

Fig. 4. The structure of the MLP model has been developed.

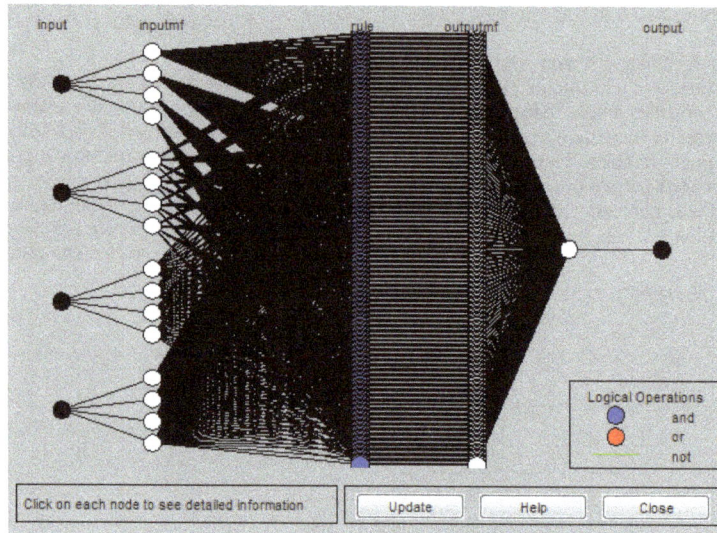

Fig. 5. The structure of the ANFIS model has been developed.

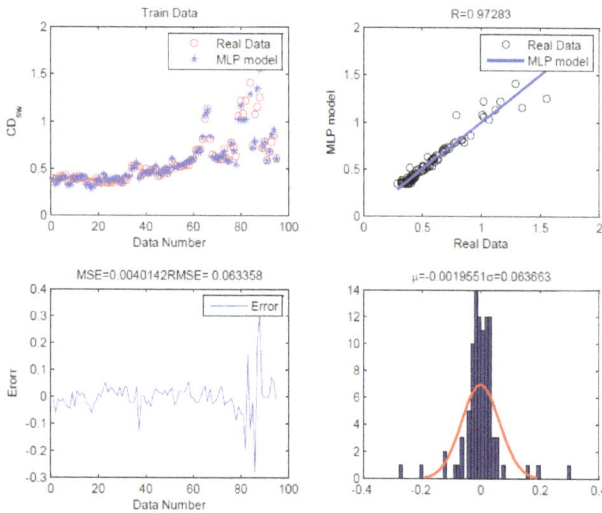

Fig. 6. Performance of MLP models during the training stage.

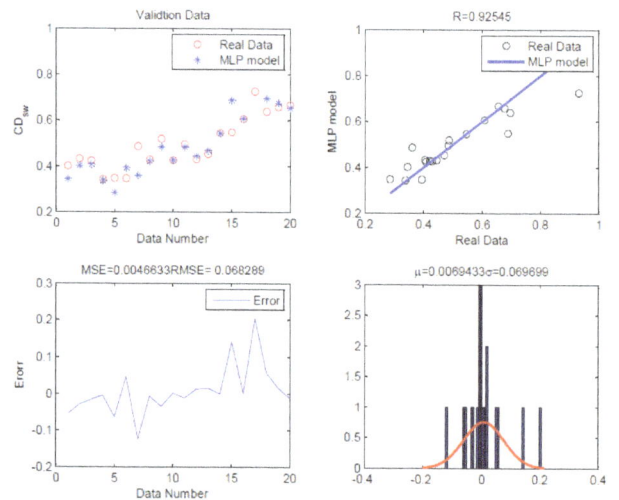

Fig. 8. Performance of MLP models during the test stage.

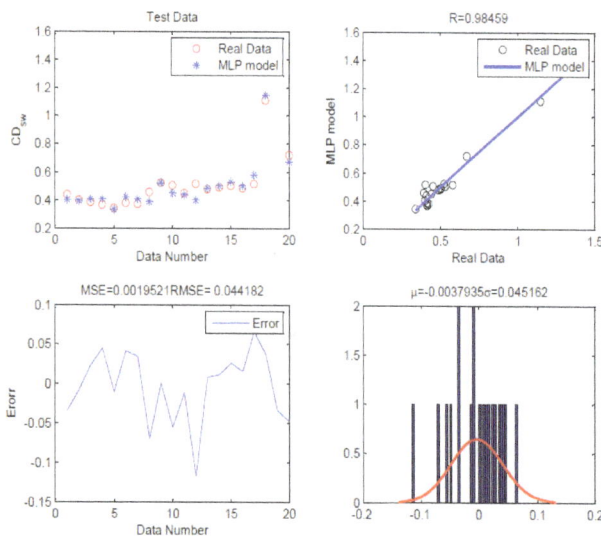

Fig. 7. Performance of MLP models during the Validation Stage.

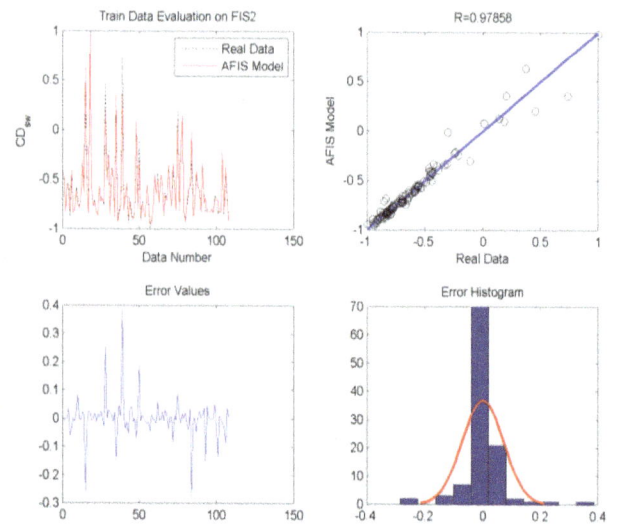

Fig. 9. Performance of ANFIS models during the training stage.

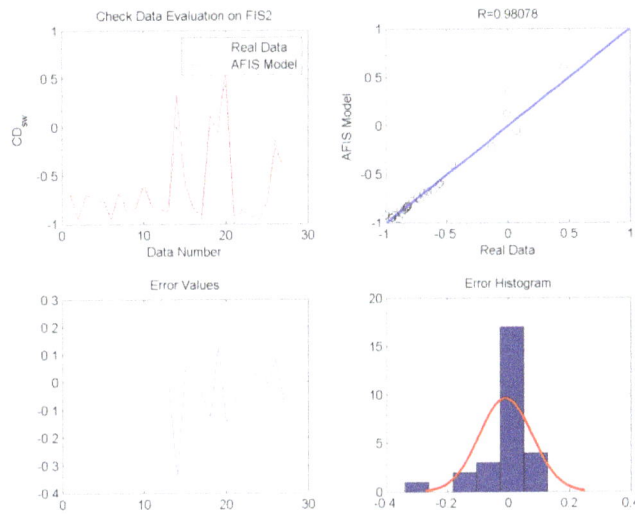

Fig.10. Performance of MLP models during the test stage.

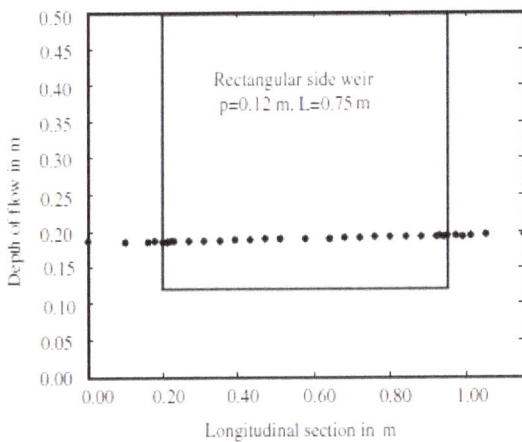

$F_l=0.28$, Surface profile along the centerline

Fig. 11. Run (1) of the Laboratory experiment of flow over side weir.

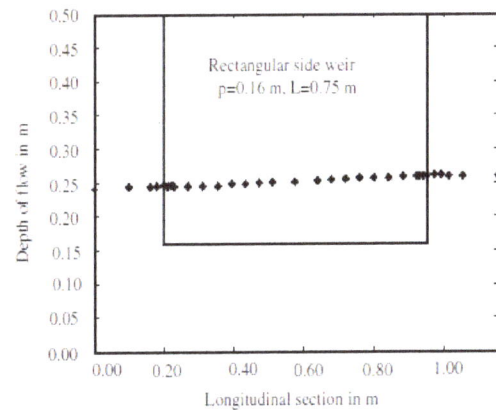

$F_l=0.39$, Surface profile along the centerline

Fig. 12. Run (2) of the Laboratory experiment of flow over side weir.

Table 4. The parameter that used in computer modeling.

Run Parameter	Fr	P	L	CD_{SW} (ANN)	CD_{SW} (ANFIS)	C_d (borghei)	Measurement
Run -1	0.28	0.12	0.75	0.52	0.54	0.97	0.58
Run -2	0.39	0.16	0.75	0.47	0.46	0.88	0.44

References

Aleksander I., Morton H., An Introduction to Neural Computing, 1995, Internat, Thomson Computer Press.

Al-Taee A.y., Theoretical analysis of flow over the side weir using runge kutta method. annals of faculty engineering hunedoara – International Journal Of Engineering 4 (2012) 47-50.

Bilhan O., Emiroglu M.E., Kisi O., Use of artificial neural networks for prediction of discharge coefficient of triangular labyrinth side weir in curved channels, Advances in Engineering Software 42 (2011) 208–214.

Borghei S.M., and Parvaneh A., Discharge characteristics of a modified oblique side weir in subcritical flow, Flow Measurement and Instrumentation 22 (2011) 370–376.

Cosar A., and Agaccioglu H., Discharge Coefficient of a Triangular Side-Weir Located on a Curved Channel, Journal ofirrigation and drainage engineering 130 (2004) 410–423.

Emiroglu M.E., Agaccioglu H., and Kaya N., Discharging capacity of rectangular side weirs in straight open channels, Flow Measurement and Instrumentation 22 (2011) 319-330.

Emiroglu M.E., Kisi O., Bilhan O., Predicting discharge capacity of triangular labyrinth side weir located on a straight channel by using an adaptive neuro-fuzzy technique, Advances in Engineering Software 41 (2010) 154–160.

Ghodsian M., Flow over triangular side weir, Journal Scientia Iranica Transaction on Civil Engineering 11 (2004) 114–120.

Haddadi H., M. Rahimpour M., A discharge coefficient for a trapezoidal broad-crested side weir in subcritical flow. Flow Measurement and Instrumentation 26 (2012) 63–67.

Hager H., lateral outflow over side weirs, Journal of Hydraulic Engineering 113 (1987) 491-504.

Honar T., Javan M., Discharge Coefficient in Oblique Side Weirs, Iran Agricultural Research 25 (2007) 27-35.

Kumar S., Discharge Characteristics of Sharp Crested Weir of Curved Plan-form. Research Journal of Engineering Sciences 1 (2012) 16-20.

Kaya N., Effect of upstream crest length on flow characteristics and discharge capacity of triangular labyrinth side weirs, Scientific Research and Essays 5 (2010) 1702-1712.

Kaya N., Emiroglu M.E., Agaccioglu H., Discharge coefficient of a semi-elliptical side weir in subcritical flow, Flow Measurement and Instrumentation 22 (2011) 25-32.

Kumar S., A New Approach to Analyze the Flow over Sharp Crested Curved Plan form Weirs, International Journal of Recent Technology and Engineering 2 (2013) 24-28.

Kumar S., Ahmad Z., Mansoor T., A new approach to improve the discharging capacity of sharp-crested triangular plan form weirs, Flow Measurement and Instrumentation 22 (2011) 175–180.

May R.W.P., hydraulic design of side weir. 2003, London: Thomas Telford publishing.

Misra S.R., Managing Canal Irrigation in India: Problems and Their Resolutions, 2000: Concept Publishing Company.

Parsaie A., Haghiabi A.H., Development and evaluating of two-neural network model (MLP and SVM) to estimate the Side weir discharge coefficient, International Journal of Agriculture and Crop Sciences 5 (2013) 2804-2811.

ahimi A., Hydraulic Design of Side Weirs by Alternative Methods, Australian Journal of Basic and Applied Sciences 6 (2012) 157-167.

ibramanya K., SC A., Spatially varied flow over side weirs. J. of the Hydraul Division 98 (1972) 1-10.

5

Flood frequency analysis using density function of wavelet

Sajad Shahabi[*], Masoud Reza Hessami Kermani

Department of Civil Engineering, Bahonar University, Kerman, Iran.

ARTICLE INFO

ABSTRACT

Keywords:
Density function
Energy function
Flood frequency analysis
Polroud river
Wavelet transform

In this paper, we present a method to perform flood frequency analysis (FFA) when the assumption of stationary is not important (or not valid). A wavelet transform model was developed to FFA. A full series was applied to FFA using two different wavelet functions, and then a combined method was investigated. In the combined method, all discharge data which were less than the lowest value of annual maximum (AM) discharge were removed. Furthermore, energy function of wavelet was used for FFA. The data was decomposed into some details and an approximation through different wavelet functions and decomposition levels. The approximation series was employed to FFA. This was performed using discharge data from of the Polroud River in Iran. This analysis was performed on the daily maximum discharge data from the Tollat station in the north of Iran. Data from 1975 to 2007 was evaluated by the wavelet analysis. The study shows that the wavelet full series model results (density function) are too small in compared with the results of combined method and they are both lesser than traditional methods (AM and PD). On the other hand, the results of energy function method were closed to the combined method when they are compared with the full series data results. These wavelet models were assessed with the AM and PD methods. The concrete result of this paper is that, the basin hydrologic conditions and data's nature are very important parameters to improve FFA and to select the best method of analysis.

1. Introduction

Despite over half a century of research on flood frequency analysis (FFA), the new methods continuously are being presented in this important branch of hydrology, which indicates its importance. Hence, increasing the accuracy in this area has been considered by many researchers. Stationary data is used in most of traditional methods such as annual maximum (AM) and partial duration (PD).

Hydrologic systems are sometimes impacted by extreme events as severe storms, floods and droughts. The magnitude of an extreme event is inversely related to its occurrence frequency; very severe events occur less frequently than more moderate events. The objective of frequency analysis of hydrologic data is to relate magnitude of extreme events to their frequency of occurrence through the use of probability distributions. The hydrologic data analyzed are assumed to be independent and identically distributed; and the hydrologic system producing them (e.g., a storm rainfall system) is considered to be stochastic space independent, and time independent in the classification scheme. The hydrologic data employed should be carefully selected so that the assumptions of independence and identical distribution are satisfied. In practice, this is often achieved by selecting the annual maximum (AM) of the variable being analyzed (e.g., the AM discharges, which is the largest instantaneous peak flow occurring at any time during the year) with the expectation that the successive observations of this variable from year to year will be independent (Chow et al. 1988).

The main aim of the FFA in hydrology is to determine the relationship hydrograph-return period. Until now, most of literatures investigated the flood peak univariate statistical procedures. However, concerning hydraulic works above all for flooding and inundation management, it is not enough to know information about flood peak only, but it is also useful to statistically estimate flood volume and duration. In order to have this information, joint cumulative distribution function (CDF) and probability density function (PDF) of involved variables is needed, and so multivariate statistical analysis has to be applied (Box et al. 1964).

Flood frequency analysis (FFA) has a major role to prevent from damages to establishment. Considering the irreparable damages of inattention to FFA, in last half of century, many different methods were presented in this branch of hydrology studies. Most of these approaches are based on statistical distributions. Such as these approaches was presented by Chow and his colleagues in applied hydrology.

A major shift in approaches to the management of flooding is now underway in many countries worldwide. This shift has been simulated by severe floods, for example on the Oder (Odra. 1997), Yangtze (1998), Elbe (Labe. 2002), Rhone (2003), in New Orleans (2005), on the Danube (2006) and in the UK (2000, 2007 and 2009) (Rossi et al. 2011).

Flood risk management is the process of decision making under uncertainty. It involves the purposeful choice of flood risk management that are intended to reduce flood risk (Rossi et al. 2011).

Traditional approaches to flood forecasting involve multi-dimensional mathematical models extensively based on underlying physical principles. In contrast, machine learning algorithms are data-driven methods whereby models are inferred directly from a database

Corresponding author Email: sajad.shahaabi@gmail.com

of training examples. Consequently, the incorporation of background knowledge, in the form of an understanding of the hydrology of the system being studied, only takes place indirectly through, for example, the choice of input variables to the artificial intelligence (AI) algorithm, or through the identification of an appropriate lead time for prediction. For this reason, data-driven models are sometimes referred to as being 'black box' (Abu and Sung. 2011).

Post-event analysis of any particular flood event will reveal that both the rainfall and snowmelt inputs that caused it and the effects in terms of areas flooded and damaged caused will be spatially variable or distributed in nature. The hydrology and hydraulics of the event will reflect the heterogeneities in the driving variables and catchment and channel characteristics. The distributed nature of the process is important, and the logical consequence is that in trying to predict flood events for flood management we should use distributed models whenever local distributed inputs interact with local nonlinear processes to produce responses where the distributed impacts might be significant (Petersen Olivier et al. 2009).

In recent 6 decades, many researches and studies were performed on FFA and its related branches. These researches and studies include several different methods from traditional methods, such as using AM and PD data, index flood, etc. to newer methods such as self-organization feature map, fuzzy clustering, regional FFA (RFFA), etc. The last subject is most considered in last decades. Below, previous researches and studies about these methods are presented.

Different types of probability distributions are one of the most usage and popular method in FFA. Most of researchers were working on developing this approach. In most of experimental projects, this method of FFA is used, in past and today. Some of researches are reviewed types of this method and was discussed about them (Jim and Edmund. 2011).

FFA in urban watersheds is complicated by non-stationary of annual peak records associated with land use change and evolving urban stormwater infrastructure. A framework for FFA is developed based on Generalized Adaptive Models for Location, Scale and Shape parameters (GAMLSS), a tool for modeling time series under non-stationary conditions. GAMLSS is applied to AM peak discharge records for Little Sugar Greek, a highly urbanized watershed which drains the urban core of Charlotte, North Carolina. It is shown that GAMLSS is able to describe the variability in the mean and variance of the AM peak discharge by modeling the parameters of the selected parametric distribution as a smooth function of time via cubic splines (Shu et al. 2008).

In another paper the joint impact of sample variability and rating curve impression in at-site FFA was considered. A novel likelihood-based framework is developed for this purpose, amusing the power-law model for the stage-discharge measurement and generalized extreme value (EV) model for the AM discharges. It shows that the two models can be pooled into one likelihood function (Guilan Regional Water Company, Research Committee. 2011).

Kale provided a synoptic view of extreme monsoon floods on all the nine large rivers of South Asia and their association with the excess (above-normal) monsoon rainfall periods. Simple techniques such as the Cramer's t-test, regression and Mann-Kendall (MK) tests and Hurst method were used to evaluate the trends and patterns of the flood and rainfall series (Adamowski. 2008, Heo et al. 2001).

At other study, the gradients of trends in the mean and the standard deviation (SD) are estimated by the weighted least squares method and the best fitting linear model of trend is with the aid of the Akaike Information Criterion (AIC). It shows that for every time series, a trend in the variance has a considerable effect on the trend estimators of the mean value. The analysis also includes seasonal peak flow series in order to obtain further insight into the detected non-stationary of the peak flows series (Adamowski and Fung Chan. 2011).

In other research, was examined the methods and approaches available in long-term flood seasonality analysis and applies them to the river Ouse (Yorkshire) in Northern England Since AD 1600. A detailed historical flood record is available for the city of York Considering of annual maximal flood level since AD 1877, with documentary accounts prior to this (Ravnik et al. 2004; Lawry et al. 2011).

RFFA has become a standard practice for determining flood quantiles at ungagged locations or at sites with short records. RFFA is the most popular method in FFA for watersheds that have not enough data for FFA. This method was used frequently and developed in last decades. Some of studies about this branch of FFA were presented below.

At first part of a study with 2 parts, a two parameter Weibull distribution with independence in both time and space was selected as a RF model and analyzed based on an index flood assumption. The method of maximum Likelihood (ML), the method of Moment (MOM), and the method of probability weighted moments (PWMs) were used to estimate flood quantiles at a site of interest (Zhang et al. 2006). In the second part of this study, flood quantile estimates determined from flood data at a single site have limited precision because ordinarily the available sample size is small. To improve the precision of such quantile estimates, an index flood technique has been employed enabling one to use available flood data at several sites in a region computer simulation experiments were performed in order to compare the sample properties of quantile estimates obtained based on the ML, MOM, and PWM methods and to determine the probability of the asymptotic variances obtained for each method for finite samples (Ozger et al. 2010).

In another study of RFFA, was compared Bayesian Generalized Least Squares (BGLS) regression approaches using a fixed and region-of-influence (ROI) framework that seeks to minimize the Bayesian model error variance (predictive uncertainty) (Haddad et al. 2012).

Shu and Ouarda presented the methodology of using adaptive neuro-fuzzy interface systems (ANFIS) for flood quantile estimation at ungagged sites (Macdonald et al. 2012). A regionalized relationship to estimate flood magnitudes for ungagged and poorly gauged catchments can be established using RFFA. RFFA was performed in this study using fuzzy c-mean, L-moment and artificial Neural network (ANN) (Lecrec and Ouarda. 2007).

Some of the other researchers performed RFFA using different methods and approaches, such as index- flood, combining self-organizing feature map and fuzzy clustering, GEV model (Pellegrini et al. 2012; Heo et al. 2001; Beven. 2011).

Up to now, we were considered to several common methods in FFA, but some other methods are being used in analysis, one of these methods is fuzzy expert system (FES). In Shu and Burn paper, the performance of the FES is improved by tuning of the membership functions of the fuzzy sets using a genetic algorithm (GA) (Quiroz et al. 2011).

We couldn't find any study in FFA by wavelet transform, but many studies have been done using wavelet transform in different branches of sciences involve hydrology and others (Patral et al. 2007; Shrinivas et al. 2008; Nourani et al. 2009; Chow et al. 2013; Vishwas et al. 2011; Strupczewski et al. 2001). One of these studies in hydrology is a large set of monthly precipitation data from 43 stations throughout Texas that was employed to investigate the spatial variability in the multi-scaling properties of wet and dry spells. Rainfall data from stations scattered across a very large size of data are analyzed by using a multi-scaling approach. Wavelet spectrum maps are interpreted considering different scale behavior of stations. It is found that stations show different scaling properties in terms of their wet and dry spells (Golizadeh et al. 2011).

Many studies and researches in FFA indicate the importance of this very important branch of hydrology. Damages of recent huge floods in the world indicate the result of inattention to this nature phenomena or mistakes in estimate of flood risk. So that many researchers were tried to make the better estimation of flood risk in last decades.

2. Case Study (Polroud basin)

FFA was performed on Polroud River in north of Iran (Guilan province). Polroud is located in the east of Guilan and it is the most important river in this region as sometimes east of Guilan basin was called Polroud basin. Fig. 1 shows the satellite picture of Polroud basin with two hydrometric stations (note: this picture is rotated about 45 deg. clockwise). Table 1 shows a summary of Polroud river characters.

Although the Manjil (Sefidroud) dam on Sefidroud River is the main source of Guilan water demand, but so it is distance to east of Guilan lands is far that so using Polroud River for supply of water demand is inevitable. In Polroud basin two main stations on two branch of this river has been established. Fig. 1 shows these two stations, Haratbar and Tollat. Table 2 shows the characters of these stations (Shu et al. 2004; Villarini et al. 2009; The Math Works. 2009.

MATLAB, Version 7.8.0.347). In this paper, 52 years discharge data of Tollat station was used.

3. Wavelet transform

The wavelet transform has increased in usage and popularity in recent years since its inception in the early 1980s, yet scientists still do not enjoy the widespread usage of the Fourier transform (Subramanya et al. 2008).

Fig. 2 shows a schematic form of wavelet transform (Gubareva et al. 2011). The time-scale wavelet transforms of a continuous-time signal, x(t), is defined as:

$$T\left(a,b\right) = \frac{1}{\sqrt{a}} \int_{-\infty}^{+\infty} g^*\left(\frac{t-b}{a}\right) x\left(t\right) dt \tag{1}$$

Table 1. Characters of Polroud river and its basin (GRWC. 2011).

Features	quantity/quality
Length	51 km
basin area	1765 km^2
origin	south of Alborz mountains
River delta	Caspian sea
River annual stream flow	472*10^6 m^3

Table 2. Characters of stations on Polroud river (GRWC. 2011).

River (branch)	Station	Height(m)	establishing date	Basin area (km^2)	Longitude	Latitude
Polroud	Tollat	113	1335	1574	50°17'30"	36°59'41"
Samoush	Haratbar	123	1336	115	50°18'11"	36°59'53"

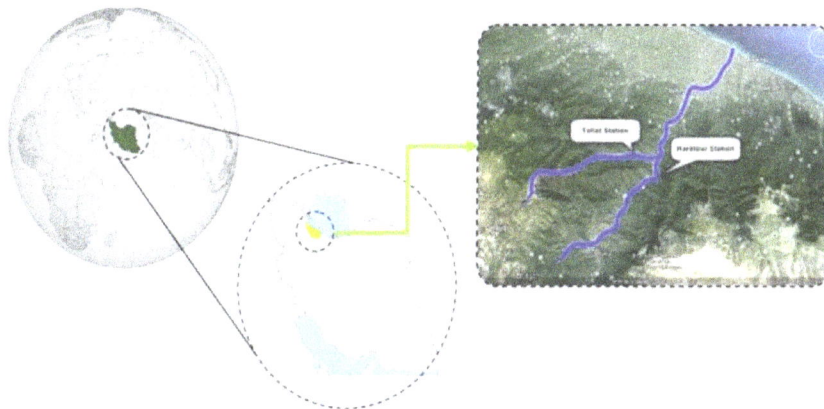

Fig. 1. The position of Polroud River.

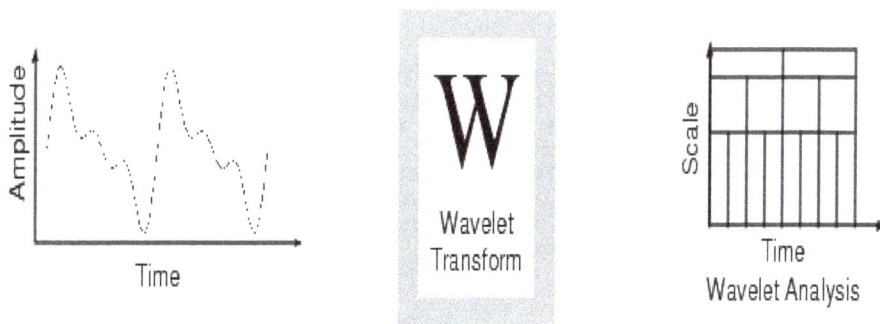

Fig. 2. Schematic showing of wavelet transform.

where, * corresponds to the complex conjugate and g(t) is called wavelet function or mother wavelet. The parameter "a" acts as adilation factor, while "b" corresponds to a temporal translation of the function g(t), which allows the study of the signal around "b". The characters of wavelet transform are to provide a time-scale localization of processes, which derives from the compact support of its basic function. This is opposed to the classical trigonometric function of Fourier analysis. The wavelet transform searches for correlations between the signal and wavelet function. This calculation is done at different scales of "a" and locally around the time of "b". The result is a wavelet coefficient (T(a, b)) contour map known as a scalogram. In order to be classified as a wavelet, a function must have finite energy, and it must satisfy the following "admissibility conditions":

$$\int_{-\infty}^{+\infty} g\left(t\right) dt = 0, \quad C_g = \int_{-\infty}^{+\infty} \frac{\left|\hat{g}\left(w\right)\right|^2}{|w|} dw < \infty \tag{2}$$

where, g^(w) is Fourier transform of g(t); i.e. the wavelet must have no zero frequency component. In order to obtain a reconstruction formula for the studied signal, it is necessary to add ''regularity conditions'' to the previous ones.

$$\int_{-\infty}^{+\infty} t^k g\left(t\right) dt = 0, \quad where \ k = 1, 2, ..., n-1 \tag{3}$$

So the original signal may be reconstructed using the inverse wavelet transform as (Subramanya. 2008):

$$x\left(t\right)=\frac{1}{C_g}\int\limits_{-\infty}^{+\infty}\int\limits_{0}^{+\infty}\frac{1}{\sqrt{a}}g\left(\frac{t-b}{a}\right)T\left(a,b\right)\frac{dadb}{a^2}\qquad(4)$$

Two functions were existed that have main role in wavelet analysis: scale function (φ) and wave function (ψ). These two functions are produced a collection of functions that is used in decomposition or reconstruction of a signal. φ and ψ called father and mother wavelet, respectively (Kjeldsen et al. 2002). Two wavelet functions that used in this study were introduced briefly.

3.1. Haar

Any discussion of wavelets begins with Haar wavelet, the first and simplest. Haar wavelet is discontinuous, and resembles a step function. It represents the same wavelet as Daubechies db1

(Gubareva et al. 2011). The simplest wavelet analysis is based on Haar scale function. The Haar scale function is shows as:

$$\varphi\left(x\right)=1\qquad\qquad if\quad x\in[0,1]\qquad(5)$$

$$\varphi\left(x\right)=0\qquad\qquad if\quad x\notin[0,1]\qquad(6)$$

Haar wavelet is discontinuous and similar step function. Haar function is like Daubechies1 function. The (7) to (9) show Haar wave function:

$$\psi\left(x\right)=1\qquad\qquad x\in[0,0.5]\qquad(7)$$

$$\psi\left(x\right)=-1\qquad\qquad x\in[0.5,1]\qquad(8)$$

$$\psi\left(x\right)=0\qquad\qquad x\notin[0,1]\qquad(9)$$

Fig. 3 and Fig. 4 show the Haar wave and scale functions, respectively.

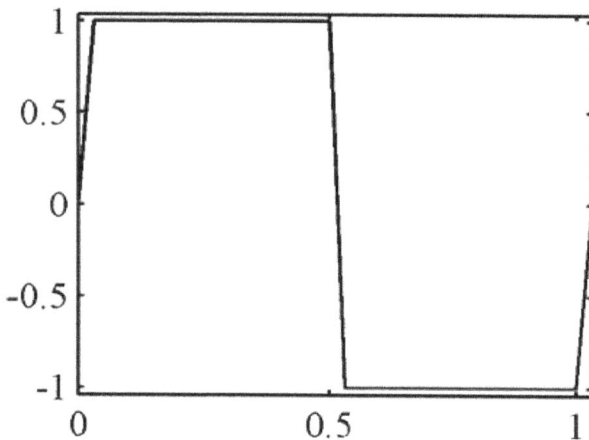

Fig. 3. Haar wave function.

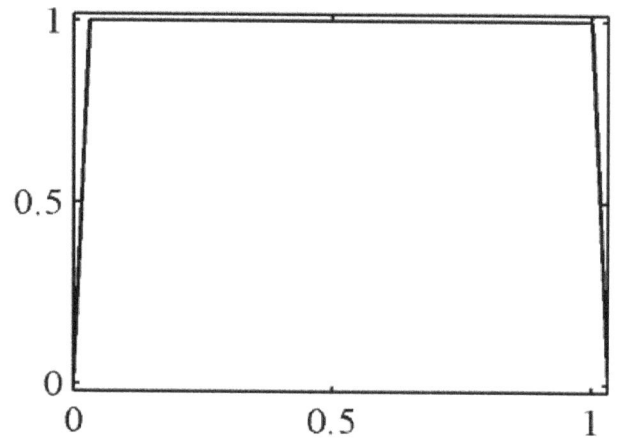

Fig. 4. Haar scaling function

3.2. Daubechies10

At first of using wavelet, Daubechies wavelets with some other by similar characteristics were only available wavelets. The simplest is Haar wavelet exactly that only discontinues wavelet in all of them. The other wavelets in this family are continues (Kjeldsen et al. 2002).

Ingrid Daubechies, one of the brightest stars in the world of wavelet research, invented what is called compactly supported orthonormal wavelet, thus making discrete wavelet analysis practicable.

The names of the Daubechies family wavelets are written dbN, where N is the order, and db is the "surname" of the wavelet. The db1 wavelet, as mentioned above, is the same as Haar wavelet (Gubareva et al. 2011).

The most of these family functions are not symmetric but dissymmetric of some these are deterministic. Functions regularity of this family is increased with increasing their orders. This family is orthogonal also (Gubareva et al. 2011).

Many researchers and scientist were believed Daubechies is the most exact functions in wavelet functions for analyzing of natural phenomena. Fig. 5 and Fig. 6 show the Daubechies10 wave and scale functions, respectively.

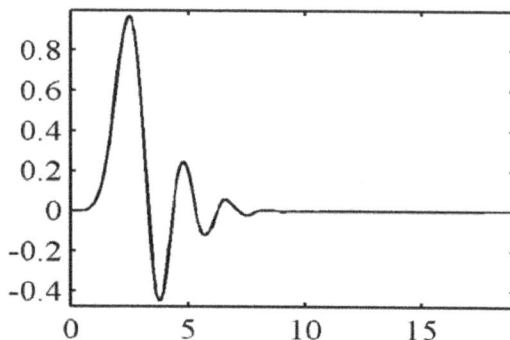

Fig. 5. Dubechies10 wave function.

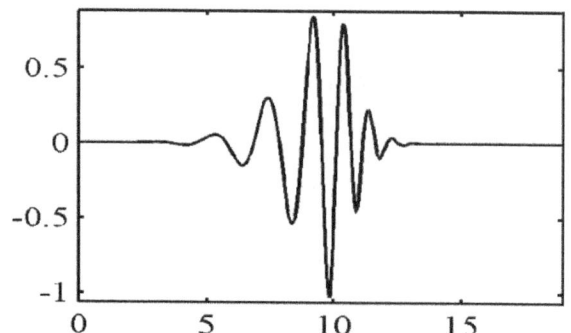

Fig. 6. Daubechies10 scaling function.

4. Materials and methods

In this study FFA using wavelet transform has been considered. For this purpose a full series of Tollat discharge data was used for making wavelet model. For this purpose, time period of study has been reduced from 52 years to about 33 years. Fig. 7 shows this signal, it contained more than 12000 data.

After making signal, next step is selection of wavelet function and its decomposition level. When is used wavelet transform, selection of wavelet function and decomposition optimized level is so important. One of the important and base points is choosing mother wavelet based on phenomena natural and series type. So each mother wavelet function pattern that make a better set adaption in geometry aspect to time series, gets better results. In this paper furthermore this important case, density function form of wavelet modeling was also considered. After choosing wavelet function, next important step is selection of decomposition level.

In theory decomposition process can continue infinitely, but really decomposition process cans perform to signal details involve one pixel only. In a signal decomposition using maximum decomposition is not correct because although it improves computations accuracy in network training have inverse result on simulated data, because over training of network pattern to training data.

In this paper the simplest and first wavelet function was used, Haar. Also Daubechies function was used in two levels.

At the end of this study, the wavelet results were compared with some traditional methods, such as AM and PD.

4.1. Analysis and investigation of data

In this part of study was investigated used data for different methods. In AM method 52 year data of Tollat station on Polroud River was used. In all of study period, the smallest data is 0.4 m^3/sec and the largest data is 537 m^3/sec. The mean and standard deviation (Std) of AM data are 119.42 and 93.84 m^3/sec, respectively.

When normal distribution was used in AM and PD methods, data have to fitness to normal distribution, this subject was investigated drawing data on normal paper (Fig. 8).

Fig. 8 clearly shows that Polroud data is not normal, so for using these data in normal distribution, must become normal. This can be done by different methods, that in this paper Box-Cox formula was used for normalization of data (Golizadeh et al. 2011). This normalization results are showed in Fig. 9. This picture clearly shows that data was normalized well. Used data in AM and PD methods was investigated for stationary, stability and homogeneity and station, stable and homogenous data was used finally.

For many days in 52 years period has not recorded any discharge and this make an incomplete series, for solving this problem two ways: at first, reproduce artificial data for no discharge days that have a special problems and mistakes. And second, used a period of time that has complete discharge data. Second method was selected because an about 33 years period was distinguished in studying period, in other hand from many researchers point a minimum 30 years period is enough (Grimaldi et al. 2006).

In this study for FFA using wavelet transform 3 TS were used: first series is almost 33 years period that contain all the data, mean and standard deviation of data are 15.75 m^3/sec and 17.85 m^3/sec, respectively. This series called 100 % or full series in this paper. Second series is produced from omitting of all the data smaller than 80 % of the least AM discharge, mean and standard deviation of this series are 40.17 m^3/sec and 26.02 m^3/sec, respectively. This series called 80 % series. And the last series produced like second series but in this case the criteria for omitting the data is the data smaller than 95 % of the least AM discharge or 25.8 m^3/sec (called 95% series). Fig. 7 shows the full series (100 %).

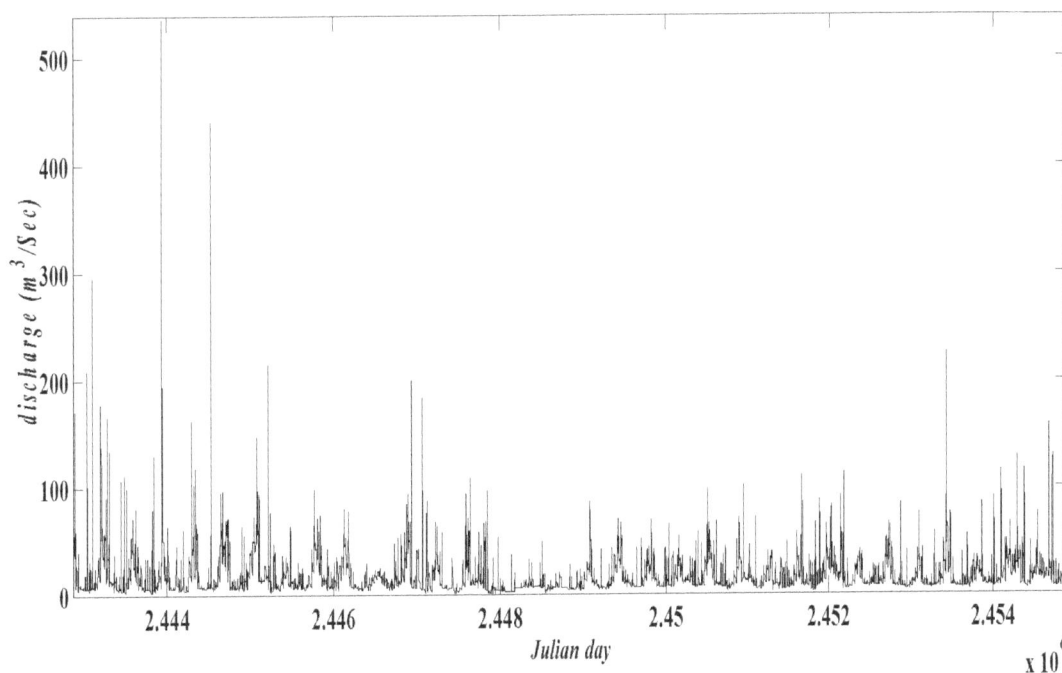

Fig. 7. Full time series for 32years period (100 % of data).

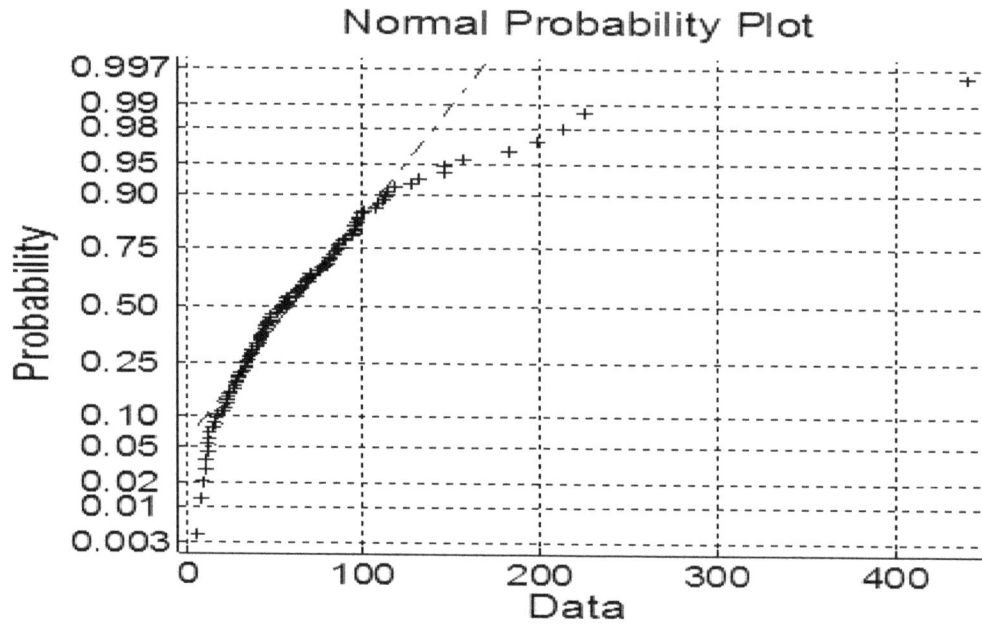

Fig. 8. Polroud discharges data on normal paper.

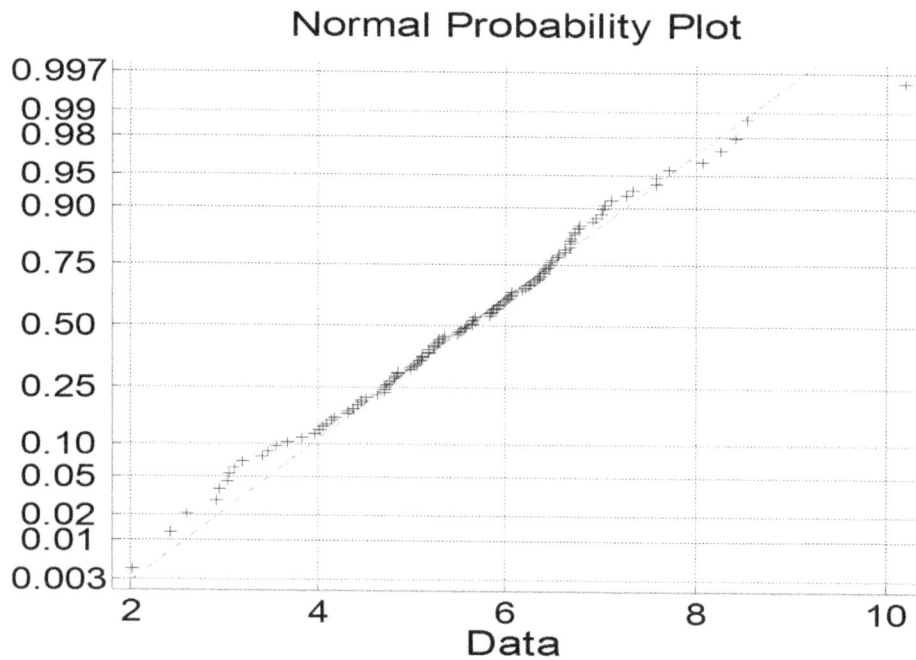

Fig. 9. Polroud normalized data on normal paper.

5. Results and discussion

In this section, we present flood frequency analyses for Polroud Basin, focusing on wavelet transform method. Fig. 10 shows the probability density distribution function related to TS that shows in Fig. 7. Also Fig. 11 shows Daubechies10 in second level of decomposition.

Table 3 shows the FFA results on full TS (100 % of data) using Haar1 and Daubechies10 with determined decomposition levels for different return periods. Table 4 and Table 5 show the results for TS 80 % and TS 95 %, respectively.

Table 3. Flood frequency analysis using haar and Duabechies10 with full series (100 % of data).

Return period	Probability	Discharge (m³/s) Haar	Discharge (m³/s) Dubechies10
2	0.5	6.63	8.13
10	0.1	32.49	31.85
25	0.04	48.38	48.32
50	0.02	61.58	60.93
100	0.01	76.44	75.63
1000	0.001	189.9	189.02

FFA results on Polroud River presented using two wavelet functions, Haar and Daubechies10, with full and combined series. Now FFA results compute using Tollat station AM and PD data.

For FFA using methods that data fitted to a special distribution, investigation of stationary, stability and homogeneity is necessary. All of these computations were performed on AM data and PD daily maximum discharge. Table 6 and Table 7 show these results, respectively. Investigation of these conditions was performed using Mann-Witeny and Wald-Wolfowitz, furthermore stationary was investigated in both cases.

Table 4. Flood frequency analysis using haar and Duabechies10 with 80 % time series.

Return period	Probability	Discharge (m^3/s) Haar	Discharge (m^3/s) Dubechies10
2	0.5	31.3	31.6
10	0.1	60.4	60.2
25	0.04	81.8	81.4
50	0.02	101.5	102.6
100	0.01	138.4	138.6
1000	0.001	293.7	293.7

Table 5. Flood frequency analysis using haar and Duabechies10 with 95 % time series.

Return period (year)	Probability	Discharge (m^3/s) Haar	Discharge (m^3/s) Dubechies10
2	0.5	36.2	36.0
10	0.1	65.5	65.3
25	0.04	89.2	89.3
50	0.02	111.2	111.5
100	0.01	160.2	159.5
1000	0.001	436.1	297.5

Table 6. Flood frequency analysis using annual maximum data and fit to LP3 and exponential distribution.

Return period (year)	Probability of occurrence	Discharge (m^3/s) LP3	Discharge (m^3/s) Exponential
2	0.5	92.3	90.6
10	0.1	215.7	241.6
25	0.04	311.6	327.6
50	0.02	400.5	392.7
100	0.01	508.5	457.7

Table 7. Flood frequency analysis using partial series and fit to normal distribution.

Return period (year)	Probability of occurrence	Discharge (m^3/s)
2	0.5	66.9
10	0.1	136.5
25	0.04	162.0
50	0.02	178.5
100	0.01	192.3
1000	0.001	234.9

5.1. Comparison of results

At the end of this section, a general comparison between results of all methods was presented. Graphical results of all presented methods in this paper are showed in Fig. 12. In this figure, AM results was presented using Log-Pearson 3 (LP3) and wavelet results in two different type: full series and TS 5 % and also mean of 5 different

wavelet functions (Haar, Daubechies3, Daubechies10, Symlet4 and Coiflet2). The time axis (horizontal axis) is logarithmic.

An exact investigation of Fig. 12 shows that:

a) AM result is the most overestimating method. This overestimation about Polroud data is clearer, because its distribution is exponential and in its AM data time series only 0.06 % of data is more than 200 m^3/sec and these numbers of data distributed in different years.

b) In PD series, used data is about three times larger than AM series and almost all the added data is smaller than AM series data, so that results of this method is smaller than AM results as predicted.

c) Above trend is continued in wavelet analysis, means with using a larger size of small data, results are becoming smaller and more accurate.

d) When 5 % series is used estimations are much larger than full series is used.

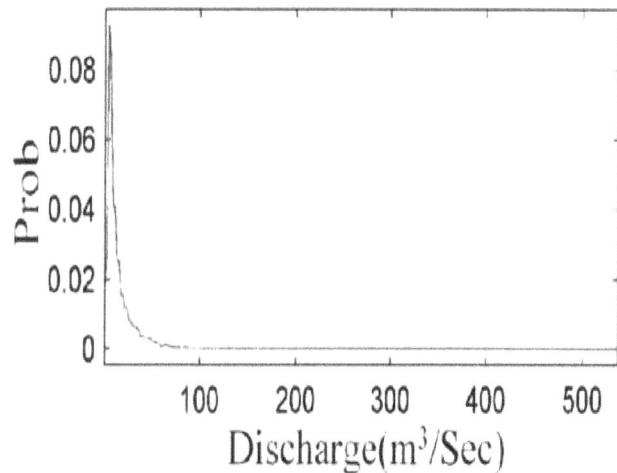

Fig.10. Probability distribution function of wavelet analysis time series (Haar1).

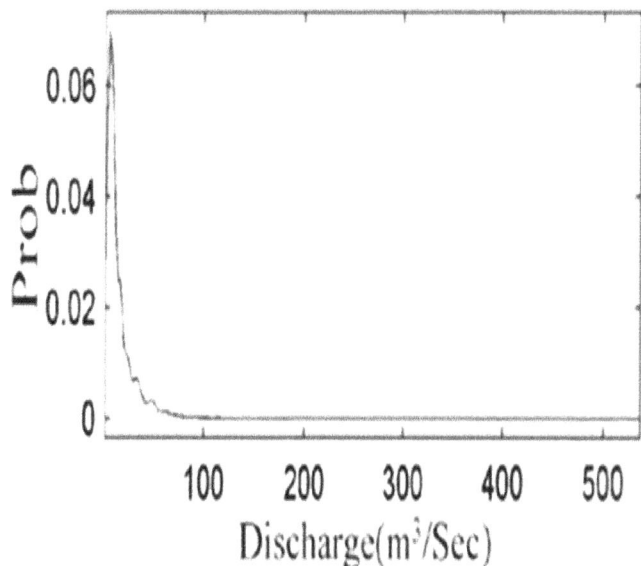

Fig. 11. Probability distribution function of wavelet analysis time series (daubechies10).

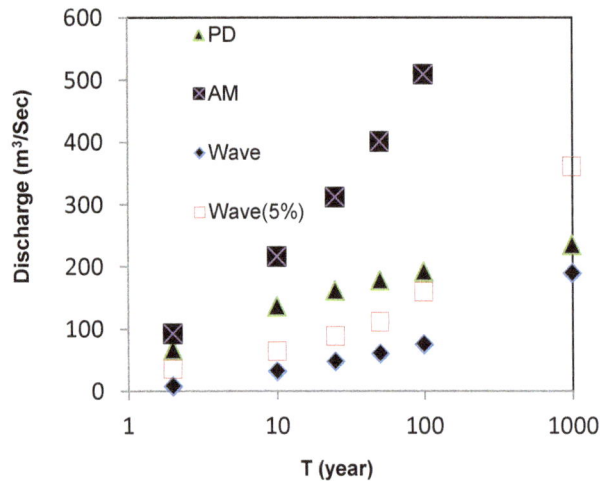

Fig. 12. Graphical comparison for diffrent methods of flood frequency analysis on Polroud river.

6. Conclusion

Presentation of variety of methods of FFA in more than last half century can't cause less consideration by researchers to this branch of hydrology, which shows the importance of FFA. Results collection of this paper shows that when numerous of small data are so going up, the model is prepared smaller analysis results. Briefly, when frequently of small data is too much and large data is occurred rarely, dependence of study aim different method can use. But when data distribution almost is uniform or has a collection of too small data only, wavelet can present the very exact results. According to last studies, statistical methods are contained too mistake hypothesis and although their large results, those ones are not closed to accurate results. In general, for every project, researcher must select the best method based on the hydrological studying of watershed and available data.

References

Abu S., Sung H.J., Wavelet Spatial Scaling for Educing Dynamic Structure in Turulent Open Cavity Flows, Journal of Fluid and Structures 27 (2011) 962-975.

Adamowski J., Development of a Short-term River Flood Forecasting Method for snowmelt Driven Floods Based on wavelet and cross-wavelet analysis, Journal of Hydrology 353 (2008) 247-266.

Adamowski J., Fung Chan H., a Wavelet Neural Network Conjunction Model for Groundwater Model Forecasting, Journal of Hydrology 407 (2011) 28-40.

Asgeir O.P., Reitan T., Accounting for rating Curve Imprecision in Flood Frequency Analysis Using Likelihood-Based Methods, Journal of Hydrology 366 (2009) 89-100.

Boggess A., Narcowich F.J., A First Course in wavelets with Fourier Analysis, John Wiley, New Jersey, 2009.

Beven K., Distributed models and uncertainty in flood risk management. In: Gareth Pender, Hazel Faulkner, editors. Flood risk science and management, Britain: Wiley-Blackwell, (2011) 291-312.

Box G.E.P., Cox D.R., An Analysis of Transformations, Journal of the Royal Statistical Society 127 (1964) 211-252.

Chow V., Maidment D., Larry Mays., Applied hydrology. 2nd ed. New York: McGraw-Hill Companies, 2013.

Golizadeh S., Samavati O.A., Structural Optimization by Wavelet Transforms and Neural Networks, Applied Mathematical Modeling 35 (2011) 915-929.

Grimaldi S., Serinaldi F., Asymmetric Coupla in Multivariate Flood Frequency Analysis, Advances in Water Resources 29 (2006) 1155-1167.

Gubareva T.S., Types of Probability Distributions in the Evaluation of Extreme Floods, Water Resources 38 (2011) 962-971.

Guilan Regional Water Company, Research Committee, 2011.

Haddad K., Rahman A., Regional Flood Frequency Analysis in eastern Australia: Bayesian GLS Regression-Based Methods whithin Fixed Region and ROI Framework-Quantile Regression vs. Parameter Regression Technique, Journal of Hydrology 430-431 (2012) 142-161.

Heo J.-H., Boes D.C., Salas J.D., regional Flood Frequency Analysis Based on Weibull Model: part 1. Estimation and Asymptotic Variances, Journal of Hydrology 242 (2001) 157-170.

Heo J.-H., Salas J.D., Boes D.C., regional Flood Frequency Analysis Based on Weibull Model: part 2. Simulations and Applications, Journal of Hydrology 242 (2001) 171-182.

Jim W.H., Edmund C.P.R., Setting the scene for flood risk management. In: Gareth Pender, Hazel Faulkner, editors. Flood risk science and management, Britain: Wiley-Blackwell, (2011) 3-16.

Jingyi Z., Hall M.J., Regional flood Frequency Analysis for the Gan-Ming River Basin in Chaina, Journal of Hydrology 296 (2004) 98-117.

Kale V., On the Link between Extreme Floods and Excess Monsoon Epoches in south Asia, Springer, 2011.

Kjeldsen T.R., Smithers J.C., Schulze R.E., Regional Flood Frequency Analysis in the KwaZulu-Natal Province, South Africa, Using the Index-Flood Method, Journal of Hydrology 255 (2002) 194-211.

Lawry J.R., McCulloch D.J., Randon N., Cluckie I., Artificial intelligence techniques for real-time flood forecasting. In: Gareth Pender, Hazel Faulkner, editors. Flood risk science and management, Britain: Wiley-Blackwell, (2011) 146-162.

Lecrec M., Ouarda T.B.M.J., Non-stationary Regional Flood Frequency Analysis at Ungaged Sites., Journal of Hydrology 343 (2007) 254-265.

Liu Y., Brown J., Demargne J., Seo D.-J., A Wavelet-Based Approch to Assessing Timing errors in Hydrologic Predictions, Journal of Hydrology 397 (2011) 210- 214.

Macdonald N., Trends in Flood Seasonality of the River Ouse (Northern England) from Archive and Instrumental, Climatic Change (2012) 901-923.

Naulet R., lang M., Ouarda T.B. M.J., Coeur D., Bobee B., Recking A., Moussay D., Flood Frequency Analysis on the Ardeche River Using French Documentary Sources from the last two Centuries, Journal of Hydrology 313 (2005) 58-78.

Nourani V., Alami M., H. Aminfar M., A combined Neural-wavelet Model for Prediction of Ligvanchai Watershed Precipitation, Engineering Applications of Artificial Intelligence 22 (2009) 466-472.

Ozger M., K.Mishra A., P.Singh V., Scaling Characteristic of Precipittion data in Conjunction with Wavelet Analysis, Journal of Hydrology 395 (2010) 279-288.

Partal T., Kisi O., A wavelet and Neuro-Fuzzy Conjunction Model for Precipitation Forecasting, Journal of Hydrology 342 (2007)199-212.

Pellegrini M., Sini F., Taramasso A.C., Wavelet – Based Automated Localization and Classification of Peaks in Streamflow Data Series, Computers and Geosciences 40 (2012) 200-204.

Quiroz R., Yarleque C., Posadas A., Mares V., W.Immerzeel W., Improving Daily Rainfall Estimation from NDVI Using a Wavelet Transform, Environmental Modelling & Software 26 (2011) 201-209.

Ravnik J., Skerget L., Hribersek M., The Wavelet Transform for BEM Computational Fluid Dynamics, Engineering Analysis with Boundary Elements 28 (2004) 1303-1314.

Rossi A., Massei N., Laignel B., A Synthesis of the Time-Scale Variability of Commonly Used Climate Indices Using Continues Wavelet Transform, Global and Planetary Change 78 (2011) 1-13.

Shu C., H. Burn D., Homogeneous Pooling Group Delineation for Flood Frequency Analysis Using a fuzzy Expert System with Genetic Enhancement, journal of hydrology 291 (2004) 132-149.

Shu C., Ouarda T.M.B.J., Regional Flood Frequency Analysis at Ungaged Sites Using the Adaptive Neuro- Fuzzy Interface System, journal of hydrology 349 (2008) 31-43.

Srinivas V.V., Tripathi S., RamachandraRao A., S. Govindaraju R., Regional Flood Frequency Analysis by Combining Self-Organizing Feature Map and Fuzzy Clustering, Journal of Hydrology 348 (2008) 148-166.

Strupczewski W.G., Singh V.P., Mitosek H.T., Non-Stationary Approch to at-site Flood Frequency Modeling. III. Flood Analysis of Polish Rivers, Journal of Hydrology 248 (2001) 152-167.

Subramanya K., Engineering Hydrology, McGraw-Hill, 2008.

The Math Works, MATLAB, Version 7.8.0.347 (R2009a), Wavelet toolbox (2009).

Villarinia G., A. Smitha J., Serinaldic F., Balese J., Batesf Paul D., F. Krajewskig W., Flood Frequency Analysis for nonstationary Annual Peak Records in an Urban Drainage Basin, advances in water resources 32 (2009) 1255-1266.

Zhang Q., Liu C., Xu C., Xu Y., Jiang T., Observed trends of annual Maximum Water Level and Streamflow During Past 130 Years in the Yangtze River Basin, China, Journal of Hydrology 342 (2006) 255-265.

6

Hydraulic influence of geometrical defects in Venturi flumes: shall we destroy and rebuild?

Matthieu Dufresne[*], José Vazquez

National School for Water and Environmental Engineering of Strasbourg (ENGEES) – ICube (University of Strasbourg, CNRS, INSA of Strasbourg, ENGEES), Mechanics Department, Fluid Mechanics Team ENGEES, Strasbourg cedex France.

ARTICLE INFO

Keywords:
Discharge
Venturi flume
Defect
Tolerance

ABSTRACT

Venturi flumes are measurement structures commonly used in water systems to measure the flow discharge. Some of them are not well installed or present some geometrical defects. The objective of this study is to investigate the hydraulic influence of a number of typical wrong installations and geometrical defects of long-throated Venturi flumes: significant positive or adverse slopes, humps and hollows on the walls of the throat, hump or hollow on the bed of the throat. The geometric tolerances corresponding to an acceptable tolerance on the discharge of 2% and 5% are calculated for each defect. A number of corrections of the head –discharge relationship are proposed to avoid the destruction of a flume if the geometric tolerance is not respected.

1. Introduction

Venturi flumes are hydraulic devices commonly used in water systems to measure the flow discharge. A Venturi flume consists of an upstream channel, a converging zone, a constricted section called the throat of the flume and then an enlargement (Fig. 1). The local constriction of cross-section in the throat is a favorable condition to critical flow to occur (Hager 1999). One of the characteristics of critical flow is that a bijective relationship Q = f(h) exists between the discharge Q and the water depth h, which makes possible the determination of the discharge with only one water level measurement. Instead of measuring the water level in the throat where the free surface is inclined and generally instable, the measurement is carried out in the approach channel (generally three to four times the maximum depth upstream of the converging zone, as recommended by ISO 4359:2012) where the free surface is nearly horizontal and quite stabilized (Fig. 1).

Our own field experience reveals that a significant number of Venturi flumes are not well installed or present some geometrical defects. Whereas the bed of the flume must be horizontal as required by ISO 4359:2012 (which may be very difficult in practice), some devices presents a positive or an adverse significant slope (Fig. 2). This may raise two problems: is the head – discharge relationship corresponding to a horizontal flume relevant? Where must the datum of the water level measurement be undertaken? Moreover, the walls of the throat are often not completely parallel; the bed, not completely flat. Indeed, they can present some humps or some hollows on their surface (Figure 3 and Figure 4), which may be problematic because the geometry of the throat is the hydraulic control of the head – discharge relationship.

The objectives of this study are the following: investigate the hydraulic influence of a number of typical wrong installations and geometrical defects of Venturi flumes; define the acceptable tolerances corresponding to each geometrical defect; if possible, propose a correction of the head – discharge relationship to take into account the geometrical default of the measurement device. The investigations are restricted to the Venturi flumes corresponding to the requirements of ISO 4359, namely the long-throated devices.

$$C_V C_D = \frac{Q}{\left(\frac{2}{3}\right)^{3/2} \sqrt{g} B_t h^{3/2}} \tag{1}$$

[*]Corresponding author E-mail: matthieu.dufresne@engees.unistra.fr

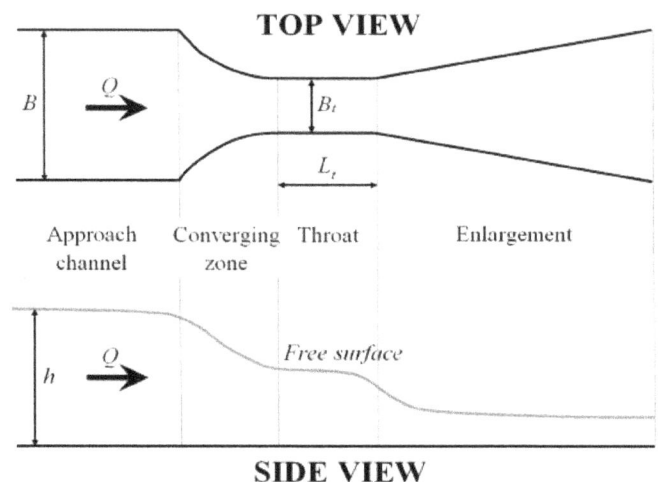

Fig. 1. Top view and side view of a Venturi flume.

2. Methods description of the approach

Since it would have been difficult to build Venturi flumes with calibrated defects, it has been decided to follow a Computational Fluid Dynamics (CFD) approach. This tool has shown its capability to simulate the flow in Venturi flumes (Dufresne and Vazquez 2013). After validation of the numerical model against experimental data from literature (Yeung 2007), the methodology consists in generating a numerical databank of simulations of flow in Venturi flumes with perfect and distorted geometries.

2.1. Numerical databank

The simulations were performed with two geometries of rectangular Venturi flume. The first one is a small-size device corresponding to the one investigated by Yeung (2007): the breadth of its throat is equal to 101 mm, which corresponds to the lower limit of the domain of validity of ISO 4359:2012. The small flume was used to study the influence of the defects for a large range of distortions. The second flume is a larger device (the breadth of the throat is equal to 480 mm) used to investigate scale effects for a limited number of defects. The whole characteristics of the two flumes are given in Table 1. The defects investigated are listed in Table 2. A total number of 158 simulations have been carried out for this study. The deformation of the walls is characterized by 2Δ/Bt since Δ is the deformation on one side (the breadth is therefore modified by 2Δ).

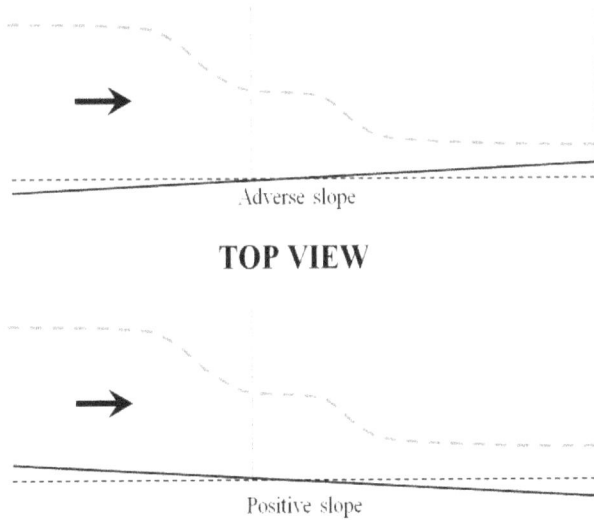

Fig. 2. Venturi flume installed with a longitudinal slope different from horizontal.

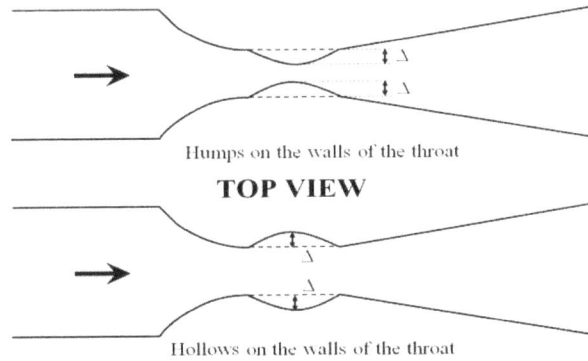

Fig. 3. Venturi flume presenting a distortion on the walls of the throat (humps or hollows) – the amplitude Δ is exaggerated in the Fig.

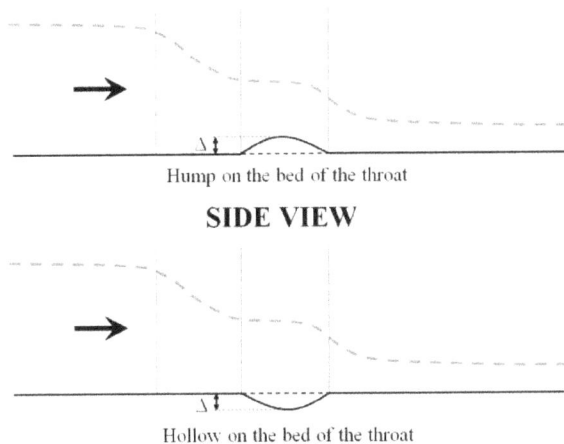

Fig. 4. Venturi flume presenting a distortion on the bed of the throat (hump or hollow) – the amplitude Δ is exaggerated in the figure.

Table.1. Characteristics of the two Venturi flumes investigated (see Fig. 1).

Characteristics	Small flume	Large flume
Breadth of the throat Bt (mm)	101	480
Breadth of the approach channel B (mm)	203	800
Length of the throat Lt (mm)	300	900

For each simulation, the global discharge coefficient CVCD is evaluated using Eq. 1, where Q is the discharge; g, the gravity acceleration; Bt, the breadth of the throat and h, the water depth in the measurement section. The discharge coefficient CVCD is the product of the velocity coefficient CV (taking into account the relationship between the water depth and the energy head) and the 'true' discharge coefficient CD (taking into account the influence of the approximations of ISO 4359:2012).

Table 2. Characteristics of the defects studied (Bt is the breadth of the throat and Δ is the amplitude of the deformation, see Figs. 3 and 4).

Defects	Values
Slope	-2.0% (adverse slope), -1.6%, -1.2%, -0.8%, -0.4%, -0.2%, -0.1%, 0%, +0.1% (positive slope), +0.2%, +0.4%, +0.8%, +1.2%, +1.6%, +2.0%
Deformation of the walls of the throat 2Δ/Bt	-20% (humps), -10%, -4%, -2%, 0%, +2% (hollows), +4%, +10%, +20%
Deformation of the bed of the throat Δ/Bt	-10% (hollow), -5%, -2%, -1%, 0%, +1% (hump), +2%, +5%, +10%

For each simulation corresponding to a Venturi flume with a defect, the error EP that is done if the flume is considered to be geometrically perfect is evaluated using Eq. 2. Here, (CVCD) perfect is the discharge coefficient of the Venturi flume with no defect; (CVCD) actual is the actual discharge coefficient of the Venturi flume with the defect.

$$E_P = \frac{\left(C_V C_D\right)_{\text{perfect}} - \left(C_V C_D\right)_{\text{actual}}}{\left(C_V C_D\right)_{\text{actual}}} \qquad (2)$$

Corrections of the head – discharge are tested using the error EC that is done when the corrected discharge coefficient (CVCD)corrected is used, as written in Eq. 3.

$$E_C = \frac{\left(C_V C_D\right)_{\text{corrected}} - \left(C_V C_D\right)_{\text{actual}}}{\left(C_V C_D\right)_{\text{actual}}} \qquad (3)$$

2.2. Settings of the numerical model

Numerical simulations were performed with the computational fluid dynamics code Open FOAM (Open FOAM 2013). The Reynolds-averaged Navier-Stokes (RANS) equations were used. In order to reproduce the non-uniformity of the water level distribution, the two-phase Volume of Fluid (VOF) model was chosen. Since the aim of the numerical investigations was to simulate the water level (and not the velocity field neither other variables maybe linked to the anisotropy of the turbulence), the k-ω SST turbulence model was chosen. The near-wall region is bridged using standard wall functions (ERCOFTAC 2000).

The main difficulty of the use of computational fluid dynamics in hydraulic applications is neither the choice of the turbulence model nor the choice of numerical schemes but the definition of the computational domain and the boundary conditions. Since the regime is subcritical in the upstream zone of the flume, the water depth in the approach channel is controlled by the critical section in the throat. Therefore, the upstream face of the computational domain was defined as a velocity inlet whose height was roughly chosen based on the value of the discharge. The approach channel was defined sufficiently long to ensure a stabilization of the water level upstream of the inlet convergence. Since the flow downstream of the throat is supercritical, the outlet boundary condition was simply defined as a pressure outlet. In order to reproduce atmospheric pressure, the top face of the domain was defined as a pressure outlet too. Free surface was defined in post-processing as the zone where the water volume fraction was equal to 50%.

2.3. Numerical uncertainty

A grid sensitivity analysis was performed with the small flume in order to evaluate the numerical uncertainty (Roache 1994). To do so, a fine mesh and a coarse mesh were built. They are respectively composed of 900,000 and 3,078,000 cells; the refinement ratio between the two grids is equal to 1.5. The Grid Convergence Index (GCI) of the discharge coefficient defined in Equation 1 (Bos 1977) for the fine mesh was then evaluated using Eq. 4 (Roache 1994).

$$\text{GCI} = \frac{3}{r^p - 1} \left| \frac{(C_V C_D)_{\text{fine}} - (C_V C_D)_{\text{coarse}}}{(C_V C_D)_{\text{fine}}} \right| \qquad (4)$$

Here, r is the grid refinement ratio; p, the order of the method (2 since second-order schemes were used); (CVCD)fine, the discharge coefficient obtained with the fine mesh; and (CVCD)coarse, the discharge coefficient obtained with the coarse mesh.
Results show a GCI on the discharge coefficient for the fine mesh of about 0.6%, which can be considered as an acceptable numerical uncertainty for the purpose of this study.

2.4. Validation of the numerical model

The comparison between the numerical simulations performed for the perfect geometry of the small flume and the experimental data of Yeung (2007) is illustrated in Figure 5. Rather than using the discharge for the abscissa of the graphics, the dimensionless parameter H/Lt is used (H is the upstream energy head; Lt, the length of the throat); H/Lt is representative of the discharge. With the exception of the two points located near a value of the discharge coefficient of 0.95 that probably correspond to errors in the experimental measurements, it can be concluded that the CFD model accurately simulates the discharge coefficient of the Venturi flume.

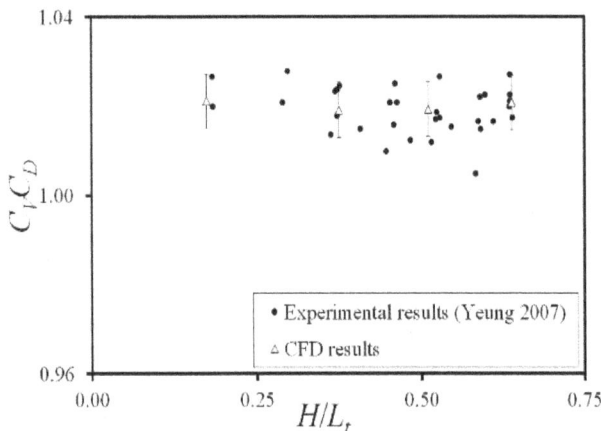

Fig. 5. Validation of the CFD model against experimental results of Yeung (2007) – H is the upstream energy head; Lt, the length of the throat.

3. Results and discussion
3.1. Influence of the slope

The relative error EP for a Venturi flume presenting a slope different from horizontal is given as a function of the H/Lt in Tables 3 and 5, for the small flume and the large flume respectively.

The comparison between the results obtained for the small flume and those obtained for the large flume (also given in Fig. 6 for a limited number of slopes for clarity reasons) shows that the order of magnitude of the error is the same, which proves that the scale effects are negligible. The conclusions drawn below for the small flume can therefore be generalized to flumes of any size.

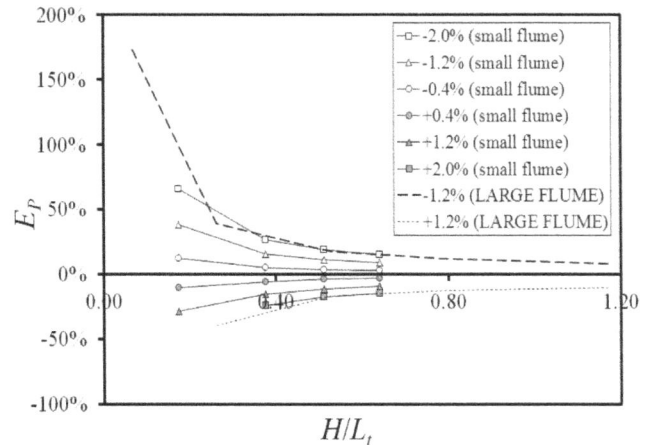

Fig. 6. Error on the discharge in small and large flumes for different slopes as a function of H/Lt.

If a tolerance of 2% is accepted for the error on the discharge, even a slope of +/-0.1% is too large and cannot be accepted (Table 3). If 5 % is acceptable, a slope lower than +/-0.2% can be accepted. It must be noticed that such slopes may be difficult to reach in practice.

A simple correction would be to choose the datum of the water level measurement not below the water level measurement in the approach channel but in the throat where the critical flow occurs, more precisely in the downstream section of the throat where the critical flow approximately occurs (ISO 4359:2012). Even if the critical flow does not always occur at the downstream section of the throat (Dabrowski and Polak 2012), the results obtained with this correction are very good, especially for adverse slopes (see Table 4). For a tolerance of 2% (respectively 5%) on the discharge coefficient, a slope of -1.2% is acceptable (respectively around -2.0 %) when the correction is applied. For positive slopes, +0.4% (respectively +0.8%) is an acceptable slope for a tolerance of 2% (respectively 5%).

3.2. Influence of the deformations of the walls of the throat

As for the results obtained for the slope, the orders of magnitude of the error due to the deformation of the walls of the throat are the same for the small and the large flumes (compare Table 7 and Table 8). Table 7 can therefore be seen as general conclusions about the influence of defects on the walls of the throat.

First, the results show that a constriction (humps on the walls) has a much greater impact than an enlargement (hollows on the walls). Indeed, the error is up to six times higher for 2Δ/Bt = -20% than for +20%. This can be hydraulically explained by the fact that a constriction probably moves the location of the critical flow in the section where the breadth is minimum (Hager 1999) whereas a local enlargement probably only creates a dead zone. If a tolerance of 2% (respectively 5%) is acceptable for the error on the discharge, a constriction of -2% (respectively -4%) and an enlargement of 4% (respectively 10%) can be accepted.

Table 3. Relative error EP on the discharge as a function of H/Lt for a small Venturi flume installed with a longitudinal slope S different from horizontal without any correction of the discharge formula – in grey errors > 2%, in dark grey errors > 5% (* presence of a hydraulic jump).

H/Lt	S (m/m)														
	-2.0%	-1.6%	-1.2%	-0.8%	-0.4%	-0.2%	-0.1%	0%	+0.1%	+0.2%	+0.4%	+0.8%	+1.2%	+1.6%	+2.0%
0.17	+66%	+53%	+38%	+26%	+13%	+6%	+4%	0%	-3%	-5%	-11%	-19%	-28%	-40%	*
0.38	+27%	+21%	+16%	+11%	+5%	+2%	+1%	0%	-1%	-3%	-6%	-10%	-15%	-19%	-24%
0.51	+19%	+15%	+11%	+7%	+4%	+1%	+1%	0%	-1%	-2%	-4%	-7%	-11%	-14%	-18%
0.64	+15%	+12%	+9%	+6%	+3%	+1%	+1%	0%	-1%	-1%	-3%	-6%	-9%	-12%	-15%

Table 4. Relative error EC on the discharge as a function of H/Lt for a small Venturi flume installed with a longitudinal slope S different from horizontal with correction of the discharge formula – in grey errors > 2%, in dark grey errors > 5% (* presence of a hydraulic jump).

H/Lt	S (m/m)													
	-2.0%	-1.6%	-1.2%	-0.8%	-0.4%	-0.2%	-0.1%	0%	+0.1%	+0.2%	+0.4%	+0.8%	+1.2%	+1.6%	+2.0%
0.17	0%	0%	0%	+1%	0%	0%	+1%	0%	0%	+1%	+1%	+4%	+6%	+4%	*
0.38	-4%	-3%	-2%	-1%	-1%	-1%	-1%	0%	0%	0%	0%	+1%	+2%	+3%	+4%
0.51	-2%	-2%	-1%	-1%	0%	0%	0%	0%	0%	0%	0%	+1%	+1%	+2%	+3%
0.64	-2%	-1%	-1%	-1%	-1%	0%	0%	0%	0%	0%	0%	0%	+1%	+1%	+1%

A "natural" correction would be to use the breadth of the throat at the location of the deformation (Bt + 2Δ for an enlargement and Bt - 2Δ for a constriction). Since the discharge is proportional to the breadth of the throat (see Eq. 1), this correction leads to directly change the discharge by using the percentage of the defect 2Δ/Bt. A look at Table 9 reveals that this correction is not relevant for enlargement (2Δ/Bt> 0); results are indeed even worse with correction than without correction, especially for large values of 2Δ/Bt! This correction is much more relevant for constrictions and can be used to accept walls with humps up to 2Δ/Bt = -4% (respectively more than -10%) for an acceptable tolerance of 2% (respectively 5%).

Table 5. Relative error EP on the discharge as a function of H/Lt for a large Venturi flume installed with a longitudinal slope S different from horizontal without any correction of the discharge formula – in grey errors > 2%, in dark grey errors > 5% (*presence of a hydraulic jump).

H/Lt	S (m/m)		
	-1.2%	0%	+1.2%
0.07	+173%	0%	*
0.26	+39%	0%	-40%
0.53	+17%	0%	-17%
0.79	+12%	0%	-13%
1.17	+8%	0%	-10%

Table 6. Relative error EC on the discharge as a function of H/Lt for a large Venturi flume installed with a longitudinal slope S different from horizontal with correction of the discharge formula – in grey errors > 2%, in dark grey errors > 5% (* presence of a hydraulic jump).

H/Lt	S (m/m)		
	-1.2%	0%	+1.2%
0.07	0%	0%	*
0.26	0%	0%	-7%
0.53	-1%	0%	-1%
0.79	0%	0%	+1%
1.17	0%	0%	-2%

Table 7. Relative error EP on the discharge as a function of H/Lt for a small Venturi flume presenting a deformation of the walls of the throat without any correction of the discharge formula – in grey errors > 2%, in dark grey errors > 5%.

H/Lt	2Δ/Bt								
	-20%	-10%	-4%	-2%	0%	+2%	+4%	+10%	+20%
0.17	+25%	+11%	+4%	+2%	0%	-1%	-1%	-3%	-4%
0.38	+20%	+8%	+3%	+1%	0%	-1%	-2%	-5%	-7%
0.51	+17%	+7%	+3%	+1%	0%	-1%	-2%	-5%	-7%
0.64	+15%	+6%	+2%	+1%	0%	-1%	-2%	-5%	-8%

Table 8. Relative error EP on the discharge as a function of H/Lt for a large Venturi flume presenting a deformation of the walls of the throat without any correction of the discharge formula – in grey errors > 2%, in dark grey errors > 5%.

H/Lt	2Δ/Bt		
	-2%	0%	+2%
0.07	+2%	0%	-1%
0.26	+3%	0%	-2%
0.53	+2%	0%	-2%
0.79	+3%	0%	-2%
1.17	+2%	0%	-2%

3.3. Influence of the deformations of the bed of the throat

As for the results obtained for the slope and the ones obtained for the deformation of the walls, the orders of magnitude of the error due to the deformation of the bed of the throat are the same for the small and the large flumes (compare Table 10 and Table 11). Table 10 can therefore be seen as general conclusions about the influence of defects on the walls of the throat.

Whereas a hollow has a small impact on the discharge coefficient (the region in the hollow is probably a dead zone), a hump on the bed may have a huge impact, especially for low discharges. For example, the error is up to seven times higher for a hump than for a hollow for a deformation Δ/Bt of 10 %. This behavior can be explained by the fact that a hump acts as a weir for small discharges, which probably moves the critical flow above the hump. If a tolerance of 2% (respectively 5%) is acceptable for the error on the discharge, a hollow of -2% of the breadth (respectively -5%) can be accepted. No hump at all can be accepted for a tolerance of 2% on the error and a hump of 1% of the breadth can be accepted for a tolerance of 5%.

Table 9. Relative error EC on the discharge as a function of H/Lt for a small Venturi flume presenting a deformation of the walls of the throat with correction of the discharge formula – in grey errors > 2%, in dark grey errors > 5%.

H/Lt	2Δ/Bt								
	-20%	-10%	-4%	-2%	0%	+2%	+4%	+10%	+20%
0.17	0%	0%	0%	0%	0%	+1%	+2%	+7%	+16%
0.38	-4%	-3%	-2%	-1%	0%	+1%	+2%	+4%	+11%
0.51	-6%	-4%	-2%	-1%	0%	+1%	+2%	+5%	+10%
0.64	-8%	-4%	-2%	-1%	0%	+1%	+2%	+4%	+10%

A "natural" correction would be to define the datum of the water level measurement at the altitude of the deformation (the top of the hump or the bottom of the hollow). The results applying this correction are given in Table 12. Since a hollow is mainly a dead zone, this correction is irrelevant for such a deformation. Nevertheless, this correction is relevant for humps and may lead to accept flumes with deformations Δ/Bt up to +2% (respectively +10%) if the acceptable tolerance is 2% (respectively 5%).

Table 10. Relative error EP on the discharge as a function of H/Lt for a small Venturi flume presenting a deformation of the bed of the throat without any correction of the discharge formula – in grey errors > 2%, in dark grey errors > 5%.

H/Lt	Δ/Bt								
	-10%	-5%	-2%	-1%	0%	+1%	+2%	+5%	+10%
0.17	-4%	-4%	-2%	-1%	0%	+3%	+6%	+15%	+30%
0.38	-7%	-4%	-2%	-1%	0%	+1%	+1%	+4%	+9%
0.51	-5%	-3%	-1%	0%	0%	+1%	+1%	+3%	+6%
0.64	-5%	-2%	-1%	0%	0%	0%	+1%	+2%	+5%

Table 11. Relative error EP on the discharge as a function of H/Lt for a large Venturi flume presenting a deformation of the bed of the throat without any correction of the discharge formula – in grey errors > 2%, in dark grey errors > 5%.

H/Lt	Δ/Bt		
	-2%	0%	+2%
0.07	-3%	0%	+23%
0.26	-4%	0%	+5%
0.53	-2%	0%	+2%
0.79	-1%	0%	+2%
1.17	-1%	0%	+1%

Table 12. Relative error EC on the discharge as a function of H/Lt for a small Venturi flume presenting a deformation of the bed of the throat with correction of the discharge formula – in grey errors > 2%, in dark grey errors > 5%.

H/Lt	Δ/Bt								
	-10%	-5%	-2%	-1%	0%	+1%	+2%	+5%	+10%
0.17	+27%	+11%	+4%	+2%	0%	0%	0%	-1%	-2%
0.38	+7%	+3%	+1%	+1%	0%	-1%	-2%	-3%	-5%
0.51	+5%	+3%	+1%	+1%	0%	0%	-1%	-2%	-4%
0.64	+4%	+2%	+1%	0%	0%	0%	-1%	-2%	-3%

4. Conclusions

The objective of this study was to investigate the hydraulic influence of a number of typical wrong installations and geometrical defects of long-throated Venturi flumes: significant positive or adverse slopes, humps and hollows on the walls and the bed of the throat.

The geometric tolerances corresponding to an acceptable tolerance on the discharge of 5% and 2% have been calculated for each defect. A number of corrections have been proposed to avoid the destruction of a flume if the geometric tolerance is not respected:

ΛChange the datum of the water level measurement for a slope significantly different from zero or a hump on the bed of the throat,

ΛUse the deformed breadth of the throat for humps on the walls of the throat. Results are summarized in Table 13.

Table 13. Geometric tolerances as a function of the defect and the acceptable tolerance on the discharge – in brackets: geometric tolerances when the proposed corrections are applied (Ø means that no defect is acceptable).

Acceptable tolerance on the discharge	2%	5%
Positive slope	Ø (+0.4%)	+0.2% (+0.8%)
Adverse slope	Ø (-1.2%)	-0.2% (-2.0%)
Humps on the walls on the throat	-2% (-4%)	-4% (-10%)
Hollows on the walls on the throat	+4%	+10%
Hollow on the bed on the throat	-2%	-5%
Hump on the bed on the throat	Ø (+2%)	+1% (+10%)

References

Bos M.G., The use of long-throated flumes to measure flows in irrigation and drainage canals, Agricultural Water Management 1 (1977) 111-126.

Dabrowski W., Polak U., Improvements in flow rate measurement by flumes, Journal of Hydraulic Engineering 138 (2012) 757-763.

Dufresne M., Vazquez J., Head–discharge relationship of Venturi flumes: from long to short throats, Journal of Hydraulic Research 51 (2013) 465-468.

ERCOFTAC, Special interest group on "quality and trust in industrial CFD" – Best practice guidelines, European Research Community On Flow, Turbulence and Combustion, (2000).

Hager W.H., Wastewater hydraulics–Theory and Practice, Springer, (1999).

ISO, Liquid flow measurement in open channels – Rectangular, trapezoidal and U-shaped flumes, International Standardization Organization, ISO 4359:2012, (2012).

Open FOAM, Open FOAM – The open source CFD toolbox – User guide, Open FOAM Foundation (2013).

Roache P.J., Perspective: a method for uniform reporting of grid refinement studies, Journal of Fluids Engineering 116 (1994) 405-413.

Yeung H., An examination of BS3680 4C (ISO/DIS 4369) on the measurement of liquid flow in open channels-flumes, Flow Measurement and Instrumentation 18 (2007) 175-182.

7

Forecasting Surface Area Fluctuations of Urmia Lake by Image Processing Technique

Javad Ahmadi, Davood Kahforoushan*, Esmaeil Fatehifar, Khaled Zoroufchi Benis, Manouchehr Nadjafi

Environmental Engineering Research Center, Faculty of Chemical Engineering, Sahand University of Technology, Iran.

ARTICLE INFO

Keywords:
Urmia Lake
Numerical Modeling
Remote sensing
Image processing

ABSTRACT

Urmia Lake's water surface area is among the most important parameters needed for water balance analysis. Periodical measurement of this parameter directly by conventional topography almost seems impossible since it is costly and time – consuming. Such limitations highlight the needs for new approaches to be taken, namely remote sensing technique which could provide a good approximation of lake's surface area in terms of some other parameters available or at least easily measured. This paper considers development of a new model for lake's surface area measurement according to available water levels and its comparison with other methods in this field as well as the calculations regarding salt-land formation and coastline changes. High resolution images provided by NASA satellites, Aqua and Terra were collected and passed an image processing stage through MATLAB software for surface area calculations. Finally, the water level and surface area values resulted from the home made code, were put together to reach relationship. The comparison between the results of proposed method and provided data by Eastern Azerbaijan Water Organization and also a similar study indicated that the proposed image processing technique has good performance to estimate the surface area of Urmia Lake. The maximum error between the results of proposed model and a similar study which was used combination method of Cellular Automata and Markov Chain was 5.96 % which indicates the good performance of image processing technique in estimation of surface area of Urmia Lake.

1. Introduction

Earth, on the whole, and all terrains on it, as an inseparable part of our world, is continuously experiencing an eternal process of transformation. Exploring and monitoring these changes, as much as possible, could lead to perception of the natural and artificial phenomenon causing them, as well as reasonable predictions about the future changes. Among these various changes, those which can be harmful to human residence are of most importance.

Lake Urmia is one of the largest saltwater lakes on earth and a highly endangered ecosystem, Monitoring Urmia Lake coastline and territory, as the second hyper saline lake in the world, and the first one in Iran has been considered in this investigation (Aghakouchak et al. 2014). Progressive decrease in size and amount of water in this lake due to the recent regional droughts and some artificial changes made in the neighborhood by human activities has made it a major concern requiring more research on this area which might lead to a proper solution for the current issue.

1.1. Study Area

Located in north-west of Iran (37° 4'–38° 17' N and 45°–46 °E), Urmia lake is ranked 20th largest and second hyper saline lake in the world (Ahmadzadeh et. al. 2009). With a semi-rectangular shape, a maximum length of 135 km, it covers an average area of 5,100 square kilometers. The maximum and average depths of this lake are 16 and 5 meters, respectively. Due to its unique biological and ecological features, it is internationally registered as a protected area as both a UNESCO Biosphere Reserve and a Ramsar site. Also, on the national

level, it is designated as a "National Park" by the Iranian Department of Environment (Eimanifar and Mohebbi. 2007).

The Lake encompasses a total of 102 islands among which is the Shahi, the biggest island, covering 250 square kilometers. About 30 main rivers, including 13 permanent and more than 17 seasonal ones, namely Zarinneh Rood, Simineh Rood, Mahabad Chai, Godarchai, Barandoozchai, etc. supply most of the inflow to the lake (4900 mm^3 out of 6900 mm^3) (Ghaheri. 1999).

1.2. Purpose

Among the effective factors considered in evaluating the annual water balance of the lake, surface area is of the most importance, since it directly influences parameters like evaporation. On the other hand, continuous falling trend in water level and progressive changes in coastlines during recent decade. Retreat of Urmia Lake from its original shoreline is not only a hydrological concern, but it also presents serious challenges for water quality, conservation, human health and economics (Sima and Tajrishy. 2013). Dried coastal salt lands, which leads to salt marshes creation, has brought to attention the importance of our knowledge about the water surface area. Since direct measurement of the lake's surface area is costly and time-consuming, therefore development or employment of new methods and tools for this purpose will be valuable. Considering some rational assumptions, surface area is a function of few parameters, among which water level is the most important one (see Fig. 1). Taking into account abovementioned, this paper aims to provide a reliable correlation between lake's surface area and level. In other words, a novel formula is proposed for area prediction in terms of lake's surface level, which is a vital relationship in practice for various hydrological and environmental analyses. To this end, remote sensing techniques have been utilized to estimate the lake area in terms of water level.

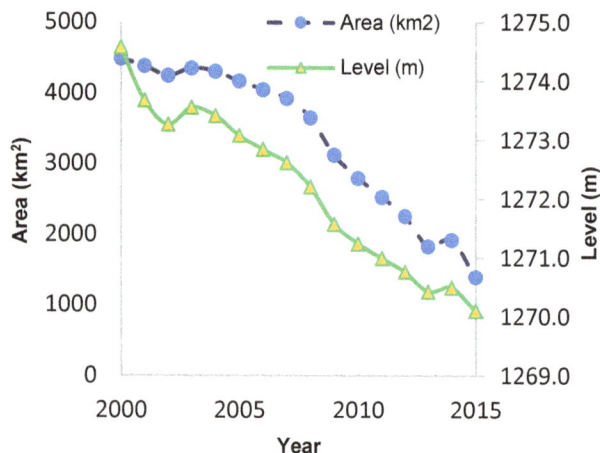

Fig. 1. Relationship between water level of Urmia Lake and its surface area, (Ahmadi. 2012).

1.3. Literature review

According to the stated purpose of this research mentioned in previous part, there is not a direct relationship between Urmia Lake level and surface area in the reported literature, except an empirical correlation presented and employed by Eastern Azerbaijan Regional Water Organization as following;

$$S_{(km^2)} = 479923 - \frac{6.05501*10^8}{L_{(meter)}} \qquad (1)$$

S and L are the predicted water surface area in km^2 and water level in meters, respectively. Moreover, since this research takes advantage of remote sensing, some of researches which have employed remote sensing for similar purposes have been reviewed. Singh et al. (1991) of Indian Bopal Research and technology association employed the remote sensing technology as a hybrid approach for evaluating surface waters and Bhopal Lake management. This research demonstrated that data collected via satellites can successfully be used for monitoring and surveying large water reservoirs.

Digital image processing is a discipline which can be applied in many areas such as astronomy, genetics, remote sensing, video communication, biomedical, transportation system (Kosesoy et al. 2015). Qudah & Harahsheh (1994) combined remote sensing and GIS as an efficient tool to determine the coastlines of Bahr ol Mayet, located in Jordan. Results from this research indicated a vast range of changes in lake coastlines and water surface area during the study period, which seemed impossible without taking advantages of image processing and precise images provided by satellites.

Seang et al. (1998) studied the coastline bound of Tonel Sap Lake of Thailand in which image processing was utilized for coastline determination during droughts. Their research was another case of successful employment of image processing and remote sensing. Furthermore, similar investigations have been done by Zavoianu et al. (2001), Kish (2002), Bayram et al. (2004), Najafi (2003), Alsheikh et al. (2007).

Rasuli et al. (2008) monitored the level fluctuations in Urmia Lake processing multi-sensor and multi-time satellite images. Their research revealed a noticeable fact that during recent decade lake area is decreased by 23 %, equal to 1200 km^2. This is an alert, calling the authority in charge for an immediate action against salt marshes development which will be crucial to the surrounding vegetation as well as human beings dwelling in neighborhood.

All above-mentioned researches have proved that remote sensing and image processing are among the best techniques for geographical and environmental investigations especially in expanded area studies.

2. Material and methods

Clearly, it could be concluded that water surface area is an indication of the lake's bed topography in various levels. Therefore, the water surface area is generally a function of two factors: lake's water level and ground topography inside the lake limits. Although, the first factor is influenced by different parameters such as evaporation, precipitation, surface and underground water entrance to the lake, tidal range, but it could be measured easily because level fluctuation due to the waves and tide is negligible. The second factor, lake topography, could also change due to natural and artificial phenomena. Natural events such as erosion, sedimentation, etc. and artificial ones like massive man-made embankments inside the lake limits, as was created in causeway construction, can impressively alter the surface area. This factor is also influenced by many parameters which could not be measured to account for in surface area estimation. The only way for this purpose, other than the costly method of surveying and mapping, is remote sensing technique which employs up-to-date images provided by satellites. Most of the time, these images are not available to public, so they could not be utilized. Furthermore, processing these images besides the water level to reach the water surface area is also a matter of question.

According to the abovementioned statement, surface area could be estimated in terms of water level and topography change detection. Since topography changes are not considerable compared to the water surface changes due to the level fluctuations, lake water surface area could be assumed only dependent on water level, of course considering the accuracy level we aim in this paper. For this purpose, holding water surface area in various lake water levels which has been recorded in the past, can lead to a correlation between these parameters. To this end, a database should be collected, including water level and surface area during the past time. These data have been gathered and been computed, as following.

2.1. Data collection and selection

Images provided by NASA website from Terra and Aqua satellites, taken from 2005 to 2012 years, were collected and selected for this study. These images were taken with resolution accuracy equal to 250 m which have been the only source for lake area calculations during the abovementioned period. On the other hand, lake level data are retrieved by daily measurements carried out and provided by Eastern Azerbaijan provincial water organization's archive. These two data sources have been put together to predict the lakes area directly and its topography indirectly to some extent.

Totally, 44 images, obtained from abovementioned satellites, beside the corresponding water level in the same date, from Mar. 2005 to Dec. 2015, measured by Eastern Azerbaijan Water Organization were used in this study. Data records include water level fluctuation from 1271 m to 1274 m above mean sea level with a mean value of 1272.75 meters. This variation during the study time period is illustrated in Fig. 1. According to this figure, Lake's water level has experienced a descending trend during the last seven years, despite some seasonal increases. This is an alert for the lake and surrounding neighborhood form the environmental point of view.

2.2. Data preprocessing and analysis

Satellite images used for this study, were processed in Photoshop graphical software and MATLAB Programming software to extract the surface area values. Thus, Lake surface and surrounding ground were precisely separated using Photoshop tools considering the different colors belonging to each part, as the distinction criteria. Extracted Lake was then exported to MATLAB and set on a white background for further calculations. MATLAB was employed for computing surface area by enumerating total pixels and white colored pixels, which represent surrounding ground as well as islands and causeway embankments. Consequently, white colored pixels number subtraction from total pixels will result in the number of pixels presenting lake surface area. As a result, pixel size and quantity multiplication will result in the gross surface area value. The whole process was carried out through a code prepared in MATLAB Programming software. Date of photos and calculated areas are as Table 1.

Table 1. Calculated areas using MATLAB and Photoshop software.

Date	Area	Level	Date	Area	Level
20050317	4437.8	1273.72	20090901	3091.5	1271.6
20050411	4508.2	1273.9	20091001	3040.8	1271.5
20050801	4216.5	1273.54	20091201	3119.1	1271.6
20050901	4198.6	1273.37	20100401	3409.6	1271.85
20051001	4165.6	1273.22	20100601	3556.9	1272.01
20051101	4145.6	1273.12	20100701	3356.4	1271.82
20051201	4185	1273.11	20100801	3090.1	1271.62
20060602	4321.1	1273.43	20100901	3024.2	1271.48
20060702	4222.1	1273.32	20101101	2819.4	1271.28
20060802	4160.9	1273.08	20101201	2852.6	1271.25
20060902	4117.4	1272.93	20110228	2813.9	1271.31
20070202	4179.1	1272.95	20110510	3468.2	1271.51
20070303	4175.2	1273	20110604	3273.8	1271.64
20070611	4275.9	1273.28	20110704	3101.5	1271.42
20070811	4160.7	1272.93	20110804	2873.3	1271.23
20080601	3860	1272.48	20110810	2977.2	1271.24
20080701	3761.9	1272.33	20110904	2738.9	1271.09
20080901	3670.4	1272.1	20110920	2779.7	1271.01
20090401	3542.4	1272.21	20121004	2664.2	1270.99
20090501	3576.1	1272.17	20131215	1270.77	2262.55
20090701	3435.8	1272	20141208	1270.43	1836.82
20090801	3299.9	1271.7	20151203	1270.50	1927.51

Throughout the above process, surface area values were obtained for all available images, then water levels (L) and corresponding water surface areas (A) were coupled to investigate the probable correlation and its reliability for future applications. This data is presented in Table 2.

Simple polynomial regression analysis resulted in a reliable relationship between these parameters as following;

$$S_{km^2} = 115665570 - 0.018623 \times - \frac{35092679000}{L^{1.5}} \qquad (2)$$

In which S is the water surface area in square kilometers predicted by the proposed correlation, and L is the water level in meters. The performance of this correlation has been evaluated in next section via appropriate criteria.

3. Results and discussion

Performance of proposed model have passed two different comparison tests; first, a comparison with Eastern Azerbaijan Water Organization's Model and second, a comparison with Ahadnejad's (2007) calculations who had used a different method to evaluate surface areas of the Urmia Lake during 2001-2007.

3.1. Comparison of proposed model and Eastern Azerbaijan Water Organization's

The proposed model performance is assessed considering three criteria namely Pearson correlation coefficient (R), Mean Absolute Error (MAE) and Root Mean Sum of Error (RMSE). The formulas and results are as following equations and Table 3.

$$R = \frac{\sum (x_i - \bar{x})(y_i - \bar{y})}{\sqrt{\sum (x_i - x)^2 \sum (y_i - y)^2}} \qquad (3)$$

$$MAE = \frac{1}{n} \sum_{i=1}^{n} |y_i - x_i| \qquad (4)$$

$$RMSE = \sqrt{\frac{1}{n} \sum (y_i - x_i)^2} \qquad (5)$$

Graphical comparison between proposed model and Eastern Azerbaijan Water Organization's has been illustrated in Fig. 2, which indicates that proposed model has a reliable performance and could substitute with the previous one.

This also has been illustrated in Fig. 3. According to the Fig., the proposed correlation performance is considerably reasonable, since data points scattered very close to the diagonal line.

3.2. Proposed Model and Ahadnejad's (2007) calculations comparison

Remote sensing technique has been used by others to predict the surface area of Urmia Lake like Ahadnejad (2007) who has used a Combination method of Cellular Automata and Markov Chain to measure the area of the lake. In spite of different methods, a simple comparison shows that maximum error of proposed model to Ahadnejad (2007) estimations is only about 6 percent which is negligible and reliable. Table 4 provides such comparison.

Table 2. Summary of statistical measures.

Parameters	Data set No.	Min	Max	Mean	Standard Deviation
Level (m)	44	1271.5	1273.9	1272.7	0.72
Area (km²)	44	3040.8	4508.2	3914.7	444.7

Table 3. Statistical Comparison between proposed model and Eastern Azerbaijan Water Organization's.

Predictor	Train Datasets	Test Datasets	R_{Test}	MAE[b]	RMSE[c]	Mean Prediction/Target	Standard Deviation of Prediction/Target
E.A.W.O.[a]	NA	5	0.978	263.5	322.4	1.08	0.063
Proposed Relationship	25	5	0.991	49.5	59.4	1.001	0.015

a. Eastern Azerbaijan Water Organization; b. Mean Absolute Error; c. Root Mean Sum of Error

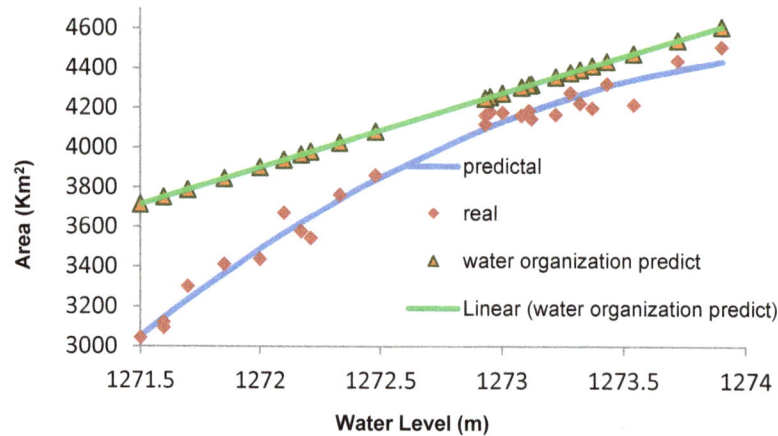

Fig. 2. Graphical comparison between proposed model and Eastern Azerbaijan Water Organization's.

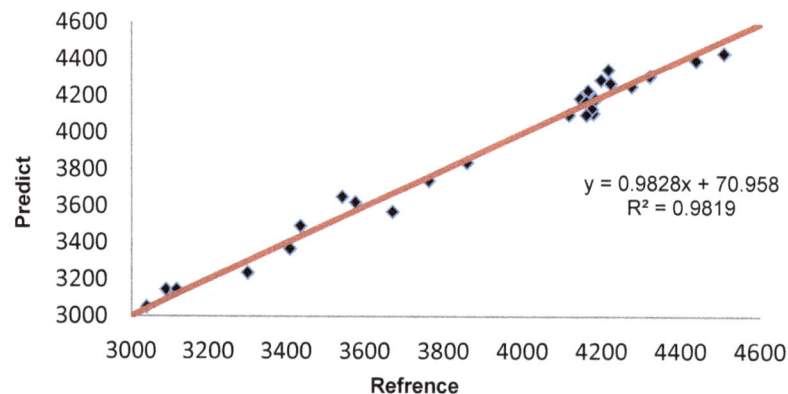

Fig. 3. Performance of proposed relationship (Predict Vs Reference Area Values).

Table 4. Comparison between Ahadnejad's calculations and proposed model.

Water year	Water Level	Model prediction	Ahadnejad (2007)	Error
2001-2002	1273.76	4406	4158	5.96%
2002-2003	1273.62	4370	4329	0.94%
2003-2004	1273.73	4399	4242	3.70%
2004-2005	1273.57	4355	4146	5.04%
2005-2006	1273.19	4219	4099	2.93%
2006-2007	1273.16	4206	4064	3.49%
2007-2008	1272.21	3653	3497	4.46%

4. Conclusion

The lake has faced extreme water loss in recent years due to overuse and mismanagement. Over the last thirty years, the population in the lake basin has been doubled and the agricultural area fed by water resources of the lake basin has tripled. To investigate the reasons of water loss of the lake, following items brought up by several specialists and authorities: 1) Shahid Kalantary causeway, which divides the lake, prevents the water circulation and this has caused the salinization of the lake, 2) vastness of lake area surface in comparison to its depth which leads to large evaporation surface, 3) construction of several dams and diversion of surface water for agriculture, in Urmia Lake basin without proper environmental risk assessments 4) reduction of precipitation and 5) warming of the region.

Urmia Lake's water surface area is one of the most important parameters needed for water balance analysis for the purpose of evaluation or modeling. Surface area measurement based on conventional methods consumes lots of energy and time for topographical works. Remote sensing technique has solved this problem and availability of water level data of the Urmia Lake has eased the way through surface area evaluation. Using Photoshop tools for extracting satellite images and processing theses photos through MATLAB software, combining these areas with available water levels; current paper has introduced a new equation for evaluating surface area of Urmia Lake using remote sensing technique, which doesn't have the former one's defects. According to abovementioned comparisons; proposed model has a reliable precision and could be used for surface area evaluations using water level data. After 40 years of Ramsar convention stating emphasized protection of Urmia Lake, Artemia Urmiana is on the edge of distinction, migrant birds have left the Lake forever, more than 52 percent of the lake surface area has become salt marsh, and the lake is transforming to a salt desert. According to proposed model 1800 square kilometers of lake surface area have been transformed to salt marshes since 2005. In case of complete desiccation of Urmia Lake, we will have 5100 square kilometers salt land. Spreading such amount of salt by wind power will destroy the economy and agriculture of the region (Nadjafi et al. 2012). We need to do something about this.

References

Abatzopoulos T.J., Baxevanis A.D., Triantaphyllidis G.V., Criel G., Pador E.L., Stappen G.V., Sorgeloos P., Quality evaluation of Artemiaurmiana Günther (Urmia Lake, Iran) with special emphasis on its particular cyst characteristics (International Study on Artemia LXIX), Aquaculture 254 (2006) 442-454.

AghaKouchak A., Norouzi H., Madani K., Mirchi A., Azarakhsh M., Nazemi A., Nasrollahi N., Farahmand A., Mehran A., Hasanzadeh E., Aral Sea Syndrome desiccates Lake Urmia: Call for action, Journal of Great Lakes Research 833 (2014) 5.

Ahadnejad M., Study of Uremia Lake Level Fluctuations and Predict Probable Changes Using Multi-Temporal Satellite Images and Ground Truth Data Period (1976-2010). New Challenge about Climate Change or Human Impact, Map Asia2010 & ISG 2010 Conference, Kuala Lumpur, Malaysia (2010).

Ahmad Zadeh T., Ahmad Zadeh B., Moradi S., Baghvand A., Calculation of effective parameters in the Urmia lake level fluctuations, Second International Symposium on Environmental Engineering, Tehran, Iran (in Persian) (2009).

Ahmadi J., Study of Urmia Lake water level and propose an optimized model for prediction of its fluctuations, M.Sc. Thesis. Sahand University of technology, Tabriz, Iran (in Persian) (2012).

Alesheikh A.A., Ghorbanali A., Nouri N., Coastline change detection using remote sensing, International Journal Environmental Science and Technology 4 (2007) 61-66.

Alipour S., Hydrogeochemistry of seasonal variation of Urmia Salt Lake-Iran, Saline Systems 2 (2006) 9.

Bakhtiari A., Zeinoddini M., Ehteshami M., Modeling flow pattern and salinity changes in Lake with a three-dimensional model, Second International Symposium on Environmental Engineering, Tehran, Iran (in Persian) (2009).

Barzegar F., Sadighian I., Study of highway construction effects on sedimentation process in Lake Urumieh (N.W. Iran) on the basis of satellites data, Geocarto International, 6 (1991) 63 - 65.

Bayram B., Bayraktar H., Helvaci C., Acar U., Coast line change detection using corona, SPOT and IRS ID Images, International archives of photogrammetry remote sensing and spatial information sciences 35 (2004) 437-441.

Eaestern Azerbaijan Water Organization, Urmia Lake Reports, Tabriz, Iran (in Persian) (2011).

Eimanifar A., Mohebbi F., Urmia Lake (Northwest Iran): a brief review, Saline Systems 3 (2007) 5.

Ghaheri M., BaghalVayjooe M. H., Naziri J., Lake Urmia, Iran: A summary review, International Journal of Salt Lake Research 8 (1999) 19-22.

Karbassi A.R., Bidhendi G.N., Pejman A., Bidhendi M.E., Environmental impacts of desalination on the ecology of Lake Urmia, Great Lakes Research 36 (2010) 419-424.

Kish S.A., Balsillie J.H., Milla K., A remote sensing and GIS Study of Long-Term water mass balance of Lake Jackson, Twelfth Annual Conference of Florida Lake Management, Tallahassee, Florida, USA (2001).

Kosesoy I., Cetin M., Tepecik A., A Toolbox for Teaching Image Fusion in Matlab, Procedia- Social and Behavioral Science 197 (2015) 525-530.

Nadjafi M., Ahmadi J., Makvandi R., Evaluation of Effective Parameters on Urmia Lake level fluctuations, First International Symposium on Urmia Lake; Challenges and solutions, Urmia, Iran (in Persian) (2012).

Najafi A., Investigation of the snowmelt runoff orumiyeh – region using modeling GIS and RA- tchniqes. M.Sc. Thesis. International institute for geoinformation science and observation, enchede the Netherland (2003).

Qudah O., Harahsheh H., Recession of Dead Sea through the Satellite Images, Royal Jordanian geographic Centre, Amman-Jordan (1994).

Rasouli A.A., Abbasian Sh., Jahanbakhsh S., Monitoring of Urmia Lake Water Surface Fluctuations by Processing of Multi- Sensors and Multi-Temporal Imageries, The Modares journal of spatial planning (in Persian)12(2008) 53-71.

Shafiei R.A., Perspective of Urmia Lake future based on water and salt balance, Second Conference on Water Resources Management, Isfahan, Iran (in Persian) (2006).

Singh A., Digital change detection techniques using remotely sensed data, International Remote sensing 10(1989) 989-1003.

Seang T.P., Murai Sh., Honda K., Schumann R., Samakaroon L., Detection of Coast Lines of Tonle Sap Lake in Flood Season using JERS-1Data for Water Volume Estimation, The 19th Asian Conference on Remote Sensing, Manila, Philippines (1998).

Sima S., Tajrishy M., Using satellite data to rxtract volume- area-evaluation relationships for Urmia lake, Iran, Journal of Great Lakes Research 39 (2013) 90-99.

Zavoianu Fl., Caramizoiu A., Badea D., Study and Accuracy Assessment of Remote Sensing Data for Environmental Change Detection in Romania Coastal Zone of the Black Sea, International Society for Photogrammetry and Remote Sensing congress, Istanbul, Turkey (2004).

Zeinoddini M., Tofighi M.A., Vafaee F., Evaluation of dike-type causeway impacts on the flow and salinity regimes in Urmia Lake, Iran, Great Lakes Research 35 (2009) 13-22.

8

Effects of curvature submerge vane in efficiency of vortex settling basin

Amin Hajiahmadi[1, *], Mojtaba Saneie[2], Mehdi Azhdari Moghadam[3]

[1]Department of Civil Engineering, University of Sistan and Baluchestan, Zahedan, Iran.
[2]Soil Conservation and Watershed Management Research Institute (SCWMRI), Tehran, Iran.
[3]Department of Civil Engineering, University of Sistan and Baluchestan, Zahedan, Iran.

ARTICLE INFO

Keywords:
Sediment
Settling basin
Submerge vane
Vortex flow
Efficiency

ABSTRACT

In this century due to population growth, the use of river water has become more complicated. Since most rivers pass across loose and erodible areas, they always act as the most important factor of transferring eroded materials from the solid crust of the earth. Vortex basins are among the solutions known for the high-speed separation of solids from liquids (filtration). One of the problems of such basins is the settling of sediments in their floor which necessitates the performance of required investigations and researches in order to present a method to exclude or reduce such sediments. The present research presents and investigates a plan to resolve this problem. This paper proceeds to perform an experimental study on the effect of a group of curvature submerged vanes in different positions at the floor of a vortex settling basin with a 90° radial section on the efficiency of the basin. The experiments were performed on a physical model with 96 cm height, 206 cm diameter, 10% floor slope, tow discharges of 45 and 37 L/S, and three orifices with 59, 46 and 36 mm diameters. Uniform aggregate (d50=0.22 mm) was applied in experiments. The efficiency of the basin was determined in six different positions of curvature submerged vanes and the values were investigated compared to each other. The results of experiments showed that the efficiency is higher when the vanes are placed more distant from the orifices while changes in orifice diameter and discharge considerably effect on the efficiency.

1. Introduction

Taking water from river has long been common for agriculture and drinking and later for industrial uses and financing energy. One of the main issues encountered by engineers while designing hydraulic structures including irrigation and hydroelectric intakes and other cases is the condition of sediments entered to the transmission system which should be controlled. Because of the harmful effects derived by the entrance of these sediments into hydraulic structures, different tools are applied in order to control a part of them. Of course, it is impossible and sometimes undesirable to eliminate all sediments. Small grained silt works as a sealant in transmission systems, modifies the soil texture and fertilizes agricultural lands as a result.

Application of settling basins, vortex tube extractors, submerged vanes, tunnel sediment extractors in the entrance of channels and vortex settling basins are among the methods of controlling sediments. In order to omit the undesirable characteristics of settling basins (such as high economical cost) and other settling methods for sedimentation of suspended solids, a cylindrical settling basin can be applied. Vortex settling basins are among the structures which control the entered sediments to water channels through vortex flows. These basins perform the separation of sediment grains in a high speed. Low water waste, being economical compared to other methods, being permanent unlike other settling systems, being needless to perform short-term dredging, and finally, smaller size compared to classic basins are all among the advantages of this sediment controlling method.

A wide range of researches have been performed on different kinds of this settling structure throughout the world. For instance, scholars such as Cecen and Akmandor (1973), Curi et al. (1979), and

Cecen & Bayazit (1975) in Istanbul Technical University (ITU) have investigated the use of vortex in sediment filtration systems as one of the possible solutions for high-speed separation of solids from liquids.

In his researches, Salakhov (1975) showed that vortex settling basins are completely efficient in controlling underneath layers of sediments (near to the basin bed) during intake operation. Ogihara & Sakaguchi (1984) have presented a model which is different in the position of the outgoing flow from spillway. In order to transmit clear water, a bell-mouth spillway was applied in this basin. Mashauri (1986) has presented a model which separates the entrance and outlet of the flows with a horizontal vane. The model presented by Paul (1991) has used a horseshoe-shaped vane (deflector) installed in the entrance channel in a distance equal to one-third of the flow depth. Ziaei (2000) has suggested that the settled sediments in the floor of settling basin can be washed through devising tow flow entrances and changing the flow direction from clockwise to anticlockwise and vice versa. Athar et al. (2002) have presented two other models of vortex settling basins. In their experiments, Niknia & Keshavarzi (2011) investigated the flow structure under deflector and compared it with the flow structure in the absence of deflector for the anticlockwise flows entered to the vortex basin.

One of the problems of vortex settling basins is the settling of sediments in their floor which necessitates the performance of required investigations and researches in order to present a method to exclude or reduce such sediments. The present research presents and investigates a model to resolve this problem. This model includes the application of a group of curvature submerge vanes in different positions placed at the floor of vortex settling basin in different positions.

*Corresponding author E-mail: amin.hajiahmadi@yahoo.com

2. Experimental equipments
2.1. Experimental model

The experiments were performed on the vortex settling basin physical model of Soil Conservation and Watershed Management

Research Institute, Tehran, Iran. Figure 1 illustrates the plan of the physical model of vortex settling basin. The characteristics of the model are mentioned in Table 1.

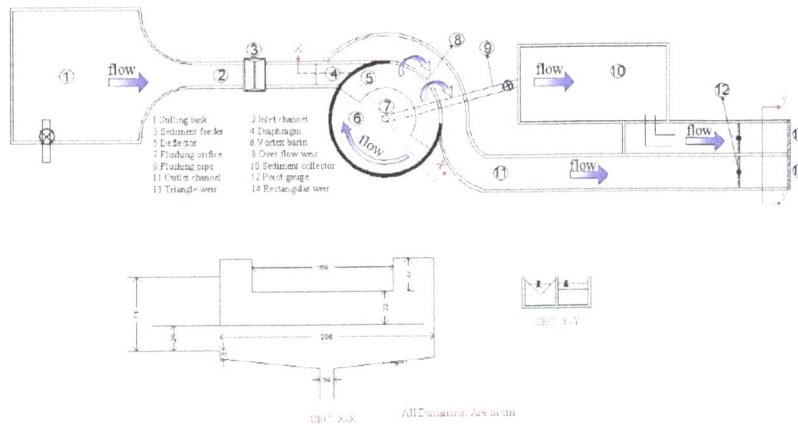

Fig. 1. The plan of the experimental model of vortex settling basin.

Table 1. The characteristics of vortex settling basin.

Basin Height (cm)	Basin Diameter (cm)	Flushing Orifice Diameter (mm)	Sloping floor (%)	basin Overflow length (cm)	basin Overflow Height (cm)	Height below the diaphragm (cm)
96	206	59	10	168	32	24

Tow outlet channels for the transmission of water excluded from the basin spillway and the flushing orifice are devised in this system. These depth and width of these channels are both 60 cm. There is also a deflector devised horizontally and flatly at the bottom of spillway, and a diaphragm on its entrance. Structure of the deflector is devised in order to prevent the sediment grains from being excluded just as they enter to the basin affected by outgoing jet flow from the spillway, which increases the duration of the settling of the sediment containing flow in the basin and finally, swirls the sediments more. This system includes a water reservoir with a content of 12.5 m³ and a centrifugal pump with a maximum discharge of 45 L/S.

2.2. Sediment injection

In this method, sediments should be injected well-distributed in the entering flow to the basin. Sediments should also be infused in a suitable area so that they have enough time for a normal distribution in flow profile and have suspended flow in the model. Scholars have used different gradations in their studies, some of which are observed in Table 2. Through investigation of different settling materials and considering issues such as providing, gradation, injection, sampling, collecting, drying, specific weight and uniform distribution, the sand with specific weight of 2.65 g/cm³ was applied in this research, The

gradation extents of the applied sand are observed in figure 2(a). d50 for this sand is equal to 0.22 mm.

The sediment injection is performed applying an infusion machine (Fig. 2(b)) placed 1.5 m far from the upstream channel of the vortex settling basin, in a dry and uniform form with a specific volume (22 g/s) and within a specific time.

Table 2. Range of sediment size used in previous studies.

Authors	Range of sediment size (mm)
Curi et al. (1977)	2.12
Mashauri M-I (1986)	0.375-1.80
Mashauri M-II (1986)	0.1875-0.75
Esen (1989)	0.320-2.7
Mashauri M-III (1986)	0.063-0.25
IPRI (1989)	0.09-0.30
Paul et al. M-I (1991)	0.175
Paul et al. M-II (1991)	0.05-1.00
Paul et al. M-III (1991)	7.64
Athar et al. (2002)	0.055-0.931
Keshavarzi et al. (2006)	0.074-0.3
Niknia et al. (2011)	0.08-2.00

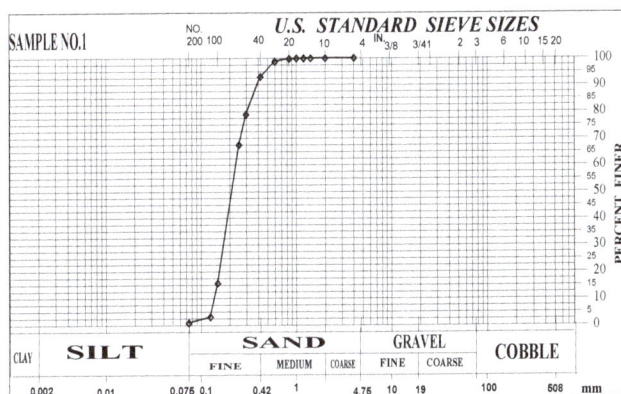

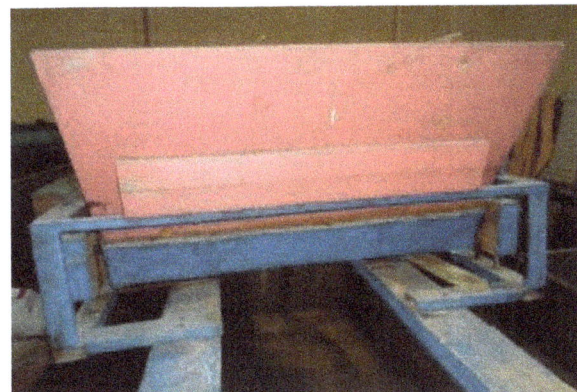

(a)

(b)

Fig. 2. (a) Gradation curve of the injected sediments to the vortex settling basin with d50=0.22mm; (b) Sediment injection machine.

2.3. Curvature submerge vanes

Considering the wide application of submerged vanes in rivers and intakes to improve efficiency of these structures, they are not that much used in vortex settling basin. Therefore, according to the suggestion of Odgaard and Kennedy (1983), the size of these vanes is illustrated in Table 3 (Fig. 3(a)).

Table 3. The Dimension of submerged vanes.

Vane Height H_v (cm)	Vane length $L_v = 3H_v (cm)$
4	12

These vanes are placed at the floor of the basin in a curvaceous form on concentric circles. The curve of each vane depends on the circle perimeter on which it is placed. Thus, considering the radial distance of vanes from each other, they have six different curves with 27.95, 40.45, 52.95, 65.45, 77.95 and 90.45 cm radiuses. Six types of vanes with different curves are observed in Fig. 3(b).

Vanes in 90° radial section (Fig. 4(a)) are placed in areas 2, 3 and 4 (Fig. 4(b)) at the floor of the basin. The vanes in 2, 3 and 4 are placed in six different positions each named considering the area in which they are devised.

These six types of arrangements (different positions) are: R2, R3, R4, R2 3, R3 4 and R2 3 4. You can see these six types in figure 5. These vanes aren't placed in area 1 because according to the experiments, without placing vanes in this area, they have no effect on the sediment flushing of basin floor.

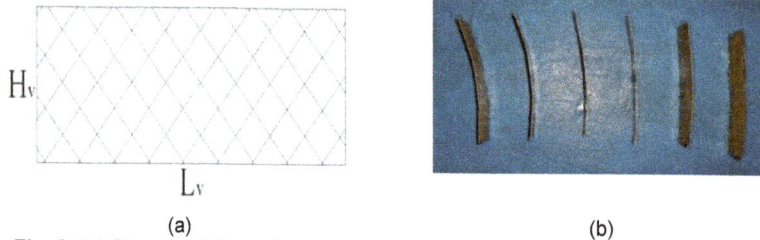

(a)   (b)

Fig. 3. (a) Characteristics submerge vane; (b) Six types of curvature submerged vanes.

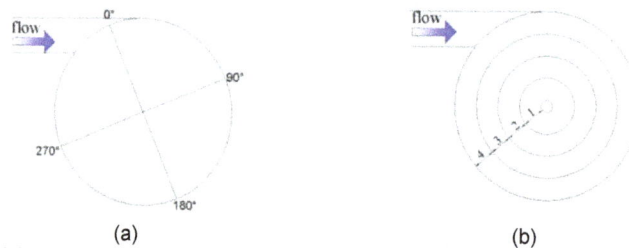

(a)   (b)

Fig. 4. (a) The radial section in which vanes are devised at the floor of basin; (b) The areas in which vanes are devised at the floor of basin.

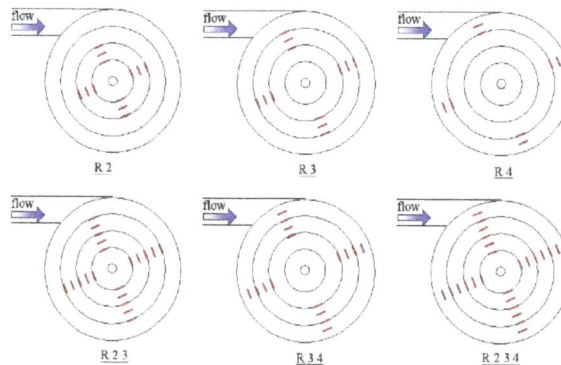

Fig. 5. Six types of arrangements for devising vanes at the floor of basin.

In order to perform a better coverage, some vanes are placed between the circles (figure 6). The traverse distance of vanes from each other is considered as a constant value of δ=3Hv (Odgaard & Kennedy, 1983).

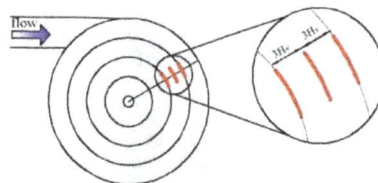

Fig. 6 The traverse distance between the curvature submerge vane.

2.4. Experimental program

Water is entered to the system through pumping and interned to the basin through the entrance of the basin and from beneath of the diaphragm. It creates a vortex flow it the basin a part of which out goes from the flushing orifice and the other part is excluded from the top of the spillway. The excluded water from the orifice goes to a channel with triangular spillway while the excluded water from the spillway goes to one with rectangular spillway. When the flow gets steady in the system, the system discharge can be calculated through reading the gauges fixed on two outlet channels and considering the type of spillways. The applied discharges are 37 and 45 L/S. Sediment nourishment gets started after stabilization of flow and adjustment of discharge. This act is performed with a specific discharge and within a specific time (20 min) in upstream channel. Afterward, as sediments are finished, the pump gets off simultaneously with closing the evacuation valve of the orifice. After settling of remained sediments in the basin, the water contained by the basin would be evacuated (figure 7) and the sediments inside the basin and those excluded from the orifice would be gathered, dried and finally weighted separately. Three flushing orifice with diameters of 36, 46 and 59 mm were used in order to determine the effect of flushing orifice on the efficiency of basin. This experimental process was first performed without and then with vanes.

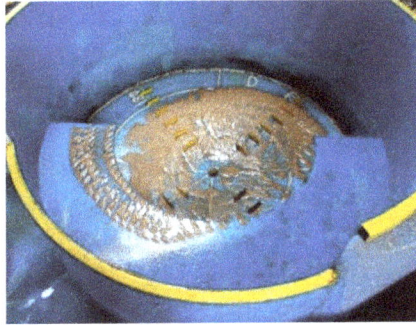

Fig. 7. Floor of the basin after the end of the experiment.

3. Results and discussion

The sediments entered to the vortex settling basin are divided into several parts. The first part of these sediments is entered to the flushing orifice, the second part settles at the floor of the basin and the third part passes from the top of the spillway. Now, we use the following relations in order to indicate the obtained efficiency:

$$\eta_T = \eta_o + \eta_B \tag{1}$$

$$\eta_o = \frac{W_o}{W_T} \tag{2}$$

$$\eta_B = \frac{W_B}{W_T} \tag{3}$$

$$W_T = W_T' - W_T'' \tag{4}$$

In the relations above, η_T represents the total settling efficiency, η_O refers to the settling efficiency of the flushing orifice, η_B is the settling efficiency at the floor of the basin, W_T represents the total weight of sediments entered to the basin, W_O refers to the weight of sediments entered to the flushing orifice, W_B represents the weight of sediments at the floor of basin, W'_T is the total weight of injected sediments in the upstream channel and W''_T refers to the weight of settled sediments under the sediment injection machine.

During the injection, a part of sediments settle under the sediment injection machine before entering to the basin. Therefore, these sediments are collected, dried after the experiment is done, then their weight would be subtracted from the weight of injected sediments.

Besides, G' parameter is used in order to indicate the rate of sediment flushing at the floor.

$$G' = \frac{(\eta_B' - \eta_B)}{\eta_B'} \tag{5}$$

In relation five, G' is the efficiency of the sediment flushing at the floor, η'_B represents the settling efficiency at the floor of the basin without curvature submerged vanes and η_B refers to the settling efficiency at the floor of the basin with curvature submerged vanes.

In order to show the value of the wasted discharge in the basin in different positions, η_O/η_t was used in which η_O is the outgoing

discharge from flushing orifice and η_t refers to the total entered discharge to the basin.

The results achieved about the effect of radial distance of the position of vanes (R2, R3, R4, R2 3, R3 4 & R2 3 4) at the floor on sedimentary parameters such as η_T, η_O and η_B, and hydraulic parameters such as η_O/η_t and G', are illustrated in figure 8. A 45 L/S discharge and a 59 mm orifice were applied in order to show this effect. R0 indicates a position in which no vane is placed at the floor of the basin.

As observed in Fig. 8 (a), the use of vanes has a small effect on total efficiency and wasted discharge while these vanes have considerably changed the value of η_O and η_B, in a way that whenever the orifice efficiency is reduced, the floor efficiency goes up in any R position.

Fig. 8 (b) shows the G' parameter in R3, R4 and R3 4 positions which indicates more sediment washing from the floor of basin. Positive values explain the successful performance of vanes n sediment flushing from the floor. On the other hand, the value of G' parameter hasn't changed much in simple and compound positions of R2 and sediment flushing from the floor of basin is low as well, i.e. the performance of vanes is not successful in flushing the sediments from the floor of basin. Fig. 8 (b) also shows that not only the sediment flushing is not performed in R2 3 4 position, but also the sediments are increased at the floor of the basin.

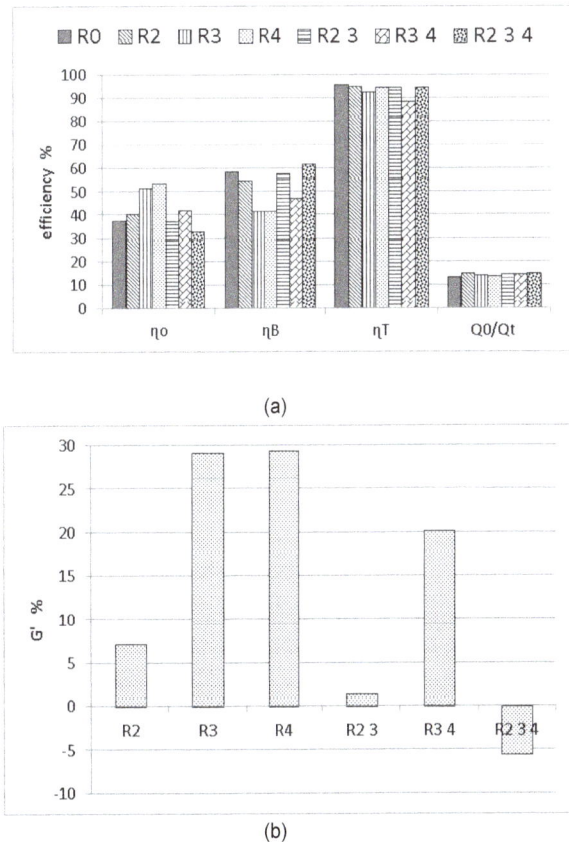

(a)

(b)

Fig. 8. (a) The effect of the radial distance of the (R) vain positions on sedimentary and hydraulic parameters; (b) The effect of the radial distance of the (R) vain positions on percentage of sediment flushing at the floor of the basin (G').

Therefore, considering the suitable and optimum performance of R3, R4 and R3 4 in sediment flushing from the floor of the basin, these positions are used in order to present the rest of the results and R2, R2 3 and R 2 3 4 arrangements would be omitted because of their low efficiency and sediment flushing from the floor of the basin.

The achieved results about the effect of orifice on sedimentary parameter η_T and hydraulic parameters such as Q_O/Q_t and G' are illustrated on Figs. 9(a), 9(b) and 9(c). A 45 L/S discharge was applied in order to show this effect. It should be mentioned that R0 explains a position in which no vane is placed at the floor of the basin.

(a)

(b)

(c)

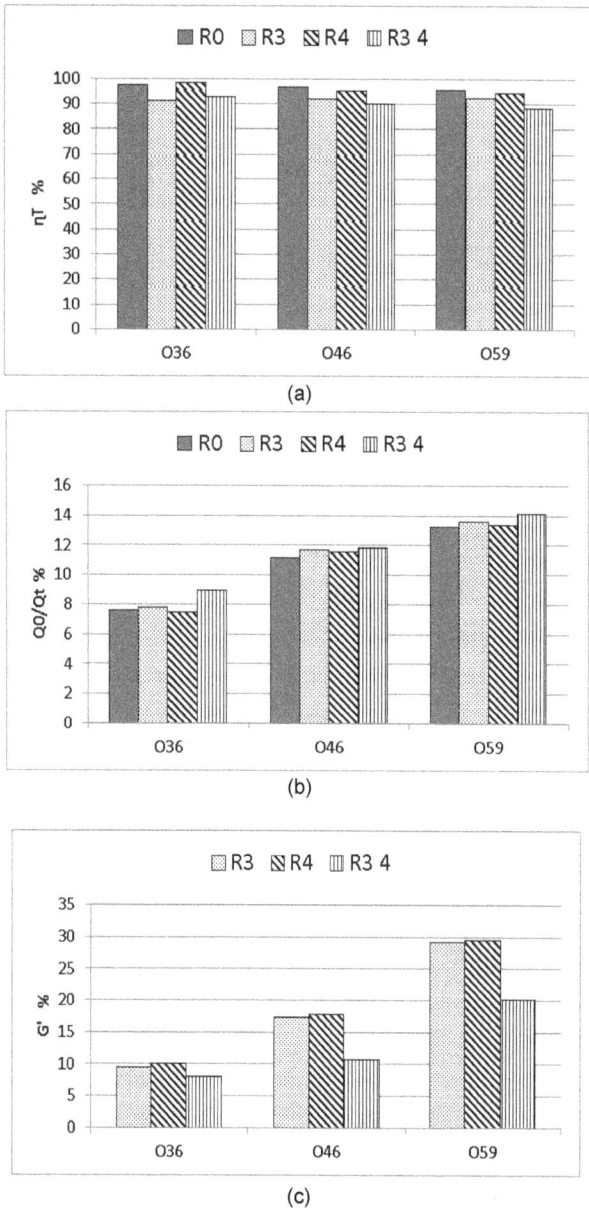

Fig. 9. (a) Mutual effect of Orifice diameter (O) and radial distance (R) on average percentage of total sediment flushing (ηT); (b) Mutual effect of Orifice diameter (O) and radial distance (R) on average percent of wasted discharge (QO/Qt); (c) Mutual effect of Orifice diameter (O) and radial distance (R) on average percent of sediment flushing from the floor of the basin.

It is observed in Fig. 9(a) that the changes of orifice have no considerable effect on ηT and it hasn't changed a lot compared to the position with no vanes. Fig. 9(b) shows that the wasted discharge goes up when the orifice diameter is increased. It also shows that the effect of these changes on wasted discharge is more compared to the changes in radial distance (R). As shown by Fig. 9(c), the sediment flushing from the floor of the basin goes up when the orifice diameter is increased. It also shows that the compared to the variations of radial distance (R), the changes in orifice diameter have a stronger effect on sediment flushing from the floor of basin.

Therefore, considering Figs. 9(b) and 9(c), we would have more sediment flushing from the floor of basin through choosing an orifice with 59mm diameter and accepting more wasted discharge. The R4 radial distance and an orifice with a 59 mm diameter were used in order to show the effect of discharge on sedimentary and hydraulic parameters. The results achieved about the effect of discharge on

As observed in figure 10(a), ηT doesn't change that much when the discharge goes up while the wasted discharge (QO/Qt) would be reduced. This hydraulic performance improvement can be caused by the increase in the power of vortex which is proportional with the increase in entered discharge. It also shows that through increasing discharge, ηO parameter is increased as well while ηB is reduced. The rate of sediment flushing from the floor of the basin for each discharge is illustrated in figure 10(b). It is completely clear that if discharge increases, the sediment flushing from the floor would be increased as well.

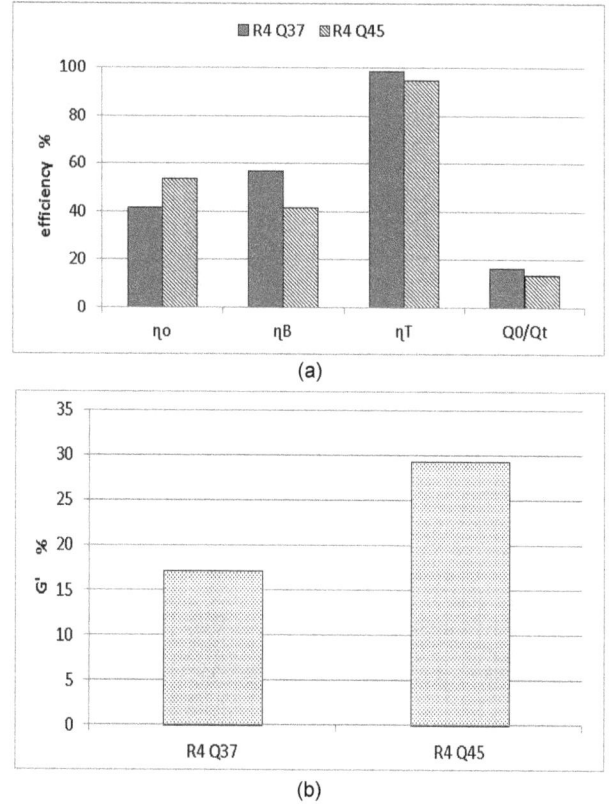

(a)

(b)

Fig. 10. (a) The effect of the entered discharge on (Q) on percentages of sedimentary and hydraulic parameters of the basin without the vanes; (b) The effect of the entered discharge on (Q) on percentage of sediment flushing from the floor of the basin.

sedimentary parameters such as ηT, ηO and ηB, and hydraulic parameters such as QO/Qt and G' are illustrated on Figs. 10(a) and 10(b).

4. Conclusions

Experimental models enable us to investigate the performance of a plan technically and economically. Considering the achieved results, the total efficiency doesn't change that much in different positions of vanes, while a considerable effect on orifice efficiency and the efficiency or the floor of basin is evident compared to the condition with no vanes. Thus, R3, R4 and R3 4 arrangements are chosen as the optimum positions for sediment flushing from the floor of the basin. This result shows that it is more efficient to place the vanes in more radial distances from the orifice. Results also indicated that when there are vanes in vortex settling basin, wasted discharge and sediment flushing from the floor goes up as the orifice diameter increases. Besides, an increase in discharge increases the wasted discharge and sediment flushing from the floor of the basin as well. By and large, the efficiency of the sediment flushing from the floor of vortex settling basin with vanes is respectively affected by orifice, discharge and radial distance of the vanes.

References

Athar M., Kothyari U.C., Garde R. J., Sediment removal efficiency of vortex chamber type sediment extractor. J. hydra. Engrg., Proc., American Society of Civil Engineers (ASCE) 128 (2002) 1051-1059.

Cecen K., Akmandor N., Circular Settling Basins with Horizontal Floor. MAG Report No 183, TETAK, Ankara, Turkey, 1973.

Cecen K., Bayazit M., Some laboratory studies of sediment controlling structures calculation of load controlling water intake structures for Mountain Rivers. In proceedings of the Ninth Congress of the ICID, Moscow, Soviet Union, (1975) 107-110.

Curi K.V., Esen I.I., Velioglu S.G., Vortex type solid liquid separator Progress in Water Technology 7 (1979) 183-190.

Esen I.I., Solid liquid separation by vortex motion, Solid Liquid Flow 1 (1989) 21–27.

Keshavarzi A.R., Gheisi A.R., Trap efficiency of vortex settling chamber for exclusion of fine suspended sediment particles in irrigation canals, Irrigation and Drainage 55 (2006) 419-434.

Mashauri D.A., Modelling of a vortex settling basin for primary clarification of water. PhD thesis, Tampere University of Technology, Tampere, Finland, 1986.

Niknia N., Keshavarzi A.R., Hosseinipour E.Z., Improvement the Trap Efficiency of Vortex Chamber for Exclusion of Suspended Sediment in Diverted Water. Proc., World Environmental and Water Resources Congress (ASCE), California, USA, (2011) 4124–4134.

Odgaard A.J., Kennedy J.F., Bed -river bank protection by submerged vanes. J. Hydra. Engrg., Proc., American Society of Civil Engineers (ASCE) 109 (1983) 1161–1173.

Ogihara H., Sakaguchi S., New system to separate the sediments from the water flow by using the rotating flow. Pros, 4th Congress of the Asian and Pacific Division, IAHR, Chiang Mai, Thailand, (1984) 753–766.

Paul T.C., Sayal S.K., Sakhuja V.S., Dhillon G.S., Vortex-settling basin design considerations, ASCE Journal of Hydraulic Engineering 117 (1991) 172–189.

Salakhov F.S., Rotational design and methods of hydraulic calculation of load-controlling water intake structures for mountain rivers, In Proceedings of Ninth Congress of the ICID, Moscow Soviet Union, (1975) 151-161.

Ziaei A.N., Study on the efficiency of vortex settling basin (VSB) by physical modeling. MSc. Thesis, Shiraz University, Shiraz, Iran, 2000.

9

On the fine sediment deposition patterns in a gravel bed open-channel flow

Akbar Safarzadeh[1], Seyed Hossein Mohajeri[2,*]

[1]Faculty of Civil Engineering, University of Mohaghegh Ardabili, Ardabil, Iran.
[2]Department of Civil Engineering, Science and Research Branch, Islamic Azad University, Tehran.

ARTICLE INFO

Keywords:
Gravel bed
Open-channel
Laboratory experiment
Fine sediment deposition
Video recording

ABSTRACT

One of the most important problems in rivers restoration and management is the problem of fine sediment deposition in the bottom of gravel beds. In fact, such phenomena can affect fauna and flora in various area. In Present study, a series of video camera measurements in an open channel with gravel bed was carried out in order to investigate the process of fine particles entry in the matrix of a gravel bed. Specifically, the present study focuses on a spatial pattern of fine sediment deposition and entrapment. The results show that deposition and entrapment of fine particles caused by the intrusion of large gravel particles and thus fine sediment deposition pattern are mostly in agreement with bed topography. Indeed, fine particles generally deposited on the downstream side of the gravel particles, while they rarely settled down in the upstream side of the gravel particles. Moreover, the results highlighted the formation of quite long longitudinal sand bars which are repeated in whole cross-section. These observations are in agreement with near bed common flow characteristics such as sweep and ejection events and strong secondary currents formation. The combined effects of sand ribbons and bed topography lead to the complex spatial pattern of deposition which questions the applicability of common transport thresholds which were developed based on the bulk properties of the flow like shear velocity.

---

1. Introduction

Gravel beds are common in nature, specifically most of the mountainous rivers are composed of gravels in different sizes (Wohl. 2013). The transport and deposition of fine sediments in gravel bed river are also common in mountainous areas (Schälchli.1992; Wohl. 2013). Improved knowledge of the distinct characteristics of fine sediments, which affects their erodibility (Grabowski et al. 2011) and the flow structures above gravel beds will further our understanding of fine sediment dynamics. This is important because fine sediments deliver benefits such as nutrient supply to biota but excessive fine sediment loads and the presence of sediment-bound contaminants can cause significant environmental impacts.

The issue of fine sediment deposition in the matrix of a gravel bed and its filtration to the deeper layer is a common topic in various engineering and science disciplines (Schälchli. 1992; Brunke and Gonser.1997; Blaschke et al. 2003). Due to this fact, there are several definitions of this phenomenon. In this regard, the term colmation is commonly used where there is more emphasis on the ecological and biological aspects of sediment deposition. In the context of the physical effects of sediment deposition, the process is called clogging (Blaschke et al. 2003; Packman and MacKay 2003; Rehg et al. 2005) Infilling is more used in the field of groundwater hydrology where the emphasis is on conductivity reduction (McCloskey and Finnemore. 1996; Li and Zhou. 1997) and finally ingress is a common term in geomorphology (Li and Zhou. 1997).

Despite many studies concerning fine sediment deposition in gravel bed, many physical aspects of this process have not been clarified properly. More precisely, it is not clear how fine particles are deposited in the matrix of gravel and what would be the spatial organization of deposited materials. In the present study, we focus on the process of fine particle entry in the matrix of gravel bed. Specifically, we aim to address spatial pattern of fine particles deposition respect to near bed flow characteristics. To this end, laboratory experiments have been conducted in open channel with an immobile gravel bed and in the presence of mobile particle. The bed topography during the process was captured using a digital camera. These measurements allow us to depict spatial deposition patterns.

2. Materials and methods
2.1. Experimental setup

The experiments were conducted in a 0.4 m wide, 0.4 m deep, and 6 m long polymethylmethacrylate rectangular tilting open channel at the Hydraulic Engineering Laboratory of the University of Trento. The flume bed was covered by a thick layer of gravel. To minimize backwater effects on the intended uniform flow conditions, an adjustable tailgate weir at the end of the flume was used. Free surface profiles were measured by an ultrasonic distance transducer to check flow uniformity. The discharge at the flume inlet was controlled by an inverter for pump speed regulation and was measured by an electromagnetic flowmeter. In our study, we employ the right-hand coordinate system, i.e., the x-coordinate is oriented along the main flow positive downstream and parallel to the mean bed; the z-coordinate refers to the vertical direction, pointing upward from the gravel tops (the z origin will be explained below); and the spanwise y-axis is directed to the left wall (Fig. 1-a).

Rough bed materials were the mixture of gravels (D_{50}=24.9 mm). Gravel-bed surface was smoothed mechanically by moving a wooden leveling table along a longitudinal guide from upstream to downstream, in order to avoid gravel clustering and produce conditions similar to water-worked gravel beds (e.g., Aberle and Nikora. 2006). In Fig. 1-b, the photo of the gravel bed in the laboratory open channel is shown. Moreover, to resemble the natural substrate in the rivers, sandy materials are also added to gravel bed (D_{50}=1.25 mm). The standard deviation of gravel bed elevations σ_1, which is a representative roughness scale (i.e., $\Delta \sim \sigma_1$, (Nikora et al. 1998) is equal to 6.1 mm.

Corresponding author Email: mohajeri@srbiau.ac.ir

(a)

(b)

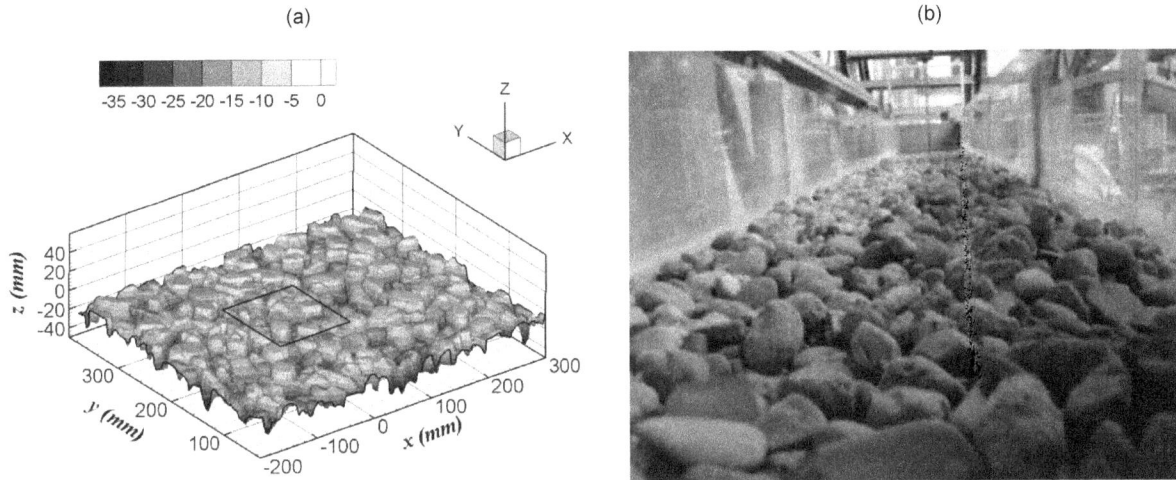

Fig. 1. (a) relative location of the PIV horizontal plane above the gravel bed, (b) photo of the gravel bed.

Two series of data have been collected during these measurements (named Series 1 and Series 2). Experimental conditions of these measurements are reported in Table 1. Hydraulic conditions were adjusted so that rough bed materials remain immobile during measurements. The aspect ratio B/H (B is channel width and H is water depth) was higher than 5 suggesting that effects of secondary currents in the central part of the flow should not be significant Nezu and Nakagawa [1993] The materials which were used as fine mobile particles were a plastic material named bakelite (phenol-formaldehyde resin) with density (ρ_s) equal to 1553 kg/lit. In Table 1, Rouse number ($R_0 = \omega_s/u_*$) is also reported. Settling velocity in Rouse number is a function of shape, size, and roughness of fine particles. For high Reynold's number, empirical formulas are available for estimation of settling velocity based on particle diameters. One of the most common forms of these empirical formulas is as below (Cheng. 1997; Julien. 2010):

$$\omega_s = \frac{\nu}{D}\left[\sqrt{\frac{1}{4}\left(\frac{A}{B}\right)^{2/m} + \left(\frac{4}{3}\frac{d_*^a}{B}\right)^{1/m}} - \frac{1}{2}\left(\frac{A}{B}\right)^{1/m}\right]^m \quad (1)$$

Where D is diameter moveable material, A, B, m are the coefficients vary according to different authors. In present study, the coefficients suggested by Cheng [1997] are used (A= 32.0, B= 1.0, m= 1.5). Also, d· is dimensionless particle diameter defined as:

$$d_* = \frac{(\rho_s - \rho)g}{\rho\nu^2}D \quad (2)$$

Where ρ and ρ_s are respectively water and fine sediment density and g is the acceleration of gravity equal to 9.81 m^2/s. When Rouse number is smaller than 1, it can be assumed that suspended sediments transport is initiated (Van Rijn. 1984). In both measurements, Rouse number is smaller than 1 (Table 1). This means that at least some parts of fine particles are transported in suspension.

To capture the deposition pattern of fine particles a video camera measurement technique is used. Based on this technique, a High-Resoloution video-camera is installed in a position where both stable gravel particles and fine particles are simultaneously recorded during a quite long/quite a long period. The video camera was Panasonic G-HVX200 DVC, with non-adjustable frame rate. The data was recorded for around 2 minutes in which the authors can clearly observe and describe those depositional patterns which were stable and remains for long period. The bed topography was captured in three difference views. In the first view, bed topography was captured from the side wall of the channel. The second view was from top where the plan of the bed topography was captured. Finally, the third view was again from top, but the zoom was wide to capture larger spatial patterns of the sand deposition. After experiments, the recorded videos are post-processing using Photoshop software, then proper photos are selected and extracted from the collected video record.

Table 1. Characteristics of the measurements in the presence of fine sediment.

	Q (lit/min)	H (mm)	S (-)	B (mm)	B/H (-)	D (μm)	U$_*$ (m/s)	ω$_s$ (m/s)	R$_0$ (-)
Series 1	420	50	0.002	400	8	500	0.031	0.027	0.87
Series 2	320	40	0.002	400	10	425-500	0.028	0.025	0.89

3. Results and discussion

In Fig. 2, deposition of fine particle near gravel bed in series 1 from side view (Fig. 2-a) and plan view (Fig. 2-b) are shown. It is clear that deposition of fine materials near gravel bed has a correlation with bed topography. Specifically, fine particles are generally deposited in the downstream side of gravel crests (see dashes curves in Fig. 2a and b). Indeed, the dashed curves in the downstream side of particles clearly show deposition of fine particles (materials with red color) in this region. In the upstream side of gravel crests, most of the fine particles are eroded. The same behavior is also observed out of the area covered by video camera.

There are two reasons for this tendency of deposition in correlation with bed topography. The first reason is due to the diversion of the water flow near gravel particles. Indeed, before gravel particles, water flow is diverted toward the lateral direction and after the particles the flow reattached. Accordingly, fine particles are brought up and down by the flow respectively in upstream and downstream of the particles. Another reason for this behavior can be related to the turbulence structure.

Ejection, which is the upward movement of low-velocity flow and sweep, which is the movement of high-velocity flow can commonly occur in rough bed flows (Grass.1971Nezu and Nakagawa. 1993). Fine particles can also be eroded and deposited by the sweep and ejection events (Mohajeri et al. 2016). In fact, previous studies show that gravel particles intrusion can affect near bed flow characteristics and prevalence of sweep and ejection events (Sambrook Smith and Nicholas. 2005; Wren et al. 2011). In this case, it can be speculated that ejection is more common in the upstream side of the particles, while sweep is more common in the downstream side of the particles. Both of the assumptions should beexamined by comparison of flow structure measurements near gravel particles.

In Fig. 2-b, it can be observed that in some cases, the fine particles are not deposited on the downstream side of the gravel particles. Indeed, fine particles cannot be deposited in the small spaces between the gravel particles. Moreover, the shape of gravel particles and their orientation can also change deposition pattern. This observation shows that the bed topography can affect the spatial deposition pattern of the fine particles.

(a)   (b)

Fig. 2. fine sediment deposition in comparison to bed topography (a) side view (b) plan view.

In Fig. 3, the sandy-grave bedsl in the wider top view for series 1 (Fig. 3-a) and series 2 (Fig. 3-b) measurements are shown. In both measurements, longitudinal bars of fine particles are formed along the main flow direction. their longitudinal bars are formed in the whole cross-section and the length is many time higher than their width. In fact, their length is fifty times higher than the water depth. As regards of the width of these longitudinal bars, it is difficult to judge about the exact values (due to the variation of their width respect to bed topography), but their width could be around water depth. As highlighted in Fig. 3 the lateral l distance between two consecutive longitudinal bars (δ) is approximately two times of water depth. The lateral spacing of these longitudinal bars and their size are in agreement with classical sand ribbons (Nezu and Nakagawa.1993). Similar straight sand ribbons are commonly notified in natural rivers (Karcz. 1966). Also, in the wide open channel (B/H>5) such sand bars can form. The main reason for the formation of sand ribbons is due to the presence of secondary currents cells. In rough bed open channel and in the presence of mobile fine materials, cellular secondary currents and sand ribbons interact mutually. The presence of sand ribbons can stabilize secondary currents, while the occurrence of upward and downward movement in secondary current cells can cause the formation of sand ribbons (Nezu and Nakagawa. 1984; Nezu and Nakagawa. 1993). The lateral spacings of sand ribbons in the present study are similar to the lateral spacing of lower longitudinal velocity zones of secondary currents observed by Kinoshita (1967); Albayrak and Lemmin. (2011); and Mohajeri et al. (2015). This observation supports the assumption that in our gravel bed straight channel even with high aspect ratio, secondary currents can form. Consequently, the particles are regularized in the longitudinal bars which are explained in the result section.

(a)   (b)

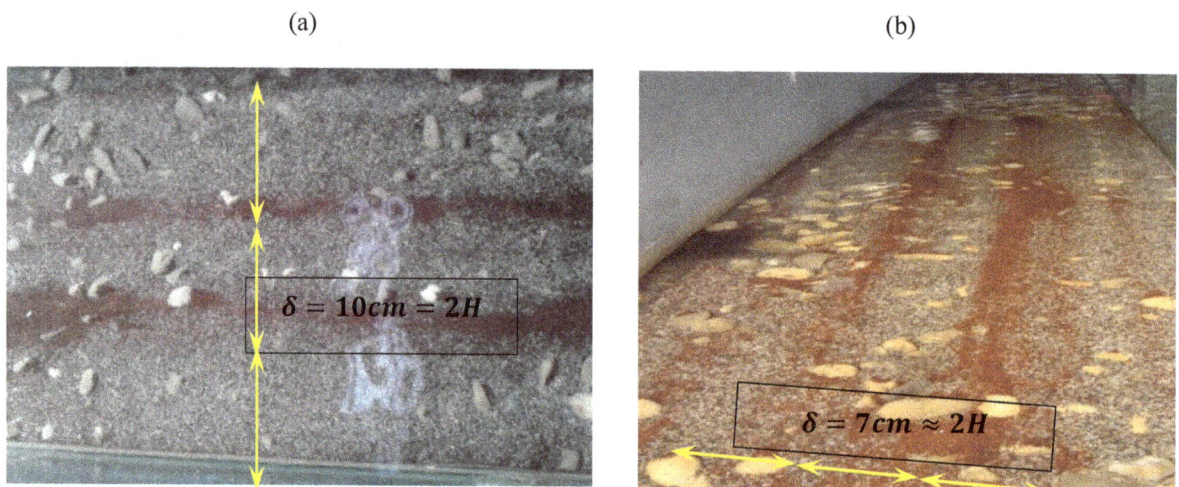

Fig. 3. Formation of sand ribbons in rough bed (a) series 1 (b) series 2.

To further examine the of the secondary currents on fine sediment deposition, measurement of the flow structure near gravel particles should be employed. This type of measurements is extracted from Mohajeri et al. 2015). In this study, the authors measure flow structure in a horizontal PIV plan just above gravel crests. In Fig. 4, an example of their measurements is shown. Water depth in plots of Fig. 4 was 4 cm and mean size of gravel bed materials was about 22 mm. The hydraulic conditions of this measurement are approximately similar to the hydraulic conditions of measurements in the present t study which allow us to use this information for better understanding of the governing process of our measurements. As shown in Fig. 4 and also properly described by Mohajeri et al. (2015), two flow structure can be depicted in near rough bed flow. The first one is the formation of secondary currents which are elongated from right to left in the whole of the contour plots. The presence of secondary currents can be better notified by the plot of stream wise averaged velocity components which are shown in the left corner of contour plots in Fig. 4. This observation can clearly support our assumption that secondary currents in gravel bed play a crucial role in the formation of sand ribbons. On the side, previous studies show that in the case that sand ribbons are formed, the secondary currents will be present and these ribbons generate a maintaining mechanism of lateral flow which results in secondary currents in time averaged velocity components (Karcz. 1966; Nezu and Nakagawa. 1993). However, comparison of Fig. 3 and Fig. 4 shows that the enhanced secondary currents in gravel bed flow can lead to sand ribbons formation. The mutual relation of sand ribbons and secondary currents in newly obtained finding which should be properly considered in future studies for better description.

It should be also highlighted that possible simultaneous effects of bed topography and secondary currents cause complex spatial distribution of the preferential zones for entrainment or deposition of fine particles. Formation of these regions confined the application of common sediment transport thresholds such as Rouse criterion, Shields criterion or any other transport threshold which does not consider spatial variations of flow characteristics. In order to consider the spatial variation of flow characteristics in sediment transport criteria, the aforementioned criteria should also be expanded in this regard. This issue can also be followed in the future studies.

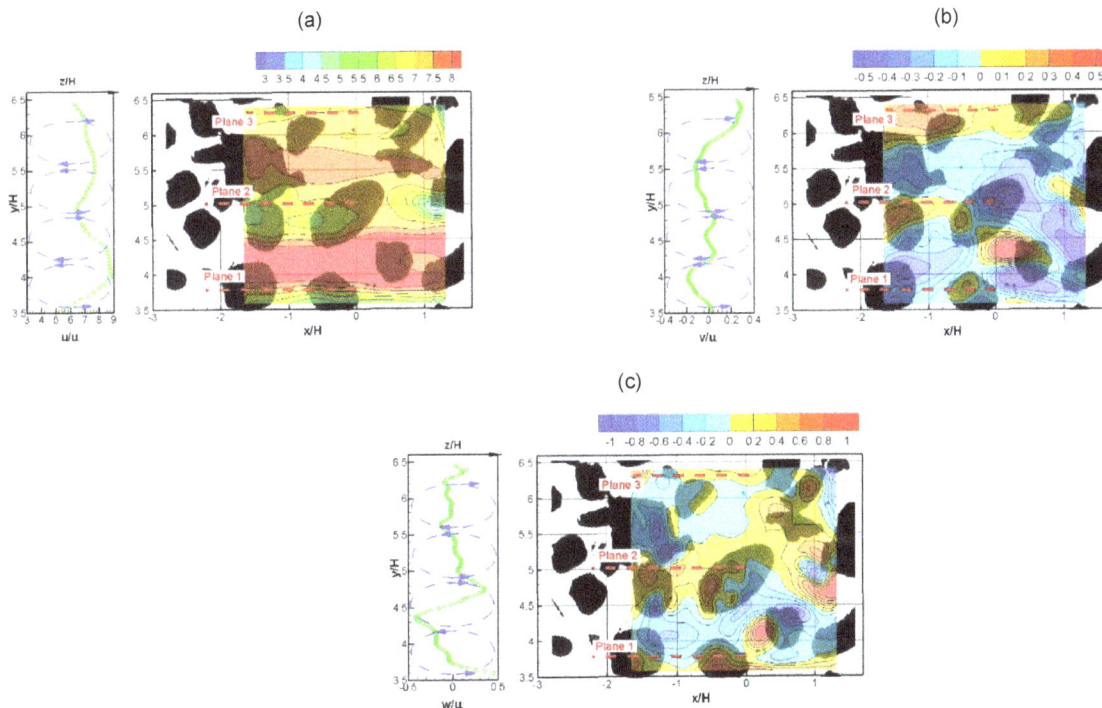

Fig. 4. Contour maps of velocity ((a) streamwise velocity (b) spanwise velocity (c) normalwise velocity) components normalized with shear velocity in the horizontal plane just above the particle crests modified from Mohajeri et al. (2015) , flowfrom left to right.

4. Conclusions

In the present study, the physical process of fine sediment deposition in gravel bed is studied using video camera measurement in a rectangular open channel. Two series of measurements were carried out in the range of shallow flow with a wide aspect ratio. The analysis of the results led to the following findings:

1- Deposition of fine particles are generally in agreement with bed topography. Fine particles rarely can be deposited on the upstream side of the gravel particles, while deposition is more common in the downstream side of the gravel particles. This observation speculates dominance of sweep and ejection events respectively in downstream and upstream sides of the gravel particles. This assumption should be examined in future studies with near bed flow field measurements.

2- It can also be observed that sand ribbons are formed in our experiments. Formation of these ribbons is mostly the result of secondary currents presence in the whole cross-section. It is interesting that in rough bed wide open channel secondary currents are present and sand ribbons are formed. Indeed, sand ribbons and secondary currents interact mutually in the smooth wide open channel. While in our experiments, sand ribbons are formed due to the secondary currents presence.

3- Combined effects of bed topography and sand ribbons formation make a very complex spatial variation of deposition zones. There are some zones where the depositin is more probable. This fact, challenge application of common sediment transport criteria. Therefore, it can also be suggested that in future studies sediment transport threshold should be somehow modified in order to take into account spatial variability of fine sediment deposition and entrapment.

Acknowledgements

The authors would like to thank Dr. Maurizio Righetti whose support in experimental data collection is truly appreciated. Moreover, we would like to appreciate the support of Trento University, especially for allowing us to use Hydraulic laboratory facilities during experimental measurements.

References

Albayrak I., Lemmin U., Secondary currents and corresponding surface velocity patterns in a turbulent open-channel flow over a rough bed, Journal of Hydraulic Engineering 137 (2011) 1318-1334.

Blaschke A.P., Steiner K.-H., Schmalfuss R., Gutknecht D., Sengschmitt D., Clogging processes in hyporheic interstices of an impounded river, the Danube at Vienna, Austria, International Review of Hydrobiology 88 (2003) 397-413.

Brunke M., Gonser T.O.M., The ecological significance of exchange processes between rivers and groundwater, Freshwater Biology 37 (1997) 1-33.

Cheng N., Simplified Settling Velocity Formula for Sediment Particle, Journal of Hydraulic Engineering 123 (1997) 149-152.

Grass A.J., Structural features of turbulent flow over smooth and rough boundaries, Journal of Fluid Mechanics 50 (1971) 233-255.

Julien P.Y., Erosion and Sedimentation, Cambridge University Press, (2010).

Karcz I., Secondary currents and the configuration of a natural stream bed, Journal of Geophysical Research 71 (1966) 3109-3112.

Kinoshita R., An Analysis of the Movement of Flood Waters by Aerial Photography Concerning Characteristics of Turbulence and Surface Flow, Journal of the Japan society of photogrammetry 6 (1967) 1-17.

Li W., Zhou L.S., edimentary facies and tectonic setting of the cretaceous in the suhongtu-yingen basin, Scientia Geologica Sinica 32 (1997) 387-396.

Mccloskey T.F., Finnemore E.J., Estimating hydraulic conductivities in an alluvial basin from sediment facies models, Ground Water 34 (1996) 1024-1032.

Mohajeri S.H., Grizzi S., Righetti M., Romano G.P., Nikora V., The structure of gravel-bed flow with intermediate submergence: A laboratory study, Water Resources Research 51 (2015) 9232-9255.

Mohajeri S.H., Righetti M., Wharton G., Romano G.P., On the of turbulent gravel bed flow: Implications for sediment transport, Advances in Water Resources 92 (2016) 90-104.

Nezu I., Nakagawa H., Cellular Secondary Currents in Straight Conduit, Journal of Hydraulic Engineering 110 (1984) 173-193.

Nezu I., Nakagawa H., Turbulence in Open-Channel Flows, A.A. Balkema (1993).

Nikora V., Goring D., Biggs B., On gravel-bed roughness characterization, Water Resources Research 34 (1998) 517-527.

Packman A.I., Mackay J.S., Interplay of stream-subsurface exchange, clay particle deposition, and streambed evolution, Water Resources Research 39 (2003) 1097.

Rehg K.J., Packman A.I., Ren J., Effects of suspended sediment characteristics and bed sediment transport on streambed clogging, Hydrological Processes 19 (2005) 413-427.

Sambrook Smith G.H., Nicholas A.P., Effect on flow structure of sand deposition on a gravel bed: Results from a two-dimensional flume experiment, Water Resources Research 41 (2005) W10405.

Schälchli U., The clogging of coarse gravel river beds by fine sediment, Hydrobiologia 235-236 (1992) 189-197.

Wohl E., Mountain Rivers, Washington D.C, American Geophysical Union, (2013).

Wren D.G., Langendoen E.J., Kuhnle R.A., Effects of sand addition on turbulent flow over an immobile gravel bed, Journal of Geophysical Research: Earth Surface 116 (2011) F01018.

10

Implementation of the skyline algorithm in finite-element computations of Saint-Venant equations

Reza Karimi[1], Ali Akbar Akhtari[1,*], Omid Seyedashraf[2]

[1]Department of Civil Engineering, Razi University, Kermanshah, Iran.
[2]Department of Civil Engineering, Kermanshah University of Technology, Kermanshah, Iran.

ARTICLE INFO	ABSTRACT
	Solving a large sparse set of linear equations is of the problems widely seen in every numerical investigation in the entire range of engineering disciplines. Employing a finite element approach in solving partial derivative equations, the resulting stiffness matrices would contain many zero-valued elements. Moreover, storing all these sparse matrices in a computer memory would slower the computation process. The objective of this study is to attain insight into Skyline solver in order to store the non-zero valued entries of large linear systems and enhance the calculations. Initially, the Skyline solver is introduced for symmetric or non-symmetric matrices. Accordingly, an implementation of the proposed solver is conducted using various grid form sets and therefore, several stiffness matrices with different sizes to evaluate the solver's capability in solving equation systems with a variety of dimensions. Comparing the obtained numerical results it was concluded that Skyline algorithm could solve the equation systems tens of times faster than a regular solver; especially in conducting iterative mathematical computations like Saint-Venant Equations.
Keywords: Skyline solver Sparse matrices Dam-break flow Saint-Venant equations	

Nomenclature

B	nodal degree of freedom
DIMMA	number of elements in the global matrix
DM	dimension of the stiffness matrix
DMSL	total number of elements in a 1D matrix
EL	number of elements
K	stiffness matrix
NO	number of nodes in an element
TOTVAL	number of non-zero cells in a matrix
TOTVALS	number of non-zero cells in a symmetric matrix
X	unknown vector
ZEVA	the number of zero valued cells in the global matrix

1. Introduction

Employing the numerical methods like finite difference and finite-element methods for numerical analysis of a hydraulic phenomenon; particularly in its three-dimensional form, can result in solving large systems of linear equations like Eq. (1). Regularly, these equation systems are symmetric and include many zero-valued entries in their equivalent matrix system form. Gauss-Jordan elimination, Atomic Triangular matrices, Cholesky-method and Strip-method are the most distinguished mathematical approaches to solve these systems of linear equations. For example, in the Gauss-Jordan elimination method, the determinants and inverse matrices must be calculated (Poole 2002). Moreover, the required numerical stability can be reached by conducting a partial diagonalization of the respective matrix (Golub and Van Loan 1996).

$$[K][X] = [B] \tag{1}$$

where, [K] is the coefficients matrix (stiffness matrix in finite-element approach), [X] is the unknown vector, and [B] is the nodal degree of freedom.

The constraints on computer storage requirements and CPU prevent using common solvers for intricate problems, like fluid flow with thousands of equations to be solved. There are two types of solvers, iterative and direct solvers. Below, we will present a concise overview of these solvers.

Iterative methods are found to be capable of solving large linear equation systems since they require less storage room and CPU time. They are inaccurate solvers, but their converged solutions can be close enough to the exact solution. The main process in an iterative method is the matrix-vector multiplication as compared to the matrix reduction in direct methods. A noteworthy benefit of iterative methods is that a given set of equations can be spit up into several subsets of equations and calculations can be performed in parallel on an array of processors. However, convergence characteristics of iterative methods depend on the condition number of the system of linear equations, and to find a suitable preconditioner is an essential to reach convergence. Normally, in Taylor-Galerkin computations of dam-break problems, less than six iterations would be enough to achieve convergence (Quecedo and Pastor 2002).

Nevertheless, solving sparse equation systems through classic methods would lead to larger computer storage requirements and therefore higher CPU costs. Nevertheless, there are various methods for re-ordering these sparse matrices. For example, Chin Shen et al. (2002) have proposed a set of parallel preconditioning approaches built up upon a multilevel block incomplete factorization method, which implements an iterative solver to complete the linear system solving process. According to their research, this technique is suitable for solving large sparse linear systems on distributed memory parallel computers that consist of a collection of independent processors. They have also proposed two new algorithms for constructing preconditioners based on the theory of block autono mous sets.

Moreover, in order to solve the large linear equations of the discretized form of shallow water equations, Fang and Sheu (2001) have made use of the bi-conjugate gradient stabilized (Bi-CGSTAB) and the generalized minimum residual (GMRES) methods, which are two Krylov subspace iterative solvers. They have evaluated these methods through comparing the obtained numerical results to the ones obtained from multi-frontal solver, which is a well-known direct solver regularly used in MATLAB software. Based on their investigations, the GMRES performed much faster than the direct solver.

The previously mentioned re-ordering methods can be employed in direct solvers too. Direct methods are those in which simultaneous linear arithmetical equations are solved by successive elimination of variables and back replacement. The Gauss elimination method is a fixed-step procedure. Furthermore, the frontal and skyline solution methods are modifications of Gauss elimination technique, which are frequently used direct solvers. These methods are common procedures in numerical simulations and especially in the finite-element analysis when we face a sparse mass matrix in the computations. The frontal solution method is faster than nearly all direct solvers as it calls for less core space as long as active variables can be kept in the core since the method completes the computational steps related to creating element matrices and assembling the global stiffness matrix in one single step (Irons 1970).

Other novel procedures include the approaches, which deal with only non-zero entries of a stiffness matrix through storing them in a new compacted form. Nour-Omid and Taylor (1984) have proposed an algorithm for assembly of K matrix. According to their studies, the data structure can be extended to be used in conjunction with any solution procedures by simply expanding the compacted form into a form appropriate for the respective solver. The scheme is similar to Skyline solver and results in considerable reduction in the storage needs during the assembly process. Alan Mathison has proposed another novel technique. In this method, the coefficient matrix must be decomposed into two lower and upper triangular matrices. Furthermore, the equation system can be solved by performing Forward Substitution and Back Substitution methods (Poole 2002). While, employing the Cholesky method, a unit valued element must be considered for the diagonal entries of the upper triangle of the matrix.

In this research, a one-dimensional dam-break test case is numerically examined to evaluate the efficiency of the Skyline solver through measuring its run time. The motivation of the investigation was to assess the necessary CPU time for different grid forms used to discretize the computational domain through a classic finite-element method and enhance the calculations. Accordingly, a shock wave induced by a dam failure will be numerically modeled and its results will be compared to the pre-existing numerical data.

2. Materials and methods
2.1. Band matrix solver

Among all conventional directs solvers; this method is more suitable to solve banded matrices. The matrix, A, of size n×n is called a banded matrix if all of its entries except the diagonal ones are zero-valued elements. The width and size of these diagonal stripes can be denoted by k_1 and k_2, which represent the width of left and right half of the diagonal strip, respectively. The global banded matrix can be constructed by storing the diagonal entries while having zeros for matrix cells except the diagonal ones. For example, the matrix, $A_{(4 \times 4)}$, with the width equal to three would be stored as a matrix like A' of the size 4×3 (Golub and Van Loan 1996). Here, the constant coefficients k_1 and k_2 are equal to one.

$$A = \begin{bmatrix} a_{11} & a_{12} & 0 & 0 \\ a_{21} & a_{22} & a_{23} & 0 \\ 0 & a_{32} & a_{33} & a_{34} \\ 0 & 0 & a_{43} & a_{44} \end{bmatrix} \tag{2}$$

$$A' = \begin{bmatrix} 0 & a_{11} & a_{12} \\ a_{21} & a_{22} & a_{23} \\ a_{32} & a_{33} & a_{34} \\ a_{43} & a_{44} & 0 \end{bmatrix} \tag{3}$$

2.2. Symmetric Band-matrix solver

This solver is similar to the Band Matrix Method, but more useful for symmetrical matrices. The following strategy can be used for positive definite matrix systems, which are banded and symmetrical. (4) and (5) demonstrates the respective arrangement of the re-ordered matrix.

$$A = \begin{bmatrix} a_{11} & a_{12} & a_{13} & 0 & 0 & 0 \\ & a_{22} & a_{23} & a_{24} & 0 & 0 \\ & & a_{33} & a_{34} & a_{35} & 0 \\ & & & a_{44} & a_{45} & a_{46} \\ & sym & & & a_{55} & a_{56} \\ & & & & & a_{66} \end{bmatrix} \tag{4}$$

$$A^* = \begin{bmatrix} a_{11} & a_{12} & a_{13} \\ a_{22} & a_{23} & a_{24} \\ a_{33} & a_{34} & a_{35} \\ a_{44} & a_{45} & a_{46} \\ a_{55} & a_{56} & 0 \\ a_{66} & 0 & 0 \end{bmatrix} \tag{5}$$

2.3. Skyline method

This method is useful for banded symmetric matrices, which are commonly seen in the finite-element analysis with an appropriate node numbering from. The scheme stores the non-zero cells of a matrix entry at the beginning and end of each column in the second-line array. However; in the case, the non-zero elements are located around the main diagonal, and the global stiffness matrix is created in a certain numerical order; an arithmetical logic can be proposed, in which the second-line array could be removed. Consequently, one can apply a series of simple mathematic rules when solving linear equations while keeping the self-determining property of the computational process. According to the following example, here the stiffness matrix can be stored in a single-dimensional array or in a single column vector. To do this, the diagonal entries of the matrix must be stored in an array.

$$A = \begin{bmatrix} a_{11} & 0 & a_{13} & 0 & 0 & 0 \\ & a_{22} & 0 & a_{24} & 0 & 0 \\ & & a_{33} & 0 & 0 & 0 \\ & & & a_{44} & a_{45} & 0 \\ & sym & & & a_{55} & a_{56} \\ & & & & & a_{66} \end{bmatrix} \tag{6}$$

$$A' = \begin{bmatrix} a_{11} & a_{22} & a_{13} & 0 & a_{33} & a_{24} & \cdots \\ 0 & a_{44} & a_{45} & a_{55} & a_{56} & a_{66} \end{bmatrix} \tag{7}$$

$$A^* = \begin{bmatrix} 1 & 2 & 5 & 8 & 10 & 12 \end{bmatrix} \tag{8}$$

where, the elements of matrix A_* represent the element values of each column up to the diagonal position.

Since there could be too many elements in each stiffness matrix, the assembly and storing steps of these sparse matrices are of the most complex steps, which engineers and scientists must take in order to discretize the respective governing equations. As a result, in most cases the analyst can manipulate the structure of the stiffness matrix to a more proper form and have a smooth computational process. For example, the banded matrix storage and the Skyline technique can be employed in order to minimize the space required for non-zero valued elements. Both methods store the non-zero cells in a narrow band near the main diagonal of the matrix. Accordingly, this can significantly reduce the computer memory required to store the matrix entries. The Skyline storage method has an efficient functionality for this; however, its way of storing and solving is complicated. Using Skyline solver, two one-dimensional matrices must be employed while its optimum output is reachable when these matrices are dense enough or in other word have fewer zero cells.

A Skyline solver can be used whenever the elements of the entire matrix are placed around the main diagonal based on a mathematical logic. The principal goal here is to store the non-zero elements and the right-hand side terms of Eq. (1). Moreover, the eliminations must be done on the previously mentioned one-dimensional array. Eq. (2) depicts the structure of a global stiffness matrix for three-node elements. According to this equation, the distance between any two consecutive cells of the main diagonal must be equal to the number of nodes in each element. Furthermore, for each column of the stiffness matrix, there is a simple mathematical relationship between the nodes and number of elements in each column, which can be used in the Gauss-Jordan solver in order to minimize the required CPU. This process would be more effective when the computational domain is partitioned into rather small elements, and the stiffness matrix has many entries. Eq. (9) depicts the structure of the respective global stiffness matrix.

$$
\begin{bmatrix}
a & a & a & 0 & 0 & 0 & 0 & 0 & 0 \\
a & a & a & 0 & 0 & 0 & 0 & 0 & 0 \\
a & a & b & b & b & 0 & 0 & 0 & 0 \\
0 & 0 & b & b & b & 0 & 0 & 0 & 0 \\
0 & 0 & b & b & c & c & c & 0 & 0 \\
0 & 0 & 0 & 0 & c & c & c & 0 & 0 \\
0 & 0 & 0 & 0 & c & c & d & d & d \\
0 & 0 & 0 & 0 & 0 & 0 & d & d & d \\
0 & 0 & 0 & 0 & 0 & 0 & d & d & d
\end{bmatrix}
\times
\begin{bmatrix}
x_1 \\ x_2 \\ x_3 \\ x_4 \\ x_5 \\ x_6 \\ x_7 \\ x_8 \\ x_9
\end{bmatrix}
=
\begin{bmatrix}
Q_1 \\ Q_2 \\ Q_3 \\ Q_4 \\ Q_5 \\ Q_6 \\ Q_7 \\ Q_8 \\ Q_9
\end{bmatrix}
\tag{9}
$$

Here, the dimension of the matrix can be calculated through a simple mathematical calculation using NO and EL parameters as the number of nodes in each element and number of elements, respectively. The following equations show the relationships between the various components of the matrix system.

$$DM = \left(EL \times (NO-1)\right)+1 \tag{10}$$

$$DIMMA = DM^2 \tag{11}$$

$$TOTVAL = \left(EL * NODE^2\right)-(EL-1) \tag{12}$$

$$TOTVALS = \left(EL * NO * \frac{NO+1}{2}\right)-(EL-1) \tag{13}$$

$$DIMSL = TOTVALS + DM \tag{14}$$

$$ZEVA = DIMMA - TOTVAL \tag{15}$$

where DM is the dimension of the stiffness matrix; DIMMA is the total number of elements in the global quadratic matrix. TOTVAL and TOTVALS are the number of non-zero valued cells in a regular and symmetric matrix formation, respectively. DIMSL is the total number of elements in a one-dimensional matrix, and ZEVA is the number of zero valued cells in the global matrix.

3. Results

Here an incompressible fluid is considered, which is assumed to have no viscosity (inviscid fluid) that is homogeneous with a constant and uniform density (ρ). Subjected to the characteristics of the flow and computational domain, some approximations are imposed to the Navier-Stokes system of equations to obtain the efficient equations capable of simulating Dam-break shock waves. Accordingly, the Saint-Venant Equations can be expressed as follows:

$$\frac{\partial A}{\partial t}+\frac{\partial Q}{\partial x}=0 \tag{16}$$

$$\frac{\partial Q}{\partial t}+\frac{\partial}{\partial x}\left(\frac{Q^2}{A}\right)+gA\frac{\partial h}{\partial x}=gA(S_0 - S_f) \tag{17}$$

where A is the cross-sectional area; t denotes the time (s); h is the water elevation (m); g is the gravity acceleration, and x is the flow direction. Sf and S0 correspond to friction and bottom slopes, respectively.

In this study, a homogeneous case is considered. Consequently, the source terms are neglected (S=0). A standard finite-element approach was used in order to have an approximate solution to the incompressible and shock-dominated flow problem. The numerical method employs the weight functions and the new parameter, α, to solve the system in an implicit approach. The numerical method approximately solves the governing equations in an iterative way so that the results obtained in the step s would be close enough to the results obtained in the computational step s+1. The written computer code uses the following equations to solve the momentum and continuity equations in an implicit approach (Karimi 2012).

$$\left[M^1\right]\left[\frac{\partial U}{\partial t}\right]+\left[K\right]\left[U\right]=\left[F\right] \tag{18}$$

$$\left[\hat{K}\right]_{s+1}\left[U\right]_{s+1}=\left[\bar{K}\right]_s\left[U\right]_s+\left[\hat{F}\right]_{s,s+1} \tag{19}$$

$$\left[\hat{K}\right]_{s+1}=\left[M^1\right]+\alpha\,\Delta t\left[K\right]_{s+1} \tag{20}$$

$$\left[\bar{K}\right]_s=\left[M^1\right]+(1-\alpha)\,\Delta t\left[K\right]_s \tag{21}$$

$$\left[\hat{F}\right]_{s,s+1}=\Delta t\left(\alpha[F]_{s+1}+(1-\alpha)[F]_s\right) \tag{22}$$

where [M1] is the coefficient matrix of time-dependent terms; [K] is the coefficient matrix of U, and [F] is the Right Hand Side (RHS) vector. In these set of equations, the parameter, α ,is varied within the range [0,1]. Accordingly, one may define $\alpha = 0, \alpha = 1.2, \alpha = 3.2$ while they act as a Backward Difference, Crank–Nicolson, Galerkin or Forward Difference schemes, respectively (Reddy 2006).

In order to assess the effectiveness of the proposed numerical scheme, results obtained from the previously mentioned mathematical approach is put side by side a finite difference model of a one-dimensional dam-break problem. Tseng and Chu (2000) have employed a predictor–corrector Total Variation Diminishing (TVD) scheme for the computation of unsteady one-dimensional dam-break flows. The presumption of an immediate and complete breach is chosen to simplify the simulation, while we are applying certain arithmetical methods for analyzing the Skyline's efficiency in solving large linear equation systems. Moreover, the experiment's presumptions illustrate a reinforced concrete arch dam failure (Seyedashraf 2012).

The test case exhibits the development of the flow in the computational domain from time t = 0 s to t = 60 s. The computational domain is a straight, rectangular channel, and the dam is located in the middle of a horizontal channel whose Manning roughness coefficient is assumed zero. Originally, the water body is motionless with uneven depths on both sides of the hypothetical barrier, 10m in the upstream and 5m in the downstream region of the dam. The obtained result is depicted in Fig. 1 and compared to its respective finite difference solution.

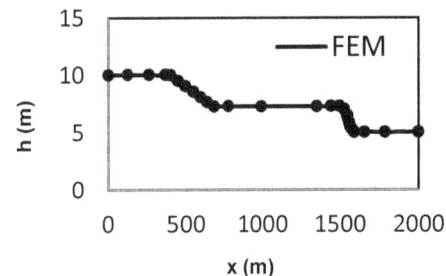

Fig. 1. Water depths at t=60(s) in a 1D dam-break problem.

Table 1. Global matrix characteristics for different element types used to discretize the computational domain.

Number of nodes on each element	3	4	5	6	7	8
Number of elements	100	100	100	100	100	100
Matrix dimension	201	301	401	501	601	701
Number of cells in the global matrix	40401	90601	160801	251001	361201	491401
Number of non-zero valued cells	801	1501	2401	3501	4801	6301
Skyline's matrix dimension	1002	1802	2802	4002	5402	7002
Number of zero valued cells	39600	89100	158400	247500	256400	485100

3. Discussion

As observed in table (1), extra zero-valued entries are produced in the global stiffness matrix, when the number of nodes in each element is increasing. To conduct the numerical computations, a QBasic code was developed. The computer code was implemented on a Pentium 4, 2.40 GHz personal computer. Tables (2) and (3) depict the numerical results obtained from different types of elements employed to discretize the computational domain while two distinctive solvers were used to solve the linear equation system, which are Gauss-Jordan and Skyline, respectively.

Table 2. Analogy of the results obtained from a finite element simulation of dam-break flow using different elements. The linear equations are solved by a Skyline technique.

Node	Δx(m)	Δt (s)	Time
3	50	1	0 min , 7 sec
3	40	1	0 min , 8 sec
3	30	1	0 min , 9 sec
3	20	0.5	0 min , 26 sec
3	10	0.25	1 min , 36 sec
3	2.5	0.1	14 min , 57 sec
5	100	1	0 min , 10 sec
5	50	0.5	0 min , 33 sec
5	40	0.5	0 min , 40 sec
5	30	0.4	1 min , 3 sec
5	20	0.2	2 min , 58 sec
5	10	0.1	11 min , 15 sec
7	100	0.5	0 min , 33 sec
7	40	0.25	2 min , 22 sec
7	30	0.25	3 min , 4 sec
7	25	0.1	9 min , 9 sec
7	60	0.5	0 min , 50 sec
7	70	0.5	0 min , 43 sec
7	50	0.4	1 min , 13 sec
7	50	0.2	2 min , 25 sec
4	50	1	0 min , 12 sec
4	40	0.5	0 min , 27 sec
4	30	0.5	0 min , 33 sec
4	20	0.25	1 min , 29 sec
4	10	0.2	3 min , 27 sec
4	5	0.1	13 min , 20 sec
6	100	1	0 min , 16 sec
6	50	0.4	0 min , 54 sec
6	40	0.4	1 min , 4 sec
6	30	0.2	2 min , 45 sec
6	20	0.2	4 min , 0 sec
6	10	0.1	15 min , 57 sec
8	100	0.5	0 min , 41 sec
8	50	0.25	2 min , 32 sec
8	40	0.25	3 min , 7 sec
8	30	0.2	4 min , 59 sec

According to these tables, the Skyline technique has enhanced the solving process, which has led to a 10-times faster computational process.

Table 3. Analogy of the results obtained from a finite element simulation of a dam-break flow using various elements. The linear equations are solved by a Gauss-Jordan technique.

Node	Δx(m)	Δt (s)	Time
3	50	1	6 min , 4 sec
3	40	1	11 min , 25 sec
3	30	1	25 min , 28 sec
3	20	0.5	>60 min
3	10	0.25	>60 min
3	2.5	0.1	>60 min
5	100	1	6 min , 6 sec
5	50	0.5	>60 min
5	40	0.5	>60 min
5	30	0.4	>60 min
5	20	0.2	>60 min
5	10	0.1	>60 min
7	100	0.5	40 min , 30 sec
7	40	0.25	>60 min
7	30	0.25	>60 min
7	25	0.1	>60 min
7	60	0.5	>60 min
7	70	0.5	>60 min
7	50	0.4	>60 min
7	50	0.2	>60 min
4	50	1	19 min , 25 sec
4	40	0.5	>60 min
4	30	0.5	>60 min
4	20	0.25	>60 min
4	10	0.2	>60 min
4	5	0.1	>60 min
6	100	1	11 min , 39 sec
6	50	0.4	>60 min
6	40	0.4	>60 min
6	30	0.2	>60 min
6	20	0.2	>60 min
6	10	0.1	>60 min
8	100	0.5	>60 min
8	50	0.25	>60 min
8	40	0.25	>60 min
8	30	0.2	>60 min

4. Conclusions

The Skyline method was discussed in this research and was employed to solve the spare linear equation systems for numerical simulation of a shock wave propagation emanating from a one-dimensional dam-break problem. Skyline is a novel approach to deal with the problems associated with sparse matrices in numerical computations. The technique only deals with the non-zero valued cells of the global stiffness matrix, and stores them in a one-dimensional matrix, which was later solved using an elimination method. The numerical results obtained from the proposed procedure were evaluated through a test case and compared to the ones acquired from a regular Gauss-Jordan solver. It has been shown that Skyline is an appropriate scheme for reducing the CPU time required for processing instructions of a computer, and therefore, computer storage costs. Moreover, using this solver, the obtainable numerical data would be the same as the ones acquired from well-known and regularly used solvers. Consequently, employing this method is recommended for problems that lead to large linear equation systems.

References

Fang C., Sheu T., Two element-by-element iterative solutions for shallow water equations, SIAM Journal on Scientific Computing 22 (2001) 2075-2092.

Golub G.H., Van Loan C.F., Matrix Computations, Johns Hopkins University Press, Baltimore, 1996.

Irons, B.A., A frontal solution program for finite element analysis, International Journal for Numerical Methods in Engineering 2 (1970) 5-32.

Karimi R., Numerical Solution of the Dam Break Problem With Finite Element Method, Thesis, Razi University, 2012.

Nour-Omid B., Taylor R.L., An algorithm for assembly of stiffness matrices into a compacted data structure, Engineering Computations 1 (1984) 312-317.

Poole D. Linear Algebra: A Modern Introduction, Cengage Learning, Thomson Brooks Cole, 2006.

Quecedo M., Pastor M., A reappraisal of Taylor–Galerkin algorithm for drying–wetting areas in shallow water computations, International Journal for Numerical Methods in Fluids 38 (2002) 515-531.

Reddy J. N., An introduction to the finite element method, McGraw-Hill Higher Education, New York City, New York, 2006.

Seyedashraf O., Development of Finite Element Method for 2D Numerical Simulation of Dam-Break Flow Using Saint-Venant Equations, MSc. Thesis, Razi University, 2012.

Shen C., Zhang J., Parallel Two Level Block ILU Preconditioning Techniques for Solving Large Sparse Linear Systems, Parallel Computing, 28 (2002) 1451-1475.

Tseng M.H., Chu C.R., The simulation of dam-break flows by an improved predictor–corrector TVD scheme, Advances in Water Resources, 23 (2000) 637-643.

11

The study and zoning of dissolved oxygen (DO) and biochemical oxygen demand (BOD) of Dez river by GIS software

Reza Jalilzadeh Yengejeh[1], Jafar Morshedi[2], Razieh Yazdizadeh[3,*]

[1]Department of Environmental Engineering, Science and Research Branch, Islamic Azad University, Khouzestan, Iran.
[2]Department of Geoghraphy, Ahvaz Branch, Islamic Azad University, Ahvaz, Iran.
[3]Department of Environmental Sciences, Science and Research Branch, Islamic Azad University, Khouzestan, Iran.

ARTICLE INFO

Keywords:
Zoning
Dissolved Oxygen (DO)
Biochemical Oxygen Demand (BOD)
Wastewater
Dez river

ABSTRACT

Pollution sources into the water, the necessity of qualitative studies of water resources is one of the most important new challenges for mankind in almost every parts of the world. Dissolved oxygen (DO) and Biochemical Oxygen Demand (BOD) are among parameters of water quality indexes which are considered as water pollution indexes. In the present research, DO and BOD of Dez River basin water (in Dezful City) were studied and zoned by applying the Geographic Information System (GIS). Nine stations were considered for sampling during six months in 2013. The results indicate that the average maximum amount of DO at an average of six months is 8.47 mg/l in S1 station and the minimum amount is 1.71 mg/l in S8 station. The average maximum and minimum amounts for BOD during an average of six months are orderly 150.83 mg/l in S8 and 3.16 mg/l in S1. By a qualitative zoning, places that are prone to pollution can be recognized and measures can be taken for monitoring and preserving such areas. Decreasing the amount of water pollution and controlling the pollution sources are possible by adoption appropriate measures.

1. Introduction

Today, by the expansion of social, economic, and developmental contexts and promotion of education level, there is a dire need to a clean and green environment (Thanh Binh et al. 2012). Continual monitoring of water quality is essential for efficient management of city rivers in order to control the pollution sources quickly (Hur & Cho 2012). As city rivers are prone to different pollution sources, they can be easily contaminated and as a result, the water quality will degrade. Instantly, the phenomenon is directly and indirectly a danger which threats the humans' health and aquatic ecosystems (Henderson et al. 2009; Su et al. 2011; Mouri et al. 2011). Generally, the density of DO is a key parameter for describing the nature and wastewaters of the environment (Keeling et al. 2002). A decrease in the level of DO in the world's oceans is increasingly on the rise which will affect the total earth ecosystem such as Carbon cycle (Joos et al. 2003), weather (Keeling et al. 2002; Keeling et al. 2010, Keeling et al. 2002; Shaffer et al. 2009; Gilbert et al. 2009). The amount of existing DO in natural waters is so essential that for doing researches related to nature, data will be gathered by considering the environment protection, hydro biological points of view and/or ecology (Nagy et al. 2008). Enough amount of DO is needed for the survival of water plants, animals and also wastewater treatment plant (Hobbs & McDonald 2010). Measuring and monitoring the density of DO is required to promote the quality of the environment (Naykki et al. 2013).

In aquatic environment, oxygen is produced by photosynthesis in plants and algae, and then plants, animals and bacteria breathe it in. The oxygen is also needed in the increase of the organic loading, in the formation of sediment which are formed by oxygen and oxidation, and aeration by exchange with the atmosphere (Radwan et al. 2003). Generally, oxygen may be added to water or be removed from it through different physical, chemical and biological actions (Fortes Lopes 2008). According to Europe standard, DO limit in a river is at least equal to 5 mg/l, for a suitable qualitative rate (Krenkel & Novotny

1980). A high amount of BOD is produced by the fast decomposition of biodegradable organic materials after which the decrease in the oxygen level will lower the water quality of the river (Jin & Cho 2012). GIS can geo reference, organize, process, and analyze the complicated information (Pedrero et al. 2011).

Pollution zoning and providing a valid image of the quality of the surface water by GIS, can help a better management decision-making which directly and indirectly, will promote the quality of surface waters (Hooshmand et al. 2008; Dunnette 1989; Curtis 2001). In a research done by Martin et al (2013) the density level of some of the effective parameters on DO, such as BOD, are surveyed for providing a model for predicting and decreasing the amount of DO in Athabasca river in Canada. BOD was considered as an index in the mentioned research. In another research conducted by Abdolhamid et al (2010) water quality index of Dokan river in Kurdistan-Iraq was determined by 10 parameters of water quality. The results of the mentioned research indicate that different human activities have affected some parameters like BOD. Kevin et al. (2000) have used GIS to study Zardchin river in order to estimate the level of relationship between pollution and hydrological and ground operation. Also, Akkoyunla & Akiner (2012) and Rosoli et al. (2012) have measured the water qualitative parameters such as BOD and DO to indicate the relationship between the level of pollution and the water quality.

The present research has focused on Dez river in Dezful city which is an important agricultural and industrial zone in Khoozestan Province to study and zone the trend of changes in the level of BOD and DO and their relation with each other.

2. Materials and methods

Since it is not possible to monitor the whole parts of the river, 9 stations along Dez river are considered which are placed before the City, in Dezful and after the City. The stations are called Gellal Dez, Ziba Shahr, Parke Dowlat, Artesh, Nabavi, Pole Panjom, Khoroojiy-e-

*Corresponding author E-mail: r.yazdizade@yahoo.com

tasfiyekhane, Koshtargah-e-Dami, and Gavmish Abad. Sampling from the determined stations was done monthly, during 6 months, from March to September in 2013. Polyethylene containers were used for estimation BOD and transferred to laboratory. Also, Horiba U-10 was used to estimate DO. Data obtained in laboratory were entered into Excel. Also, ArcGIS software was used to provide zoning map. It should be noted that since the river is not wide, borders are considered around it to have a clear zoning image and easier

analyzing process. The positions of the sampling stations are shown in Fig. 1.

3. Results and discussion

The present research has attempted to determine the status of DO and BOD indexes in all of the stations by doing a comparison between spring data and summer data. The six-month averages are indicated in Table 1.

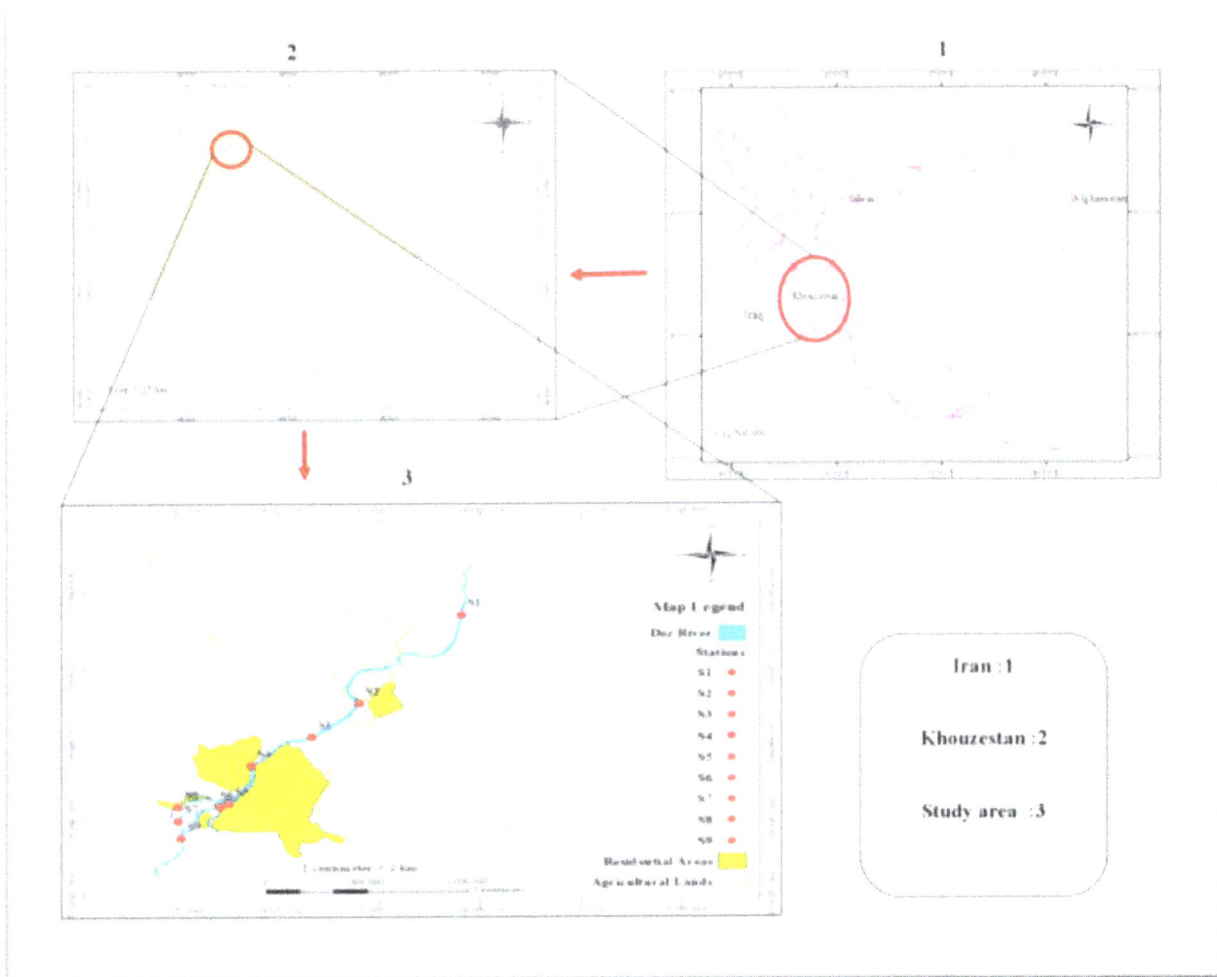

Fig. 1. The position of sampling stations in the map of administrative divisions.

Table 1. The level of DO and BOD in spring and summer.

Six-month average		Summer average		Spring average		
BOD	DO	BOD	DO	BOD	DO	Stations
3.16	8.47	3.76	8.31	2.56	8.36	S_1
43.38	7.28	7.76	7.2	79	7.36	S_2
20.05	7.06	6.86	5.43	33.23	8.7	S_3
26.13	5.45	7.93	4.23	44.33	6.66	S_4
7.83	7.65	7.33	6.83	8.33	8.46	S_5
17.55	4.95	19.1	3.63	16	6.26	S_6
27.83	4.28	24.3	3.76	31.33	4.8	S_7
150.83	1.71	93	0.57	208.66	2.86	S_8
10.93	5.56	9.86	4.96	12	6.16	S_9

The maximum amount of DO in spring was for S3 (central station) because of favorable temperature, flow, and wind turbulence. The maximum amount of DO in summer was for S1 (the first station located at upstream) because of low input organic loading and microbial population of the river. The minimum amounts of DO in both spring and summer were for S8 (located at the downstream) because of an increase in the volume of raw animal waste entered into the river

which itself, increases the level of river pollution. So, the maximum amount of BOD in spring and summer is for S8 and the minimum amount of BOD is spring and summer is for S1.

Figures 2 and 3 have indicated the changes in the level of DO in a 6-month average. The maximum amount of DO was for S1 (8.47 mg/l) and the minimum amount of it was for S8 (1/71 mg/l). The maximum amount of BOD was for S8 (150.33 PPM) and the minimum amount it was for S1 (3.16 mg/l). So, it can be said that because of an increase in the volume of raw and untreated animal waste, existence of organic materials, and microbial agents, DO has decreased and BOD has increased. Since S1 is located at an upper level from industrial, agricultural and residential areas, it enjoys a high level of DO.

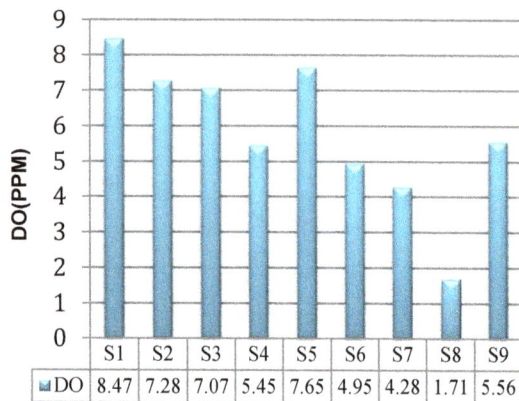

Fig. 2. Position-dependant changes in DO density (mg/l) in determined stations (an average of 6 months)

	S1	S2	S3	S4	S5	S6	S7	S8	S9
DO	8.47	7.28	7.07	5.45	7.65	4.95	4.28	1.71	5.56

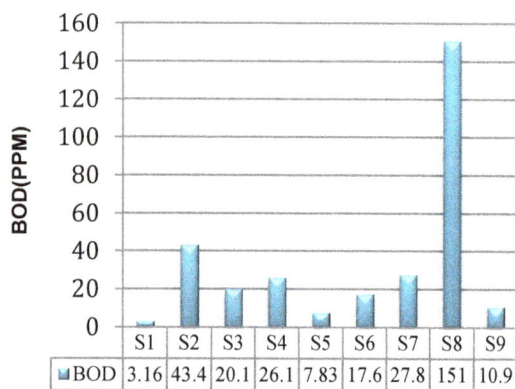

	S1	S2	S3	S4	S5	S6	S7	S8	S9
BOD	3.16	43.4	20.1	26.1	7.83	17.6	27.8	151	10.9

Fig. 3. Position-dependant changes in BOD density (mg/l) in determined stations (an average of six months).

In Fig. 4, monthly DO changes are shown. S1, as it is located at a higher place, has the maximum level of DO, and S9 which is located in the lowest part has the minimum level of DO. S5 is a central station (between S1 and S9). With an overview to the DO changes during spring and summer, it can be concluded that DO level in spring is higher than the summer. The findings are similar to what Lehman (2002) and Fortes Lopes (2006) indicate in their studies regarding rivers' sediments.

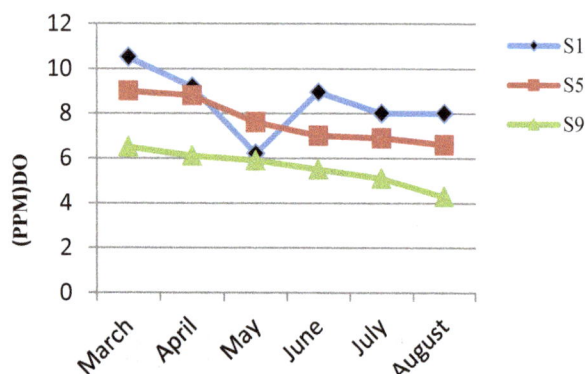

Fig. 4. Monthly changes at the level of DO in S1, S5, and S9.

The low level of DO in S1 during May can be the result of increased level of rainfall and erosion, the area around the river being washed and entering the microbial agents and organic materials into the river.

Fig. 5 has indicated the monthly changes of BOD in determined stations. According to the diagram, it can be said that S9 has the maximum amount of BOD, S1 has the minimum amount of BOD, and S5's BOD is in between. With an overview to the diagram, it is clear that in comparison with summer, BOD in spring has increased because of increased level of rainfall and organic material. Aerobic biodegradation includes biologic oxidation of organic materials. During the process, microorganisms change the organic materials into microbial biomass (Jouanneau et al. 2013). So, in spring, microbial population, especially Fecal Coliform, will be increased due to the increased level of rainfall and as a result, BOD will increase too. Romas et al. (2006) and Moungi et al. (2003) indicate a strong relationship between the level of rainfall and its runoff and the level of microbial pollution loading of surface waters. Kim et al. (2005), has also claim that E. coli has increased 7 times in rainfall months more than that of in dryer months. Therefore, in summer, due to the decreased level of rainfall, its runoff, and suspended particular matters, E. coli will be lessening in the rivers' waters. According to the provided zoning in figures 6 and 7, dark blue indicates the maximum amount of DO which is for S1, while red indicates the minimum amount of DO which is related S8. In S1 discharge of wastewater into the river has not been observed and the pollution loading is lesser in comparison with the other stations which is a reason of high level of DO. Also, BOD level for the average of 6 months in S8 is higher in comparison with the other stations and is shown in blue. Usually, surface flow of the residential areas will be increased intensely in spring. Consequently, microbial loading in surface waters will be increased too (Jaglass 1997; Venter et al. 1997).

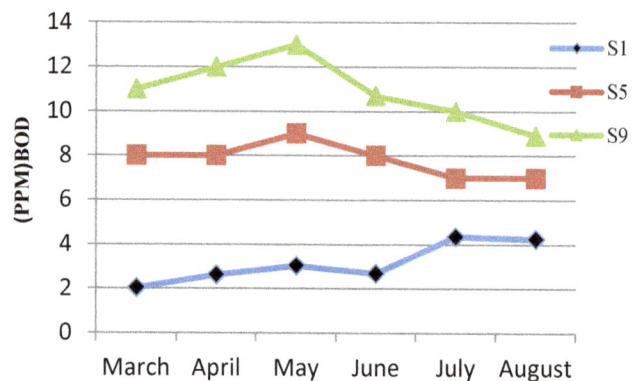

Fig. 5. Monthly BOD changes in S1, S5 and S9.

4. Conclusions

Monitoring of DO and BOD provides us with enough information about a river conditions (Tyagi et al. 1999). If BOD is low, the water is clear and without organism. In the case of existence of any organism they do not use Oxygen (Tchobanoglous et al. 2003).

So, in the stations with low amount of BOD, we observe a smaller microbial population. Where there is a low level of DO and naturally, a high level of BOD, this is the place of entrance of wastewaters, and raw and untreated wastewaters.

An increased level of rainfall in spring will cause the entrance of organic materials and expands the microbial population which all of these factors affect the level of BOD.

The present research states that the reason for the low density of DO in some stations is due to the entrance of wastewaters, having stagnant water in some places, and existence of microorganisms or algae. Using a shared database for boosting the relationship between monitoring and reconstruction of environmental projects is essential in the promotion of technology transferring (Quinn et al. 2005). By the help of GIS, critical points of pollution in the rivers can be recognized. So, online monitoring of rivers for promotion of water quality is possible. Therefore, GIS software, and controlling and monitoring the pollution sources entered into the aquatic ecosystems of the rivers are suggested for having precise information about the rivers' qualitative status and studying the effective and important estimation parameters in the rivers' waters quality, as it is used of Dez river.

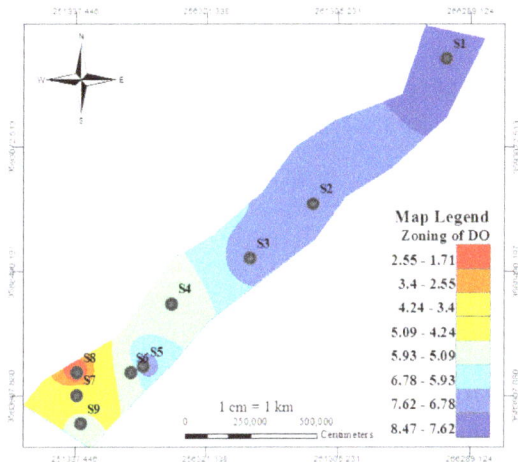

Fig. 6. Zoning the average DO in six months.

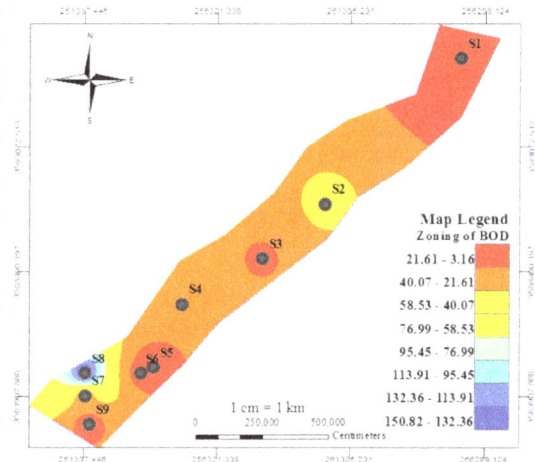

Fig. 7. Zoning the average BOD in six months.

References

Abdul Hameed M., Alobaidy J., Abid H.S., Maulood B.K., Application of water quality index for assessment of Dokan Lake Ecosystem, Kurdistan Region, Iraq, Journal of Water Resources Protection (2010)792-798.

Akkoyunlu A.E., Akiner M., Pollution evaluation in streams using water quality indices: A case study from Turkey's Sapanca Lake Basin, Ecological Indicators (2012) 501-511.

Arlyapov V., Kamanin S., Ponamoreva O., Reshetilov A., Biosensor analyzer for BOD index express control on the basis of the yeast microorganisms Candida maltosa, Candida blankii, and Debaryomyceshansenii, Enzyme Microbial Technology 50 (2012) 215-220.

Chang Chen Y., Chung Yeh H., Wei C., Estimation of river pollution index in a tidal stream using Kriging analysis, International Journal of Environmental Research and Public Health 9 (2012) 3085-3100.

Cox B.A., A review of dissolved oxygen modelling techniques for lowland rivers, The Science of the Total Environment 316 (2003) 303-334.

Curtis G., Oregon water quality index: A tool for evaluating water quality management effectiveness, Journal of American Water Resource Association 37 (2001) 76-83.

Dunnette D.A., A geographically variable water quality index used in Oregon, Journal of Water Pollution Control Federation, 51 (1989) 53-61.

Evanson M., Ambrose F., Sources and growth dynamics of fecal indicator bacteria in a costal wetland system and potential impacts to adjacent water, Water Research 40 (2006) 475-486.

Fortes Lopes J., Silva C., Temporal and spatial distribution of dissolved oxygen in the Ria de Aveiro lagoon, Ecological modelling 197 (2006) 67-88.

Gilbert D., Rabalais N.N., Diaz R.J., Zhang J., Evidence for greater oxygen decline rates in the coastal ocean than in the open ocean, Biogeosciences Discussion 6 (2009) 9127-9160.

Helm I., Jalukse L., Leito I., Measurement uncertainty estimation in amperometric sensors: A tutorial review, Sensors 10 (2010) 4430-4455.

Henderson R.K., Baker A., Murphy K.R., Hambly A., Stuetz R.M., Khan S.J., Fluorescence as a potential monitoring tool for recycled water systems: A review, Water Research 43 (2009) 863-881.

Hobbs J.P.A., McDonald C.A., Increased seawater temperature and decreased dissolved oxygen triggers fish kill at the Cocos (Keeling) Islands, Indian Ocean, Journal of Fish Biology 77 (2010) 1219-1229.

Hooshmand A., Delgandi M., Sied Kaboli H., Zoning of water quality on Karoonriver bases on WQI index with GIS. 2nd Congress on Environmental Eng. Proceedings. Tehran University (2008) (Persian).

Hur J., Cho J., Prediction of BOD, COD, and Total nitrogen concentrations in a typical urban river using a fluorescence excitation-emission matrix with PARAFAC and UV absorption indices, Sensors 12 (2012) 972-986.

Jagals, P., Storm water runoff from typical developed and developing south Africa urban development: definitely not for swimming, Water Science and Technology 35 (1997) 133-140.

Joos F., Plattner G.K., Stocker T.F., Körtzinger A., Wallace D.W.R., Trends in marine dissolved oxygen: Implications for ocean circulation changes and the carbon budget. EOS Trans. AGU 84 (2003) 197–204.

Jouanneau S., Recoules L., Durand M.J., Boukabache A., Picot V., Primault Y., Lakel A., Sengelin M., Barillon B., Thouand G., Methods for assessing biochemical oxygen demand (BOD): a review, Water Resaerch 49 (2013) 62-85.

Keeling R.F., Garcia H.E., The change in oceanic O2 inventory associated with recent global warming. Proc. Natl. Acad. Sci. USA99 (2002) 7848-7853.

Keeling R.F., Körtzinger A., Gruber N., Ocean deoxygenation in a warming world, Annual Review Material Science 2 (2010) 199-229.

Kevin G., Whealer K., Strzapek M., A method for rapid water quality assessment in developing countries; A case study of Chinese Yellow river, 4th International Conference on Integrating GIS and Environmental Modeling, Canada, GIS/EM, No, 89 (2000).

Kim G.T., Choi E., Lee D., Diffuse and point pollution impacts on the pathogen indicator organism level in the Geum river, The Science of the Total Environment 350 (2005) 94-105.

Krenkel P A, Novotny V., Water Quality Management", Academic Press, New York, NY (1980).

Kumar A., Dhall P., Kumar R., Redefining BOD:COD ratio of pulp mill industrial wastewaters in BOD analysis by formulating a specific microbial seed, International Biodeterioration & Biodegradation 64 (2010) 197-202.

Lehman P., Oxygen demand in the San Joaquin river deep water channel, Fall 2001. CALFED 2001 San Joaquin Low Dissolved Oxygen Studies. California Department of Water Resources (2002).

Mahler B.J., Personne J.C., Lods G.F., Drogue C., Transport of free and particulate-associated bacteria in karst, Journal of Hydrology 238 (2000) 179-193.

Martin N., Mc Eachern P., Yu T., Zhu D.Z., Model development for prediction and mitigation of dissolved oxygen sags in the Athabasca River, Canada, The Science of the Total Environment 443 (2013) 403-412.

Mouri G.,Takizawa S., Oki T., Spatial and temporal variation in nutrient parameters in stream water in a rural-urban catchment, Shikoku, Japan: Effects of land cover and human impact. Journal of Environmental Management 92 (2011) 1837–1848.

Mvungi A., Hranova R.K., Love D., Impact of home industries on water quality in a tributary of the Marimba River, Harare: implications for urban water management. Physics and Chemistry of the Earth, Parts A/B/C (2003) 1131-1137.

Nagy S.A., Dévai G.Y., Grigorszky I, Schitchen C.S., Tóth A., Balogh E., Andrikovics S., The measurement of dissolved oxygen today-Tradition and topicality, Acta Zoology Academic Science Hung 54 (2008) 13-21.

Naykki T., Jalukse L., Helm I., Leito I., Dissolved oxygen concentration Inter laboratory Comparison: What Can We Learn?, Water 5 (2013) 420-442.

Pedrero F., Albuquerque A., Monte H.M., Cavaleiro V., Alarcon J.J., Application of GIS-based multi-criteria analysis for site selection of aquifer recharge with reclaimed water, Resources, Conservation and Recycling 56 (2011) 105-116.

Quinn N.W.T., Jacobs K., Chen C.W., Stringfellow W.T., Elements of a decision support system for real-time management of dissolved oxygen in the San Joaquin River Deep Water Ship Channel, Environmental Modelling & Software 20 (2005) 1495-1504.

Radwan M., Willems P., El-Sadek A., Berlamont J., Modelling of dissolved oxygen and biochemical oxygen demand in river water using a detailed and a simplified model, International Journal of River Basin Management 1 (2003) 97-103.

Ramos M.C., Quinton J.N., Tyrrel S.F., Effects of cattle manure on erosion rates and runoff water pollution by fecal coliforms, Journal of Environmental Management 78 (2006) 97-101.

Rosli N.A., Zawawi M.H., Bustami R.A., Salak river water quality identification and classification according to physico-chemical characteristics, International Conference on Advances Science and Contemporary Engineering (2012) 69-77.

Shaffer G., Olsen S.M., Pedersen J.O.P., Long-term ocean oxygen depletion in response to carbon dioxide emissions from fossil fuels, Natural Geoscience 2 (2009) 105-109.

Su S., Li D., Zhang Q., Xiao R., Huang F., Wu J., Temporal trend and source apportionment of water pollution in different functional zones of Qiantang river, China, Water Research 45 (2011) 1781-1795.

Tchobanoglous G., Burton F.L., Stensel H.D., Wastewater Engineering: treatment disposal reuse. 4th ed. NewYork: McGraw-Hill (2003).

Thanh B.N., Tuan A.N., Ngoc Q.V., Thi T.H.C., Minh Q.C., Management and monitoring of air and water pollution by using GIS technology, Journal of Viet. Environment 3 (2012), 50-54.

Tyagi B., Gakkhar S., Bhargava D.S., Mathematical modelling of stream DO–BOD accounting for settleable BOD and periodically varying BOD source, Environmental Modelling & Software 14 (1999) 461–471.

Venter S.N., Steynberg M.C., De Wet, C.M.E., Hohls D., Du Plessis G., Kfir R., A situational analysis of the microbial water quality in a peri-urban catchment in South Africa. Water Science and Technology 35 (1997) 19-124.

12

Developing finite volume method (FVM) in numerical simulation of flow pattern in 60° open channel bend

Azadeh Gholami[1,2], **Hossein Bonakdari**[1,2,*], **Ali Akbar Akhtari**[1,2]

[1]*Department of Civil Engineering, Razi University, Kermanshah, Iran.*
[2]*Water and Wastewater Research Center, Razi University, Kermanshah, Iran.*

ARTICLE INFO

Keywords:
Numerical simulation
Finite volume method (FVM)
Developed FLUENT software
60° bend

ABSTRACT

In meandering rivers, the flow behavior is very complex due to topography and flow depth changes. In general, effective forces on bend flow pattern include centrifugal force due to non-uniformity of the vertical velocity profile and radius pressure gradient induced by the lateral slope of water surface. In this paper, the 60° bend flow pattern is simulated by developing FLUENT computational software based on finite volume method (FVM), numerically. The k-ε (RNG) turbulence model and volume of fluid (VOF) method are used for turbulence and flow depth modeling. The FVM numerical results are verified by existing experimental data in velocity and flow depth. The results illustrate that the FVM model has high accuracy in prediction flow variables in the bend. As the average value of root mean square error (RMSE) and mean absolute percentage error (MAPE) values between the observational and numerical results for depth-averaged velocity (DAV) in the different transverse profile are 4.5 and 9%, respectively, which is an acceptable error percentage. The advanced software can well simulate the both major and minor secondary current cells with opposite rotation direction in the vicinity of channel bed and vicinity the water level in the outer wall, respectively. By the development of the major and minor secondary currents in sections located 40 (cm) after the bend, longitudinal velocity shift, and the high-velocity zone moves further to the outer wall (and the channel bed) in depth. Therefore, it can be said the developing FLUENT software can be utilized in practical cases in design and execution of curved channel.

1. Introduction

For Meandrous Rivers, flow patterns are complex, so that flow mechanics has specific characteristics at bends, not observed within straight channels. Combining between field experiments, laboratory experiments, and numerical simulations is the most effective way to enhance flow knowledge and turbulence structure in curved channels. There have been extensive researches undertaken into the flow in bend channels and this research can be distributed into two types, experimental and Numerically studies; Leschziner and Rodi (1979) using a numerical model simulated the flow schema in the 180° sharp bend. They found that the longitudinal pressure gradient is the main factor of the maximum velocity component transmission into the outer wall at the final of the sharp bend while in the mild bends secondary current is the main factor of displacement. De Vriend and Geoldof (1983) did extensive studies on changing the water level and velocity experimentally and declared that in the beginning of the bend, the maximum velocity is located near the inner wall and transmitted to the outer wall in the final cross section. Bergs (1990) examined the flow and topographical changes of the bed in a laboratorial curved flume with trapezoidal cross section and mobile bed. The result demonstrates that the flow is spiral in the primary cross sections of the bend. Jung and Yoon (2000) investigated the flow velocity field in any kind of bend. They declared that in mild bend the maximum velocity is located at the inner wall and gradually transmitted to the outer wall from the internal cross section of the bend. Blanckaert and Graf (2004) examined the velocity component changes, boundary shear stress and bed

topography variation in curved flumed. Sui et al. (2006) measured the local scour at the 90° bend experimentally. They stated that the Froude number, the properties of protective wall and the size of bed particles have a high effect on the scour amount in the vicinity of channel bed. Naji et al. (2010) conducted vast experimental and numerical studies on flow properties at 90° mild bend. They declared that the maximum velocity is placed in the inner wall in half of the bend and is transferred to outer wall from the 45° cross section. Barbhuiya and Talukdar (2010) tested the 3D flow variables such velocity component, turbulent tension, shear stress and scour in different section at the 90° bend. Uddin and Rahman (2012) measured the velocity components in three directions by ADCP velocity meter and erosion in the channel bed in the river bend by experimental and numerical models. They compared the experimental and numerical model result and pointed to the good agreement between data. Liaghat et al. (2014) examined the flow hydraulic in a curved flume with variable width by SSIIM model. Ramamurthy et al. (2013) simulated three-dimensional flow properties at 90° sharp bend by different turbulence models associate with the numerical model. The researchers said that the RSM (Reynolds stress model) turbulence model and volumes of fluid method are in acceptable adaptation with experimental results. Vaghefi et al. (2014) analyzed the shear stress amount in the 180° sharp bend channel bed. In this bend, the maximum shear stress in channel bed is located at the 40° sections close to the inner wall. Gholami et al. (2014) studied the hydraulic of flow in 90° sharp curved channel experimentally and numerically. They stated that the maximum velocity till the final sections remains in the inner wall and is moved to the channel axis in the final cross sections.

Corresponding author Email: bonakdari@yahoo.com.

Ajeel Fenjan et al. (2016) modeled the flow variables in the 60° bend using computational fluid dynamic (CFD) and artificial neural network (ANN) soft computing model. They compared the numerical and ANN results with experimental data and with each other. Their results illustrate that the ANN model acts more accurately than CFD model with low error indices. Gholami et al. (2015 a, b and 2016 a, b) studied the flow hydraulic of sharp curved channels by different soft computing methods. They pointed to study complex bend flow type and presented noticeable results. In this study, the accuracy and application of the FLUENT software based on finite volume method (FVM) in prediction of total flow variables in 60° bend is evaluated. Vast experimental study was done by author to test the velocity and water surface depth (Akhtari et al. 2009). Also, the numerical simulation is done for predicting flow variables and the results are verified in comparison of available experimental data. Also, the advancing the numerical software is examined in the prediction of velocity contour and secondary cell development. The different statistical indices are used to examine model efficiency.

2. Experimental model

Experimental Model Akhtari (2009) measured the flow variables in a 60° sharp bend in the hydraulics laboratory of Ferdowsi University of Mashhad, Iran. Channel test was included of three parts: the straight inlet channel 360 cm before the point in which the bend begins, the curved channel, central radius of this channel (R_c) is 60.45 cm. The ratio of the bend central radius to the channel width is 1.5 (R_c/b) and since this ratio is lesser than 3, the under- study bend is considered a sharp bend. The straight outlet channel after the bend is 180 cm long. The cross sections of the intended flume are square-shaped with a width of 40.3 and a height of 40.3 cm. channel bed and walls are made of Plexiglas with rigid fix bed. Figs. 1 and 2 show the laboratorial and geometrical shape of the flume, respectively.

In this paper, the experiments were carried out on a discharge of 25.3 l/s. The Froude number and Reynolds number was 0.34 and 44705, respectively. The hydraulic characteristic is shown in Table 1.

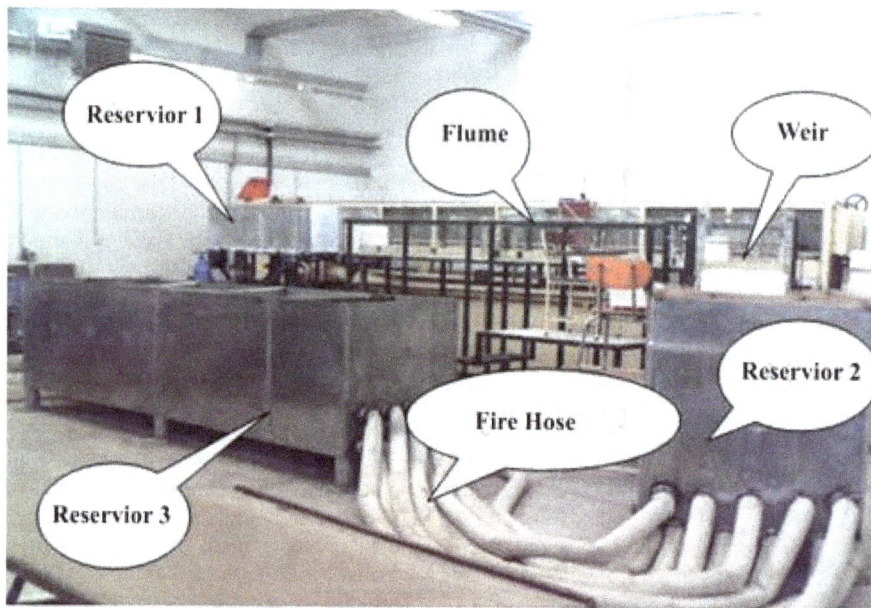

Fig. 1. The laboratorial flume of the 60° bend.

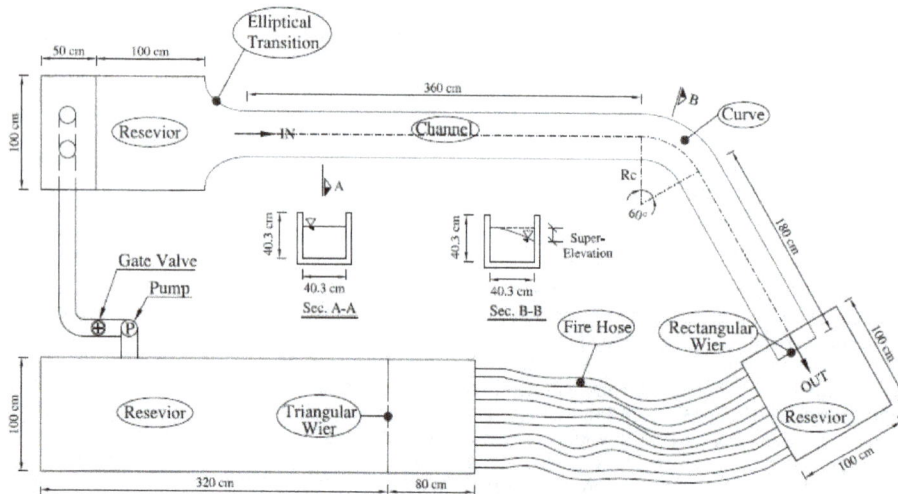

Fig. 2. The plan and geometry of the experimental flume related to the 60° bend.

Table 1. Hydraulic properties of test.

Flow discharge Q (lit/s)	Flow depth Y (cm)	Flow velocity V (m/s)	Reynolds number	Froude number	Flow regime
25.3	15	0.418	44705	0.34	Turbulent and subcritical

2.1. Numerical model

In this paper, the software Fluent was used for the numerical modeling. The software used the Finite Volume Method (FVM) for solving the flow equation (Manual. 2005). The FVM as a numerical method according to the integral conservation is utilized to compute the partial differential equations (e.g., Navier-Stokes equation) and then to calculate the flow variables values, averaged across the volume. The law of integral conservation is applied on small control volumes which are defined by the computational mesh.

2.2. The equations governing the flow

The equation governing the motion of a viscous in the compressible fluid at a turbulent state lays down down by average Navier-Stokes equations known as Reynolds (RANS). A state of the art CFD package; FLUENT, was employed to solve the governing equation, which used a finite volume method (FVM) to approximate the equations. The continuity equation (conservation of mass) and motion equation (momentum conservation) are stated as:

- Continuity equation

$$\frac{\partial \rho}{\partial t} + \frac{\partial}{\partial x_i}(\rho u_i) = 0 \tag{1}$$

- Momentum equation

$$\frac{\partial}{\partial t}(\rho u_i) + \frac{\partial}{\partial x_i}(\rho u_i u_j)$$
$$= -\frac{\partial P}{\partial x_i} + \frac{\partial}{\partial x_j}\left[\mu\left(\frac{\partial u_i}{\partial x_j} + \frac{\partial u_j}{\partial x_i} - \frac{2}{3}\delta_{ij}\frac{\partial u_l}{\partial x_l}\right)\right]$$
$$+ \frac{\partial}{\partial x_j}\left(-\rho\overline{u_i'u_j'}\right) \tag{2}$$

The k- ε (RNG) turbulence model as an exact and actuarial method and is utilized to the disjunction of Navier-Stokes equations.

In this paper, two methods are applied to determine the accuracy and suitability of the models. First, two relative and absolute error relations as the root mean square error (RMSE, Eq. (3)) and the mean absolute percentage error (MAPE, Eq. (4)) are used to evaluate model efficiency. Both are expressed in percentages and have an ideal value of zero.

$$RMSE = \left[\frac{\sum\limits_{i=1}^{N}(Y_{i(actual)} - Y_{i(model)})^2}{N}\right]^{1/2} \times 100 \tag{3}$$

$$MAPE = \frac{1}{N}\sum_{i=1}^{N}\left(\frac{|Y_{i(actual)} - Y_{i(model)}|}{Y_{i(actual)}}\right) \times 100 \tag{4}$$

In this case, N is the number of measurement points, Y_{actual} is experimental measured data and Y_{model} is predicted by the model.

3. Results and discussion
3.1. Depth distribution of longitudinal velocity

In general, when flow approaches the bend, the secondary flow is generated gradually and causes the change in the velocity components at three directions x, y, z.

The maximum amount of the longitudinal velocity component is placed beneath the water surface and about the flow mid depth. In addition, the longitudinal velocity distribution is not logarithmic in the bend. In Fig. 3(a), (b), (c), and (d) show typical comparisons of the longitudinal computed and experimental velocity at 0°, 30°, 60° and 80 cm after bend cross sections, respectively. The mean difference (RMSE and MAPE errors) between the observational and FLUENT data have been shown in the top of the graphs. Fig 3 (b) which related to 30° bend section illustrates that the position of maximum velocity is close to the water surface. In Fig. 3 (c) the effect of secondary flow on the bend flow is noticeable at 60° section. In general, in sharp bends, by moving along the bend, the maximum velocity position is changed as it moves from the inner wall to the central channel axis and outer wall in the final cross sections of the bend. According to error values, it could be seen that the error values are acceptable. Relative error value in the 0° and 80 cm after the bend is more and less than the other cross section, respectively (1.53 % and 2.4 %). So, it can be said that model accuracy in the first and exit cross sections is high and low, respectively. Gholami et al. (2015a) Simulated hydraulic of bend flow using CFD and ANN models in 90° sharp bend. And their results represented that the CFD models can forecast flow hydraulic with no experimental data by RMSE of 2.33.

30 degree
RMSE= 4.7, MAPE= 2%

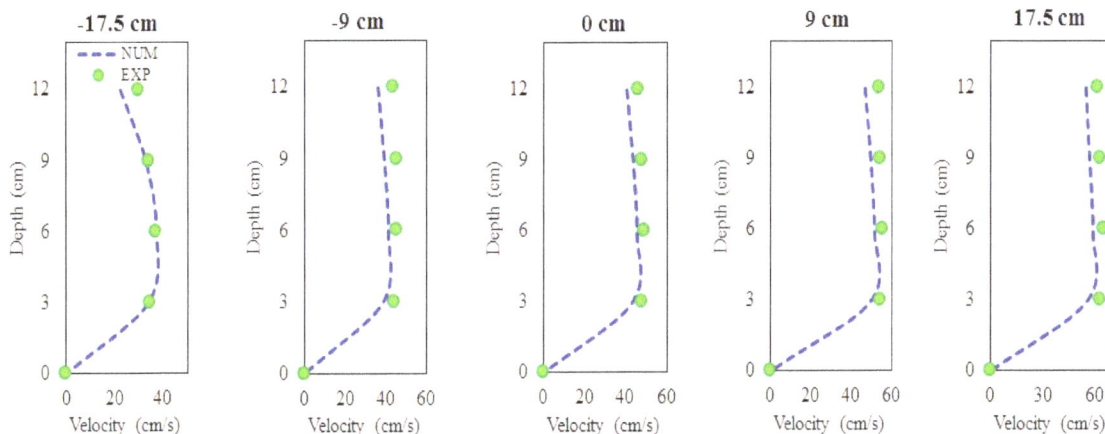

60 degree
RMSE= 4.57, MAPE= 2%

80 cm after bend
RMSE= 5.58, MAPE= 2.4%

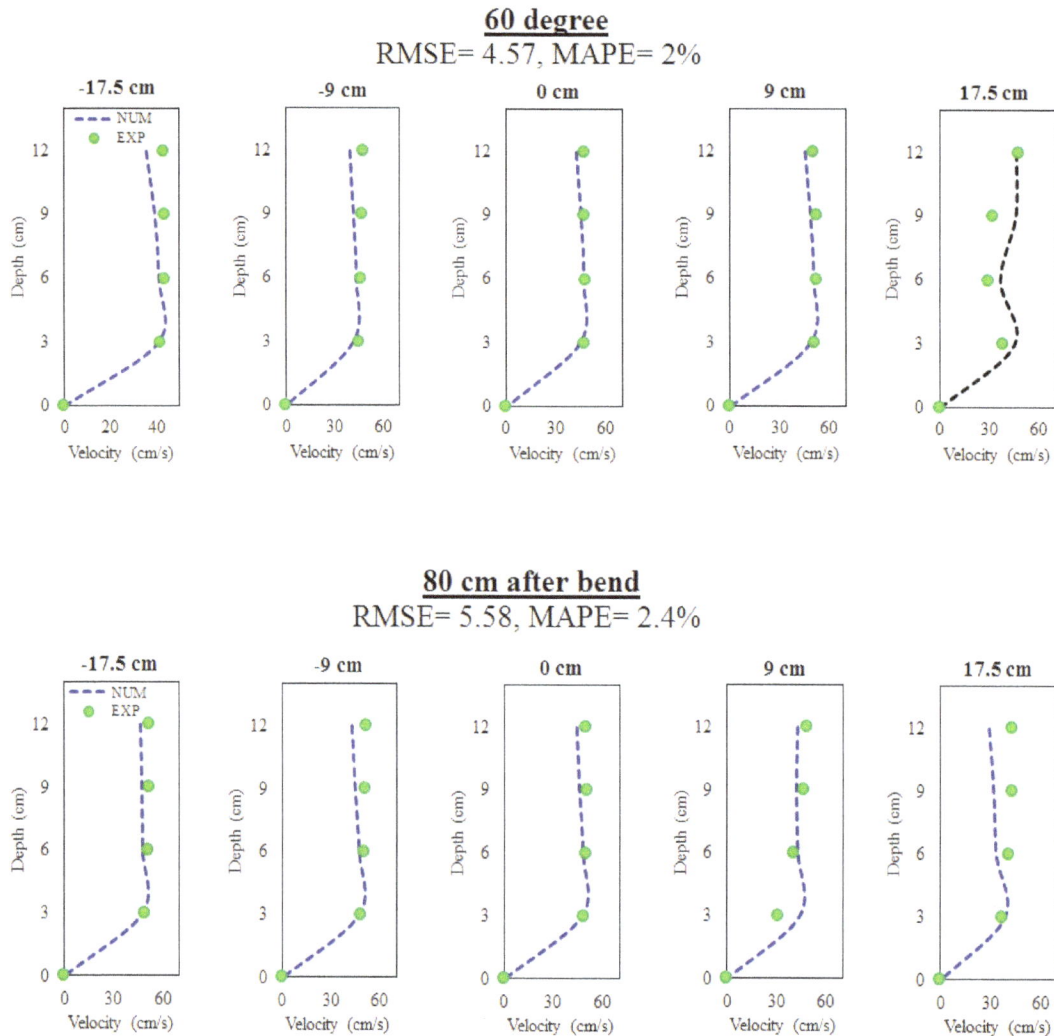

Fig. 3. Depth distributions of longitudinal velocity at cross sections of: (a) 0°, (b) 30°, (c) 60°, (d) 80 cm after bend.

Figs. 4 and 5 show the transverse profile of velocity predicted by numerical FLUENT model in comparison of observational values in different distance from the channel bed (Z= 3, 6, 9 and 12 cm from the channel bed) in 30° and 80 cm after the bend cross sections, respectively. The numerical simulation results match the experimental values with a great degree of compliancy. The RMSE and MAPE error values in Z= 3, 6, 9, and 12 (cm) and also the error values of the depth averaged velocity have been presented in Table 2 for all different cross sections.

The difference between the numerical and experimental results (RMSE) of the transversal distribution of axial velocities is at most approximately 16 % and the rest of these values range between 2 and 9 %. These error values are very much satisfactory. The lateral velocity profile represents the maximum level of inconsistency between the numerical and observational results at the input cross sections (40 cm before the bend and 0°) and the 50° and 60° cross sections especially in the layers near the channel bed which indicates that the numerical model is unable to suddenly increase the velocity and the flow undergoes separation in these cross sections (such Fig. 5).

Table 2. The RMSE and MAPE error values of velocity values predicted by numerical and experimental data at different cross sections, distance from the channel bed and averaged level.

Cross sections	Z= 0.03 m		Z= 0.06 m		Z= 0.09 m		Z= 0.12 m		Average velocity	
	RMSE	MAPE	RMSE	MAPE	RMSE	MAPE	RMSE	MAPE	RMSE	MAPE
40cm before	2.34	4.73	3.08	6.00	4.27	8.22	6.24	12.69	3.96	7.88
0°	2.52	9.54	2.90	5.56	4.65	8.78	6.03	11.94	3.97	7.81
10°	4.66	9.54	4.46	8.06	4.71	8.60	6.69	13.72	5.04	9.95
20°	4.34	8.93	3.75	6.91	5.14	9.84	6.44	13.61	4.81	9.74
30°	4.84	9.37	4.03	7.19	5.17	9.84	6.83	14.55	5.12	10.14
40°	4.64	8.69	3.72	6.89	5.27	10.54	9.78	16.35	4.97	10.06
50°	3.19	6.11	2.57	5.08	4.43	8.76	6.81	13.78	4.03	8.29
60°	3.04	5.78	2.78	5.50	5.56	10.64	6.97	13.27	4.13	8.52
40 cm after	5.79	12.17	2.90	6.10	6.91	13.20	7.67	15.06	4.04	8.07
80 cm after	5.15	9.45	4.18	8.08	7.85	15.01	8.38	16.11	4.96	9.12

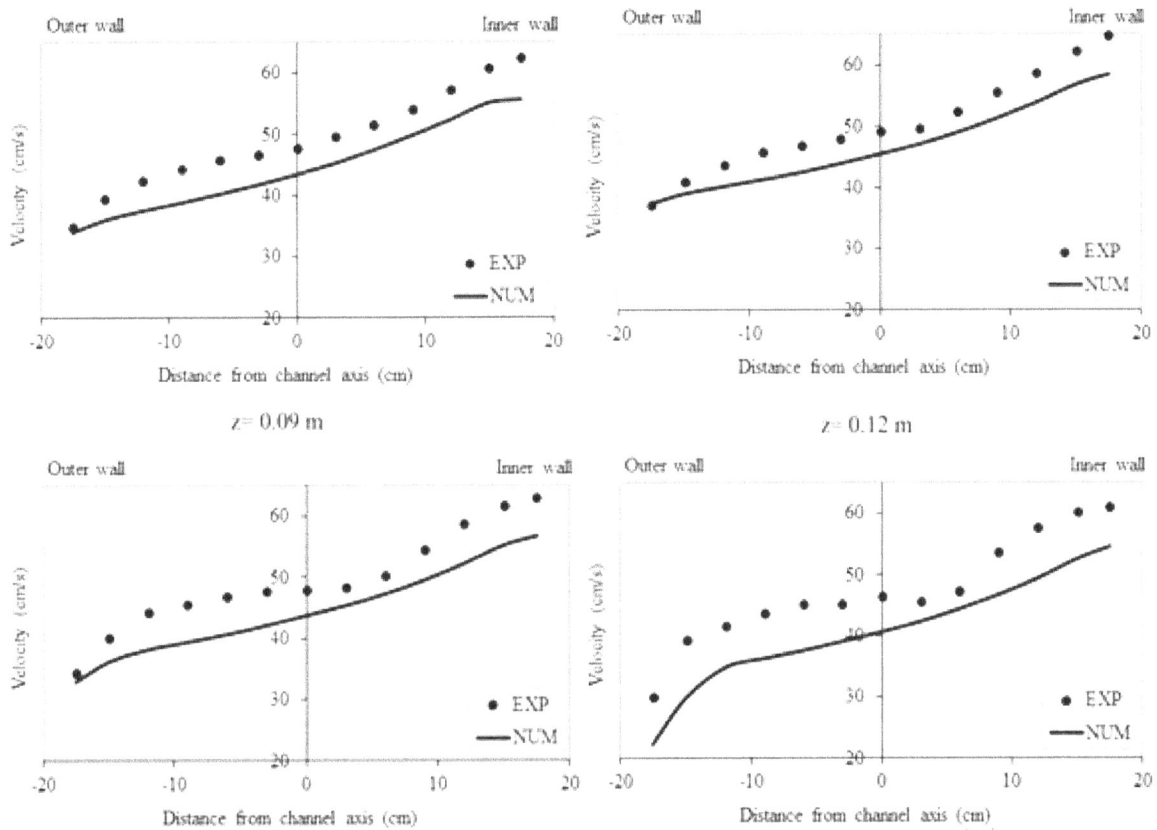

Fig. 4. The experimental and numerical results comparison of the lateral velocity distribution in the 30° cross section at the different distance from the channel bed (z= 0.03, 0.06, 0.09, and 0.12 m).

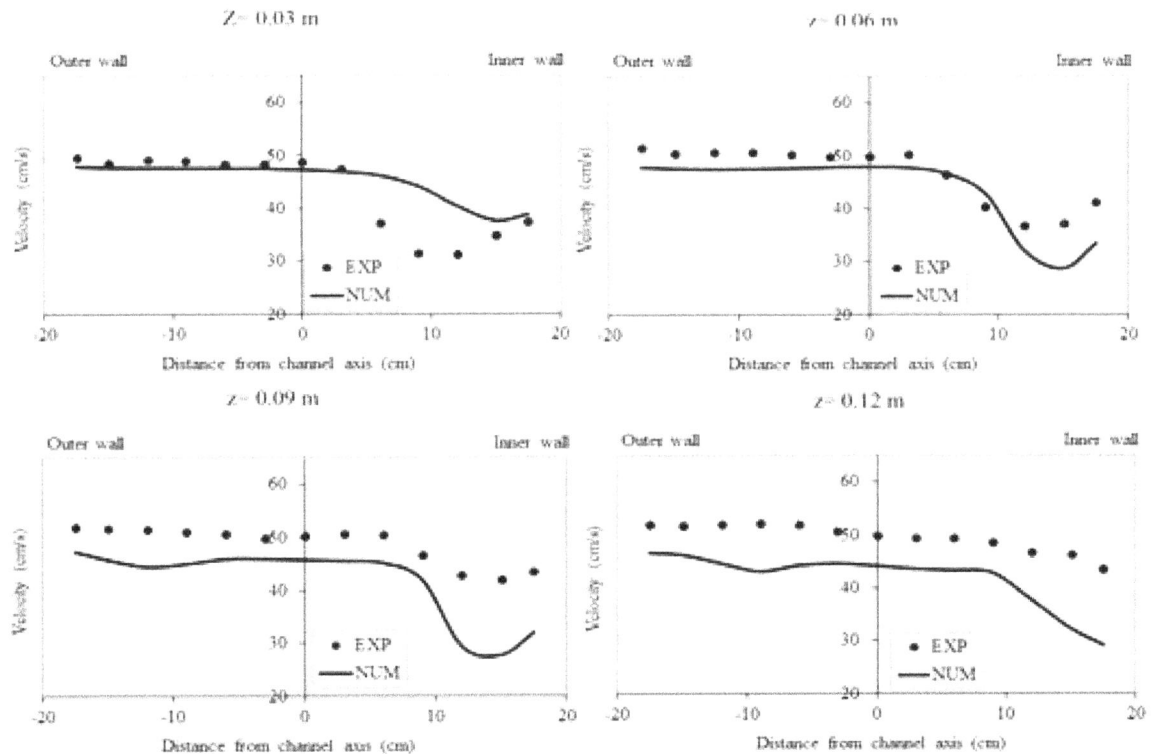

Fig. 5. The experimental and numerical results comparison of the lateral velocity distribution in the cross section of 80 cm after the bend at the different distance from the channel bed (z= 0.03, 0.06, 0.09, and 0.12 m).

3.3. Longitudinal distribution of flow depth

Fig. 6 shows the numerical results of longitudinal flow depth distribution through the inner wall, the axis, and the channel outer wall in comparison with the corresponding experimental values. The values of RMSE and MAPE between these results have been presented in Table 3. The figure and the values make it clear that the fairly consistent results along with 0.12 and 0.61 % respectively for RMSE and MAPE

mean values through the bend are acceptable. MAPE error value in the outer wall is higher than another longitudinal axis (1.03 %). The increase in the water level before the bend in order to gain the required energy to enter the bend could be clearly seen in both the experimental and numerical models. Therefore, it can be said that the numerical model enables to untangle the governing equation into the flow type in bends and has high accuracy in flow variables prediction.

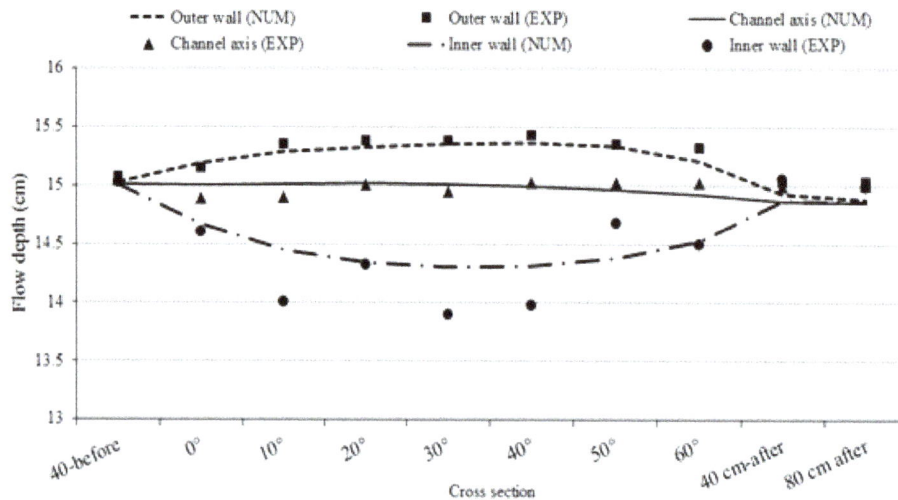

Fig. 6. Comparing the longitudinal distribution of the flow depth in the inner wall, the channel axis, and the outer wall in experimental and numerical models.

Table 3. RMSE and MAPE values between the flow depth values in the experimental and the numerical models through the channel length.

Channel length	RMSE	MAPE (%)
Outer wall	0.22	1.03
Channel axis	0.081	0.42
Outer wall	0.07	0.37
Mean RMSE	0.12	0.61

3.4. Velocity contours prediction

Fig. 7 shows three dimensional contours of the velocity magnitude in different cross section of 0°, 10°, 20°, 30°, 40°, 50°, 60°, 40 and 80

cm after the bend. When flow enters the bend, the maximum velocity zone locates in the inner wall. This trend continues up to the other sections. Since the presented 60° bend is sharp, the maximum velocity takes place in the inner wall and reminds in this wall.

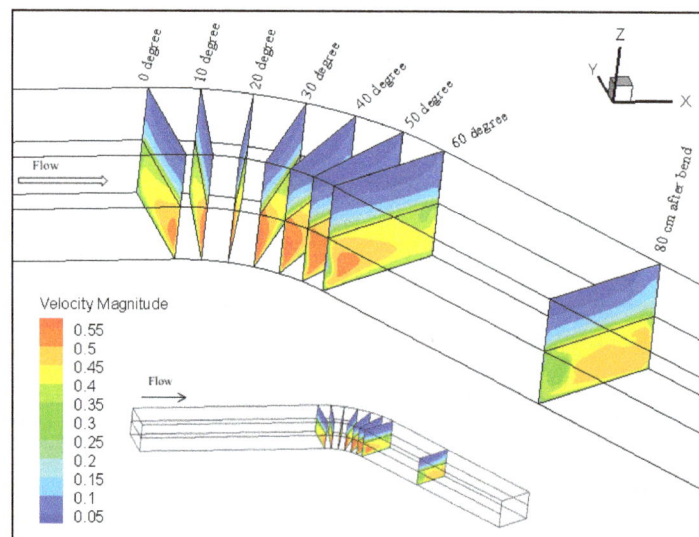

Fig. 7. Longitudinal velocity contours at different sections of 0°, 10°, 20°, 30°, 40°, 50°, 60°, and cross sections located 40 and 80 cm after the bend.

By moving along the bend, the secondary flow is generated which is important hydraulic phenomena in bend channels. By the enlargement of the major and minor secondary flows, it can be seen that in the section located 40 cm after the bend longitudinal velocity shifts and the high-velocity zone moves further to the outer wall and the bed of channel. And at the section of 80 cm after the bend, the high-velocity zone is completely separated from the inner wall and transfers to the outer wall.

3.5. Secondary currents cells simulation

Fig. 8 shows the secondary flow at the different cross sections. As seen, at the bend entrance section (0°) a one-way radial flow towards the inner wall is seen which is caused by the longitudinal pressure gradient created in the bend. In the later cross sections, the secondary flow is produced and continued up till the cross section after the bend. As seen in figures, at all sections, the rotation of the major secondary flow with clockwise rotation direction is seen. The center of the major secondary flow transfers from the outer wall towards the inner wall by advancing along the bend. In this state, the major secondary flow causes a surface flow to be pushed to the outer wall; this flow comes up after disorienting with the outer wall, and then goes back towards the inner wall. This process causes another secondary flow cell to be formed in the outer wall and close to the water surface, which is called the minor secondary flow and its rotation direction is contrary to the major secondary flow. Therefore, at a 60° sharp bend, the minor secondary flow begins from the angle of 20° and continues to some distances following the bend. By going through the bend in the cross sections located after the bend, both cell secondary flows decrease because of the weakness of their power.

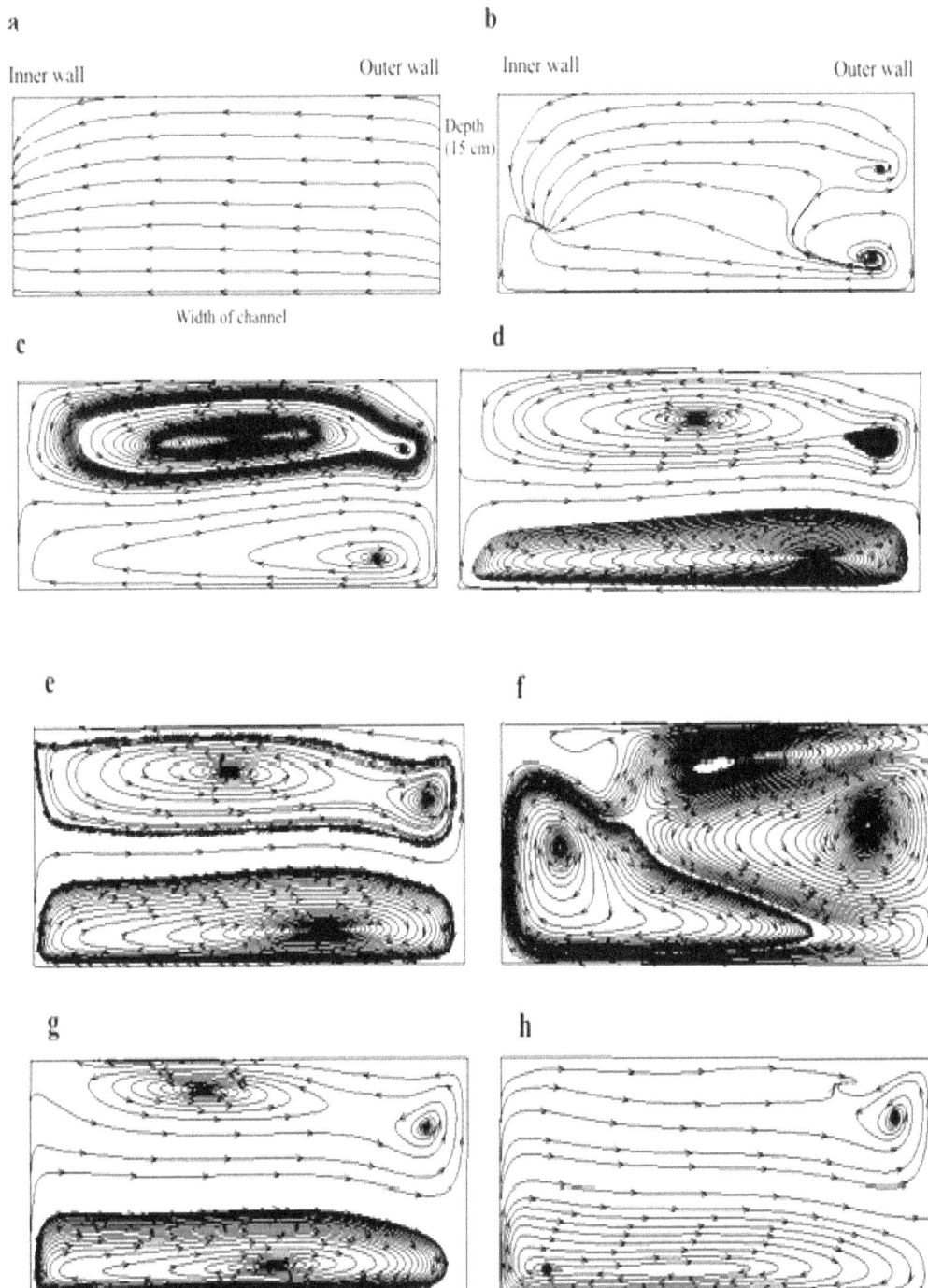

Fig. 8. The secondary flows at different sections: (a) 0°, (b) 10°, (c) 20°, (d) 30°, (e) 40°, (f) 50°, (g) 60°, (h) 80 cm after the bend cross section.

4. Conclusions

In this paper, the application of numerical software is evaluated through flow hydraulic in a curved channel. The used numerical model is based on finite volume method (FVM) that can simulate the complicated flow pattern in bends. The results illustrate that the FVM model has high accuracy in prediction of flow variables and also can well predict the powerful secondary current cells that this cell simulation is so difficult sometimes (Naji Abhari et al. 2010). The important results show that:

1. In the sharp bends, the maximum velocity through the exit bend cross sections are located in the inner wall and later with the advent and the growth of the secondary flow is moved to the channel axis and in the section after the bend in the outer wall occurs.

2. The error values of velocity component between numerical and observational data are more (1.53 %) and less (2.4 %) than the other cross section at the section of 0° and 80 cm after the bend, respectively.

So, it can be said the model accuracy in the first and exit cross sections is high and low, respectively.

3. The MAPE, when comparing the longitudinal distribution of the flow depth in the inner wall, the channel axis, and the outer wall in experimental and numerical models was acceptable and in the outer wall is higher than another longitudinal axis (1.03 %).

4. The fairly consistent results along with 0.12 and 0.61 % for RMSE values through the bend are acceptable.

5. The FLUENT software simulates the major and minor secondary cell as well as other flow variables in channel width and outer wall, respectively.

6. The proposed numerical model can well predict the high and low velocity zones and subsequently the erosion and sedimentation place in sharp bends.

References

Akhtari A.A., Abrishami J., Sharifi M.B., Experimental investigations water surface characteristics in strongly-curved open channels, Journal of Applied Sciences 9 (2009) 3699-3706.

Barbhuiya A. K., Talukdar S., Scour and three dimensional turbulent flow fields measured by ADV at a 90degree horizontal forced bend in a rectangular channel, Flow Measurement and Instrumentation 21(2010) 312-321.

Bergs M. A., Flow processes in a curved alluvial channel, PhD Dissertation, Iowa University (1990).

Blanckaert K., Graf W. H., Momentum transport in sharp open-channel bends, Journal of Hydraulic Engineering 130 (2004) 186-198.

De Vriend H. J., Geldof H.J., Main flow velocity in short river bends, Journal of Hydraulic Engineering 109 (1983) 991-1011.

Fenjan S.A., Bonakdari H., Gholami A., Akhtari A.A., Flow variables prediction using experimental, computational fluid dynamic and artificial neural network models in a sharp bend, International Journal of Engineering-Transactions A: Basics 29 (2016) 14-22.

Gholami A., Akhtari A. A., Minatour Y., Bonakdari H., Javadi A. A., Experimental and numerical study on velocity fields and water surface profile in a strongly-curved 90° open channel bend, Engineering Applications of Computational Fluid Mechanics 8 (2014) 447-461.

Gholami A., Bonakdari H., Zaji A.H., Ajeel Fenjan S., Akhtari A.A., Design of modified structure multi-layer perceptron networks based on decision trees for the prediction of flow parameters in 90° open-channel bends, Engineering Applications of Computational Fluid Mechanics 10 (2016a)194-209.

Gholami A., Bonakdari H., Zaji A. H., Akhtari A.A., Simulation of open channel bend characteristics using computational fluid dynamics and artificial neural networks, Engineering Applications of Computational Fluid Mechanics 9 (2015a) 355-369.

Gholami A., Bonakdari H., Zaji A.H., Akhtari A.A., Khodashenas S.R., Predicting the velocity field in a 90° open channel bend using a gene expression programming model, Flow Measurement and Instrumentation 46 (2015b) 189-192.

Gholami A., Bonakdari H., Zaji A.H., Michelson D.G., Akhtari A. A., Improving the performance of multi-layer perceptron and radial basis function models with a decision tree model to predict flow variables in a sharp 90° bend, Applied Soft Computing 48 (2016b) 563-583.

Jung J. W., Yoon S. E., Flow and bed topography in a 180-degree curved channel, In 4th international conference on hydro-science and engineering, Korea Water Resources (2000) Association.

Leschziner M. A., Rodi W., Calculation of strongly curved open channel flow, Journal of the Hydraulics Division 105 (1979) 1297-1314.

Liaghat A., Mohammadi K., Rahmanshahi M., 3D Investigation of Flow Hydraulic in U Shape Meander Bendswith Constant, Decreasing and Increasing Width, Journal of river engineering 2 (2014) 12-23.

Manual, Fluent, Manual and user guide of Fluent Software, Fluent Inc, (2005).

Naji Abhari M., Ghodsian M., Vaghefi M., Panahpur N., Experimental and numerical simulation of flow in a 90° bend, Flow Measurement and Instrumentation 21 (2010) 292-298.

Ramamurthy A., Han S., Biron P., Three-Dimensional simulation parameters for 90° open channel bend flows, Journal of Computing in Civil Engineering- ASCA 27 (2013) 282-291.

Sui J., Fang D., Karney B.W., An experimental study into local scour in a channel caused by a 90 bend, Canadian Journal of Civil Engineering 33 (2006) 902-911.

Uddin M. N., Rahman M.M., Flow and erosion at a bend in the braided Jamuna River, International Journal of Sediment Research 27 (2012) 498-509.

Vaghefi M., Akbari M., Fiouz A.R., Experimental Investigation on Bed Shear Stress Distribution in a 180 Degree Sharp Bend by using Depth-Averaged Method, International Journal of Scientific Engineering and Technology 3 (2014) 962-966.

13

Developing a framework for compatibility analysis of predictive climatic variables distribution with reference evapotranspiration in probabilistic analysis of water requirement

Ehsan Fadaei Kermani*, Gholam Abbas Barani, Mohamad Javad Khanjani

Department of Civil Engineering, Shahid Bahonar University, Kerman, Iran.

ARTICLE INFO

Keywords:
Compatibility analysis
Reference evapotranspiration
Chaw analysis
Probability distribution function

ABSTRACT

In this paper, a new framework has been developed for compatibility analysis of predictive climatic variables distribution with reference evapotranspiration (ETo) in probabilistic analysis of water requirement. Initially, measured monthly meteorological data of four cities of Iran including Kerman, Shiraz, Ramsar and Babolsar synoptic weather station recorded from 1961 to 2003 were considered based on De Martonne climate classification. Then monthly ETo was calculated using FAO-Penman-Moentith (FAO-PM), and optimum Probability distribution function (PDF) was determined. The Chow method has been used for frequency analysis, and compatibility analysis was implemented on results. Based on the results, the Generalized Pareto (GP) was selected as optimum PDF for ETo. Results showed that the optimum PDF for minimum and maximum temperature, solar radiation and relative humidity is GP which had compatibility with EToPDF. Eventually, obtained results in compatibility analysis framework were confirmed using Correlation analysis. The proposed methodology developed in this research has application capability in probability scheduling of design water requirement, and can be utilized to optimize probability estimate of water requirement.

1. Introduction

Given the problems associated with the development of new water resources due to environmental impact and costs, the effective methods of water resources management need the correction of water use efficiency. The first step in this way is to understand the water requirement based on the time and location (Yoo et al. 2008). Evapotranspiration data are permanently required for irrigation design or planning at the desired probability level. Therefore this study attempted to develop a new framework for analyzing the distribution compatibility of predictive climatic variables of reference evapotranspiration for the probabilistic analysis of water requirement. To evaluate the efficiency of the present method and other methods of plant reference evapotranspiration under different climatic conditions, a basic research has been performed under the supervision of American Society of Civil Engineers (ASCE).In this study, to assess the validity of these methods, the efficiency of twenty different methods has been studied and analyzed in comparison with the lysimeter data from eleven regions with variable climate conditions. The results from analyzing the performance of various methods have proposed the FAO Penman Moentith (FAO-PM) method as the only standard method (Kin et al. 2010); moreover, this method has also been suggested as the reference method by the International Committee on Irrigation and Drainage (ICID) and the World Meteorological Organization (WMO).
In the standard design of agricultural water resources development, locations–plot formulas including California and Weibull have been proposed to find the irrigation requirements and alternate design (Yoo et al. 2008).
Probability distribution functions have been mostly used in the hydrological discussions and less of them have been used in the

*Corresponding author Email: ehsanhard@gmail.com

debates of ETo probability determination and water requirements. The two-parameter gamma distribution has been used for daily rainfall and flooding (Oksoy 1999). The review of the studies in this area (Ricciardi et al. 2003; Abida and Ellouze 2008; Wright et al. 2000) shows that only limited research has been done to investigate the distribution governing ETo; however, no research has been done on the compatibility of predictive variables of water requirements with ETo, despite its importance. Moreover, there is no framework on the use of probabilistic distributions governing the water requirements and ETo as a suitable tool in the analysis. Therefore, the main objective of this study is to develop a multi-stage framework for analyzing the compatibility of predictive variables with water requirements in probabilistic estimates and the calibration results of the proposed method.

2. Materials and methods

In this study, at first the values of ETo are computed using FAO-PM method. Then the best statistical distribution function of ETo and meteorological variables of monthly time series are specified using Kolmogorov Smirnov method and probability plot correlation coefficient. Finally, the compatibility analysis and assessment of the methodology are performed.

2.1. The study area

In this study, it has been used from data of synoptic stations located in four cities of Iran with different climates based on Domarten climate classification method from 1961 to 2003. Details of the study stations have been listed in Table 1. Table 2 also shows the characteristics of the variables used in this study.

Table 1. Characteristics of the studied stations.

Station	Longitude	Latitude	Elevation above sea level (m)	Climate
Kerman	56o 58′ E	30o 15′ N	1749	Arid
Shiraz	52o 36′ E	29o 32′ N	1484	Semi-Arid
Babolsar	52o 39′ E	36o 43′ N	-21	Semi-Humid
Ramsar	50o 40′ E	36o 54′ N	-20	Humid

Table 2. Annual average (43 years) of meteorological parameters at the studied stations.

Station	Min temperature (oC)	Max temperature (oC)	Humidity (%)	Sunshine hours	Wind speed (m/s)
Kerman	3.5	21	32.1	7.4	1.9
Shiraz	6.5	22.4	38.7	8.3	1.7
Babolsar	13.2	20.9	82.8	4.9	0.9
Ramsar	12.6	19.4	84.2	3.6	1.1

2. 2. Period Analysis

Probabilistic estimation of water demand with different return periods requires period analysis. Usually in water resources and hydrological studies, Chow (1951) method is mainly used. In this paper, in order to determine the optimal probability distribution function of ETo and meteorological variables, 15 Typical probability distribution functions including Beta (B), Johnson SB (JSB), Gumbel Min (GMin), Gumbel Max (Gmax), Normal (NOR), Gamma (GAM), Gen Gamma (GG), Log Gamma (LG), Lognormal (LN), Pareto (PAR), Person 5 (P5), Weibull (WBU), Gen Extreme Value (GEV), Gen Pareto (GP) and Log Person 3 (LP3) have been used. It is worth noting that in this study, period analysis was used to estimate the meteorological variables and ETo values.

Period analysis method used is as follows: (1) Creating monthly time series for each station, (2) estimation of parameters in each PDF using the method of moments or square error, (3) ETo estimation using optimized PDF and Chow Periodic factor method, (4) selecting and defining optimized PDF using Goodness-of-Fit tests including Kolmogorov – Smirnov (K-S) tests and Probability Plot Correlation Coefficient (PPCC) (Chow 1951; Vogel and McMartin 1991; Vose 2010; Temesgen et al. 2005).

3. Results and discussion

The results of this study have been presented in this section. At first the tests were carried out on Evapotranspiration variables are provided, and then the results of predictive variables are presented. Finally the results of Harmonic Analysis and validation of methodwill be presented.

3.1. The results of period analysis of reference Evapotranspiration

In this study, ETo values have been calculated for each region using FAO-PM method. The statistical characteristics of monthly mean values of ETo have been given for all stations on Table 3. According to results, the maximum value of ETo happens in Kerman station which has dry climate, and its minimum value happens in Ramsar station located in a very humid climate. Parameters of 15 probability distribution functions also have been estimated using the moment method, and related K-S values have been calculated and ranked for each distribution. The probability distribution functions related to K-S values have been given on Table 4.

Table 3. Statistical characteristics of monthly ETo.

Station	Minimum	Maximum	Mean	Standard Deviation	Variation Coefficient	Skewness
Kerman	0.7	7.4	3.6	1.7	0.1	0.2
Shiraz	1.1	7.4	3.8	1.7	0.1	0.1
Babolsar	0.6	5.4	2.4	1.4	0.1	0.4
Ramsar	0.8	5.4	2.1	1.2	0.1	0.5

Table 4. Values of KS index in distribution functions.

Station	GP	GG	GEV	LP3	WBU	LN	P5	NOR	GAM
Kerman	0.028*	0.052	0.070	0.081	0.078	0.078	0.075	0.082	0.090
Shiraz	0.048*	0.070	0.088	0.099	0.098	0.099	0.099	0.097	0.108
Babolsar	0.065	0.064	0.101	0.085	0.089	0.089	0.091	0.117	0.106
Ramsar	0.046*	0.070	0.084	0.076	0.087	0.087	0.084	0.117	0.086

*critical values at 5 % significance level

According to Table 4, the Probability Plot Correlation Coefficient (PPCC) values are calculated which have been shown on Table 5. To determine the optimized PDF, mean and standard deviations values have been calculated for each PDF. If the mean value is the highest and standard deviation is the lowest, PDF can accurately create sample data. According to table 5, maximum value of mean and minimum value of standard deviation are related to the GG function, but because this function was not significant based on the K-S test, therefore, the GP function is selected as the optimal function. Figure 1 shows the probabilities for the GP distribution function for all stations which show high correlations and estimations for all stations except Babolsar. In Tables 6 and 7, respectively, the statistical properties of GP distribution for all stations have been given for ETo data and GP parameters. As indicated on Table 6, the GP distribution on ETo data at the Babolsar station has the highest variation coefficient and

skewness coefficient, representing that this distribution cannot create acceptable results in this climate but is the best distribution function.

The GP probability distribution function has been chosen as optimized PDF is calculated as follows (Castillo and Hadi 1997):

$$F(x, k, \sigma) = \begin{cases} 1 - \left(1 - \dfrac{kx}{\sigma}\right)^{\frac{1}{k}}; & k \neq 0, \sigma > 0 \\ 1 - \exp\left(-\dfrac{x}{\sigma}\right); & k \neq 0, \sigma > 0 \end{cases} \quad (1)$$

where, σ and k are scale and shapeparameters of the distribution function.

Table 5. Optimized probability distribution functions based on PPCC index.

Station	GP	GG	GEV	LP3	WBU	LN	P5	NOR	GAM
Kerman	0.999	0.982	0.677	0.282	0.963	0.781	0.663	0.978	0.945
Shiraz	0.997	0.974	0.776	0.462	0.957	0.817	0.762	0.969	0.941
Babolsar	0.861	0.973	0.786	0.434	0.975	0.847	0.774	0.959	0.963
Ramsar	0.997	0.978	0.788	0.419	0.984	0.859	0.776	0.963	0.974
Mean	0.964	0.977	0.757	0.399	0.970	0.826	0.744	0.967	0.956
Standard Deviation	0.068	0.004	0.053	0.080	0.012	0.035	0.054	0.008	0.015

Table 6. Statistical characteristics of optimized probability distribution functions based on ETo (mm per month).

Station	Minimum	Maximum	Mean	Variance	Standard Deviation	Variation Coefficient	Skewness
Kerman	1	7	3.6	2.8	1.7	0.6	0.2
Shiraz	1	7	3.8	2.8	1.7	0.4	0.1
Babolsar	0	5.2	1.3	3.6	1.9	1.4	7.1
Ramsar	0.5	5.6	2.1	1.3	1.2	0.6	0.6

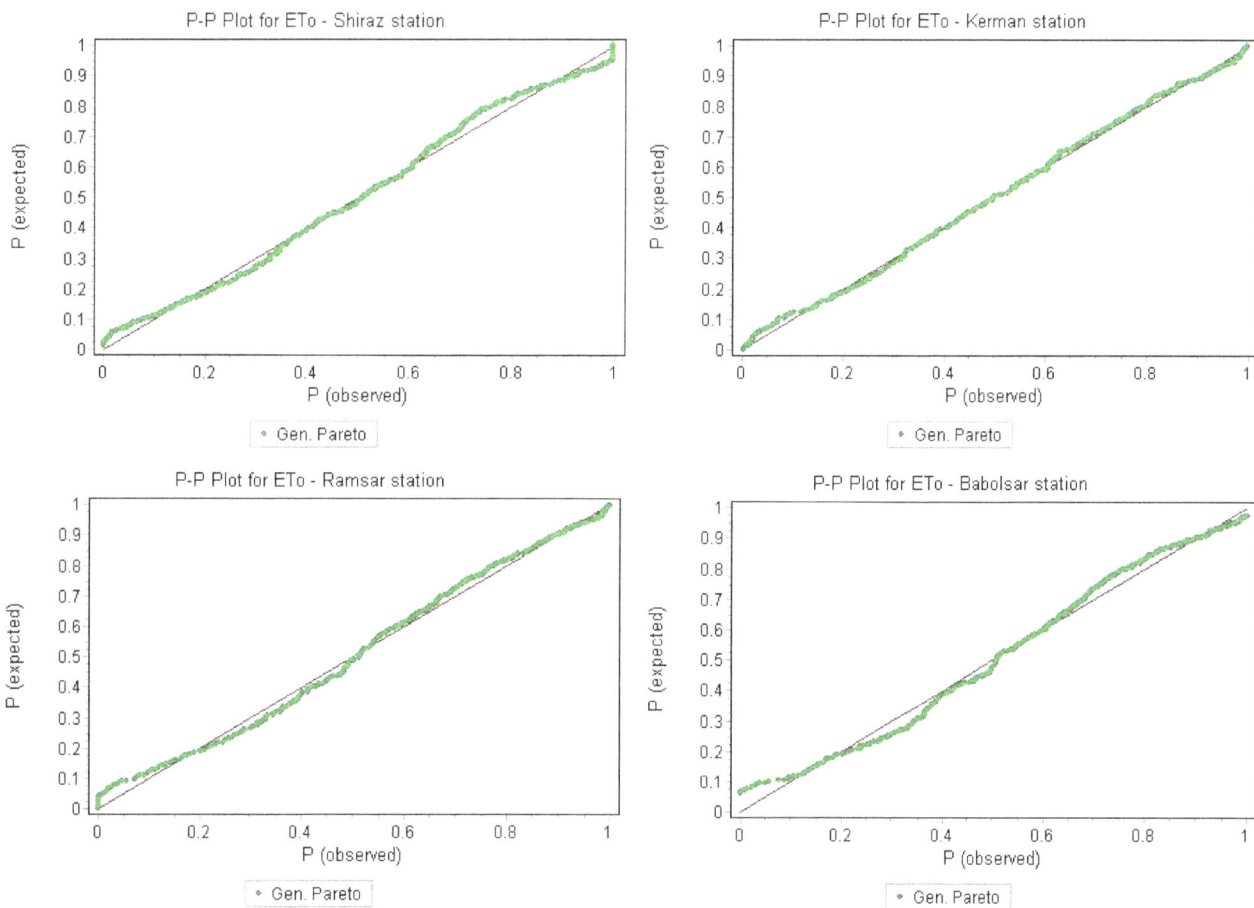

Fig. 1. Probability Plots of GP in study stations.

Table 7. Specifications of parameters for optimal distribution function at each station.

Station	Distribution parameters		
	k	σ	μ
Kerman	-0.8	4.8	1
Shiraz	-0.9	5.4	1
Babolsar	0.3	1	0
Ramsar	-0.5	2.3	0.5

3.2. Period analysis results of predictive variables

In the next step, the probability distribution functions of different variables in minimum temperature, maximum temperature, mean relative humidity, radiation and wind speed were also investigated. In this section, the same mentioned framework used in the ETo analysis was used and the Framework was developed for each of the variables. Table 8 presents the K-S values of significant functions, and the PPCC values for the optimized functions have been given on Table 9.

According to the mean and standard deviation values, it can be find out that the GP function for two variables of maximum temperature and light intensity has better answer than two variables of minimum temperature and relative humidity. The PDF parameters of GP and GEV are given on Tables 10 and 11. GEV Probability distribution function is calculated as follows:

$$t(x) = \begin{cases} \left(1 + k\left(\frac{x-\mu}{\sigma}\right)\right)^{-\frac{1}{k}} & \text{if } k \neq 0 \\ e^{-\frac{x-\mu}{\sigma}} & \text{if } k = 0 \end{cases} \quad (2)$$

where μ is the location parameter, σ is the scale parameter, and k is the shape parameter. Figure 2 shows the probability plot graph (PP plot) for variable wind speed and efficiency of the GEV cumulative distribution function for all stations. It can be seen that the results of the GEV distribution function for the variable of wind speed are reliable which can create better results than other distribution functions for wind speed.

Table 8. K-S values of distribution for the variables.

Station	Minimum temperature (GP)	Maximum temperature (GP)	Mean relative humidity (GP)	Radiation (GP)	wind speed (GEV)
Kerman	0.050	0.039	0.041	0.046	0.039
Shiraz	0.029	0.034	0.050	0.047	0.025
Babolsar	0.054	0.041	0.066	0.048	0.024
Ramsar	0.047	0.034	0.067	0.049	0.055

Table 9. PPCC values of distribution for the variables.

Station	Minimum temperature (GP)	Maximum temperature (GP)	Mean relative humidity (GP)	Radiation (GP)	wind speed (GEV)
Kerman	0.851	0.995	0.998	0.998	0.994
Shiraz	0.998	0.996	0.996	0.997	0.836
Babolsar	0.995	0.997	0.821	0.996	0.855
Ramsar	0.996	0.998	0.789	0.997	0.981
Mean	0.960	0.996	0.901	0.997	0.916
Standard Deviation	0.072	0.001	0.111	0.0008	0.082

Table 10. Parameters of the GP distribution function for the variables.

Station	Minimum temperature			Maximum temperature			Mean relative humidity			Radiation		
	K	σ	μ	k	σ	μ	K	σ	μ	k	σ	μ
Kerman	0.25	1	0	-0.009	23.81	9.08	-0.48	30.22	11.87	-1.12	24.52	6.005
Shiraz	-0.88	15.91	-1.98	-1.004	23.38	7.59	-0.53	36.80	14.73	-1.25	26.85	6.95
Babolsar	-0.96	22.78	1.49	-0.99	24.24	8.64	0.25	1	0	-0.64	13.70	5.05
Ramsar	-0.98	22.73	1.03	-1.004	23.38	7.58	0.25	1	0	-0.52	10.43	4.91

Table 11. Parameters of the GEV distribution function for variable of speed wind.

Station	Distribution parameters		
	k	Σ	μ
Kerman	-0.32	0.95	1.64
Shiraz	0.5	1	0
Babolsar	0.5	1	0
Ramsar	-0.02	0.43	0.87

3. Harmonic analysis results

It was found from the results obtained in the previous sections that optimized PDF values related to ETo for different climates has been the GP function. Moreover, the optimized PDF for four meteorological variables including minimum temperature, maximum temperature, relative humidity and radiation intensity is GP function that shows the probability distribution of these variables is harmonic with ETo. But the probability distribution of wind speed (GEV) was different from the ETo. Fig. 3 shows the harmonic variables process with ETo for Kerman station located in the dry climate. It can be seen that the variables crated similar trends wit ETo values so that the ETo values increase with increasing of these variables values and decrease with decreasing of them. It is to be mentioned that the variable of wind speed does not have similar trends with ETo as shown on Fig. 4.

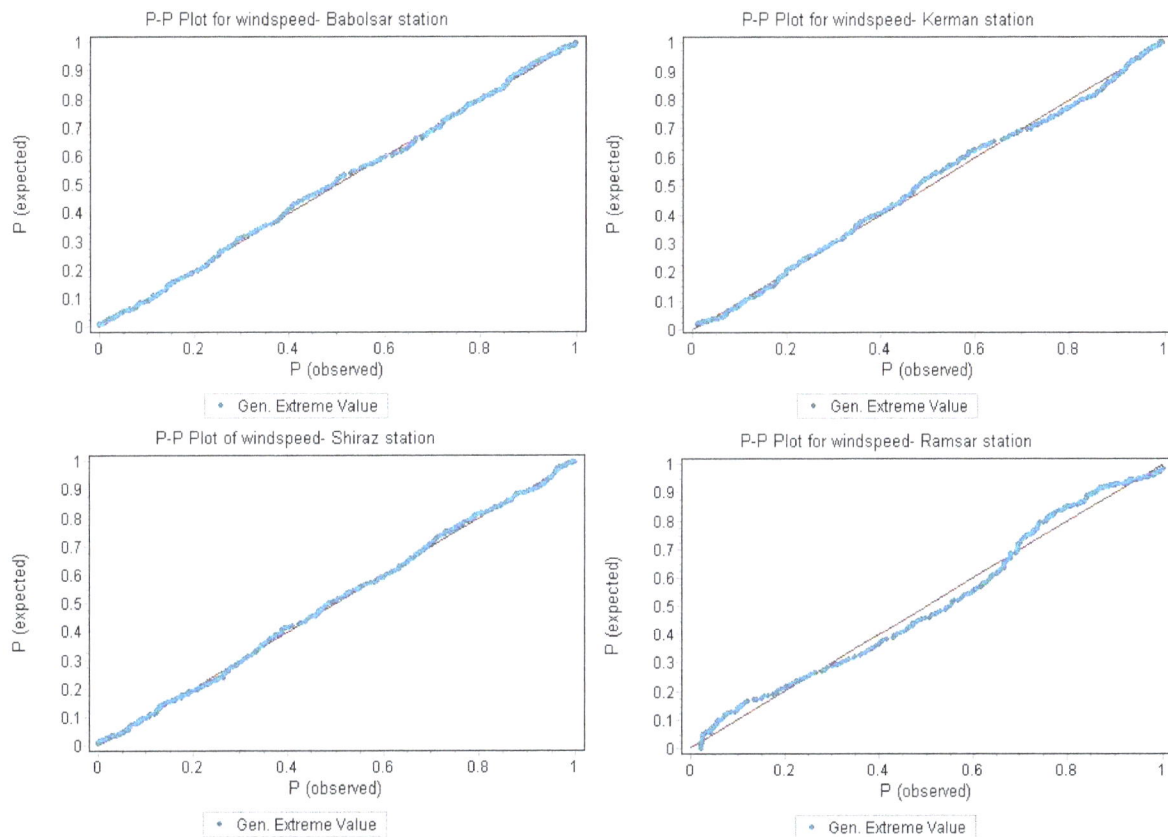

Fig. 2. GEV probability plots of variable wind speed for the study stations.

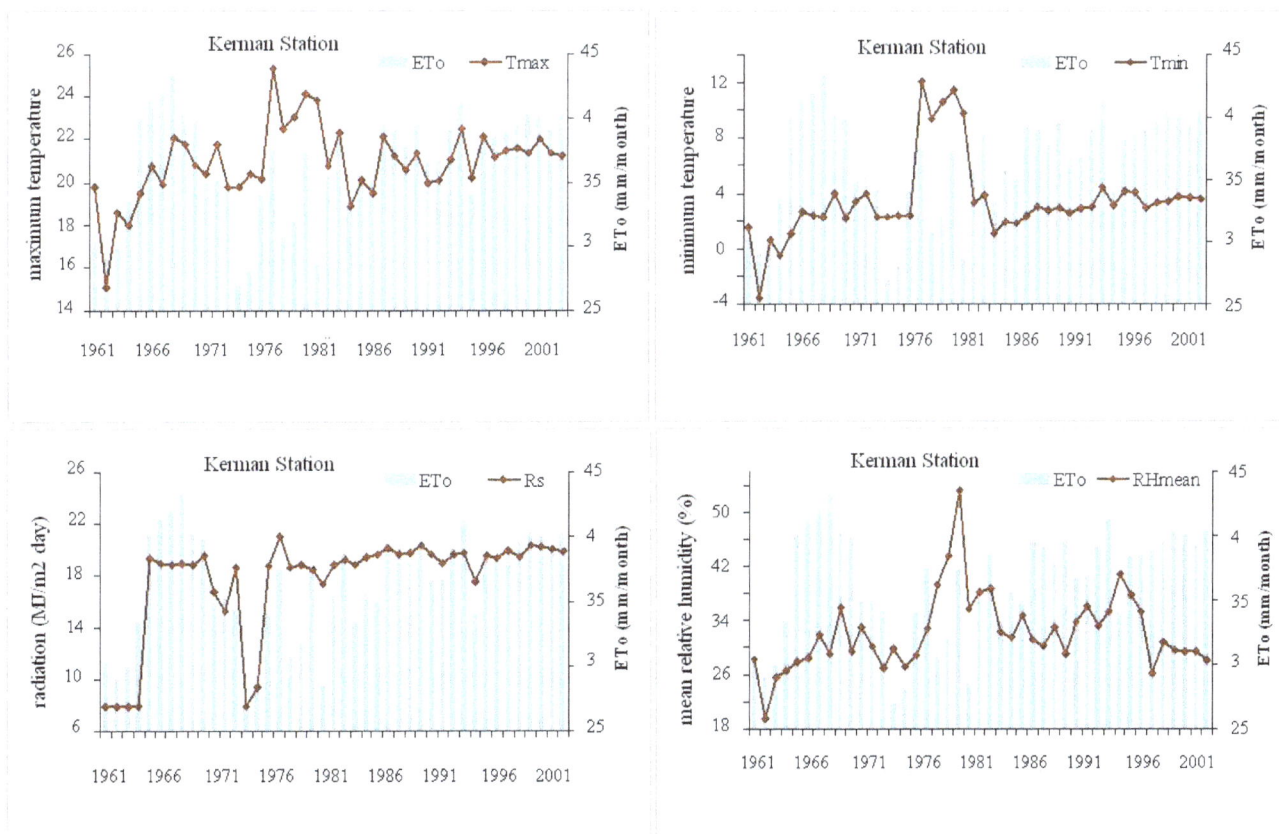

Fig. 3. Process of variables with harmonic distribution of ETo.

3.4. Results validation

Using the present results in practical aspects require confirmation and verification with other independent methods. In this section the results of the harmonic analysis of predictive variables for water requirement have been validated. Accordingly, it is expected to obtain a better correlation between the parameters of the harmonic probability distribution with different distribution parameters compared to ETo distribution. Therefore, the verification of the results was done by calculating the coefficients of determination (R2) between measured parameters at stations and calculated ETo values. The relationship of reference Evapotranspiration with the variables is presented in Figs. 5 to 8 for Kerman station. It can be seen that the three variables including minimum temperature, maximum temperature and radiation intensity have reasonable coefficient of determination associated with ETo values. The R2 value of the variable wind speed and ETo is very low (R2= 0.22) which represented no compatibility between the two variables in terms of the probability distribution. It can be find out that the results of the analysis variables are compatible with ETo to approve and confirm the framework presented in this research.

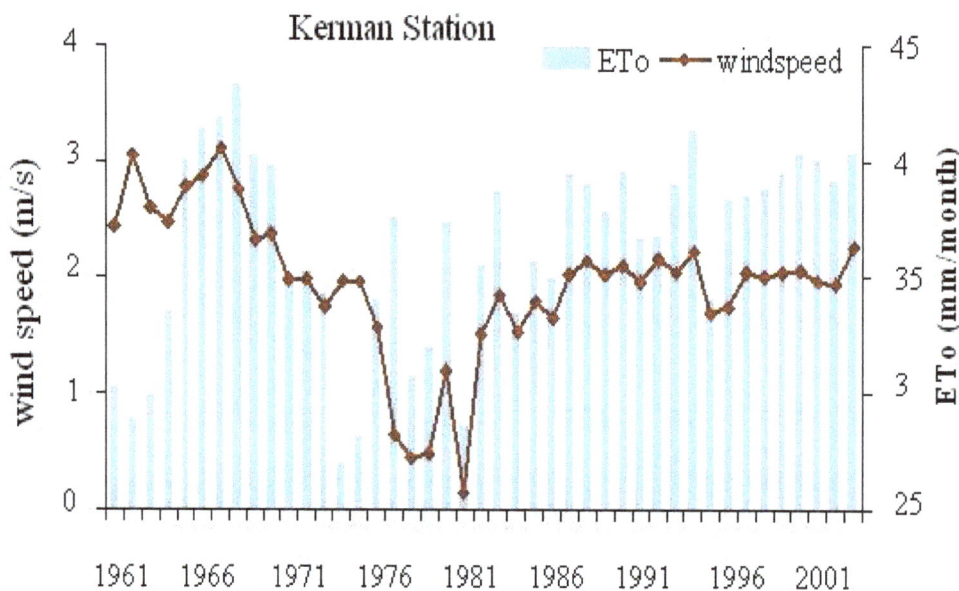

Fig. 4. Process of wind speed variable with non-harmonic distribution of ETo.

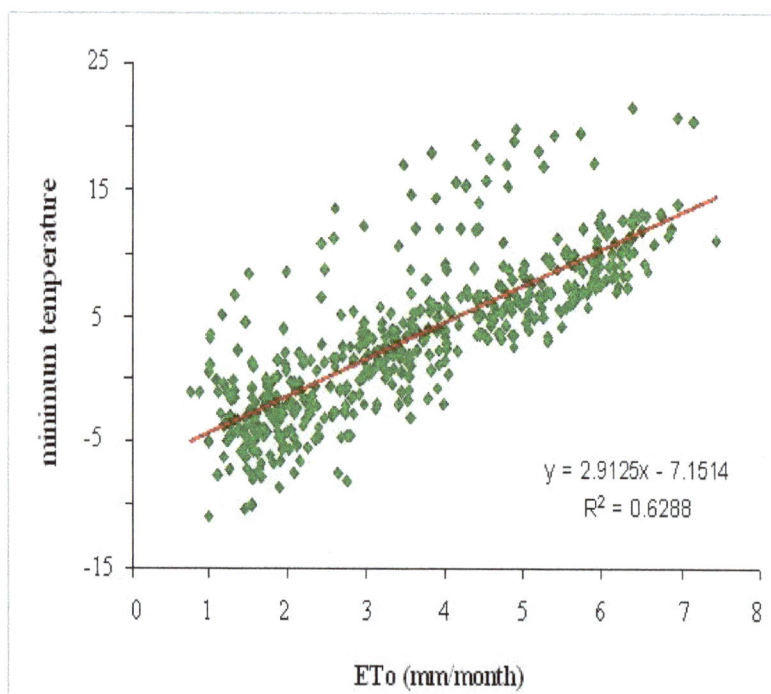

Fig. 5. Relationship of reference evapotranspiration with minimum temperature variable.

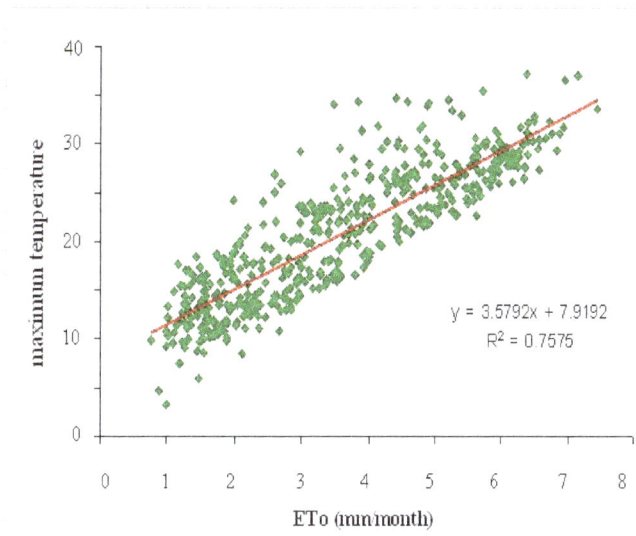

$y = 3.5792x + 7.9192$
$R^2 = 0.7575$

Fig. 6. Relationship of reference evapotranspiration with maximum temperature variable.

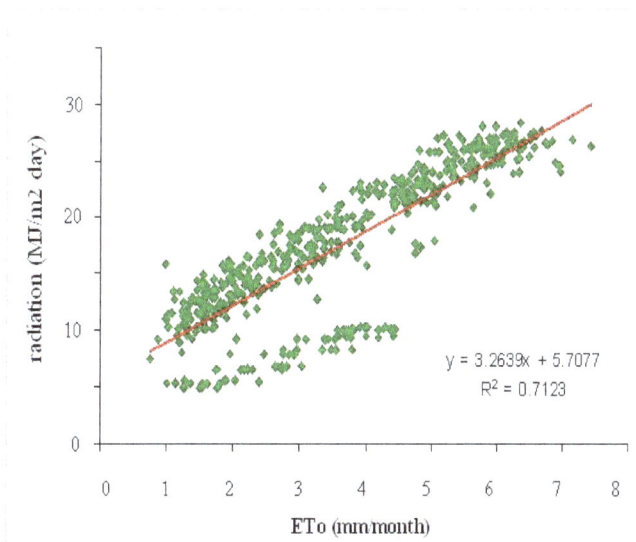

$y = 3.2639x + 5.7077$
$R^2 = 0.7123$

Fig. 7. Relationship of reference evapotranspiration with radiation variable.

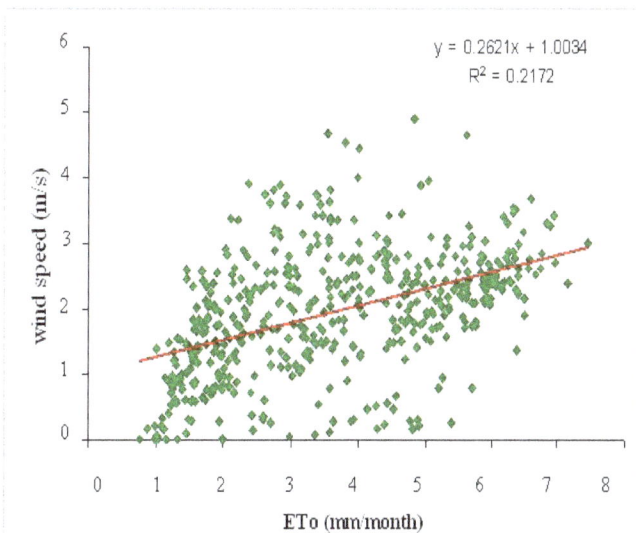

$y = 0.2621x + 1.0034$
$R^2 = 0.2172$

Fig. 8. Relationship of reference evapotranspiration with wind speed variable

4. Conclusions

Given the importance of probabilistic estimates of reference evapotranspiration (ETo) in irrigation management, in this paper, it has been attempted to develop a framework for finding the best probability distribution function (PDF) in determining the average monthly ETo and the most important meteorological variables affecting ETo including minimum and maximum temperature, relative humidity, wind speed, sunshine hours and radiation in four different climates of Iran. Results showed that the best PDF for monthly ETo values is the GP function. Moreover, it was found that the best PDF for variables of minimum and maximum temperature, relative humidity, radiation intensity is also the GP function which was compatible with the distribution function of ETo. Finally the validity of the presented framework was evaluated using correlation analysis which showed the developed framework can be utilized as a suitable tool in the probabilistic analysis of water requirement. The results of this study can be used as a guide for calculation of ETo at different climates, and the approach developed in this study can be applied in probabilistic water demand planning projects.

References

Abida H., Ellouze M., Probability distribution of flood flows in Tunisia, Hydrology and Earth System Sciences 12 (2008) 703-714.

Castillo E., Hadi A.S., Fitting the Generalized Pareto distribution to data, American Statistical Association 92 (1997) 1609-1619.

Chow V.T., A general formula for hydrologic frequencyanalysis. Transactions, American Geophysical Union 32, 1951.

Kim S., Shin H., Kim T., Heo J.H., Derivation of the probability plot correlation coefficient test statistics for the generalized logistic distribution, International Workshop Advances in Statistical Hydrology, Taormina, Italy, May (2010) 23-25.

ksoy H., Use of Gamma distribution in hydrological analysis, Turkish Journal of Engineering and Environmental Sciences 24 (1999) 419-428.

icciardi K.L., Pinder G.F., Belitz K., Comparison of the lognormal and beta distribution functions to describe the uncertainty in permeability, Hydrology 313 (2005) 248-256.

Temesgen B., Eching S., Davidoff B., Frame K., Comparison of some referenceevapotranspiration equations for California, Journal of Irrigation and Drainage Engineering 131 (2005) 73–84.

Vogel R.M., McMartin D.E., Probability plot goodness-of-fit and skewness estimation procedures for the Pearson type distribution. Water Resources Research 27 (1991) 3149-3158.

Vose D. Fitting distributions to data and why you are probably doing it wrong, 2010, Available in www.voesoftware.com.

Wright J.L., Allen, R.G., Howell, T.A. Conversion between evapotranspiration references and methods.Proceedings of the 4th National Irrigation Symposium, ASAE, Phoenix, AZ, 2000.

Yoo S.H., Choi, J.Y., Jang, M.W., Estimation of design water requirement using FAO Penman–Monteith and optimal probability distribution function in South Korea, Agricultural Water Management 95 (2008) 845-853.

14

Time series analysis of water quality parameters

Abdollah Taheri Tizro[1,*], Maryam Ghashghaie[1], Pantazis Georgiou[2], Konstantinos Voudouris[3]

1Department of Water Engineering, College of Agriculture, Bu-Ali Sina University, Iran.
2Deptartment of Hydraulics, Soil Science & Agriculture Engineering, School of Agriculture, Faculty of Agriculture, Forestry and Natural Environment, Aristotle University of Thessaloniki,Thessaloniki, Greece.
3Laboratory of Engineering Geology & Hydrogeology, Department of Geology, Aristotle University, Egnatia St., Thessaloniki, Greece.

ARTICLE INFO

Keywords:
Water quality
Time series
ARIMA
ACF
PACF

ABSTRACT

Water quality is a worldwide problem which affects human beings lives fundamentally. Water scarcity is intensified in result of quality deterioration. Different factors such as population increase, economic development and water pollution could be considered as the origins of the problem. The study and forecasting of water quality is necessary to prevent serious water quality deteriorations in future. Different methodologies have been used to predict and estimate the quality of water. In present study using time series modeling, the quality of Hor Rood River is studied at Kakareza station using time series analysis. 9 parameters of water quality are studied such as: TDS, EC, HCO_3^-, SO_4^{2-}, Mg^{2+}, Ca^{2+}, Na^+, pH and SAR. Investigation of observed time series show that there is an increasing trend for all parameters unless Na^+, pH and SAR. The order of model for each parameter was determined using auto correlation function (ACF) and partial auto correlation function (PACF) of time series. ARIMA (autoregressive, integrated, moving average) model was found suitable to generate and forecast the quality of river water. AIC, R^2, RMSE and VE % criteria were used for evaluating the generation and forecasting results. Results show that time series modeling is quite capable of water quality forecasting. For all generated and forecasted parameters the value of R^2 was greater than 0.66 Except for SO_4^{2-}. The value of R^2 for generated $SO4^{2-}$ was 0.48 and this value was 0.43 for forecasting this parameter. Also the study show that the quality of water is deteriorating based on an increasing trend for the majority of parameters and needs serious managerial actions.

1. Introduction

Water pollution is one of the most important environmental issues which the world faces. Universal problem of water deficiency and shortage of safe and healthy water require the investigation of problem. Water pollution, from a local river and basin to regional water pollution, from single to complex pollution, from surface water to groundwater, has been a serious restraint to sustainable economic development. Water quality could be affected by salinity, overdraw of ground water, urban and domestic wastewater entrance into surface streams as well as agricultural drainage. The purpose of most water quality and stream flow studies is to point out the information and necessary knowledge to manage water resources as well as their use, control and development. Time and money are saved through these studies and future development of water resources becomes inexpensive. The main objectives of water quality modeling could be to: (i) imply cause and effect relationships, (ii) identify impacts of pollutant sources, (iii) assess necessary levels of monitoring, (iv) evaluate planning and management alternatives, (v) focus on additional monitoring and management objectives and (vi) assess and evaluate future water quality conditions. Time series analysis is one of the useful methods which are applied in water quality modeling and forecasting.

Nowadays time series analyses are used in different aspects of science such as physics, economy and engineering. Water resources engineering lies within this category as there are many characteristics of water bodies, streams and groundwater resources as well as lakes and seas which are defined using time series of data. This procedure is useful in understanding and modeling the process of a phenomenon through which the past observations are generated.

It is also helpful in forecasting the future values based on the past memory. Time series is a string of data over time and there is an equal interval between all data. The interval can be defined as daily, weekly, monthly as well as yearly time steps. Time series analyzing is used in decision making in many hydrological processes and operation systems. The aim of time series analysis in hydrology can be defined as two main goals: at first it is used to understand and model the stochastic mechanism of hydrologic phenomena and at the second stage it is used to forecast the future values of the phenomena. Many works have been accomplished on hydrological components modeling using time series analysis. The application of this method for water quality forecasting is possible as well. Also evaluation of existing water resources, determining the quality of discharge as well as its quantity, identifying its variation on a watershed scale and forecasting these variables, could be a main step in integrated water resources management. Also stochastic characteristics of hydrological phenomena lead the hydrologists and water resources engineers towards benefiting from time series concepts in modeling and forecasting the future of water resources.

1.1. Applied time-series analysis

Time-series analysis using ARIMA approaches have been used to examine runoff and river discharge (Rao et al. 1982, Papamichail and

*Corresponding author E-mail: ttizro@yahoo.com

Georgiou 2001; Yurekli et al. 2005), water levels in lakes (Irvine and Eberhardt 1992; Sheng and Chen 2011), sediment yield and erosion (Hanh et al. 2010), and water quality (Papamichail et al. 2000; Lehmann and Rode 2001; Faruk 2010; Hanh et al. 2010; Voudouris et al. 2010).

ARIMA models are capable of reproducing the main statistical characteristics of a hydrologic or environmental time series. These models also provide information about system dynamics and could be used to forecast a time series for the future. Thomas and Fiering, (1962) used auto correlated models in their studies on stream analyzing. Chow and Kareliotis (1970) analyzed univarite time series of rainfall and temperature. They discovered yearly strict and 6 months slight periodic components in time series. The main step of time series application in hydrology was performed by McKerchar and Delleur (1974) as they used ARMA (Autoregressive Integrated Moving Average) and seasonal modeling in analyzing seasonal characteristics of stream parameters.

Zhang (2003) applied a hybrid of ARIMA and ANN model to take the advantages of the ARIMA and ANN models in linear and nonlinear modeling. Results showed that the combined model was capable of forecasting the real data sets more accurately in comparison with the separately applied methods of ANN and ARIMA.

Jalal Kamali (2006) also used time series models for monthly inflows to Jiroft dam. The results of this study showed that time series modeling is capable of identifying and forecasting monthly stream pattern and integrated water resources management. Also Komornk et al. (2006) studied hydrological time series in Czech through which high capability of forecasting by this kind of modeling was proved. Dalme and Yalcin (2007) applied time series analyzing in Mississippi River to forecast the values of flood. The results of their study showed the capability of time series modeling application in generating daily discharge as well as validity of forecasting. There are many studies which have been focused on water quality parameters mentioned as follow. Khashei and Bijari (2010) applied an artificial neural network (pdq) model to estimate time series forecasting. In this paper, a new hybrid model of ANN was introduced using ARIMA models in order to achieve a more accurate forecasting model than artificial neural networks. The empirical results with three well-known real data sets showed that the proposed model can be an effective way to improve forecasting accuracy achieved by artificial neural networks. The research proposed the application of model as a convenient alternative method to forecast accurately thanks to times series capabilities.

1.2. Applied time series analysis on surface water quality

Hirsch et al. (1982) introduced techniques to analyze monthly water quality data for monotonic trends. The first procedure is a non-parametric test to detect trend, which is used for seasonal time series. The second method of seasonal Kendall estimator estimates the magnitude of trends. The third procedure provides a tool to test temporal changes in correlation of constituent concentration and stream flow. Also El-Shaarawi et al. (1983) studied temporal changes in water quality parameters such as PH, Alkalinity, total Phosphorous and Nitrate concentrations using a 5-year data series of Niagara (on Ontario lake). Results showed that PH and Alkalinity were decreasing while Nitrate was increasing. Yu et al., (1993) examined surface water quality data of the Arkansas, Verdigris, and Neosho as well as Walnut river basin to study trends in 17 major constituents using 4 different nonparametric methods. Robson and Neal (1996) studied the trend of ten-year upland stream and bulk deposition water quality data from Plynlimon, mid wales through the seasonal Kendall test, the stream water dissolved organic carbon was increasing over time. However, any increase for PH was not found. It was suggested that long term monitoring programs should be applied for several decades.

In a study accomplished by Turner et al. 1995, Long-term simulations results of Lake Bosumtwi in Ghana, showed that stochastic climatic variations very similar to those observed in this century could produce the full range of lake levels observed in terrace deposits. The low salinity of about 1% suggests that dissolved solutes were removed by Lake Overflow in the recent geological past.

Primarily Graphical and statistical time series techniques have been used to analyze the trends and specified time changes, in river water quality data. The information obtained may be associated with some socio-economic variables, such as industrial or agricultural development, urban increase and wastewater discharge around or upstream of the measure station. Such a study may now be applied to more rural stations in order to compare the evolution of water quality and perhaps, historical monthly average values to evaluate the seasonality effect on annual trends (Gun and Vilagines 1997).

Papamichail et al. (2000) examined stochastic models to improve understanding and forecasting of monthly flow and some water quality parameters of Strymon River (Greek part) in an effort to reduce the negative impacts of incurred by interests using the river. Especially, they developed seasonal and nonseasonal ARIMA models for Strymon River using the time series of monthly measurements of flow and some water quality parameters. The selected models for each parameter data set can be used to forecast monthly values of one or more time periods ahead.

Antonopoulos et al. (2001) analyzed the time series of water quality parameters and the discharge of Strymon River in Greece from the 1980 to1997. The nonparametric Spearman's criterion was used to detect the trends for: discharge, EC_w, DO, SO_4^{2-}, Na^+, K and NO_3^{3-}. The Verification of the best fitted models was performed using χ^2 and Kolmogorov-Smirnov tests. The relationships between concentration and loads of constituents of both with the discharge were investigated as well. In spite of the relation between loads and discharge (r> 0.9), the correlation between concentrations and discharge is not good (r< 0.59). Ahmad et al., 2001, accomplished a study to analyze water quality data collected from Ganges River in India. Three approaches of stochastic modeling such as: multiplicative ARIMA model, deseasonalised model and Thomas–Fiering model were applied to model the observed time series of water quality. The multiplicative ARIMA model having non- seasonal and seasonal components we identified as a convenient model. The de seasonalised modeling approach was recommended to forecast water quality parameters of the river.

Through a water quality monitoring program (New Zeland) Stansfield (2001) illustrated the importance of considering detection limits of variables and sampling frequencies through analyzing the trends in water quality time series using nonparametric seasonal Kendall test and Sen Slope test. Result showed that if the sampling frequency was changed from monthly to quarterly fewer trends were detected. What is more results showed that the quarterly data present with a different magnitude in terms of slope in comparison with monthly data. Gangyan et al. (2002) investigated the temporal sediment load characteristics of the Yangtze River using the turning point test, Kendall's rank correlation test and the Anderson correlogram test to prove randomness and determine the trend. The annual sediment load data from 1950 to 1990 and the monthly sediment load data from 1950 to 1969 were used. The stochastic component was modeled using autoregressive model. Using the AR (1) model for the dependent stochastic component, 100 years of monthly sediment data were generated and the observed and generated data matched well.

Jassby et al. (2003) developed a time series model for Secchi depth in Lake Tahoe, USA considering an understanding of inter annual variability. The Secchi depth was found sometimes over 40 m. however the mean annual Secchi disk depth has declined about 10 m since 1967 inspiring a large scale restoration program. Yearly variability was extremely high, obscuring restoration actions and conformance with water quality standards. The model suggested a tool to determine the compliance with water quality standards when precipitation anomalies may persist for years. Also some studies have focused on water temperature time series such as Webb et al. (2003) who showed that when discharge is below the annual median, correlations between air and water temperature is high. Kurunc et al. (2005) applied time series analysis for water quality constituents and stream flow of the Yesxilirmak River at Durucasu which is a monitoring station. Two modeling approaches, ARIMA and Thomas–Fiering were evaluated in this study. A 13-year monthly time series records were used to obtain the best model of each water quality constituent and stream flow from both modeling approaches. The results of study showed that that between two approaches, for Yesxilirmak River Thomas–Fiering model presents more reliable forecasting of water quality constituents and stream flow than ARIMA model.

Panda et al. (2011) studied trends in sediment load of the tropical river basin of India and explored the influence of climate and human forcing mechanisms on the land ocean fluvial system. Sediment time series of different timescales during the period 1986- 87 to 2005- 06 from 133 gauging station were analyzed. Results showed significant diminishing in sediment load in the tropical river basins. The rainfall characterized by the non-significant decreasing trends and frequent drought years was found to be the reason of sediment load reductions for most of the river basins. Also results showed that the maximum reduction in sediment loads was referred to Narmada River among the tropical rivers (2.07×106 t / yr) because of construction of the dam. Also Irvine et al. (2011) accomplished a study on temporal variability of turbidity, dissolved oxygen, conductivity, temperature, and fluorescence in the lower Mekong River. Results showed that A strongly developed vertical variation of turbidity, DO, and conductivity in the flooded forest fringe may be related to a combination of factors, including dissolved material release from bed sediment and a floating organic-rich particulate layer near the bottom of the lake.

Halliday et al. (2012) studied two hydrochemical time-series derived from stream samples taken in the Upper Hafrencatchment, Plynlimon, Wales. A subset of determinants such as: aluminium, calcium, chloride, conductivity, dissolved organic carbon, iron; nitrate, pH, silicon and sulphate were examined within a framework of non-stationary time-series analysis to identify determinant trends, seasonality and short-term dynamics. The results demonstrate that both long-term and high frequency monitoring provide valuable and unique insights into the hydrochemistry of a catchment. Such studies moving forward demonstrate the need for both long-term and high-frequency monitoring to facilitate a thorough understanding of catchment hydro chemical dynamics.

Different studies on time series analyzing in water resources management demonstrate the efficiency and necessity of this kind of modeling as it takes into account the stochastic nature of hydrological processes such as discharge and climatology. Present study aims to apply this methodology on chemical water quality properties. In present study water quality parameters of Hor Rood is investigated at Kakareza station. The methodology and the study area are presented at the following stages.
The objectives of this study are to:
(i) plot time series of data to find any possible trend,

(ii) Omit the trend which demonstrates a deterministic nature of data,
(iii) Obtain the best model fits for each time series of parameters using a stochastic modeling approach including ARIMA, and
(iv) Evaluate the performance of modeling approach using five-year observed data vs. forecasted data

2. Methodology and the study area
2.1. The study area

The study area is located in the west of Iran in Kuhdasht region which is shown in Fig. 1. Kakareza station of Hor Rood River lies at 48o 15' E and 33o 43' N. The upstream area of the station is about 1148 Km2 located at Kashkan sub-basin of Karkheh watershed. The topography in the Karkheh watershed varies spatially with elevation ranging from 3 to more than 3500 masl. The average value of temperature varies from -3 to 21 ^{0}C across Kashkan basin. The mean annual precipitation of the basin varies between 345 mm/yr and 849 mm/yr. Also evaporation of the basin varies from 25 to 2922 based on a yearly average. The contribution of rainfall, snow and Karstic springs is significant to discharge of Kashkan River. Also Kakareza River is the main tributary of this river.

The land use of the study area is composed of nearly 26% agriculture, 16.9% pasture, 39.33 % forest, 0.12 % residential area and about 2.5 % of the area is composed of other kinds (Jamab Consulting Engineers 2005). Based on field studies agricultural drained waters and industrial waste waters are either directed or flow through surface streams as well as residential litters and swages. The time series of 9 water quality parameters such as TDS, EC, HCO_3^-, SO_4^{2-}, Mg^{2+}, Ca^{2+}, Na^+, pH and SAR of Kakareza station at Hor Rood River were studied in this research.

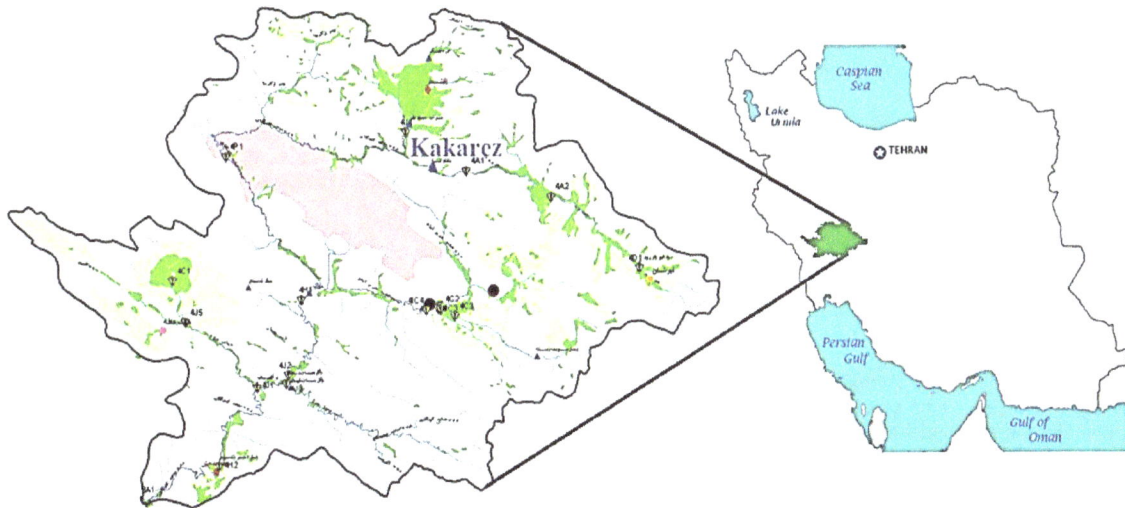

Fig.1. The location of the study area.

2.2. Methodology

Stochastic models reveal time series characteristics in terms of correlation as well as consider the randomness of phenomenon although they do not consider their physical nature. This method analyzes the past of the time series with respect to successive correlation which is used as system input in other words. The present or the future is then predicted as the system output. Two main applications of the time series models are generating simulated samples and forecasting hydrologic events. Forecasted time series are used as input for analyzing complex water resources systems. Generated series could show many possible hydrologic conditions that do not appear in the historic series explicitly. Consequently, using simulated time series, different designs and operational strategies can be tested under various conditions. Forecasted data from known historic observations can help to assess and evaluate options for a real system operation.

Although time series modeling originated from different scientific fields, it has demonstrated its capability and reliability in stochastic hydrology, and the applications of time series analysis in water resources management are important. Development of stochastic modeling in hydrology began at the beginning of the 1960s, when time

series analyses of hydrologic phenomena were extended to the synthetic generation of stream flow using a table of normal random numbers. Thomas and Fiering (1962) were the first to propose a first-order Markov model to generate stream-flow data. The classic book on time series analysis by Box and Jenkins (1976) presents the foundation of modern hydrologic stochastic modeling eq. (1):

$$y_t = f(x_t, x_{t-1}, \ldots; y_{t-1}, y_{t-2}, \ldots; 1, 2, \ldots) + \varepsilon \qquad (1)$$

where f is the selected mathematical function; y_t is the predicted output at time t; y_{t-1}, y_{t-2}, … , are the successive members of the output time series recorded at corresponding time intervals t −1, t −2; x_t, x_{t-1}, x_{t-2}, … , are the successive members of the input time series recorded at time intervals t, t −1, t −2; 1, 2, … , are the model parameters found by mathematically minimizing the differences between estimated (calculated) and observed y_t values; ε is the model error (residual) given as the difference between the calculated and the recorded value of the output series at time t.

Stochastic modeling generally follows the approach proposed by Box and Jenkins, (1976), who introduced autoregressive moving average (ARMA) models. The mathematical formulation of ARMA models is written as Eq. (2):

$$Z_t = \varphi_1 Z_{t-1} + \ldots + \varphi_p Z_{t-p} + \alpha_t - \theta_1 \alpha_{t-1} - \ldots - \theta_q \alpha_{t-q} \qquad (2)$$

where Z_t represents the time dependent series, φ_i, i=1,2,..., p are nonseasonal AR parameters, θ_i, i=1,2,..., q are the nonseasonal MA parameters,Time series models used to generate synthetic time series can be classified into autoregressive models (AR (p)), moving average models (MA (q)), and their combination, autoregressive moving average (ARMA (p, q)) with variations, such as autoregressive, integrated moving average (ARIMA) models (p, d, q) and others, where p and q are the orders of autoregressive and moving average terms, respectively, and "d " is an order of differencing.

An autoregressive model estimates values for the dependent variable, Z_t, as a regression function of previous values Z_{t-1}, Z_{t-2}, ..., Z_{t-n}. A moving average model is conceptually a linear regression of the current value of the series against the white noise or random shocks of one or more prior values of the series.

An autoregressive (AR) model, which is called a Thomas-Fiering model, has been applied extensively in hydrology for modeling annual and periodic hydrologic time series. Autoregressive (AR) models basically estimate values for the dependent variable, Z_t, as regression function of previous values, Z_{t-1}, Z_{t-2} ... Z_{t-n}. An AR model of order 1 (i.e. an AR (1) model) can be expressed as eq. (3):

$$Z_t = \varphi_1 Z_{t-1} + \alpha_t \qquad (3)$$

where, Z_t and Z_{t-1} are the deviations from the mean of the time series, φ_1 is the first-order AR coefficient describing the effect of a unit change in Z_{t-1} on Z_t, and α_t represent random shock errors or white noise. Values for α_t are assumed normally and independently distributed with zero mean and constant variance. Model stationarity requires that the variance of Z_t be non-negative and finite (Vandaele 1983) and for these conditions to be met, $|\varphi_1|$ must be less than 1. Higher order AR models are possible, much like a multiple regression, and in this case, the absolute value of each AR coefficient should be less than 1.

Moving average (MA) models incorporate past random fluctuations to represent the time series and an MA model of order 1 (i.e. an MA (1) model) can be expressed as eq. (4):

$$Z_t = \alpha_t - \theta_1 \alpha_{t-1} \qquad (4)$$

where, θ_1 is the MA coefficient to be estimated and the random shocks (αt) are assumed normally and independently distributed with mean 0 and constant variance. The model structure requires the condition of reversibility to be met and | θ1 | therefore must be less than 1. Values greater than 1 indicate that observations further in the past have a greater influence on Z_t than more recent observations which is unlikely in hydrologic time series. Higher order of MA models is possible, and like the AR model coefficients, the absolute value of each MA coefficient should be less than 1.

A parsimonious model can be achieved using a mixed ARMA model as a combination of a moving average process and an autoregressive process rather than a merely AR or MA model. Therefore, low-order of ARIMA models has been widely used in hydrological practice (Salas et al., 1982; Weeks and Boughton 1987; Padilla et al., 1996; Montanari et al. 2000).

The statistical structure of a time series should be represented by a parsimonious model, and in some cases, parsimony can be achieved using a mixed (ARMA) model rather than a pure AR or MA model. As such, it would be more parsimonious to represent a time series with an ARMA (1, 1) model than an AR (3) model because fewer model parameters need to be estimated. It is possible to mix models because these models theoretically can be rewritten as pure AR or MA models of infinite order (Vandaele 1983). Furthermore, a hydrologic time series is the result of several interactive processes that may have both a seasonal and a random fluctuation component. The mixed model structure can provide additional flexibility in describing the result of the interaction between the processes (Salas et al. 1980).

2.3. Modeling water quality time series of Hor Rood River

The main goal of a time series analysis may be to understand seasonal changes and/or trends over time. Plotting the data against time was accomplished as the first step of analyzing time series. Time

plot show a lot of information about the time series. Trends and seasonal variations are often evident in time plots. Hydrologic time series frequently exhibit a regular seasonal pattern that can be removed by standardizing the data for the seasonal mean and standard deviation and then retrending the forecasts using the inverse of the deseasonalizing transformation. Also time plots indicate the presence of outliers in the time series which are observations that are not consistent with the rest of the data.

However, another goal that is often of primary importance is to understand and model the correlational structure in the time series. This type of analysis is generally done on stationary processes. A stationary process is one that looks basically the same at any given time point. That is, a stationary time series is one without any systematic change in its mean and variance and does not have periodic variations.

Many studies have been written about the theory of ARIMA modeling as well as its applications (Pankratz, 1983; Vandaele, 1983; Nelson, 1973; Box and Jenkins, 1976). Here a brief description is presented to point out main stages accomplished in this study.

The basic stages in ARIMA modeling are composed of: (1) identifying the autocorrelation and partial autocorrelation of time series, (2) estimating the orders of the identified model, and (3) verify the model through standard tests. The results of time series analyzing of this study are explained in the following stage.

In this study 8 river quality parameters were studied. All series showed trend line. MINITAB 14 was used to analyze these 9- time series. Also Normality test of series was investigated using Easy fit. All series were normal. Then ACF and PACF of time series were plotted at next stage. Fig. 2 shows the ACF and PACF for TDS time series. Based on this Figure p= 1 and q= 3 is suggested for TDS time series. Fig. 3 shows the ACF and PACF of EC time series. For EC p= 1 and q=2 is offered. ACF and PACF of HCO_3^- time series are shown in Fig. 4 and p=1 and q=2 is suggested for HCO_3^-.

Also ACF and PACF of SO_4^{2-} are demonstrated in Fig. 5. P=1 and q=2 is proposed for the series. Fig 6 presents the ACF and PACF of Ca^{2+} and p=1 and q=2 is suggested for the series. Fig 7 shows the ACF and PACF of Mg^{2+}. As it is clear in the Fig. 7 there is not any autocorrelation between data. ACF and PACF of Na^+ are shown in Fig. 8. P=1 and q=2 is proposed for the series. Fig. 9 shows the ACF and PACF for pH series. P= 1 and q= 1 is suggested for this series. And finally ACF and PACF of SAR are shown in Fig. 10. P=1 and q=2 is offered for SAR series.

All parameters suggest the value of p= 1 except Mg^{2+}. At the following steps generation of data are accomplished and the results of it are demonstrated for TDS, EC, HCO_3^-, SO_4^{2-}, Ca^{2+}, Na^+, pH and SAR respectively. The standard (Z) time series of all parameters were plotted at the second stage, which are shown in Fig. 11. Using one difference the data were transformed to make a yearly stationary time series.

After identifying the ACF and PACF and removing the trend of each time series first the order of model was determined and then 4 criteria such as Akaike Information Criterion (AIC), Determination Coefficient (R^2), Root Mean Square Error (RMSE) and Mean Absolute Percentage Error (MAPE %) (Karamouz and Araghinejad 2005) were used to compare the results of series generation through suggested models. The value of AIC is estimated through eq. (5):

$$AIC = n \times Ln\ (\sigma^2) + 2 \times (p + q) \qquad (5)$$

where, σ denotes the standard error of residuals; n shows the sample size; p and q show the order of AR and MA, respectively.
Also the value of VE % is calculated using eq. (6):

$$MAPE = \frac{\sum_{i=1}^{n} \left| \frac{y_t - \hat{y}_t}{y_t} \right| 100}{n} \qquad (6)$$

where, y_t and $\hat{y}_t$ show the observed and estimated values respectively and n is the sample size. A thirty-five (35) year time series were generated for each parameter.

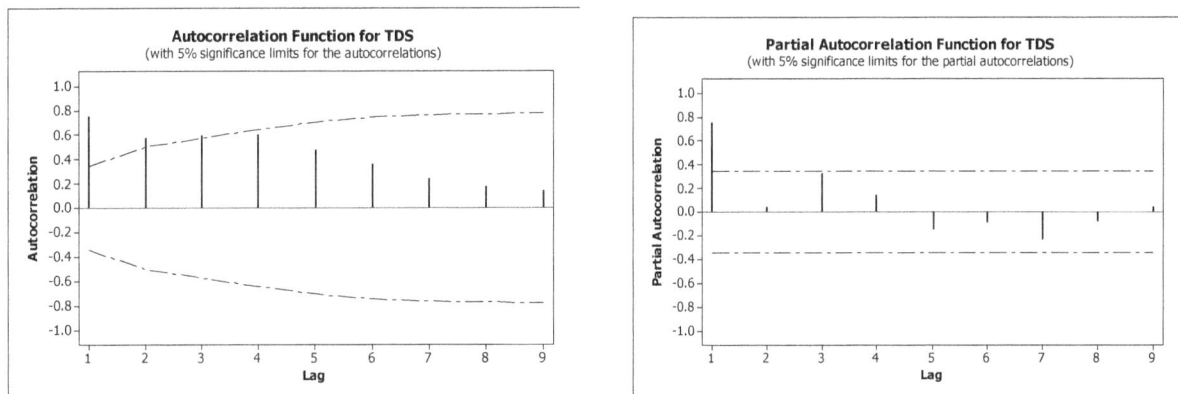

Fig. 2. ACF and PAC of TDS time series.

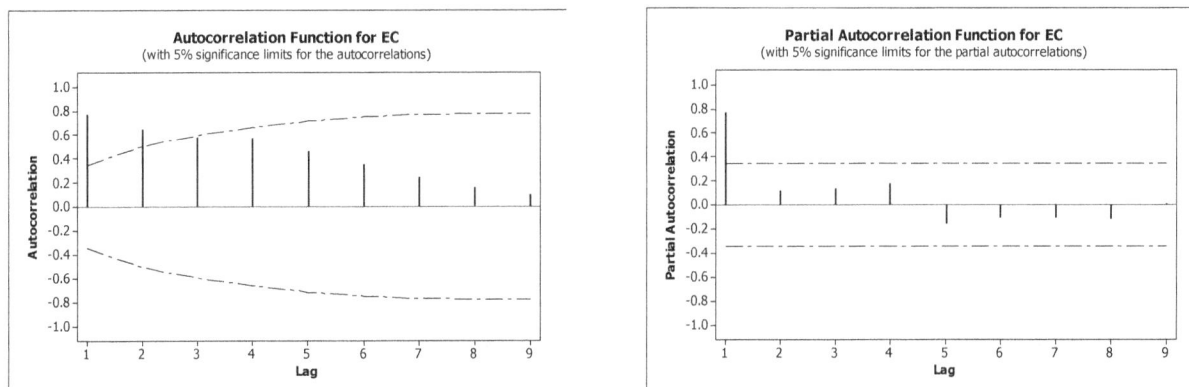

Fig. 3. ACF and PAC of EC time series.

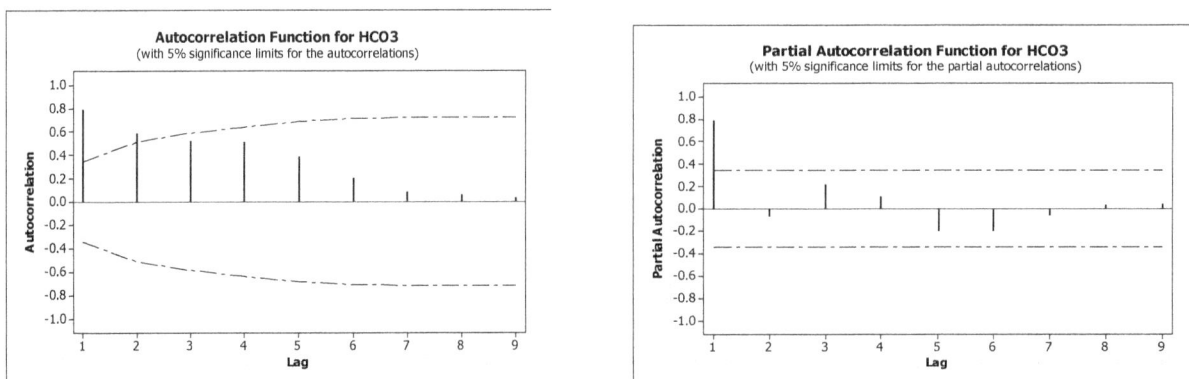

Fig. 4. ACF and PAC of HCO_3^- time series.

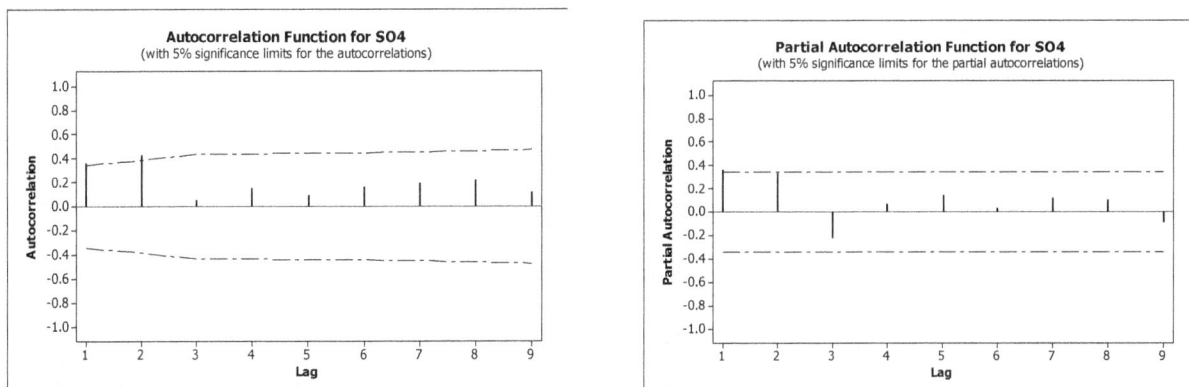

Fig. 5. ACF and PAC of SO_4^{2-} time series.

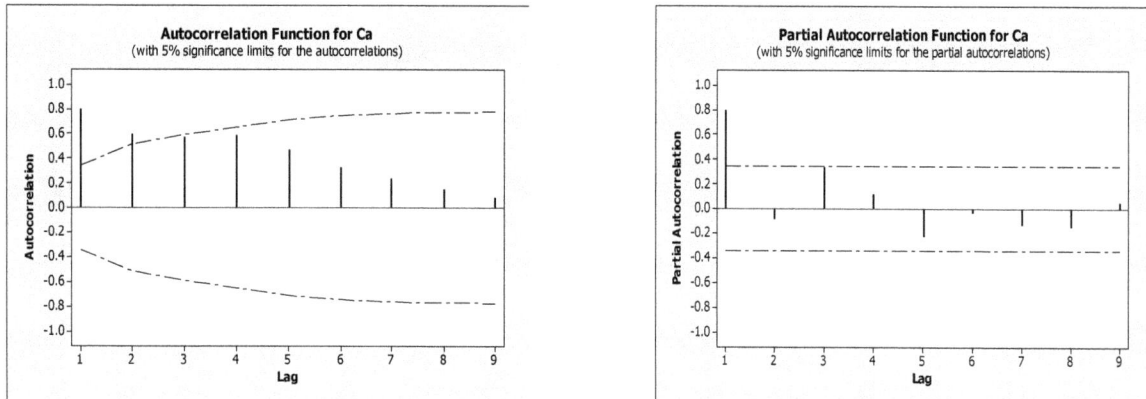

Fig. 6. ACF and PAC of Ca^{2+} time series.

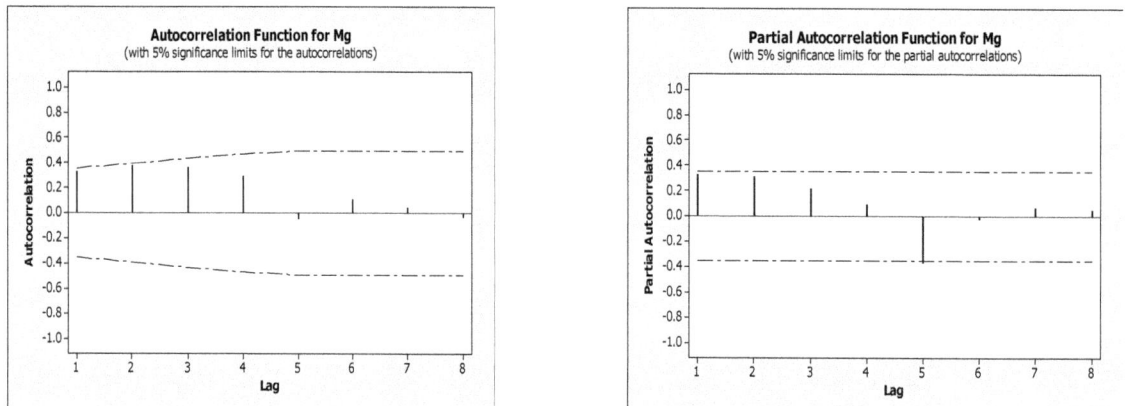

Fig. 7. ACF and PAC of Mg^{2+} time series.

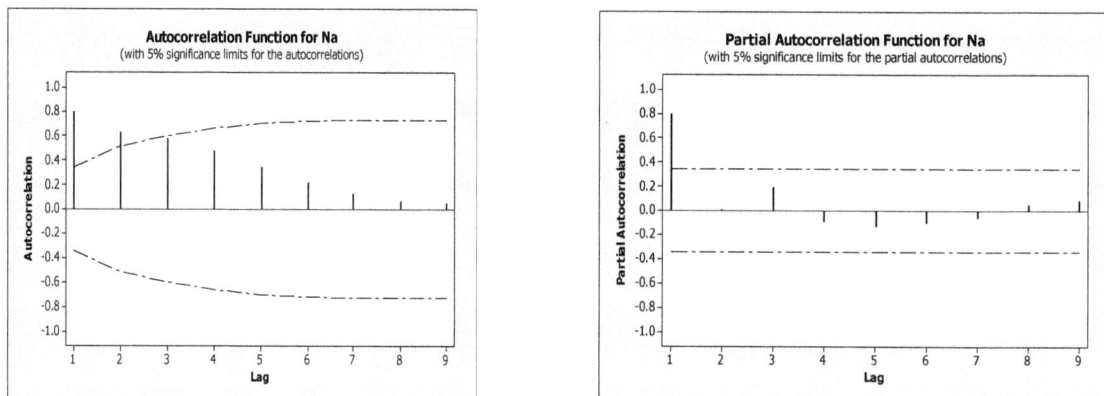

Fig. 8. ACF and PAC of Na^+ time series.

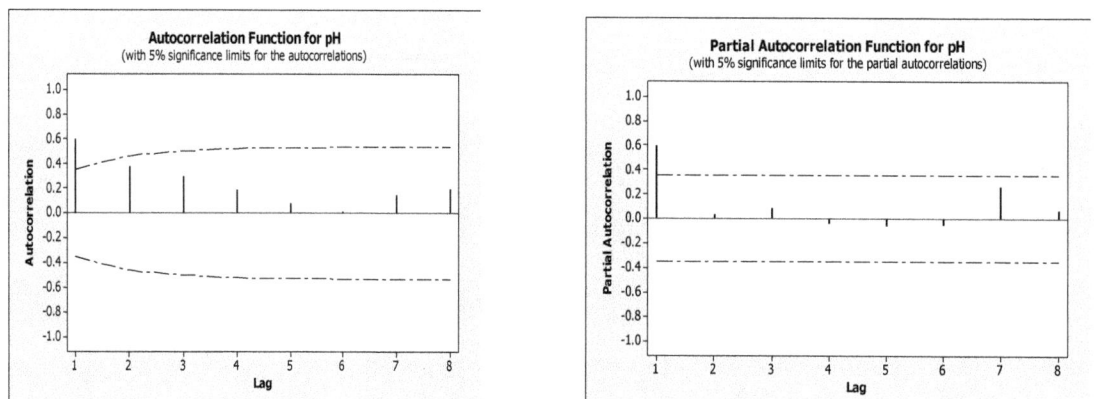

Fig. 9. ACF and PAC of pH time series.

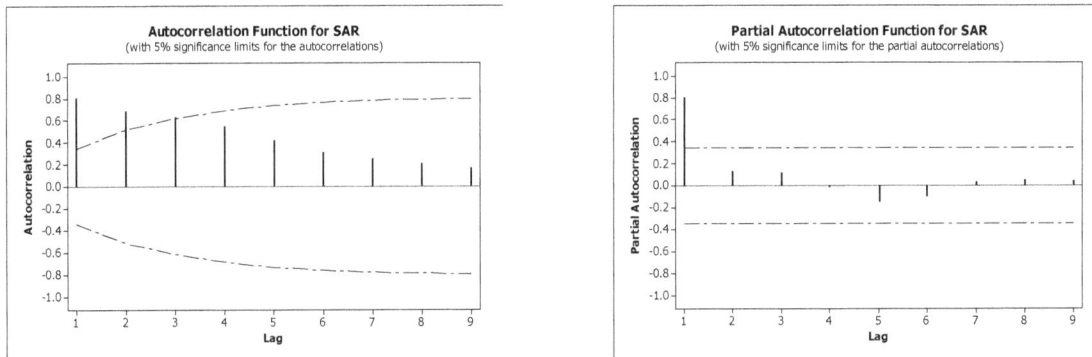

Fig. 10. ACF and PAC of SAR time series.

4. Results
4.1. Results of data generation

Time series of TDS values shows a positive trend which was removed through one difference. The best fitted model of the series based on ACF and PACF, AIC, R^2, RMSE and MAPE is demonstrated in Table 1. Results show that the model is capable of modeling the time series well.

Table 1. The results of TDS generation, order (1, 3).

MODEL	R^2	AIC	RMSE	MAPE %
(1,1,3)	0.73	-45.66	0.07	0.64
(2,1,3)	0.78	-49.13	0.07	0.86
(3,1,3)	0.76	-45.85	0.07	0.63

For the second parameter, as it is clear from Fig. 11, EC time series follows an increasing trend. Table 2 shows the results for generating and choosing the best fit after removing the trend.

Standardized time series of HCO_3^- is presented in Fig. 11. HCO_3^- follows an increasing slope. Modeling was accomplished after trend elimination. Table 3 shows the results of modeling time series for this parameter.

Table 2. The results of EC generation, order (1, 2).

MODEL	R^2	AIC	RMSE	MAPE %
(1,1,3)	0.72	-46.00	0.07	8.11
(2,1,3)	0.81	-54.11	0.06	3.90
(3,1,3)	0.79	-54.55	0.06	1.28

Table 3. The results of HCO_3^- generation, order (1, 2).

MODEL	R^2	AIC	RMSE	MAPE %
(1,1,2)	0.67	-29.22	0.10	4.16
(2,1,1)	0.61	-21.73	0.11	3.19
(1,1,3)	0.75	-35.69	0.08	3.17
(2,1,3)	0.76	-35.50	0.08	3.20
(3,1,3)	0.76	-33.61	0.08	3.21

Also Z time series of SO_4^{2-} is shown in Fig. 11. The results of generation after trend elimination and the best fit are shown in table 4. Results show that ARIMA modeling for this parameter is rather capable of predicting as well as previous parameters.

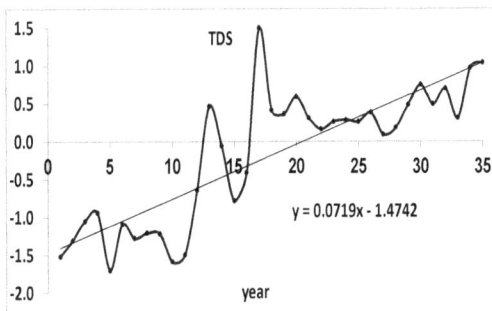

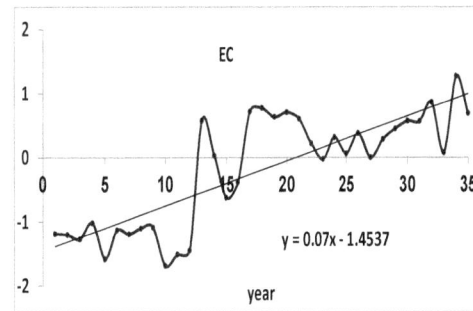

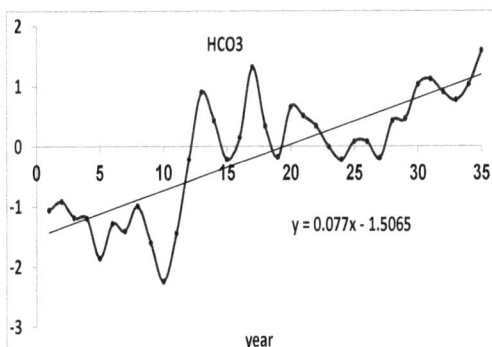

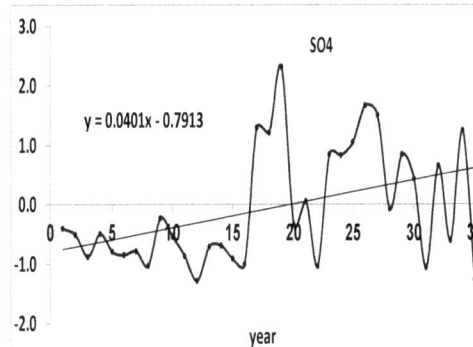

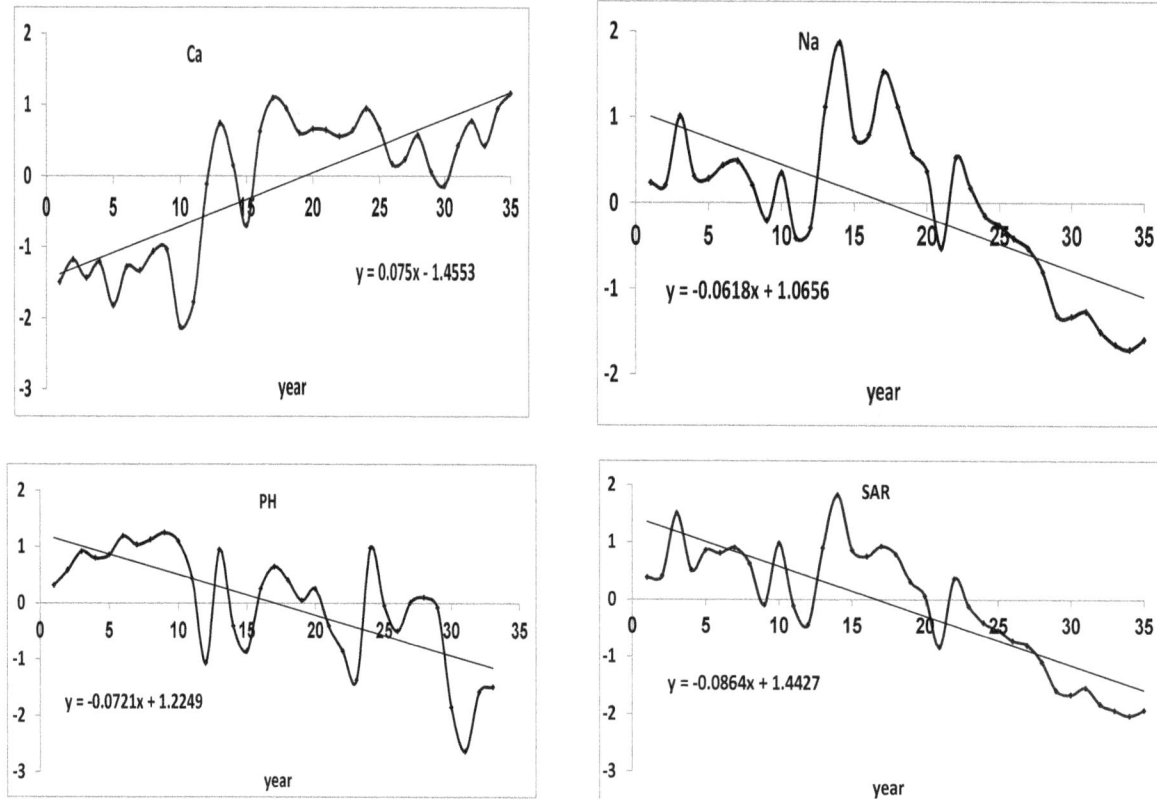

Fig. 11. The standard time series of TDS, EC, HCO_3^-, SO_4^{2-}, Ca^{2+}, Na^+, pH and SAR.

Table 4. The results of SO_4^{2-} generation, order (1, 1).

MODEL	R^2	AIC	RMSE	MAPE %
(1,1,2)	0.19	-1.01	0.15	1.08
(2,1,2)	0.35	-6.89	0.13	1.47
(1,1,3)	0.43	-11.37	0.12	1.11
(2,1,3)	0.39	-6.92	0.13	1.26
(1,1,1)	0.28	-1.95	0.15	2.04
(2,0,1)	0.29	-7.12	0.14	1.48
(2,1,1)	0.28	-0.01	0.15	2.04

Also time series of Z for Ca^{2+} is presented in Fig. 11. The results of generating presented in Table 5 show that the selected model, shown in Table 5, is capable of modeling the series quite well.

Table 5. The results of Ca^{2+} generation, order (1, 1).

MODEL	R^2	AIC	RMSE	MAPE %
(1,0,1)	0.59	-61.39	0.06	1.32
(1,0,2)	0.67	-66.36	0.06	1.33
(1,1,1)	0.60	-61.90	0.06	1.15
(1,1,2)	0.70	-69.94	0.05	1.16
(2,1,1)	0.76	-76.41	0.05	1.23
(2,1,2)	0.76	-74.52	0.05	1.22

Z time series of Na^+ show that series follow a decreasing trend. Modeling was done for Na^+ series after trend elimination. Also the results of modeling are presented in Table 6. The results show that the selected model is capable of modeling the series. Time series of pH values shows a decreasing trend which was removed through one difference. Table 7 shows the results of generating for the series. The results show that the model is quite capable of generating the data. Finally modeling the SAR series was done after trend elimination. Table 8 shows that selected model is capable of modeling the series properly.

4.2. The results of forecasting

Fig. 12 shows the results of forecasting 5 data for the 5 last years. to evaluate the selected models. These results are shown for time

series of TDS, EC, HCO_3, $SO4^{2-}$, Ca^{2+}, Na , pH and SAR, respectively.

Table 6. The results of Na+ generation, order (1, 2).

MODEL	R^2	AIC	RMSE	MAPE %
(1,0,1)	0.69	-42.63	0.09	0.66
(1,1,1)	0.70	-42.62	0.08	0.61
(1,0,2)	0.71	-42.87	0.08	0.73
(2,0,1)	0.70	-41.23	0.09	0.68
(2,0,2)	0.72	-41.55	0.08	0.70
(1,1,2)	0.72	-43.14	0.08	0.64
(2,1,2)	0.74	-42.89	0.08	0.61
(2,1,1)	0.74	-44.75	0.08	0.66
(1,1,3)	0.72	-40.54	0.08	0.66

Table 7. The results of pH generation, order (1, 2).

MODEL	R^2	AIC	RMSE	MAPE %
(1,0,1)	0.40	-12.67	0.13	1.91
(1,0,2)	0.32	-5.99	0.14	1.60
(2,0,2)	0.45	-11.97	0.13	1.26
(1,1,1)	0.53	-17.38	0.12	3.12
(2,1,1)	0.52	-14.94	0.12	3.18

Table 8. The results of SAR generation, order (2, 2).

MODEL	R2	AIC	RMSE	MAPE %
(1,1,1)	0.76	-39.42	0.09	0.92
(1,1,2)	0.77	-38.36	0.09	0.93
(2,1,2)	0.84	-50.05	0.07	0.80
(1,1,3)	0.77	-36.86	0.09	0.94
(2,1,3)	0.78	-36.13	0.08	0.86

Table 9 shows the results of forecasting water quality parameters. AIC, RMSE and VE %, criteria were used at this stage to show the capability of each model to forecast the value of data. Results show that the selected model for each parameter is quite able to estimate the future values of water quality parameters.

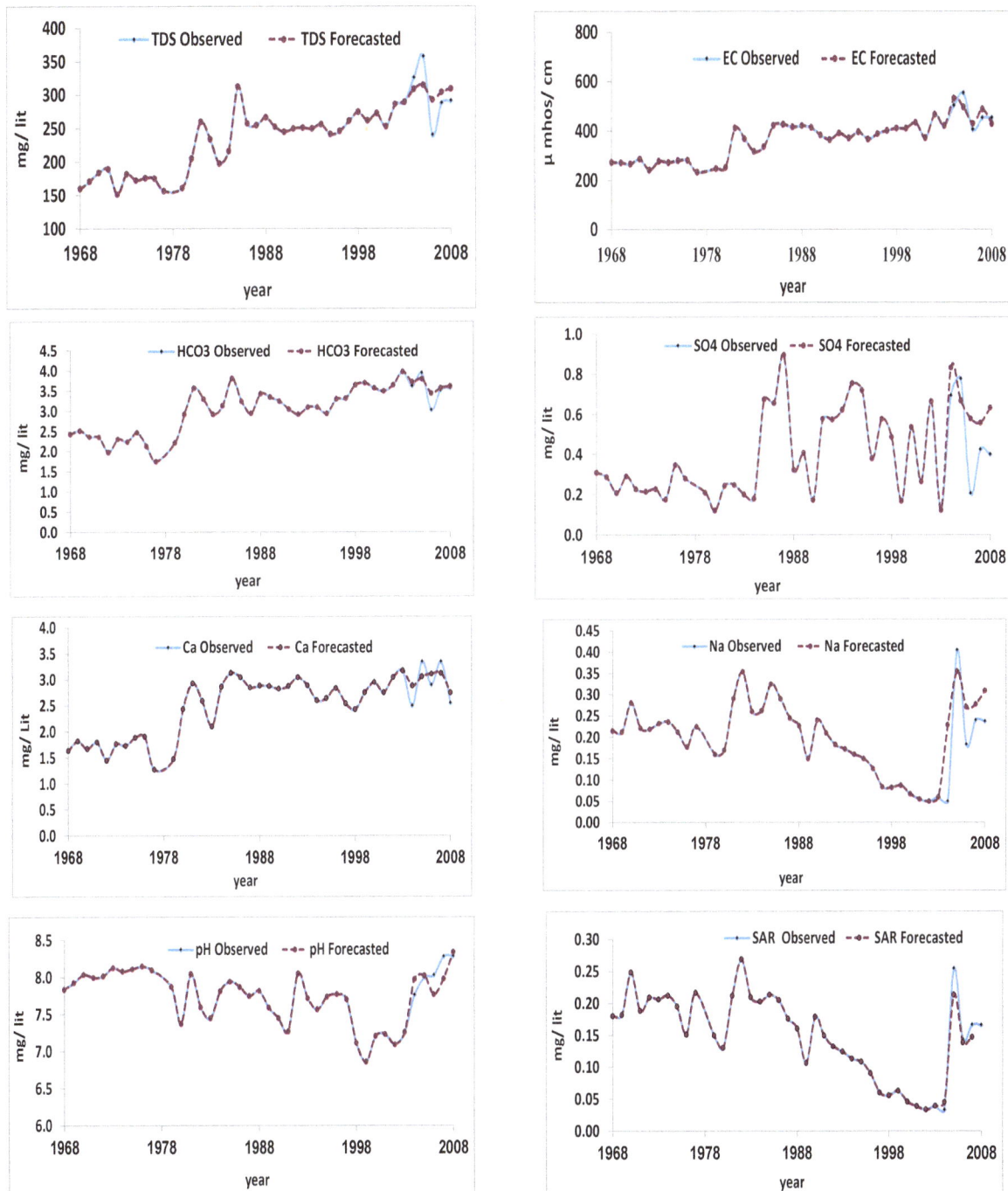

Fig. 12. Forecasted values of 5 last years for TDS, EC, HCO$_3^-$, SO$_4^{2-}$, Ca^{2+}, Na$^+$, pH and SAR.

5. Conclusions

In this study, nine water quality parameters of Hor Rood River were studied at Kakareza station. First the normality of series was examined. All parameters showed that they follow a normal distribution. Then the ACF and PACF of each series were estimated to guess the best model for generating the value of series. Also standard time series of all parameters were plotted. Na$^+$, pH and SAR show decreasing trend in spite of other elements of water quality which show an increasing trend.

Investigation of observed time series proves that EC, Ca^{2+}, SO$_4^{2-}$ and HCO$_3^-$ show a significant increasing trend which is a sign for water quality deterioration in the region. Before data generation using one difference to eliminate the trend, stationary time series were prepared to work on. The results of modeling show that ARIMA modeling process is suitable in generating and forecasting the parameters.

Table 9. Results of forecasting 5 years of parameters.

Parameter	RMSE	VE %	R2
TDS	18.87	0.1	0.87
EC	16.19	0.07	0.49
HCO$_3^-$	0.09	0.04	0.91
SO$_4^{2-}$	0.10	0.61	0.48
Ca^{2+}	0.12	0.09	0.66
Na$^+$	0.04	0.32	0.9
pH	2.08	1.45	0.92
SAR	0.02	0.23	0.7

References

Ahmad S., Khan I.H., Parida B.P., Performance of stochastic approaches for forecasting river water quality, Water Research 35 (2001) 4261–4266.

Antonopoulos V.Z., Papamichail D.M., Mitsiou K.A., Statistical and trend analysis of water quality and quantity data for the Strymon River in Greece, Hydrology and Earth System Sciences 5 (2001) 679- 691.

Box G.E.P., Jenkins G.M., Time Series Analysis, Forecasting and Control. Revised ed. Toronto: Holden-Day (1976).

Chow W.T., Kareliotis S.J., Analysis of stochastic hydrologic systems, Water Resources Research 16 (1970) 1569-1582.

Dalme C., Yalcin A., Flood prediction using time series data mining, Journal of Hydrology 333 (1970) 305-316.

El-Shaarawi A.H., Esterby S.R., Kuntz K.W., A statistical evaluation of trends in the water quality of the Niagara river, Journal of Great Lakes Research 9 (1983) 234- 240.

Faruk D.O., A hybrid neural network and ARIMA model for water quality time series prediction, Engineering Applications of Artificial Intelligence 23 (2010) 586–594.

Gangyan Z., Goel N.K., Bhatt V.K., Stochastic modeling of the sediment load of the upper Yangtze river (Chaina), Hydrological Sciences Journal 47 (2002) 93-105.

Gun C., Vilagines R., Time series analysis on chlorides, nitrates, ammonium and dissolved oxygen concentrations in the Seine, The Science of the Total Environment 208 (1997) 59-69.

Halliday S.J., Wade A.J., Skeffington R.A., Neal C., Reynolds B., Rowland P., Neal M., Norris D., An analysis of long-term trends, seasonality and short-term dynamics in water quality data from Plynlimon, Wales, The Science of the Total Environment 434 (2012) 186–200.

Hanh P.T.M., Analysis of variation and relation of climate, hydrology and water quality in the lower Mekong river, Water Science and Technology 62 (2010) 1587–1594.

Hirsch R.M., Slack J.R., Smith R.A., Techniques of Trend analysis for monthly water quality data. Water Resources Research 18 (1982) 107- 121.

Irvine K.N., Richey J.E., Holtgrieve G.W., Sarkkula J., Sampson M., Spatial and temporal variability of turbidity, dissolved oxygen, conductivity, temperature, and fluorescence in the lower Mekong River–Tonle Sap system identified using continuous monitoring, International Journal of River Basin Management 9 (2011) 151-168.

Irvine K.N., Eberhardt A.J., Multiplicative, seasonal ARIMA models for Lake Erie and Lake Ontario water levels, Water Resources Bulletin 28 (1992) 385–396.

Jalal Kamali N., Forecasting the variations of inflow to Jiroft Dam using Time Series Theories, 6th international seminar on River Engineering, ShahidChamran University, Ahvaz, Iran, 2006.

Jamab Consulting Engineers, Integrated Program of Adaptation to Climate Study, Karkhe Watershed 1 (2005).

Jassby A.D., Reuter J.E., Goldman C.R., Determining long term water quality change in the presence of climate variability, Lake Tahoe (USA), Canadian Journal of Fisheries and Aquatic Sciences 60 (2003) 1452- 1461.

Karamouz M., Araghinejad S.H., Advanced Hydrology. Industrial University of Amir Kabir (Poly Technics), Tehran, Iran, Publication Centre of Amir Kabir University (2005).

Khashei M., Bijari M., An artificial neural network (p,d,q) model for time series forecasting, Expert Systems with Applications 37 (2010) 479–489.

Kim J.-H., Lee J., Cheong T.-J., Kim R.H., Koh D.-C., Ryu J.-S., Chang H.–W., Use of time series for theidentification of tidal effect on groundwater in the coastal area of Kimje, Korea, Journal of Hydrology 300 (2005) 188- 198.

Komornık J., Komornıkova M., Mesiar R., Szokeova D., Szolgay J., Comparison of forecasting performance of nonlinear mod mparison of forecasting performance of nonlinear mod hydrological time series, Physics and Chemistry of the Earth 31 (2006) 1127–1145.

Kurunc A., Yurekli K., Cevik O., Performance of two stochastic approaches for forecasting water quality and stream flow data from Yesilirmak River, Turkey, Environmental Modeling & Software 20 (2005) 1195–1200.

Lehmann A., Rode M., Long-term behavior and cross-correlation water quality analysis of the River Elbe, Germany, Water Research 35 (2001) 2153–2160.

McKerchar A.I., Delleur L.W., Application of seasonal parametric linear stochastic models to monthly flow data, Journal of Water Resource Reservoir 10 (1974) 246-255.

Montanari A., Rosso R., Taqqu M.S., A seasonal fractional ARIMA model applied to the Nile River monthly flows at Aswan. Journal of Water Resource Reservoir 36 (2000) 1249–1259.

Nelson C.R., Applied Time Series Analysis for Managerial Forecasting, San Francisco: Holden-Day,1973.

Padilla A., Pulido-Bosch A., Calvache M.L., Vallejos A., The ARMA models applied to the flow of karstic springs, Journal of Water Resource Reservoir 32 (1996) 917–928.

Panda D.K., Kumar A., Mohanty S., Recent trends in sediment load of the tropical (Peninsular) river basins of India, Global and Planetary Change 75 (2011) 108- 118.

Pankratz A., Forecasting with Univariate Box-Jenkins Models, New York: John Wiley & Sons, (1983).

Papamichail D.M., Georgiou P.E., Seasonal ARIMA inflow models for reservoir sizing, Journal of American Water Resources Association 37(2001) 877-885.

Papamichail D.M., Antonopoulos V.Z., Georgiou P.E., Stochastic models for Strymon river flow and water quality parameters. Proc. of International Conference "Protection and Restoration of Environment V", I (2000) 219-226.

Rao A.R., Kashyap R.L., Mao L.-T., Optimal choice of type and order of river flow time series models, Water Resources Research 18 (1982) 1097–1109.

Robson A.J., Neal C., Water quality trends at an upland site in Wales, UK, (1983- 1993), Hydrological Processes 10 (1996) 183- 203.

Salas J.D., Boes, D.C. and Smith, R.A., Estimation of ARMA models with seasonal parameters. Water Resources Research 18 (1982) 1006–1010.

Salas J.D., Applied Modeling of Hydrologic Time Series, Littleton, CO: Water Resources Publications. 1980.

Sheng H., Chen Y.Q., FARIMA with stable innovations model of Great Salt Lake elevation time series, Signal Processing 91 (2011) 553–561.

Stansfield B., Effects of sampling frequency and laboratory detection limits on the determination of time series water quality trends, New Zeland, Journal of Marine and Freshwater Research 35 (2001).

Thomas H.A., Fiering M.B., Mathematical synthesis of stream flow sequences for the analysis of river basin by simulation, Harward University Press, Cambridge 1962.

Turner B.F., Gardner L.R., Sharp W.E., The hydrology of Lake Bosumtwi, a climate-sensitive lake in Ghana, West Africa, Journal of Hydrology 183 (1996) 243-261.

Vandaele W., Applied Time Series and Box-Jenkins Models. New York: Academic Press, Inc. (1983).

Voudouris K., Georgiou P., Stiakakis E., Monopolis D., Comparative analysis of stochastic models for simulation of discharge and chloride concentration in Almyroskartsic spring in Greece. e-Proceedings of the 14th Annual Conference of the International Association of Mathematical Geosciences, IAMG, Budapest, Hungary (2010) 1-15.

Webb B.W., Clack P.D., Walling D.E., Water- Air Temperature Relationships in a Devon River System and the Role of Flow, Hydrological processes 17 (2003) 3069- 3084.

Weeks W.D., Boughton W.C., Tests of ARMA model forms for rainfall-runoff modeling, Journal Hydrology 91(1987) 29–47.

Yu Y.-S., Zou S., Whittemore D., Non parametric trend analysis of water quality data of rivers in Kansas, Journal of Hydrology 260 (1993) 161-175.

Yurekli K., Kurunc A., Performance of stochastic approaches in generating low streamflow data for drought analysis, Journal of Spatial Hydrology 5 (2005) 20–32.

Zhang G.P., Time series forecasting using a hybrid ARIMA and neural network model, Neurocomputing 50 (2003)159–175.

15

Long-term assessment of water quality and soil degradation risk via hydrochemical indices of Gharasoo River, Iran

Akram Fatemi*

Soil Science Department, Razi University, Kermanshah, Iran.

ARTICLE INFO	ABSTRACT
Keywords: Agriculture water types Soil degradation risk Nutrition disorders Irrigation systems	The suitability of Gharasoo River water for irrigation uses was evaluated in Kermanshah city, Iran. Long-term datasets including major cations, anions and other parameters such as electrical conductivity (EC), total dissolved solids (TDS) were analyzed. Sodium absorption ratio (SAR), magnesium ratio (MR), % sodium (%Na), residual sodium carbonate (RSC), permeability index (PI) and Ca^{2+}/Mg^{2+} ratio were calculated to evaluate the suitability of Gharasoo River water for irrigation purposes. Piper trilinear diagram reveals that the water is the alkaline earth than alkaline type. Based on the SAR values plotted in the U.S. Salinity Laboratory Staff diagram, Gharasoo River water belongs to class medium-salinity hazard and low-sodium hazard (C_2S_1) which indicates that there is no limitation to use water for irrigation. According to FAO method, soil degradation risk was low in the study area and potential plant nutritional disorders will not be expected. Different indices showed the regional sodicity problems: the high risks for %Na, PI, Ca^{2+}/Mg^{2+} and magnesium ratios for soil and clogging of irrigation systems only at one station.

1. Introduction

Rivers play an important role in human development. They are important natural potential sources of irrigation water particularly for agricultural production in semiarid regions where rainfall is not sufficient to uphold crop growth (Sundaray, Nayak et al. 2009, Jafar Ahamed, Loganathan et al. 2013). Agriculture is the single largest user of water throughout the world. The productivity of irrigated agriculture is significantly higher than the productivity of rained agriculture, particularly in arid and semiarid regions. The water quality for irrigation plays a very important role for agricultural production and environment protection (Ağca 2014). Water quality or suitability for agricultural use is evaluated by the potential severity of problems that may be occurred during long-term use (Ayers and Westcott 1994, Ağca 2014). The water quality is depend on the concentration and composition of solutes, which can not only enrich soil with soluble salts, but also cause to precipitate insoluble salts and affect its exchangeable cation composition or even increase sodicity (Keren 2012, Keren and Levy 2012). In order to avoid plant toxicity problems, the presence of potentially toxic elements, nitrate amount, should also be evaluated. An imbalanced N supply to crops or algal development in irrigation reservoirs should be also considered. These factors are all included in the FAO practical guidelines for assessing irrigation water quality (Ayers and Westcott 1985, Peragóna, Delgadob et al. 2015).

In irrigated agriculture especially in arid climatic conditions, irrigation water with poor quality is of concern due to constant threat of the hazard of salt water. Crop yield, soil physical conditions, fertilizers needs, irrigation system performance and longevity as well as how the water can be applied are affected by irrigation water quality (Ayers and Westcot 1994). Natural and anthropogenic pollution processes cause rapidly decline water quality globally and mostly in developing countries (Zhou, Wang et al. 2013, Ağca 2014). Rezaei and Sayadi (2014) reported the point and non-point source pollutions for Gharasoo River water. Sayadi, Rezaei et al. (2015) with reference to multivariate statistical analyses revealed that there are some sources which release the pollutants into the Gharasoo River. Despite these researches indicated water quality of Gharasoo River for drink purpose, there is no detailed information about the water quality assessment for agricultural purposes based on physico-chemical parameters. Since water quality for irrigation is not a minor problem in Kermanshah city, virtually no local research has been conducted. This work was undertaken in order to evaluate Gharasoo River suitability for irrigation. In addition, attempts were made to identify potential degradation risk of soil irrigated by Gharasoo River via some calculated hydrochemical parameters and graphical representations.

2. Materials and methods
2.1. Study area

The Gharasoo River in length 20.7 km runs through the Kermanshah city after joining two its branches of Merek and Raz Avar Rivers. The study area lies between latitudes 47° 36´- 47° 47´ N and longitudes 34°00´- 34°91´ E with the height of 1322 meters above sea level. The mean annual temperature and rainfall are 13.7°C and 424.4 mm, respectively.

2.2. Datasets

Datasets of this study consist of four sampling stations data which included 10 water quality characteristics monitored monthly over a period over than 15 years (Anonymous 2009). Different durations of available data series rose from various establishment times of the stations. The monitoring stations geographical positions are presented in Table 1. The 10 analyzed parameters according to standard methods (Eaton, Clesceri et al. 1994) are presented in Table 2.

In this study focused on hydrochemical properties of water and the subsequent indices can be appropriate explaining the following undesirable effects which categorized by (Peragóna, Delgadob et al. 2015) as below:

Corresponding author Email: a.fatemi@razi.ac.ir

(a) Soil degradation through accumulation of soluble salts and sodicity, and infiltration risks. Salinity problem can be evaluated by electrical conductivity (EC) (U.S. Salinity Laboratory Staff 1954, Ayers and Westcott 1985). While sodicity and attributed infiltration risks can be estimated from EC, adjusted sodium adsorption ratio (SAR_{adj}) and residual sodium carbonate (RSC) (Suárez 1981, Ayers and Westcott 1985), percent sodium (%Na) and permeability index (PI). According

to piper method (Fig.1), it was revealed that in one station water type was Mg^{2+} dominant. Although the potential effect of Na^+ may be slightly in Mg-dominated water ($Ca^{2+}/Mg^{2+}< 1$), but sufficiently enhances a higher than normal soil exchangeable sodium percentage (ESP) at a given SAR (Rahman and Rowell 1979, Ayers and Westcott 1985, Peragóna, Delgadob et al. 2015). In addition to Ca^{2+}/Mg^{2+} ratio, the magnesium ratio (MR) can also show the Mg hazard.

Table 1. Contributed stations detailed information.

Station	Latitude	Longitude	Height (m)	Data series of different durations
Station 1 (S₁)	47° 44′ N	34° 30′ E	1338	1992-2009
Station 2 (S₂)	47° 47′ N	34° 33′ E	1290	1972-2009
Station 3 (S₃)	47° 00′ N	34° 29′ E	1300	1978-2009
Station 4 (S₄)	47° 15′ N	34° 14′ E	1380	1972-2009

Table 2. The average of chemical composition and irrigational quality parameters of water of stations.

Parameter	EC	pH	TDS	HCO_3^-	Cl^-	SO_4^{2-}	Ca^{2+}	Mg^{2+}	Na^+	SAR	SAR_{adj}	pH_c	RSC	%Na	PI	Ca/Mg	MR	French degrees	Langelier index (Is)
Stations	dS m⁻¹	-	mgL⁻¹			meqL⁻¹				(meqL⁻¹)^½		-	meqL⁻¹			%		°fH	-
S₁	0.54	8.0	346.8	0.46	0.7	1.11	2.19	3.65	1.69	0.99	1.29	8.1	-5.38	28.9	86.3	0.6	62.5	2.1	-0.1
S₂	0.46	7.8	299.0	0.16	0.3	0.68	2.96	1.29	0.36	0.25	0.17	8.7	-4.09	8.5	35.4	2.3	30.3	1.3	-0.9
S₃	0.35	7.9	225.7	0.27	0.2	0.23	2.33	0.96	0.35	0.27	0.25	8.5	-3.02	10.6	45.6	2.4	29.2	1.0	-0.6
S₄	0.51	7.8	325.4	0.23	0.5	0.63	2.82	1.51	0.76	0.52	0.41	8.6	-4.10	17.6	54.9	1.9	34.9	1.3	-0.8

(b) Nutritional disorders. Elements such as B^-, Na^+ and Cl^- can be toxic to plants. Also, bicarbonate, HCO_3^- can decrease micronutrients uptake (specially Fe^{2+} and Zn^{2+}) (Ayers and Westcott 1985). Zn^{2+} uptake and translocation to the shoot are inhibited by high concentrations of HCO_3^- (Marschner 1995). Also, a Ca^{2+}/Mg^{2+} ratio < 1 can decrease Ca^{2+} uptake via an antagonistic effect (Ayers and Westcott 1985, Peragóna, Delgadob et al. 2015).

(c) Clogging of irrigation systems. French degrees (°fH), Langelier index (Is), RSC and Ca^{2+}/Mg^{2+} ratio will be considered as useful indices to assess the participation of Ca^{2+} and Mg^{2+} compounds and carbonate precipitation (Peragóna, Delgadob et al. 2015).

To assess the suitability of water for irrigation uses, the following irrigational quality parameters were computed by the ensuing equations:

$$SAR = \frac{Na^+}{\sqrt{\frac{(Ca^{2+} + Mg^{2+})}{2}}} \tag{1}$$

$$SAR_{adj} = SAR \times [1 + (8.4 - pH_c)] \tag{2}$$

$$pH_c = (pK_2' - pK_s') + p(Ca^{2+} + Mg^{2+}) + p(Alk)$$

$$(pK_2' - pK_s') = f(Ca^{2+} + Mg^{2+} + Na^+);$$

$$p(Alk) = f(CO_3^{2-} + HCO_3^-) \tag{3}$$

$$RSC = (HCO_3^- + CO_3^{2-}) - (Ca^{2+} + Mg^{2+}) \tag{4}$$

$$\% Na = \frac{(Na^+ + K^+)}{\sqrt{(Ca^{2+} + Mg^{2+} + Na^+ + K^+)}} \times 100 \tag{5}$$

$$PI = \frac{(Na^+ + \sqrt{HCO_3^-})}{\sqrt{(Ca^{2+} + Mg^{2+} + Na^+)}} \times 100 \tag{6}$$

$$MR = \frac{Mg^{2+}}{Ca^{2+} + Mg^{2+}} \times 100 \tag{7}$$

$$°fH = (2.5\,Ca^{2+} + 4.12\,Mg^{2+})/10 \tag{8}$$

$$Is = pH - pH_c \tag{9}$$

All the ionic concentrations in the above equation are expressed in meq L⁻¹, and % Na and PI in %.

3. Results and discussion

The results of physico-chemical parameters and calculated irrigation water quality parameters are given in Table 2. The pH of water in the study area ranges from 7.8-8.0. EC in all stations is lower than permissible limit of FAO classification (Table 3). Total dissolved solid (TDS) values varied from 225.7 to 346.8 mg L⁻¹.

Measured cations and anions, include Ca^{2+}, Mg^{2+}, Na^+, K^+, HCO_3^-, Cl^- and SO_4^{2-} were plotted in piper trilinearpiper by AquaChem (2011) shown in Fig. 1. The plot showed that the dominant water types for station 1 was Mg^{2+}-Ca^{2+}-Na^+, while stations 2 and 3 were identified as Ca^{2+}-Mg^{2+} and station 4 classified as Ca^{2+}-Mg^{2+}-Na^+ type. This result revealed that water was mostly alkaline earth ($Ca^{2+}+Mg^{2+}$) than alkaline (Na^++K^+).

3.1. Salinity problems

Irrigation water has relatively small amounts salts. The origination of water salts is from dissolution or the rocks and soil weathering, including dissolution of lime, gypsum, and other soil minerals. These salts are carried with the water to agricultural lands. Salts remain behind in the soil as irrigation water evaporates or is used by the crops (Ayers and Westcott 1994). The appropriate range of EC for agricultural uses is defined as 0.25 to 3 dS m⁻¹. If EC is greater than 3 dS m^{-1}, water intake by the plant significantly decreases and then crop productivity is affected very much (Ayers and Westcott 1985). In the study area, EC varied from 0.46 to 0.54 dS m^{-1}, is present in the medium-salinity hazard class (C$_2$), indicated the moderate-to-low crop productivity (Jafar Ahamed, Loganathan et al. 2013). Therefore, Gharasoo River water can be used without any special practices for salinity control.

3.2. Sodicity problems
3.2.1. SAR and USSL diagram

The excessive Na$^+$ content in water reduces the permeability of soils, consequently, the available water for the plant. Na$^+$ replacing adsorbed Ca^{2+} and Mg^{2+} causes damage to the soil structure resulting in compact and impervious soil (Arveti, Sarma et al. 2011). The excess of Na$^+$ with Ca^{2+} and Mg^{2+} is evaluated by SAR. According to the SAR values plotted in the USSL diagram (Fig. 2), all stations fall under C$_2$S$_1$ (moderate-salinity hazard and low- Na$^+$ hazard) class. Al-Bassam and Al-Rumikhani (2003) reported that the relatively low to medium sodicity level of water is due to chemical amendments which are used frequently. The previous reports about Gharasoo River pollution sources (Rezaei and Sayadi 2014, Sayadi, Rezaei et al. 2014) showed industrial and domestic waste, agricultural runoff, as anthropogenic activities and hydro-geochemical sources.

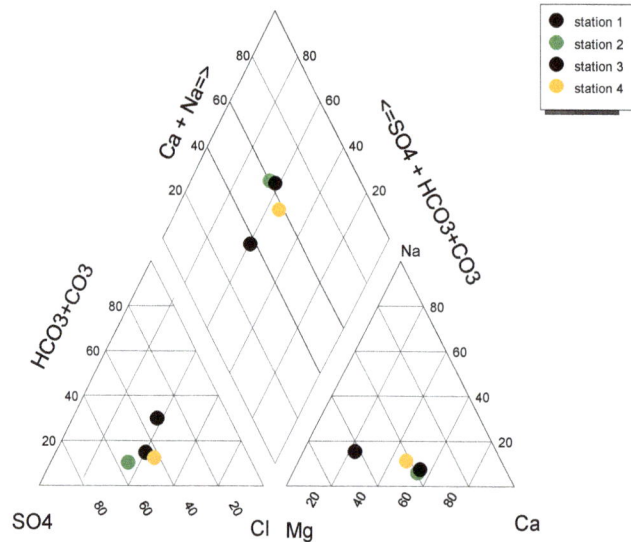

Fig. 1. Piper diagrams for water type classification by (Back and Hanshaw 1965).

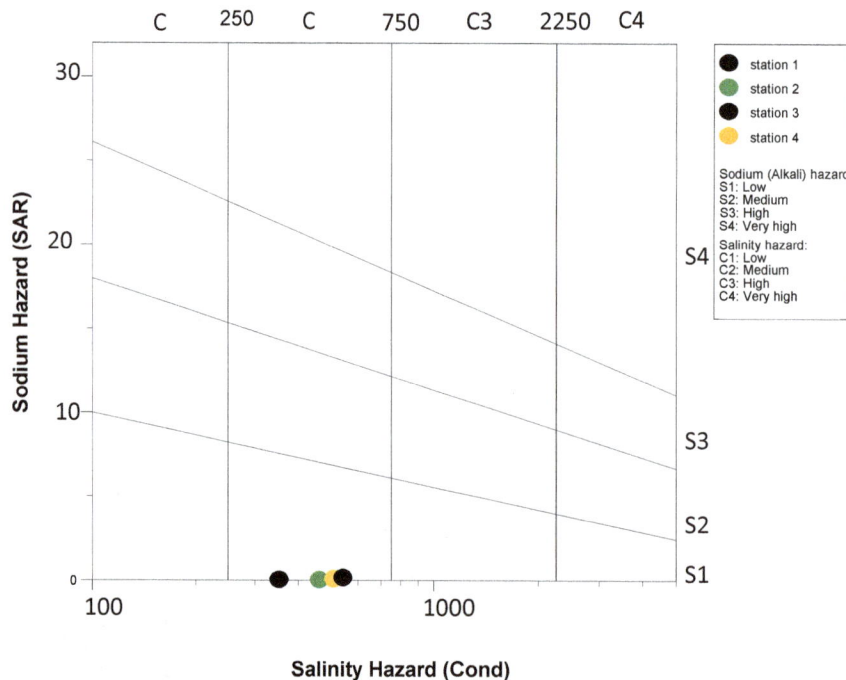

Fig. 2. Plotting SAR against Electrical Conductivity (U.S. Salinity Laboratory Staff 1954).

3.2.2. Residual sodium carbonate (RSC)

The RSC values were calculated to determine the hazardous effect of CO_3^{2-} and HCO_3^- on the water quality for agricultural purpose. Water quality may be diminished when total CO_3^{2-} and HCO_3^- levels exceed the total amount of Ca^{2+} and Mg^{2+}. At high magnitude of excess (residual) CO_3^{2-} concentration, CO_3^{2-} combine with Ca^{2+} and Mg^{2+} and form a solid material. The suitability of water for irrigation is affected by the relative abundance of Na^+ with respect to alkaline earths and the quantity of CO_3^{2-} and HCO_3^- in excess of alkaline earths. This excess is denoted by 'residual sodium carbonate' (RSC) and is determined as suggested by (Richards 1954). According to Wilcox, Blair et al. (1953), an RSC value <1.25 meq L^{-1} is probably safe for irrigation. If it is >2.5 meq L^{-1}, it is not suitable for irrigation (Table 3). Hence, in the study area, all the values fall in the safe zone.

3.2.3. Percent sodium (%Na⁺)

For irrigation purpose, the percentage of Na^+ is highly important. % Na^+ ranges from 8.5 to 28.9 % which is attributed to stations 2 and 1, respectively. These results show that all stations fall in low-risk gradation (Table 3). Based on average %Na^+, water is classified as excellent (< 20 %), good (20-40 %), permissible (40-60 %), doubtful (60-80%) and unsuitable (> 80 %) (Wilcox (1955). According to this classification, only station 1 fall in the good class and the rest stations fall in the excellent category. Thus, Gharasoo River water is suitable for irrigation purposes (Table 3).

Table 3. Classification of the Gharasoo River water quality according to the FAO method and other water quality indices.

Potential irrigation problems	Units	Risk-gradation		
		Low	Medium	High
Soil degradation				
Salinity (effects on crop water availability)				
EC	dS m⁻¹	<0.7	0.7-3.0	>3.0
Infiltration (effects on infiltration rate of water into the soil)				
SAR and EC=				
0-3	dS m⁻¹	>0.7	0.7–0.2	<0.2
3–6	dS m⁻¹	>1.2	1.2–0.3	<0.3
6–12	dS m⁻¹	>1.9	1.9–0.5	<0.5
12-20	dS m⁻¹	>2.9	2.9–1.3	<1.3
20-40	dS m⁻¹	>5.0	5.0–2.9	<2.9
RSC	meq L⁻¹	<1.25	1.25-2.5	>2.5
% Na	%	<20	4-60	>80
PI	%	<25	25-75	>75
MR	%			>50
Ca²⁺/Mg²⁺ ratio	-	>1		<1

3.2.4. Permeability index (PI)

Continuous usage of water for irrigation will affect the soil permeability quality. Na^+, Ca^{2+}, Mg^{2+}, and HCO_3^- contents in soil influence PI. Doneen (1964) results showed the suitability of water for irrigation considering the permeability index (PI). According to PI values as classified by Ragunath (1982), the Gharasoo River in station 1 fall under class III (>75 %), in rest stations falls under class II (25–75).

3.2.5. Ca²⁺/Mg²⁺ ratio

The Ca^{2+}/Mg^{2+} ratio will be also considered in the evaluation of sodicity problem. Since at similar °fH or RSC values, high Ca^{2+}/Mg^{2+} ratios can increase precipitation of Ca^{2+} phosphates and carbonates, which are less soluble than their Mg^{2+} counterparts (Peragóna, Delgadob et al. 2015). The Ca^{2+}/Mg^{2+} ratio of Gharasoo River water in all stations showed no special problems (Table 2) with the exception of station 1 (Table 3).

Table 4. Different risks related to the irrigation water quality for Gharasoo River water according to the FAO method and water quality indices.

Potential irrigation problems	Units	Risk-gradation		
		Low	Medium	High
Nutritional disorder				
Sodium (Na⁺)				
Surface Irrigation	meq L⁻¹	<3.0	3.0–8.7	>8.7
Sprinkler Irrigation	meq L⁻¹	<3.0	>3.0	
Chloride (Cl⁻)				
Surface Irrigation	meq L⁻¹	<4.0	4.0–10.0	>10.0
Sprinkler Irrigation	meq L⁻¹	<2.9	>2.9	
Bicarbonate (HCO₃⁻)				
Overhead Sprinkling	meq L⁻¹	<1.5	1.5–8.5	>8.5
Ca²⁺/Mg²⁺ ratio		>1	-	<1
Clogging irrigation systems				
French degrees	° fH	<1. 7	1.7–12	≥ 12
RSC	meq L⁻¹	<1.25	1.25–2. 5	> 2.5
Langelier index (pH-pHc)		<0		>0
Ca/Mg ratio		>1	-	<1

3.2.6. Magnesium ratio (MR)

Generally, Ca^{2+} and Mg^{2+} maintain a state of equilibrium in waters. In equilibrium, more Mg^{2+} in the water will adversely affect crop yields (Sundaray, Nayak et al. 2009). MR>50 is considered as unsuitable for irrigation purposes. In all stations, MR ranges from 29.18 to 62.50 which are lower than the permissible limit with the exception of station 1. High MR may be due to the passage of surface water through limestone formation in the study area.

3.3. Nutritional disorder

In addition to assessment of irrigation water quality on soil properties, potential nutritional disorders derived from Cl^-, HCO_3^- and Na^+ concentration or Ca^{2+}/Mg^{2+} ratio of Gharasoo River water were also evaluated. Nutritional disorders will not be expected for all stations (Table 4) with the exception of station 1 due to the high Ca^{2+}/Mg^{2+} ratio.

3.4. Potential irrigation problems

The surface water-irrigated lands of study area were under low-risk of irrigation problems by considering of Langelier index (Is). Although, negative Langelier index in all stations indicated a low-risk of clogging irrigation systems through irrigation, other properties such as °fH, RSC, and Ca^{2+}/Mg^{2+} ratio exhibited the moderate risk rating for °fH and high risk for Ca^{2+}/Mg^{2+} ratio in station 1 (Table 4). Therefore, a moderate risk of precipitation of Ca^{2+} and Mg^{2+} compounds will be expected (Peragóna, Delgadob et al. 2015). The precipitation of Ca^{2+}(and/or Mg^{2+}) with HCO_3^-, will result in a relative increase of Na^+ in solution accompanied by an increase in pH (due to the hydrolysis process where OH^- ions dominate) (Al-Bassam and Al-Rumikhani 2003).

To sum up, Gharasoo River water quality evaluation according to USSL and FAO methods is excellent for irrigation purposes. However, more indices showed the high risks for %Na, PI, Ca^{2+}/Mg^{2+} and MR in station 1. Therefore, based on the information to diminish the sodicity problem for land irrigated with water from station 1 specific practices are required. Leaching requirements (LR) should be considered to avoid deleterious salt accumulation with application of water amendments (e.g. gypsum, Ca^{2+}-containing fertilizers) and manure application instead of fertilizer chemicals to lower the risk of infiltration.

4. Conclusion

Different methods and indices were evaluated for the assessment of the Gharasoo River quality for irrigation usage purposes. While USSL and FAO methods classified water for all stations as C_2S_1 and unrestricted. Water quality indices introduced more precise definition to categorize water quality in regional scales. Water quality indices indicated that water in the station 1 has sodicity problems. Soil degradation risk was low in the study area and potential nutritional plant disorders arising from the use of irrigation will not be expected. To avoid soil degradation and plant disorders in this place by continuous irrigation, the use of water amendment and manure application may be suggested.

References

Ağca N., Spatial variability of groundwater quality and its suitability for drinking and irrigation in the Amik Plain (South Turkey), Environmental Earth Sciences 72 (4115-4130) (2014) 1-16.

Al-Bassam A.M., Al-Rumikhani Y. A., Integrated hydrochemical method of water quality assessment for irrigation in arid areas: application to the Jilh aquifer, Saudi Arabia, Journal of African Earth Sciences 36 (2003) 345-356.

Anonymous, Comprehensive studies for development of agriculture in Kermanshah province, Natural landscape and exploit of productive factors. Agricultural Organization of Kermanshah Province (2009) 138 (in Persian).

AquaChem A., professional application for water quality data analysis, plotting, reporting, and modeling (2011).

Arveti N., Sarma M.R.S., Aitkenhead-Peterson J.A., Sunil K Fluoride incidence in groundwater: a case study from Talupula, Andhra Pradesh, India, Environmental Monitoring and Assesment 172 (2011) 427-443.

Ayers R.S., Westcott D.W., Water quality for agriculture, FAO Irrigation and Drainage Paper 29 Rev. 1. Food and Agricultural Organization Rome (1985).

Back W., Hanshaw B.B., Chemical geohydrology, Advances in Hydroscience 2 (1965) 49-109.

Doneen L.D., Notes on water quality in agriculture, Davis, CA: Water Science and Engineering, University of California (1964).

Eaton A.D., Clescerim L.S., Greenberg A.E., Standard methods for the examination of water and wastewater, 18th ed. Washington, DC: American Public Health Association, American Water Works Association, Water Environment Federation (1994).

Jafar Ahamed A., Loganathan K., Ananthakrishnan S., Manikandan K., Assessment of groundwater quality for irrigation use in Alathur Block, Perambalur District, Tamilnadu, South India. Applied Water Science 3 (2013) 763 -771.

Keren R., Levy G. J., Saline and boron affected soils. In: Huang PM, Li, Y., Sumner, M.E. (ed) Handbood of Soil Sciences, Resource Management and Environmental Impacts, 2nd ed. CRC Press, Boca Ratón (2012).

Marschner H., Mineral nutrition of higher plant. 2nd edition. Academic Press, New York (1995).

Pandian K., Sankar K., Hydrogeochemistry and groundwater quality in the Vaippar River basin, Tamilnadu, Geological Society of India 69 (2007) 970-982.

Peragóna J.M., Delgadob A., J., Pérez-Latorre F. A., GIS-based quality assessment model for olive tree irrigation water in southern Spain, Agricultural Water Management 148 (2015) 232-240.

Ragunath H.M., Groundwater. Wiley Eastern Limited, New Delhi (1982).

Rahman W.A., Rowell D.L., Influence of magnesium in saline and sodic soils-specific effect or a problem of cation-exchange, Journal of Soil Science 30 (1979) 535-546.

Rezaei A., Sayadi M.H., Long-term evolution of the composition of surface water from the River Gharasoo, Iran: a case study using multivariate statistical techniques, Environmental Geochemistry and Health 37.2 (2015) 251-61.

Richards L. A., Diagnosis and improvement of saline and alkaline soils. In: Agricultural handbook 60 US department of Agriculture, Washington DC (1954).

Sayadi M., Rezaei A., Rezaei M., Nourozi K., Multivariate statistical analysis of surface water chemistry: A case study of Gharasoo River, Iran, In: Proceedings of the International Academy of Ecology and Environmental Sciences 3 (2014) 114-122.

Suárez D.L.,   Relation between PHC and sodium adsorption ratio (SAR) and an alternative method of estimating SAR of soil or drainage waters, Soil Science Society of America Journal 45 (1981) 469-475.

Sundaray S.K., Nayak B.B., Bhatta D., Environmental studies on river water quality with reference to suitability for agricultural purposes: Mahanadi river estuarine system, India–a case study, Environmental monitoring and assessment 155 (2009) 227-243.

U.S.Salinity Laboratory Staff Diagnosis and Improvement of Saline and Alkali Soils, U.S. Department of Agriculture Handbook 60, Washington (1954).

Wilcox C.V., Blair G.Y., Bower CA Effect of bicarbonate on suitability of water for irrigation, Soil Science Society of America Journal 77 (1953) 259-266.

Wilcox L., Classification and use of irrigation waters, US Department of Agriculture, Washington DC (1955).

Zhou Y., Wang Y., Li Y., Zwahlen F., Boillat J., Hydrogeochemical characteristics of central Jianghan Plain, China, Environmental Earth Sciences 68 (2013) 765-778.

16

Developing turbulent flows in rectangular channels: A parametric study

Hossein Bonakdari[1,*], Gislain Lipeme-Kouyi[2], Girdhari Lal Asawa[3]

[1]*Department of Civil Engineering, Razi University, Kermanshah, Iran.*
[2]*Department of Civil and Environmental Engineering, University of Lyon, Villeurbanne, France.*
[3]*Department of Civil Engineering, GLA University, Mathura, India.*

ARTICLE INFO

Keywords:
Fully developed flow
Establishment length
Open channel flow
Numerical modelling
Velocity field

ABSTRACT

The developing turbulent flow in an open channel is a complex three-dimensional flow influenced by the secondary currents and free surface effects and is, therefore, not amenable to analytical solution. This paper aims to study the impact of three key hydraulic parameters (relative roughness, the Froude number and the Reynolds number) on the establishment length using computational fluid dynamic (CFD) analysis. CFD analysis is based on the use of the ANSYS-CFX commercial code. The CFD strategy of modelling is validated against experimental velocity distribution in a cross-section and a good agreement is achieved. A dimensionless length is suggested for predicting the length of the developing flow zone for rectangular open channel. A linear relationship has also been developed for assessing the establishment length.

Nomenclature

g	gravitational acceleration	(m/s²)
Le	establishment length	(m)
H	flow depth	(m)
P	pressure	(Pa)
$u*$	shear velocity	(m/s)
U	flow velocity	(m/s)
Umax	maximum velocity	(m/s)
y	width	(m)
z	height	(m)
Z	vertical elevation	(m)
ζ	channel roughness	(m)
δ	boundary layer thickness	(m)
ρ	density of fluid	(kg/m³)
ν	kinematic viscosity	(m²/s)
$\overline{u_i u_j}$	Reynolds stress tensor	(Pa)
F	Froude number	
R	Reynolds number	
lec	non dimensional index	

$$z+= \frac{u*z}{\nu}$$

1. Introduction

Turbulent flows in ducts and open channels are often encountered in engineering. The most basic requirement for the experimental and numerical study of fully developed three-dimensional open channel flows is the knowledge of the channel length necessary for its establishment. Indeed, in open channels, the velocity gradients in the entrance and near the channel bed are high due to the growing boundary layer (see Fig. 1).
The boundary layer thickness (δ) increases with distance from the entrance of the channel. After a certain distance (establishment length, Le), the boundary layer reaches the free surface where after the velocity profile remains invariant. Information about this establishment length is required for any experimental study and numerical modelling of free surface flows. Beyond this distance, the

flow is fully developed. The turbulent structure of open channel flows in the developed zone can be divided into two sub-regions, (Nezu and Nakagawa 1993; Cebeci 2004): inner region (δ_i): z/h<0.2 and outer region (δ_o): 0.2≤z/h≤1.0.

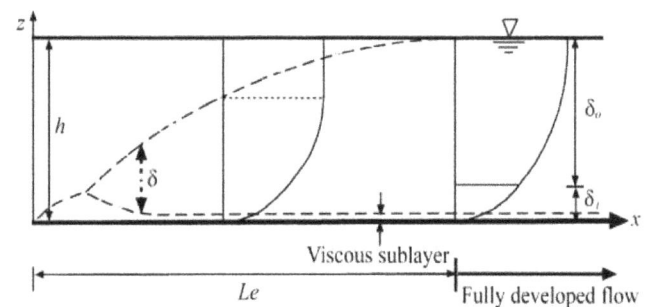

Fig. 1. Boundary layer and fully developed open channel flow.

Most experimental facilities for open channel flows are cited as being sufficiently long, yet no formal definition or documentary evidence of the fully developed condition is widely accepted. Determination of the establishment length (Le) under different hydraulic and geometric conditions is of vital importance in open channels as meaningful flow studies must be carried out in the developed flow zone. The establishment length or development length in an open channel has been studied in many ways over the years. While considerable amount of information is available regarding the development of flow in a circular pipe, there is limited information available in this respect for open channels to enable one to decide the length required for the establishment of fully developed turbulent flow in channels.

Nikuradse (1933) compared the mean velocity profiles at successive stream-wise lengths and concluded that the fully developed zone in a pipe was achieved in his experiments at a distance of fifty times the diameter from the entrance. Wang and Tullis (1974) reported similar results for the developing turbulent flow in a pipe. Since the hydraulic radius of a circular pipe is equal to one fourth of its diameter, the length required for establishment of the fully developed flow in an open channel can be taken as 200 times the hydraulic radius (or the depth of flow in a wide channel) using the above criterion. This length appears to be rather large in the context of lengths of the existing flumes in different laboratories.

Flow in a rectangular channel is characterized by the presence of corners and the free surface. Thus, the flow structure in any developing turbulent flow in channels is different from that in circular pipes. The secondary currents are induced mainly due to the inequality of the normal turbulent stresses at different locations in the cross-sectional plane (Perkins 1970). These secondary currents affect the primary mean flow field and cause the maximum velocity to occur below the free surface (dip phenomenon). It is important to investigate their influence in the developing flow region of a channel. The developing flow is a complex three-dimensional flow influenced by the secondary currents and free surface effects. CFD modelling can, probably, be used to analyze data in the developing flow region of a rectangular open channel for different flow conditions and wall roughnesses.

In the present study, the experimental data obtained by Tominaga et al. (1989) were processed to validate CFD modelling strategy and, then, the validated numerical approach was used to study the length of flow establishment in open channels as a function of wall roughness (the same roughness value is considered for both the side walls and channel bed) and relevant flow parameters.

2. Establishment length in open channel

The studies (either numerical or experimental) on the flow should be made only in the fully developed zone. Indeed, upstream of this zone, the parameters in the vicinity of the entrance are under the influence of a boundary condition that is rather uncertain. The choice of a short channel length induces the risk of the test section not being in the developed zone. However, the choice of a very large length increases the experimental cost or enhances the computing time in numerical studies. Several studies have been carried out in the past on the prediction of the establishment length. From these studies, the establishment length Le in an open channel is seen to vary from 50h to about 150h. Here, h is the depth of flow in the channel. Table 1 recapitulates the results obtained from the previous studies.

Table 1. Summary of selected studies for prediction of the establishment length value.

Reference	Type of channel	Le
Grass (1971)	Smooth & rough	130h
Dean (1978)	Smooth	55h
Nezu and Rodi (1985)	Smooth	60h-90h
Cordaso et al. (1989)	Smooth	69h-108h
Graf (1991)	Rough	140h
Ranga Raju et al. (2000)	Smooth & rough	50h-100h
Zanoun et al. (2003)	Smooth	115h
Lien et al. (2004)	Smooth	130h

These previous investigations, however, do not allow quantitative and precise evaluation of the establishment length. Indeed, in these studies, the main parameter studied was the hydraulic radius (or the flow depth). There are some other parameters such as velocity, channel width, and channel roughness which too affect the establishment length. However, these were not considered in the above-mentioned studies. Kirkgoz and Ardichoglu (1997) have, from their experimental data in a smooth channel, found that the dimensionless length of the developing flow region (Le/h) is related to the Reynolds number (R) and Froude number (F) as follows (R/F<500000):

$$\frac{Le}{h} = 76 - 0.0001\frac{R}{F} \qquad (1)$$

The measured velocity profiles of 12 tests of their investigation showed that the length of the boundary layer development varies between 50h to 70h. There is no clearly defined criterion for flow establishment length in channels. One could possibly define the length of establishment, Le, as the distance beyond which: (1) a characteristic boundary layer thickness like the displacement or momentum thickness becomes constant, or (2) the mean velocity profile along the center line of the channel remains the same, or (3) the mean velocity profile along the whole of the cross section remains the same. There is evidence to suggest that the lengths obtained using these different criteria are not the same. In order to take into account, the wall effect on the velocity field in the cross section, this study is based on criteria (3) to determine the establishment length.

3. Methodology
3.1. CFD modelling strategy

The present numerical study makes use of the Ansys - CFX software package for solving three-dimensional fundamental flow equations and the package enables one to calculate the velocity fields in any cross section.

3.1.1. Governing fluid flow equations

Two key equations for the fluid motion in an open channel are: (i) the law of conservation of mass for an incompressible fluid in Eulerian form,

$$\frac{\partial U_i}{\partial x_i} = 0 \qquad (2)$$

and (ii) the Reynolds' time-averaged Navier-Stokes equations for an incompressible turbulent fluid flow

$$\frac{\partial U_i}{\partial t} + U_j\frac{\partial U_i}{\partial x_j} = -\frac{1}{g}\frac{\partial Z}{\partial x_i} - \frac{1}{\rho}\frac{\partial p}{\partial x_i} + \frac{\partial}{\partial x_j}\left(\nu\frac{\partial U_i}{\partial x_j} - \overline{u_i u_j}\right) \qquad (3)$$

where x_i's, represent the coordinate axes, U_i's are the mean velocities in the x (stream-wise), y (lateral) and z (vertical) directions, Z is the vertical elevation, p is the pressure, ρ is the fluid density, and $\overline{u_i u_j}$ are the components of the Reynolds stress tensor.

3.1.2. Turbulence closure

Since most of the practical engineering flows are large Reynolds number flows, the simulation of turbulence is of great importance in order to obtain accurate numerical results. Before Eqs. 2-3 can be solved, a turbulence model must be introduced for determining the Reynolds stresses ($\overline{u_i u_j}$) appearing in the momentum equations, Eq. 3. The software provides various turbulence models. Stovin et al. (2002) have shown the influence of the turbulence model on the ability to represent the complexity of turbulent flows. Bonakdari (2006) and Bonakdari and Zinatizadeh (2011) have shown that the isotropic models are unable to represent the 3-D behavior of the velocity fields in narrow channels. Therefore, the anisotropic Reynolds Stress Model (RSM) has been used.

3.1.3. Boundary conditions

Boundary conditions around the solution domain are required in order to determine the solution of the governing fluid flow equations. The common boundary conditions in the open channel flow problems are:
(1) The water level at the inlet boundary needs to be given and this should be consistent with the mean velocity of flow in the channel.
(2) At the outlet, pressure is specified at the center of the cell face, while Cartesian velocity components and turbulence quantities are extrapolated from the interior of the cell using a second-order extrapolation.
(3) The interaction between the fluid and boundary walls is of great importance in the turbulent flows. Due to the strong velocity gradients occurring near the walls, a large amount of turbulence is generated there. This turbulence plays a very important role in several physical phenomena such as heat exchange and reattachment of the separated regions. For the present analysis, the wall function approach outlined by Launder and Spalding (1974) was used. This, in effect, means that the boundary conditions are not specified right at the wall but at a point, outside the viscous sub-layer, where the logarithmic law of the wall prevails.
(4) Assuming that the atmosphere exerts no shear and no inertia, equality of forces on the two sides of the interface is enforced. The

free surface location is determined by the Volume of Fluid (VOF) method (Hirt and Nicholas 1981) which has been incorporated into the solution of the Reynolds' time-averaged Navier-Stokes equations in order to take into account the free surface effects on the Reynolds stresses and, thus, velocity distribution in the cross section. In the commercial package Ansys-CFX, the water flow and some of the air flow above the water are solved simultaneously. The governing fluid flow equations for air and water are expressed in a single form but with different physical properties. A second order implicit scheme is used to discretise the governing equations i.e., the continuity equation, the momentum equations, and the turbulence model.

3.1.4. Computational meshes

To design a computational mesh suitably spaced for the solution domain described previously, different structured meshes with various cell concentrations were tested (Bonakdari, 2006). A prismatic mesh with rectangular base is used because such a mesh can be conveniently generated for the geometry of the channel; it also gives the highest accuracy (Bonakdari, 2012). The cell density should be increased in the flow regions with large gradients in velocity and free surface profiles (Akoz et al., 2009). Relatively finer local meshes in the vertical direction were used particularly near the channel bed in the water zone so that the first mesh point remained within the viscosity-affected wall region for $z+ (= \frac{u*z}{v}) \leq 30$, where high-gradient velocity profiles occur. The mesh density is decreased in the air zone. In the water phase, this grid is composed of cells with rectangular sections with its smaller side in the vertical direction so as to correspond with a higher gradient of flow characteristics in this direction; in the air phase, stepping away from the water surface, compression decreases linearly. A value of 10-6 has been employed as the convergence criterion at each time step for the sum of each of the normalized residuals over the whole fluid domain for all the governing fluid flow equations and also the mass residuals. The maximum number of iterations was equal to 30000. However, if the convergence criterion is reached for all the residuals, the simulation was stopped before reaching 30000 iterations.

3.2. Experimental data

The experimental results obtained from the study of Tominaga et al. (1989) were used in the present study. The bed of the channel was a painted iron plate and the side-walls were of glass. The experimental study was carried out in smooth and rough rectangular open channels. In rough cases, the roughness elements were glass beads with l2 mm diameter and they were densely attached to the wall. The results were obtained in a channel with 12.5 m length, and a square cross section (0.40 m × 0.40 m). The experiments were performed with the Reynolds number equal to 1.9×104 and Froude number equal to 0.19 for a depth of flow equal to 0.10 m and the mean velocity equal to 0.187 m/s. Velocity measurements were performed using a hot-film anemometer. Velocities were measured at the mid-section at about 100 locations. In this case, the flow width to depth ratio is smaller than 5. In such hydraulic contexts, the maximum velocity is below the free surface (dip phenomenon) unlike the usual accepted situation for river flows (Stearns 1883, Nezu and Nakagawa 1993).

3.3. Methodology to assess the impact of the relative roughness, Reynolds number and Froude numbers

Dimensional analysis is a powerful tool for deriving the dimensionless relationships among the variables governing any fluid flow. The variables that affect the establishment length Le are the mean velocity in the stream-wise direction (U), depth of flow in the channel (h), channel roughness (ζ), mass density (ρ) and kinematic viscosity (v) of the fluid, and gravitational acceleration (g). The dimensional analysis yields the following functional relation:

$$Le/h = f(\zeta/h, \frac{U}{\sqrt{gh}}, \frac{Uh}{v}) \tag{4}$$

$$Le/h = f(\zeta/h, F, R) \tag{5}$$

To study the effect of variations of the relative roughness, Reynolds number, and Froude number, it is necessary to vary one while keeping the other dimensionless numbers constant.

4. Results
4.1. Experimental validation of the CFD modelling strategy based on experimental data

The validity of the proposed model was evaluated by the experimental results of Tominaga et al. (1989). For the simulated case, the average water depth is h = 0.1 m and the bulk mean velocity at the entrance is equal to 0.187 m/s corresponding to the mean velocity measured experimentally by Tominaga et al. (1989). This channel is narrow with aspect ratio (width/depth) less than 5. Therefore, the secondary currents move the fluid with relatively high stream-wise momentum towards the central portion of the channel and cause the observed depression of the maximum velocity below the free surface called the dip phenomenon (Nezu and Nakagawa, 1993). Computational Fluid Dynamics (CFD) modelling was able to simulate the iso-velocity field in the whole of the cross section. Figure 2 illustrates the comparison between the experimental data and numerical results. In order to quantitatively compare the performance of the numerical model in the entire cross section with the measured results, two indicators, relative error (Rerr) and root-mean-square error (RMSE) were used and which are as follows:

$$R_{err} = \left| \frac{U_{mo} - U_{exp}}{U_{exp}} \right| \tag{6}$$

$$RMSE = \sqrt{\frac{1}{n} \sum_{i=1}^{n} \left(\frac{U_{mo} - U_{exp}}{U_{exp}} \right)^2} \tag{7}$$

where, n is the total point measurements in the cross section, U_{mo} the estimated velocity according to the numerical modelling and U_{exp} the measured velocity. The values of Rerr = 3 % and RMSE = 4.5 % obtained showed good agreement between the modelled and experimental data. CFD results show the dip phenomenon centered in the middle of the main channel at 70% of the water level from the bed.

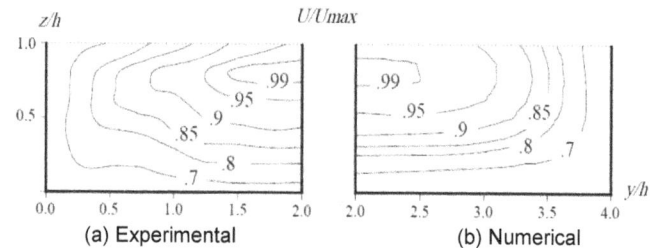

Fig. 2. Comparison of the experimental results of Tominaga et al. (1989) with the numerical results.

4.2. Establishment length

Fig. 3 shows the distribution of velocity for four different sections located at various distances from the inlet. In this example, it was observed and concluded that the flow can be treated as fully developed beyond the distance of 6.5 m from the entrance. This means that for these data, the length of boundary layer, i.e., the establishment length is 6.5 m. It represents 65 times the water depth.

Use of a structured mesh provides an identical positioning of the mesh nodes from one section to another. For this feature, the velocity distribution at all points of the successive sections were compared. For the estimation of the establishment length, an indicator of variation was proposed. This is based on the comparison of the velocity values at the corresponding points of the mesh grid in the two consecutive sections as follows:

$$I_{ec}(n, n+1) = \frac{\sqrt{\sum_{i,j\,section} \left(U_{i,j}^{n+1} - U_{i,j}^n \right)^2}}{U} \tag{8}$$

where, n and n+1 are the numbers of sections and the summation is performed for any differences between the values of all points Pij of the two sections. The longitudinal development of the mean flow velocity was investigated by calculating this indicator as a function of the distance from the entrance of the channel. This indicator was found to decrease with increase in the distance. The establishment length was defined as the distance from the entrance of the channel where the variation of this indicator values for the two consecutive sections remained within 0.5%.

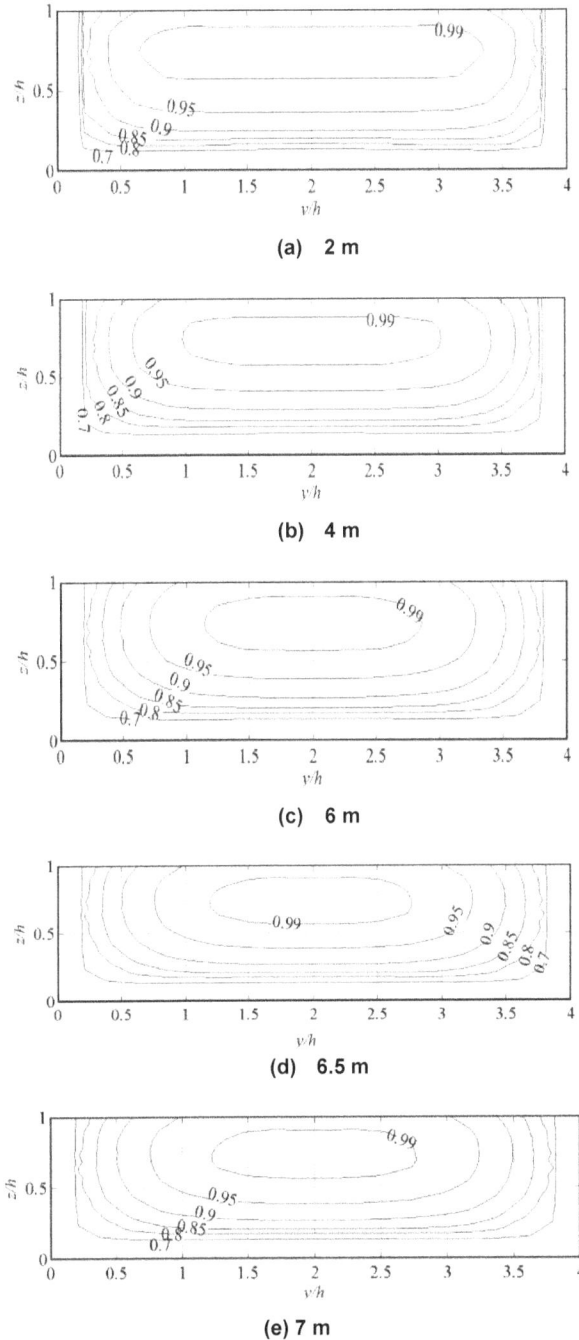

(a) 2 m

(b) 4 m

(c) 6 m

(d) 6.5 m

(e) 7 m

Fig. 3. Distribution of velocity (U/Umax) at various sections near the entrance of the channel.

4.3. Impact of the relative roughness, Reynolds and Froude numbers on the establishment length

Eq. (5) shows that the establishment length may be a function of the roughness of the walls (the same roughness value is used for the side wall and channel bed), Reynolds number, and Froude number. In the following section, the influence of these three variables on the length of establishment will be presented.

4.3.1. Influence of the roughness

The changes in roughness of wall produce considerably higher secondary currents than the smooth wall (Demuren and Rodi 1984). Fig. 4 shows the distribution of velocity at different distances from the lateral wall for smooth as well as rough walls. The value of y = 0.20 m presents the centerline of cross section. The surface velocity defect (i.e., the difference between the maximum velocity and the surface velocity) near the wall is larger in case of the rough wall compared with that for the smooth wall. Near the rough wall, higher turbulent stresses are generated in contrast to the region over the smooth wall. Therefore, fairly strong gradients of velocity and stresses exist. The flow in the boundary layer is, probably, affected by the development of the turbulent structure produced by the roughness elements.

In general, the roughness slowed the velocity gradient and slightly changed the velocity profiles close to the bed and wall. As the roughness increases, the friction at the wall surface also increases. This increase affects the boundary layer, mainly in the viscous sub-layer and near the walls. The components of the velocity and Reynolds stresses in the outer region ($z/h \geq 0.2$) should be affected very little by the roughness (Hinze 1975). Figure 5 indicates the influence of increasing roughness on the velocity field in the developed flow. The effect of roughness on the velocity field remains confined to the region near the wall. However, the average velocity in the outer layer is affected only marginally.

Variation of the velocity gradient with the variation of roughness showed to be more significant near the wall and free water surface. It might be attributed to the rate of dissipation of turbulent kinetic energy.

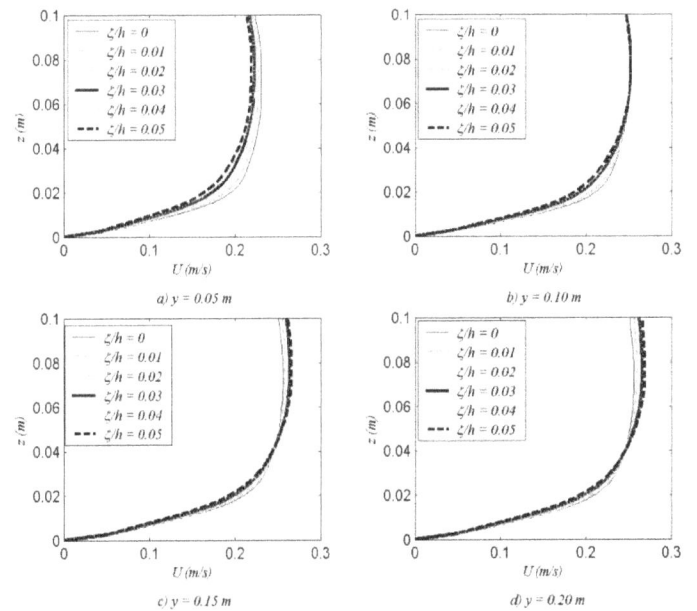

Fig. 4. Variation of velocity at different distances from the lateral wall.

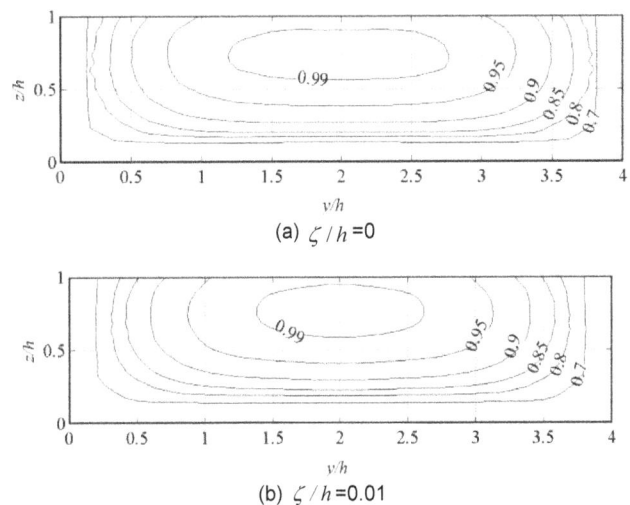

(a) ζ / h =0

(b) ζ / h =0.01

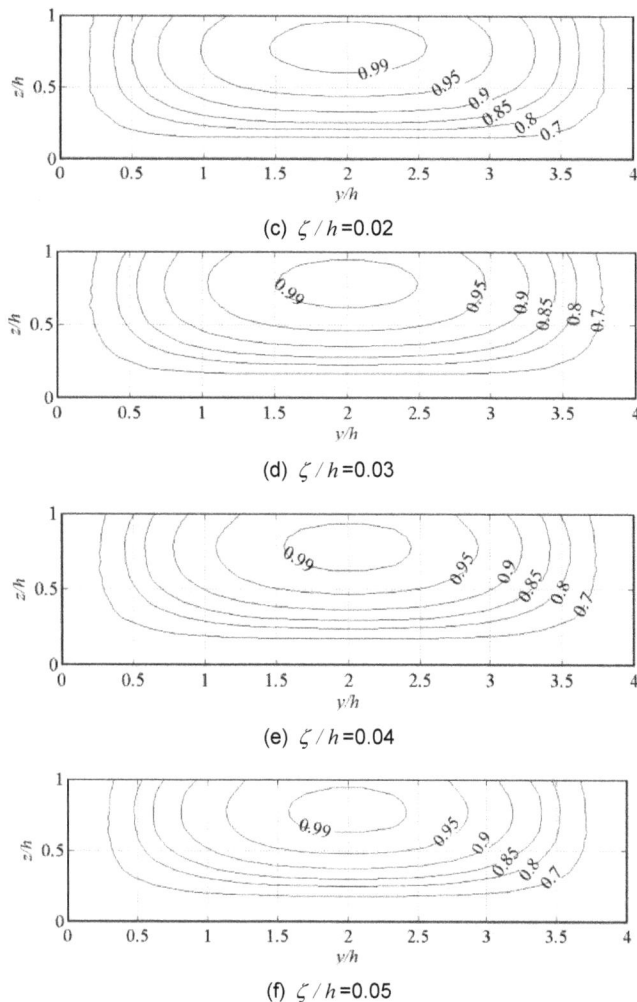

(c) $\zeta / h = 0.02$

(d) $\zeta / h = 0.03$

(e) $\zeta / h = 0.04$

(f) $\zeta / h = 0.05$

Fig. 5. Velocity fields (U/U_{max}) for different roughness in the developed flow (x = 6.5 m).

4.3.2. Influence of the Froude number

As per Eq. 5, the establishment length is likely to be affected by the Froude number. Therefore, the effect of the Froude number is investigated keeping the Reynolds number constant (Eq.5) for the various roughness values. The results indicated that the Froude number has negligible influence on the establishment length. This result is in conformity with the findings of the investigations carried out by Ranga Raju et al. (2000).

4.3.3. Influence of the Reynolds number

Figure 6 presents the dimensionless establishment length as a function of the Reynolds number for different wall roughnesses and for a constant Froude number (= 0.19). As can be seen from Fig. 6, the establishment length increases with a decrease in roughnesses for a specified Reynolds number. Figure 6 also shows that an increase in the Reynolds number causes a decrease in the establishment length.

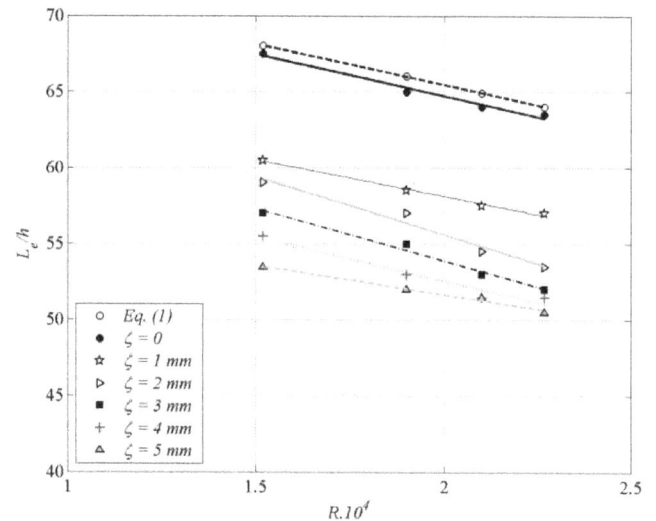

Fig. 6. Variation of Le/h with the Reynolds number for different values of relative roughness (F constant).

Present results, Fig. 6, are almost similar to those given by Kirkgoz and Ardichoglu (1997), Eq.(1), for smooth channels. In the rough conditions, however, one may accept a linear relationship between the dimensionless length of the flow developing zone and the Reynolds number which is as follows:

$$\frac{Le}{h} = aR/F+b \qquad (9)$$

Table 2 shows the evaluated values for the parameters determining the establishment length.

Table 2. Values for the parameters a and b for Eq.(9).

Relative roughness	a×10⁻⁵	B
$\zeta / h = 0$	-4.8	67.7
$\zeta / h = 0.01$	-7.5	70.7
$\zeta / h = 0.02$	-6.8	67.5
$\zeta / h = 0.03$	-3.8	60
$\zeta / h = 0.04$	-5.7	63.9
$\zeta / h = 0.05$	-5.5	75.6

5. Conclusions

The entrance length for the fully developed turbulent flow in an open channel was investigated numerically. The establishment length was determined by comparing the modelled velocity fields at successive sections that was substantially different in the establishing zone when compared with the corresponding threshold velocity field of the developed flow zone. The results obtained in the study proved that the effect of the Froude number on the establishment length is negligible. The establishment length was found to decrease with increase in the channel roughness. The dimensionless establishment length showed a linear relationship with the Reynolds number and can be computed using Eq.(9) for which a and b are dependent on the relative roughness, Table 2.

References

Akoz M.S., Kirkgoz M.S., Oner A.A., Experimental and numerical modeling of a sluice gate flow. Journal of Hydraulic Research 47(2) (2009) 167-176.

Bonakdari H., Modélisation des écoulements en collecteur d'assainissement – Application à la conception de points de mesures. Thesis (PhD), University of Caen, France, 2006.

Bonakdari H., Zinatizadeh A.A.L., Determination of Doppler flow meters position in sewers using computational fluid dynamics, Flow Measurement and Instrumentation 22 (2011) 225-234.

Bonakdari H., Effects of 3D structured grids on computational modeling of flow. LAP LAMBERT Academic Publishing, Saarbrücken, Germany, 2012.

Cebeci T., Analysis of turbulent flows. ELSEVIER Ltd, Oxford, U.K. 2004.

Cardoso A.H., Graf W.H., Gust G., Uniform flow in a smooth open channel, Journal of Hydraulic Research 27(5) (1989) 603-616.

Dean R., Reynolds number dependence of skin friction and other bulk flow variables in two-dimensional rectangular duct flow. Journal of Fluids Engineering 100 (1978) 215-223.

Dean R., Reynolds number dependence of skin friction and other bulk flow variables in two-dimensional rectangular duct flow. Journal of Fluids Engineering 100 (1978) 215-223.

Demuren A.O., Rodi W., Calculation of turbulence-driven secondary motion in non-circular ducts. Jounal of Fluid Mechanics 140 (1984) 189-222.

Graf W.H., Turbulent characteristics in rough uniform open channel flow. Annual Report, Lausanne, Switzerland, 1991.

Grass A.L., Structural features of turbulent flow over smooth and rough boundaries, Journal of Fluid Mechanics 50 (1971) 233-255.

Hinze J.O., Turbulence. McGraw-Hill Book Company, Second edition, New York, US, 1975.

Hirt C.W., Nicholas B.D., Volume of fluid (VOF) method for dynamics of free boundaries, Journal of Computational Physics 39 (1981) 201-225.

Kirkgoz S., Ardichoglu M., Velocity profiles of developping and developed open channel flow, Journal of Hydraulic Engineering 115 (1997) 1099-1105.

Launder B.E., Spalding D.B., The numerical computation of turbulent flows, Computer Methods in Applied Mechanics and Engineering 3 (1974) 269-89.

Lien K., Monty J.P., Chong M.S., Ooi A., The entrance length for fully developed turbulent channel flow. 15th Australasian Fluid Mechanics Conference, Ausralia 2004.

Nezu I., Rodi W., Experimental study on secondary currents in open channel flow. 2lst IAHR congress, Melbourne, Australia, 2 (1985) l9-23.

Nezu I., Nakagawa H., Turbulence in open-channel flows. IAHR-Monograph, A. A. Balkema Publishers, Rotterdam, The Netherlands, 1993.

Nikuradse J., Laws of flow in rough pipes. NACA Technical Notes, No. 1295, Washington D. C., US. 1933.

Perkins H.J., The formation of streamwise vorticity in turbulent duct flow. Journal of Fluid Mechanics 44 (1970) 721-740.

Ranga Raju K.G., Asawa G.L., Mishra H.K., Flow establishment length in rectangular channels and duct. Journal of Hydraulic Engineering 126(7) (2000) 533-539.

Stearns E.P., A reason why the maximum velocity of water flowing in open channels is below the surface, Transactions of the American Society of Civil Engineers 7 (1883) 331-338.

Stovin V.R., Grim J.P., Buxton A.P., Tait S.J., Parametric studies on CFD models of sewerage structures. 9th International Conference on Urban Drainage, Portland, US, 2002.

Tominaga A., Nezu I., Ezaki K., Nakagawa H., Three dimensional turbulent structure in straight open channel flows, Journal of Hydraulic Research 27 (1989) 149-173.

Wang J.S., Tullis J.P., Turbulent flow in the entry region of a rough pipe. Journal of Fluids Engineering 96 (1974) 62-68.

Zanoun E.S., Durst F., Nagib H., Evaluating the law of the wall in two-dimensional fully developed turbulent channel flows, Physics of Fluids 15 (2003) 3079–3089.

17

A TVD-WAF finite volume model for roll wave simulation

Ali Mahdavi

Department of Civil Engineering, Arak University, Arak, Iran.

ARTICLE INFO	ABSTRACT
Keywords: Roll waves Numerical method WAF scheme Steep channel	Roll waves appear as successive transitions from super- to sub-critical flows by passing through moving hydraulic jumps. Such discontinuous periodic waves propagate in a staircase pattern at constant wave celerity. On the basis of nonlinear shallow water (NLSW) equations, a finite volume model is presented to study the evolution of roll waves in inclined steep channels. The numerical model exploits the total variation diminishing version of weighted average flux (TVD-WAF) explicit method to solve the homogeneous NLSW equations. An implicit trapezoidal time integration operator is implemented for the treatment of source term which includes contributions from both the channel slope and frictional resistance. The simulated surface profile and flow velocity are in well agreement with available analytical solution for roll waves. The time evolution of wave amplitude under different undisturbed Froude numbers are investigated numerically and compared with theoretical predictions. Temporal decay of initial disturbance is discussed in which case roll waves no longer form. The observed agreement implies the efficiency and accuracy of the present scheme for roll wave modeling.

1. Introduction

Considering a steady uniform flow down an inclined steep channel with bed friction, once the Froude number exceeds a definite critical value, the flow reveals some sort of hydrodynamic unstability and the disturbances on the free surface, if exist, may eventually grow forming a series of successive hydraulic bores connected by smooth profiles of gradually varied flow. Such discontinuous periodic waves propagate in a staircase pattern at constant wave celerity and are known as roll waves.

From practical point of view, it is important to consider the role of roll waves in the design of channels, since these water carrying structures may be overtopped by formation of roll waves (Montes. 1998). Dressler (1949) was the first who mathematically explained the occurrence of periodic roll waves in an open channel of constant slope and obtained a quasi-steady solution for roll waves. He assumed a uniformly moving frame of reference to formulate traveling waves. The critical value of Froude number to onset instabilities was found to be $F_0=2$ based on mathematical analysis of linearized theory where a small perturbation is being imposed on a steady uniform state (Whitham. 1974; Que and Xu. 2006). This threshold condition for the formation of roll waves has been also confirmed by experiments (Brock. 1967).

In addition to man-made canals such as draining systems and dam spillways, roll waves have also been observed in natural water courses such as ice channels (Carver, Sear and Valentine. 1999) and lakes (Fer, Lemmin and Thorpe. 2003). Balmforth and Mandre (2004) reported disturbances in a variety of physical situations that behaves, in some aspects, similar to roll waves in open channel flows. These include perturbations in multi-phase fluid, mudflow, granular layers and flow down collapsible tubes and elastic conduits. It is worth mentioning that the phenomenon of roll wave is precisely analogous to the stop-start waves encountered in freeway models of traffic theory (Kühne. 1984).

The nonlinear shallow water (NLSW) equations have been broadly accepted for modeling long wave evolution in open channels as well as coastal areas (Mahdavi and Talebbeydokhti. 2009, 2011; Mahdavi et al. 2012). The hyperbolic nature of this system of partial differential equations admits discontinuous solutions (e.g., hydraulic jump, bore, breaking wave and roll wave) even when the initial conditions are spatially smooth. Based on the one dimensional NLSW, Zanuttigh and Lamberti (2002) found satisfactory results in reproducing the Brock's experimental roll waves in a laboratory flume. The same authors further studied the stability of viscoplastic fluid in uniform flow, proving that debris flows become unstable even for $F_0<1$ (Zanuttigh and Lamberti. 2004). Di Cristo et al. (2008) established a formula on the minimum channel length required for roll wave development. Their results were found reliable when compared to available experimental data, regardless of the channel slope. Que and Xu (2006) developed a high resolution scheme based on the gas-kinetic Bhatnagar–Gross–Krook (BGK) model to study roll-waves down an inclined open channel.

In the present study, the generation and propagation of roll waves are numerically investigated in the framework of an initial value problem for NLSW equations. The TVD-WAF scheme is considered as an approximate Riemann solver to provide the model with shock capturing property in presence of discontinuities across the front of roll waves and to allow the formation of smooth segments of gradually varied flow linking each two consecutive bore faces. The accuracy of the scheme is verified through comparisons with the available analytical solution and with the numerical results obtained by the BGK kinetic scheme of Que and Xu (2006).

2. Materials and methods

Governing equations

Referring to the definition sketch (Fig. 1), the generation and evolution of roll waves are considered here. The relevant process can be appropriately described by the non-linear shallow water equations written in the conservative form as:

$$\frac{\partial \mathbf{U}}{\partial t} + \frac{\partial \mathbf{F}}{\partial x} = \mathbf{S} \tag{1}$$

Corresponding author Email: a-mahdavi@araku.ac.ir

The conserved variables U, the flux F, and the source term S are defined respectively by:

$$\mathbf{U} = \begin{bmatrix} h \\ hu \end{bmatrix}, \quad \mathbf{F}(\mathbf{U}) = \begin{bmatrix} hu \\ hu^2 + \frac{1}{2}gh^2 \end{bmatrix}, \quad \mathbf{S}(\mathbf{U}) = \begin{bmatrix} 0 \\ ghS_o - C_f u^2 \end{bmatrix}. \quad (2)$$

In the above equations, t denotes time, x is the distance along the channel, h (x, t) is the water depth, u (x, t) is the depth-averaged velocity along x-direction, C_f is the bed roughness coefficient, S_o is the bottom slope and g is the gravitational acceleration.

In the context of numerical schemes, it is usual to approximate the spatial derivative of non-linear flux F in Eq. (1) at x_i (the center of cell i) by the conservative difference as:

$$\left(\frac{\partial \mathbf{F}}{\partial x} \right)_{x=x_i} = \frac{\mathbf{F}_{i+1/2} - \mathbf{F}_{i-1/2}}{\Delta x} \quad (3)$$

where Δx is the spatial cell size and $F_{i+1/2}$ and $F_{i-1/2}$ are the inter-cell numerical fluxes that will be explained in the next paragraphs.

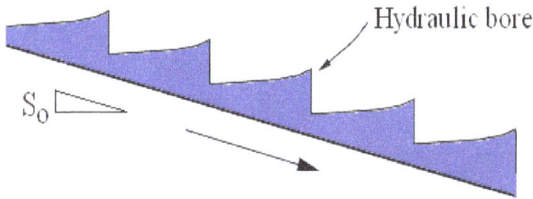

Fig. 1. Definition sketch for propagation of roll waves.

Evaluation of numerical flux

To precisely handle the shock-like wave front of roll waves, the inter-cell numerical fluxes are computed by the weighted average flux (WAF) shock capturing method. A limiter function enforces the total variation diminishing (TVD) constraint on the scheme and, thereby, adding sufficient dissipation to the scheme to ensure the monotonicity near large gradients of the solution. The TVD-WAF scheme preserves second order accuracy in spatial and temporal coordinates and is oscillation-free across discontinuities. The numerical flux may be written as (Toro. 2001):

$$\mathbf{F}_{i+1/2} = \frac{1}{2}(\mathbf{F}_i + \mathbf{F}_{i+1}) - \frac{1}{2}\sum_{k=1}^{N} \text{sign}(c_k)\phi_{i+1/2}^{(k)}\Delta\mathbf{F}_{i+1/2}^{(k)} \quad (4)$$

where N denotes the number of waves in the solution of Riemann problem, $\Delta\mathbf{F}_{i+1/2}^{(k)} = \mathbf{F}_{i+1/2}^{(k+1)} - \mathbf{F}_{i+1/2}^{(k)}$ is the flux jump across wave k with fluxes defined by $\mathbf{F}_{i+1/2}^{(1)} = \mathbf{F}(\mathbf{U}_L)$, $\mathbf{F}_{i+1/2}^{(2)} = \mathbf{F}(\mathbf{U}^*)$ and $\mathbf{F}_{i+1/2}^{(3)} = \mathbf{F}(\mathbf{U}_R)$, c_k is the Courant number for wave k and $\phi_{i+1/2}^{(k)}$ is the limiter function. Some suitable choices for limiter function are reported by Toro (2001). The SUPERBEE limiter is preferred for the present applications. It is given by:

$$\phi_{i+1/2}^{(k)} = 1 - \left(1 - |c_k|\right).\max\left[0, \min\left(1, 2r^{(k)}\right), \min\left(2, r^{(k)}\right)\right] \quad (5)$$

where $r^{(k)}$ is the ratio of the upwind change to the local change in flow depth for which the details can be found in Toro(2001). In (4), the numerical flux $\mathbf{F}_{i+1/2}^{(2)} = \mathbf{F}(\mathbf{U}^*)$ refers to the Harten-Lax-van Leer (HLL) approximate Riemann solver. Based on the HLL approach, the Riemann problem with data U_L and U_R is characterized by three constant states separated by two waves. The numerical flux in the intermediate region of the wave structure can be determined as follows:

$$\mathbf{F}(\mathbf{U}^*) = \frac{S_R \mathbf{F}(\mathbf{U}_L) - S_L \mathbf{F}(\mathbf{U}_R) + S_R S_L (\mathbf{U}_R - \mathbf{U}_L)}{S_R - S_L} \quad (6)$$

where S_L and S_R represent the wave speed estimates on the left and right sides of the cell interface, respectively. Several options are available for these wave speeds. The wave speed expressions derived by Toro (1992) is implemented in the present model:

$$S_R = \max(u_R + \sqrt{gh_R}, u^* + \sqrt{gh^*})$$
$$S_L = \min(u_L - \sqrt{gh_L}, u^* - \sqrt{gh^*}) \quad (7)$$

In above expressions $h^*$and u^*denote, respectively, the flow depth and flow velocity in the intermediate region of the wave structure. According to the two-rarefaction approach, these flow variables can be evaluated by the following closed-form solutions:

$$h^* = \frac{1}{g}\left[\frac{1}{2}(\sqrt{gh_L} + \sqrt{gh_R}) + \frac{1}{4}(u_L - u_R)\right]^2$$
$$u^* = \frac{1}{2}(u_L + u_R) + \sqrt{gh_L} - \sqrt{gh_R} \quad (8)$$

Treatment of source term

A numerical procedure for dealing with the homogeneous NLSW equations was explained in foregoing section. In presence of source term, the TVD-WAF scheme can be applied unchanged, if the source term is treated by additional integration steps. To this end, a three step splitting scheme is implemented that relies on successive solution of following system of initial value sub-problems.

$$\text{ODEs:} \quad \frac{d\mathbf{U}}{dt} = \mathbf{S}(\mathbf{U}) \left.\right\} \xrightarrow{\Delta t'} \mathbf{U}^{(1)}$$
$$\text{ICs:} \quad \mathbf{U}^n$$

$$\text{PDEs:} \quad \frac{\partial \mathbf{U}}{\partial t} + \frac{\partial \mathbf{F}}{\partial x} = \mathbf{0} \left.\right\} \xrightarrow{\Delta t} \mathbf{U}^{(2)}$$
$$\text{ICs:} \quad \mathbf{U}^{(1)} \quad (9)$$

$$\text{ODEs:} \quad \frac{d\mathbf{U}}{dt} = \mathbf{S}(\mathbf{U}) \left.\right\} \xrightarrow{\Delta t'} \mathbf{U}^{n+1}$$
$$\text{ICs:} \quad \mathbf{U}^{(2)}$$

where superscript n denotes the current time level, Δt is the time step size and $\Delta t' = \Delta t/2$. In present work, the source term parts are treated by the trapezoidal time integration method. This implicit operator is second-order accurate and can be expressed for the first sub-problem in Eq. (9) as:

$$\left[\mathbf{I} - \frac{\Delta t'}{2}\left(\frac{\partial \mathbf{S}(\mathbf{U})}{\partial \mathbf{U}}\right)_i^n\right]\Delta\mathbf{U}_i = \Delta t' \mathbf{S}(\mathbf{U}_i^n) \quad (10)$$

where I denotes the identity matrix and $\Delta U_i = U_i^{(1)} - U_i^n$ is the temporal jump in conserved variables. The second term of the right-hand side of Eq. (10) is the Jacobian matrix of the source term. The first intermediate value $U^{(1)}$ can be calculated from Eq. (10). Taking $U^{(2)}$ as the initial condition, the TVD-WAF scheme is then implemented in an explicit conservative scheme to obtain the second intermediate value $\mathbf{U}^{(2)}$ as:

$$\mathbf{U}^{(2)} = \mathbf{U}^{(1)} - \frac{\Delta t}{\Delta x}(\mathbf{F}_{i+1/2} - \mathbf{F}_{i-1/2}) \quad (11)$$

The above conservative formula results from integrating the homogeneous NLSW equations over a suitable control volume in the x-t plane. Finally, applying the source term operator to $U^{(2)}$ will give the conserved variables at the new time level, n+1. Because the TVD-WAF is an explicit scheme, the magnitude of time interval Δt is dynamically adjusted according to the Courant- Friedrichs-Lewy (CFL) criterion, defined as:

$$\Delta t = C_n \min_i \frac{\Delta x}{|u_i| + \sqrt{gh_i}} \tag{12}$$

with C_n being the CFL number ($0 < C_n \leq 1$).

Initial and boundary conditions

In order to constitute a mathematically well-posed problem, it is necessary to prescribe the initial and boundary conditions that are consistent with the true behavior of the physical phenomenon under consideration. To specify the initial conditions for modeling roll waves, the approach proposed by Que and Xu (2006) is followed which provides a simple and robust procedure to model the generation and subsequent evolution of roll waves. According to their formulation, the initial sinusoidal disturbance and corresponding initial flow velocity are respectively designated by:

$$h(x,0) = h_0 \left[1 + \varepsilon \sin(k_w x)\right]$$
$$u(x,0) = u_0 + r_p \varepsilon \sin(k_w x + \theta_p) \tag{13}$$

In the above equations, ε is an amplification factor accounting for the amplitude of the initial disturbance which was suggested to be $\varepsilon=0.005$ in the original paper, k_w is the 2π-wave number, h_0 and u_0 are the water depth and flow velocity of the undisturbed initial flow, respectively. θ_p is the phase lag between the initial water depth and associated flow velocity and r_p represents a function of angular velocity, wave number and initial flow velocity for which an expression was introduced by Que and Xu (2006). The existence of initial uniform flow down a sloping channel requires a situation in which the gravitational and frictional forces approach equilibrium such that:

$$S_o = C_f F_0^2 \tag{14}$$

where $F_0 = u_0/(gh_0)^{1/2}$ is the Froude number of initial undisturbed flow and the variables with subscript 0 refer to those associated with initial uniform flow. The adoption of periodic boundary condition at either ends of the computational domain enables efficient representation of the periodic properties of roll waves. This type of boundary condition simultaneously equates the conserved variables at two boundaries of the model.

3. Results and discussion

The numerical values of the physical parameters adopted herein mostly follow those of Que and Xu (2006). Fig. 2 shows the snapshots of depth profiles for a uniform flow ($q_0 = h_0 u_0 = 0.001$ m^2 s^{-1}, $F_0 = 2.5$) primarily disturbed by imposing a sinusoidal perturbation ($k_w = 10\pi$ and $\varepsilon = 0.005$) over its free surface. The flow domain has a length of 2 m ($0 < x < 2$m) and is discretized by 1000 computational cells ($\Delta x = 0.002$ m). The numerical stability of the roll wave simulation is guaranteed by setting $C_n = 0.65$. The bed friction is included in the computation by a roughness coefficient of $C_f = 0.006$ and the channel slope is $S_0 = 0.0375$ as determined by Eq. (14). Obviously, the amplitude of initial disturbance grows up with time until a permanent form wave is established around $t=20$s. At this stage, the wave train experiences successive transitions from super- to sub-critical flows through moving hydraulic jumps. This situation is shown in Fig. 3. The moving reference Froude number is defined as $F_{mr} = (c-u)/\sqrt{gh}$ with c being the constant wave celerity which is evaluated as $c = 0.55\,\text{ms}^{-1}$ in this case. This definition is equivalent to an observer moving with roll wave for which the wave appears to be stationary. In this manner, the problem may be simplified to steady-flow formation of hydraulic jumps. It can be seen that the difference between the maximum F_{mr} and minimum F_{mr} increases by increasing F_0 (Fig. 4). These maximum and minimum Froude numbers correspond to conjugate depths associated with moving hydraulic jumps. An increase of F_0 obviously leads to greater maximum F_{mr} which in turn increases the specific force before jump. Therefore, the specific force after the jump should also grow to maintain the balance between the specific forces acting on either sides of the jump (taking into account the bottom friction force). The threshold

Froude number $F_0 = 2$ implies the dominance of critical flow regime throughout the domain which is indicated by $F_{mr} = 1$ in Fig. 4.

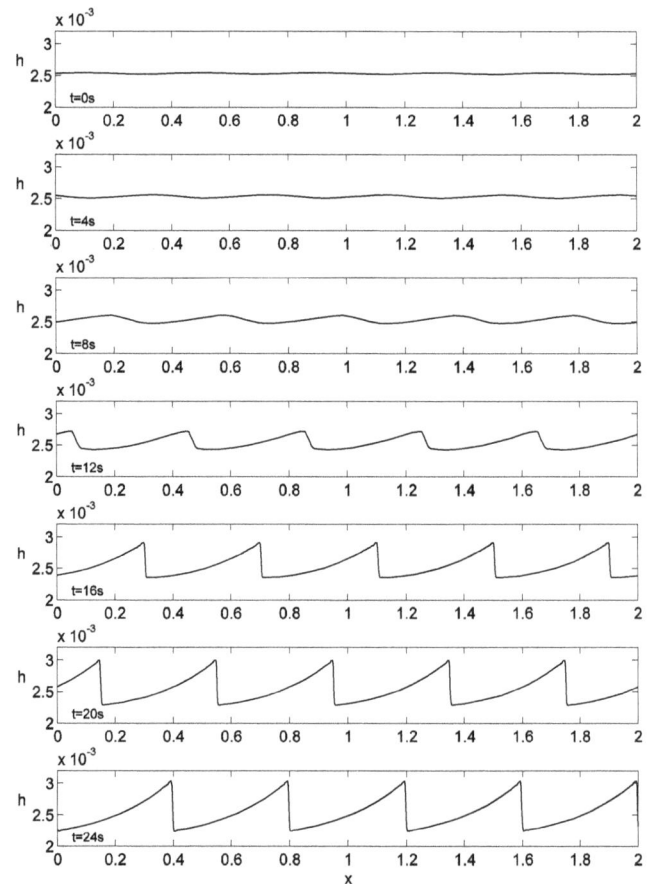

Fig. 2. Snapshots of depth profiles representing the generation and evaluation of roll waves at different times. The channel has a length of 2m; the steady state flow is designated by: $q_0 = 0.001$ m^{2}S^{-1}, $F_0 = 2.5$, $C_f = 0.006$; and the initial disturbance takes these parameter values: $k_w = 10\pi$ and $\varepsilon = 0.005$ (The axes are in meter).

In Fig. 5, depth and velocity profiles of emerged roll waves are checked against the analytical solution of Dressler (1949). The agreement is quite satisfactory in terms of predicting the locations of bore faces and simulating the smooth connecting profiles. The time history of flow variables at the middle of the channel i.e., $x=1$m are depicted in Fig. 6 along with the results obtained by a high resolution BGK kinetic scheme developed by Que and Xu (2006). The results of these two models reveal a similar pattern for both amplitude growth and temporal variation of flow velocity at this location.

Lukáčová-Medviďová and Teschke (2006), performing a comparison study of various numerical shallow water models, noted that the CPU-efficiency needs to be considered relatively since it depends on the optimality and robustness of a code. Therefore, comparisons are also made between simulation run times of the two models. In the cases studied, it was found that the CPU time consumed by the TVD-WAF scheme is only about 30–33 % of that needed by the BGK scheme, implying the efficiency of the present model with respect to its CPU performance.

The natural logarithm of maximum wave amplitude as a function of time under several Froude numbers ranging from $F_0=1.5$ to $F_0=3.7$ is presented in Fig. 7. Also depicted in this figure are the linear theory predictions (Que and Xu, 2006). At the first stages of wave evolution, the amplitude growth rate is in agreement with theoretical results obtained by the linear theory, but as time continues, the wave amplitude is no longer small and the wave form undergoes a change towards successive hydraulic bores leading to appearance of nonlinear effects and thus divergence from linear theory. However, the validity range of linear theory tends to extend when the Froude number of initial flow is reduced.

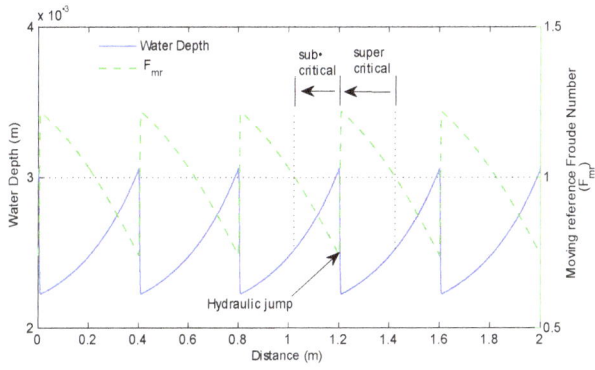

Fig. 3. Transition from super- to sub-critical flow regimes in roll waves. The waves propagate toward right.

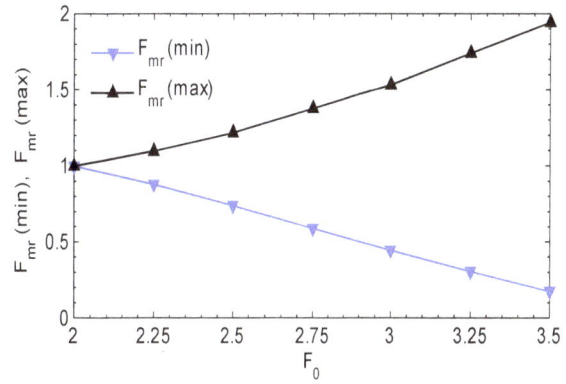

Fig. 4. The maximum and minimum values of moving reference Froude number versus initial Froude number.

Fig. 8, shows the variation of natural logarithm of wave amplitude versus 2π-wave number, k_w, for different values of Froude number. For a given Froude number, the wave amplitude descends in response to an increase in k_w. Conversely, the wave amplitude ascends as the Froude number increases for a given value of k_w.

It was remarked earlier that the linearly unstable conditions of flow are dominated as the Froude number exceeds the limit value $F_0=2$. In conformance with this fact, the numerical results demonstrate a relatively time invariant wave amplitude for $F_0=2$ while a substantial decay in wave amplitude is apparent when $F_0<2$ (Fig. 7). To further illustrate the latter case, the time evolution of free surface is considered in Fig. 9 for $F_0=1.5$. Under such a condition, the amplitude of initial disturbance would theoretically approach zero if the time goes to infinity. The simulated free surface profiles in Fig. 9 reveal this pattern of amplitude decay and the flow practically recovers its undisturbed steady state at the end of simulation time.

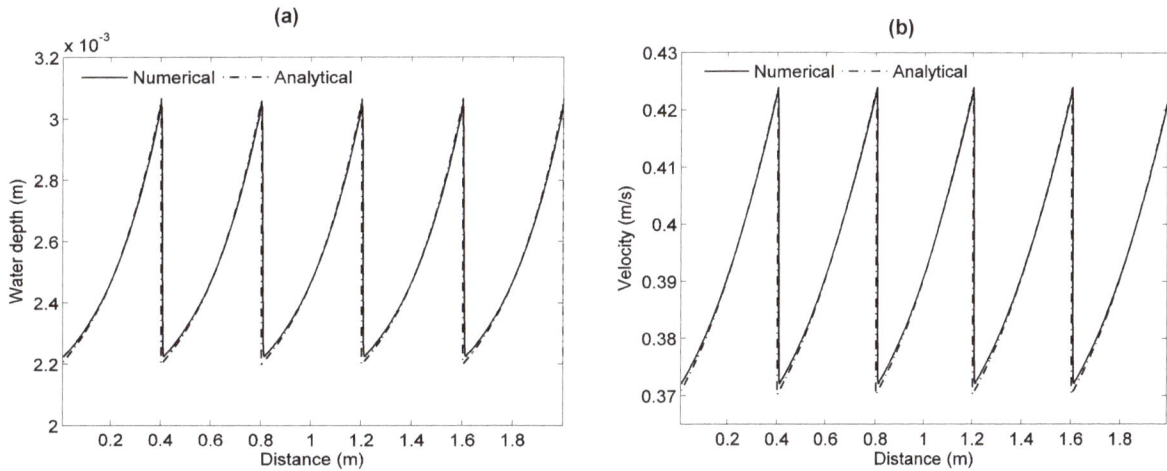

(a)

(b)

Fig. 5. Comparison between simulated roll waves and analytical solution of Dressler (1949) for a: flow depth and b: flow velocity.

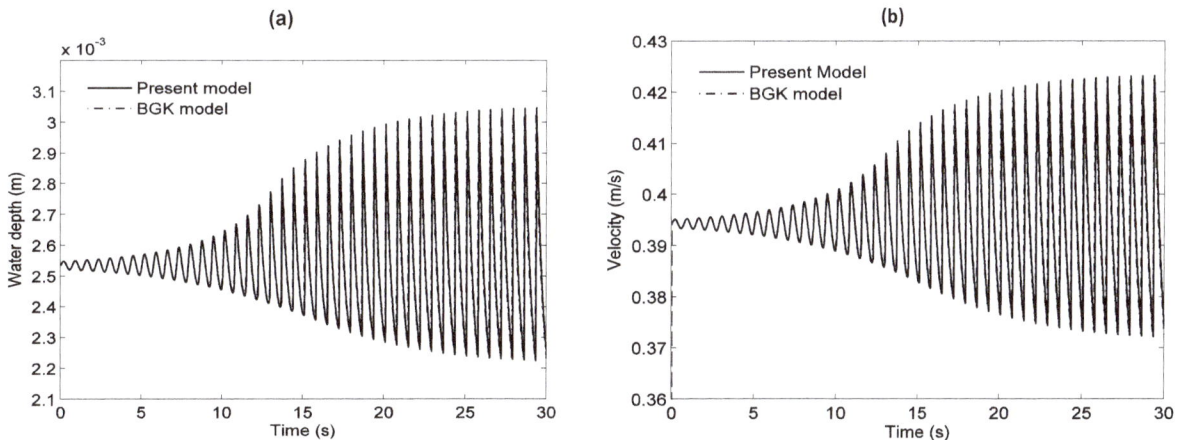

(a)

(b)

Fig. 6. Time histories of flow variables at the middle of channel ($x = 1\,\text{m}$) for a: flow depth and b: flow velocity.

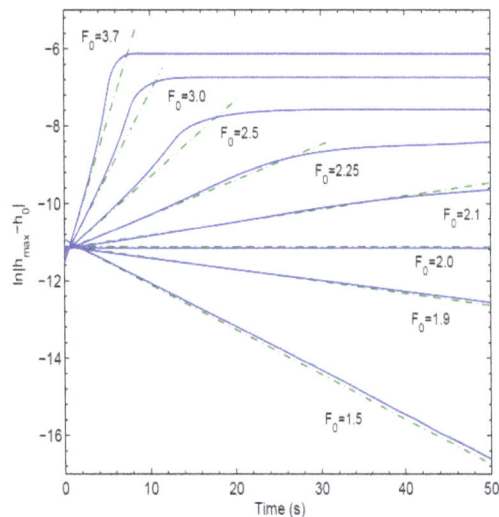

Fig. 7. Time evolution of the natural logarithm of wave amplitude under different Froude numbers and comparison with linear theory (Que and Xu, 2006). (Solid line: Present study; Dash-doted line: linear theory).

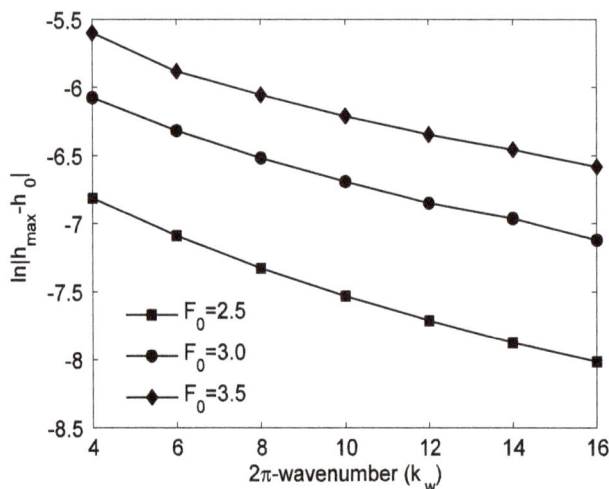

Fig. 8. Natural logarithm of wave amplitude as affected by changes in for different Froude numbers.

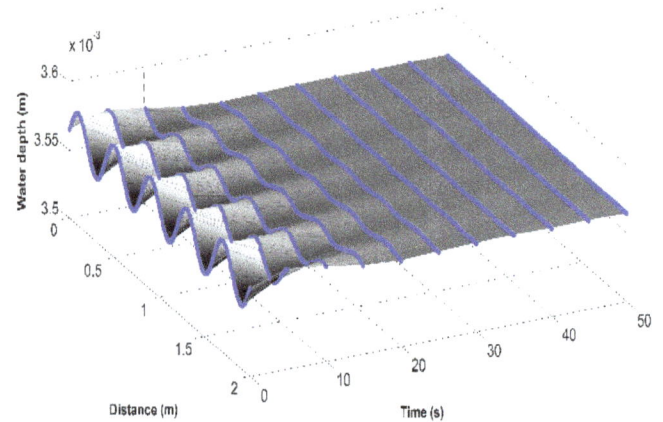

Fig. 9. Temporal decay of initial disturbance for $F_0 = 1.5$.

4. Conclusions

The TVD–WAF scheme has been combined with a three step splitting scheme to numerically solve the conservative form of one dimensional nonlinear shallow water equations. The model was adopted to investigate the generation and evolution of roll waves down an inclined steep channel. It is accomplished by introducing appropriate initial and boundary conditions in the framework of an initial value problem. The numerical results confirm the capabilities of model as a robust tool to eliminate the numerical instabilities due to small water depths usually encountered in roll wave simulation. Moreover, the present model requires a reduced amount of CPU time when compared to that of a high resolution shock capturing scheme. The amplitude growth of emerged roll waves for $F_0 > 2$ together with amplitude decay of initial disturbance for $F_0 < 2$ are simulated by the scheme and compared satisfactorily with a theoretical approach. The present numerical model requires only modest computational effort and, nevertheless, has the ability to handle free surface flows in cases where shocks and smooth profiles are to be considered simultaneously in the computational domain.

Acknowledgment

The author gratefully acknowledges Prof. Kun Xu and Mr. Que Yin-Tik (Mathematics Department; Hong Kong University of Science and Technology) for providing the BGK model source code for the numerical comparison in this study.

References

Balmforth N.J., Mandre S., Dynamics of roll waves, Journal of Fluid Mechanics 514 (2004) 1–33.

Brock R.R., Development of roll waves in open channels, Technical report, W. M. Keck Lab. of Hydraul.and Water Resources, California Institute of Technology, Report KH-R-16 (1967).

Carver S., Sear D., Valentine E., An observation of roll waves in a supraglacial meltwater channel, Harlech Gletscher, East Greenland, Journal of Glacial Archaeology 40 (1994) 75–78.

Cristo C., Iervolino M., Vacca A., Zanuttigh B., Minimum channel length for roll-wave generation, Journal of Hydraulic Research 46 (2008) 73–79.

Dressler R.F., Mathematical solution of the problem of roll waves in inclined channel flows, Communications on Pure and Applied Mathematics 2 (1949) 149–194.

Fer I., Lemmin U., Thorpe S.A., Winter cascading of cold water in Lake Geneva, Journal of Geophysical Research 107 (2003).

Kühne R.D., Macroscopic freeway model for dense traffic-stop-start waves and incident detection, In Proceedings of the 9th International Symposium on Transportation and Traffic Theory (1984) 21–42.

Lukáčová-Medviďová M., Teschke U., Comparison study of some finite volume and finite element methods for the shallow water equations with bottom topography and friction terms, ZAMM - Journal of Applied Mathematics and Mechanics 86 (2006) 874–891.

Mahdavi A., Hashemi M.R.,Talebbeydokhti N., A localized differential quadrature model for moving boundary shallow water flows, Journal of Hydraulic Research 50 (2012) 612–622.

Mahdavi A., Talebbeydokhti, N., Modeling of non-breaking and breaking solitary wave run-up using FORCE-MUSCL scheme Journal of Hydraulic Research 47 (2009) 476–485.

Mahdavi A., Talebbeydokhti N., Modeling of non-breaking and breaking solitary wave run-up using shock-capturing TVD-WAF scheme, KSCE Journal of Civil Engineering, 15 (2011) 945–955.

Montes S., Hydraulics of open channel flow. ASCE Press, USA (1998).

Que Y.-T., Xu K., The numerical study of roll-waves in inclined open channels and solitary wave run-up, International Journal for Numerical Methods in Fluids 50 (2006) 1003–1027.

Toro E.F., Riemann problems and the WAF method for solving the two-dimensional shallow water equations, Philosophical Transactions of the Royal Society A 338 (1992) 43–68.

Toro E.F., Shock-capturing methods for free-surface shallow flows. Wiley, Chichester UK (2001).

Whitham J., Linear and nonlinear waves. Wiley, New York (1974).

Zanuttigh B., Lamberti A., Roll waves simulation using shallow water equations and weighted average flux method, Journal of Hydraulic Research 40 (2002) 610–622.

Zanuttigh B., Lamberti A., Analysis of debris wave development with one-dimensional shallow-water equations, Journal of Hydraulic Engineering 130 (2004) 293–304.

18

Combination of gradually varied flow theory and simulated annealing optimization in of manning roughness coefficient

Majid Heydari[1*], Jalal Sadeghian[2], Milad Faridnia[1], Saeid Shabanlou[3]

[1]Department of Water Science and Engineering, Faculty of Agriculture, Bu-Ali Sina University, Hamadan, Iran.
[2]Department of Civil Engineering, Faculty of Technical and Engineering, Bu-Ali Sina University, Hamadan, Iran.
[3]Department of Water Engineering, Kermanshah Branch, Islamic Azad University, Kermanshah, Iran.

ARTICLE INFO

Keywords:
Manning roughness
Simulated annealing algorithm
Gradually varied flow
Nonlinear optimization

ABSTRACT

Manning roughness coefficient is one of the most important parameters in designing water conveyance structures. Unsuitable selection of this coefficient brings up some mistakes. This research aims to present a method to determine the Manning roughness coefficient based on a combination of optimization algorithm of simulated annealing (SA) with gradually varied flow equations. Therefore, in a lab rectangular flume of 12 m, 60 cm and 65 cm in length, width and height with fixed channel bed slope of 0.0002, nine series of water level profiles were carried out. Then, an objective function based on observed and calculated water level gradient was defined to decide on manning roughness coefficient while it was minimized with simulated annealing optimization method. The values of objective function parameters were discussed by sensitivity analysis and the most optimal objective function was obtained. To measure the accuracy of coefficient obtained, Statistics indices of R^2, Root mean square error (RMSE), Mean bias error (MBE), d were used. The results showed that manning roughness coefficient has a great accuracy.

1. Introduction

Gradually varied flow is a permanent non-uniform flow in which the change of depth is so small that the pressure distribution can be considered to be hydrostatic, this helps up to take the flow one dimensional without pressure gradient (except for what is applied in normal gravity direction) (Chow, 1959). Regarding Fig .2, the dynamic equation of gradually varied flow is as equation (1) (Abrishami and Hosseini, 2011).

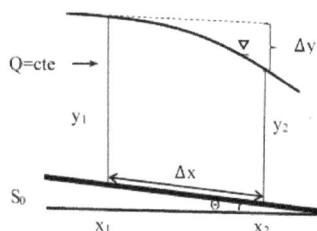

Fig. 1. Schematic diagram of GVF

$$\frac{dE_s}{dx} = S_0 - S_f \tag{1}$$

where, S_o is the channel bed slope, S_f is the friction slope, E_s is the specific energy and X is the distance from the origin. This equation is used to calculate the water surface profile in open channels. In Eq. (1), instead of friction slope, the uniform flow equations such as Manning equation with resistance coefficient of uniform flow can be used as in Eq. (2) (Abrishami and Hosseini, 2011). The dynamic equation of GVF is seen in Eq. (3).

$$S_f = \frac{n^2 Q^2}{A^2 R^{4/3}} \tag{2}$$

$$\frac{dE_s}{dx} = S_0 - \frac{n^2 Q^2}{A^2 R^{4/3}} \tag{3}$$

where, n is the Manning roughness coefficient, Q is discharge, A is cross section and R is the hydraulic radius.

Manning roughness coefficient is one of the most important parameters in designing hydraulic structures such as water conveyance open channels, so that this coefficient covers all the factors affecting the channel bed resistance against the flow varying with depth, velocity and type of lined (Kochakzadeh and Maghsoudi, 2011). Therefore, there have been various researches on the conditions of flow in open channels to estimate this coefficient. The engineering judgment plays a role to estimate roughness coefficient of Manning. If this coefficient is not taken properly, there will be plain mistakes. To determine the roughness coefficient of manning, different experimental methods such as using slides and pictures (Tadayonfar, 2009) and experimental relationships (Gazer Zadeh, 2010) are presented, due to the constraints of analytical methods, optimization becomes necessary. Also, some researchers have performed limited researching on estimation of Manning roughness coefficient using optimization, including Ramesh et al (2000) to estimate the manning coefficient in multi branch channels. In this research, SQP algorithm was used in which objective function minimizes the sum of squares of relative differences (observed values) between estimated values and observed values of water height in channel. The result showed that optimization model does not have enough convergence toward an optimum solution. Ding et al (2004) carried out a research on determining manning roughness coefficient in shallow flows with different optimization algorithms. The results showed that for the roughness coefficient of manning, the algorithms in which the dominant constraints are upper and lower bounds have greater convergence speed, and the algorithm of problem-solving process can be used to solve other problems in hydraulics. Neguyan and Fenton (2005) investigated the determination of roughness coefficient in mixed channels using pawell algorithm. Objective function is calculated from

*Corresponding author E-mail: mheydari@basu.ac.ir

sum of difference square of observed and calculated water level. The results showed that in main channels and plain channels, it is different but it can be accepted with a permissible error percentage. This algorithm has low convergence velocity to find optimum solution [8]. Gazer Zadeh (2010) estimated Manning roughness coefficient of rivers using nonlinear optimization and Genetic algorithm and gave it as an input to program Mike II to get the values of water level and discharge (calculated), The objective function was obtained using a function based on the difference square of water level and discharge values with minimization. The output of this function is the optimum roughness coefficient in each interval along the river.

The present research aimed to study Manning roughness coefficient based on Simulated Annealing using gradually varied flow relationship. The benefit of this algorithm is the greater convergence velocity in getting optimum solution. We estimate Manning roughness coefficient with numerical method and Simulated Annealing. The value of Manning roughness coefficient obtained from optimization is given to HEC-RAS software as input and water level profile is obtained.

2. Materials and methods
2.1. Governing equations

Regarding Fig .2, the discrete from of finite difference of dynamic equation in gradually varied flow (Eq. (2)) is as the following:

$$\frac{\Delta E_s}{\Delta x} = S_0 - \frac{n^2 Q^2}{\bar{A}^2 \bar{R}^{4/3}} \tag{4}$$

$$\frac{E_{i+1} - E_i}{\Delta x} = S_0 - \frac{n^2 Q^2}{(\frac{A_i + A_{i+1}}{2})^2 (\frac{R_i + R_{i+1}}{2})^{4/3}} \tag{5}$$

in which (j) is the average of specific energy variation to distance and index (i) is the number of cross-section.

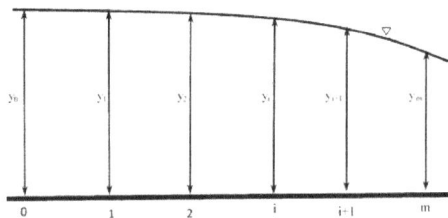

Fig. 2. Schematic figure of target function.

In Fig .2, the all interval of channel is divided in to m subinterval while $\Delta E / \Delta x$ and $\Delta E / \Delta x$ are the gradient of calculated specific energy (Eq. (5)) and observed specific energy gradient, respectively. The objective function can be the sum of square of difference between observed and calculated $\Delta E / \Delta x$ in which the decision-making is roughness coefficient of Manning (Eq. (6)).

$$OF = \sum_{j=1}^{m} [\left(\frac{\Delta E}{\Delta X}\right)_{j_{cal}} - \left(\frac{\Delta E}{\Delta X}\right)_{j_{obs}}]^2 \tag{6}$$

$$OF = \sum_{j=1}^{m} [S_0 - \frac{n^2 Q^2}{(\frac{A_j + A_{j+1}}{2})^2 (\frac{R_j + R_{j+1}}{2})^{4/3}} - \left(\frac{\Delta E}{\Delta X}\right)_{j_{obs}}]^2 \tag{7}$$

in which (j) is the number of reach. In this research, to minimize the objective function, the problem was written as a computer program in MATLAB R2015 software.

2.2. Simulated Annealing Algorithm (SAA)

Simulated Annealing (SA) algorithm is a numerical optimization method with an intelligent random structure. The idea of mathematical principles for Simulated Annealing was first introduced by Metropolis in 1953. Then, Kirkpatrick (1983) and Cereni (1985) proposed it as an optimization algorithm. The idea of Simulated Annealing algorithm is taken from annealing the metals to solid state (environment temperature) so that crystal structure of metal is form regularly in least energy level. In optimization of mathematical functions, the minimized value of objective function corresponds lower energy levels of a material in the freezing state. Simulated Annealing algorithm is simple and powerful used to solve optimization (minimization) with a large search space. The most important feature of Simulated Annealing algorithm is not being located in local optima. Also, the rate of temperature reduction in Simulated Annealing method is very important. According to Simulated Annealing, to get these minimum values, the least variations are considered in problem solutions stepwise. The most important parameters which must be examined in Simulated Annealing method are T_0 Initial temperature, B Temperature update function, I_t Max iterations, EPOCH Reannealing interval and EBS Function tolerance and Annealing function. The temperature reduction function consists of linear functions, exponential function and logarithmic function. The annealing function consists of Baltzman and fast function. To get the optimal solution of this algorithm, the objective function obtained underwent the sensitivity analysis. Table 1 shows the variation of parameters change and the values for sensitivity analysis.

2.3. Tests

The tests were carried out using nine discharges of 17.99, 28.19, 38.14, 49.26, 61.74, 70.86, 79.37, 83.55, 84.58 lit/sec in a glass rectangular flume of 12 m length and 60 cm width, 65 cm height with constant bottom slope of 0.0002 (Fig. (3)). To measure the flow discharge from a calibrated rectangular weir located in flume downstream. To measure the depth of flow profiles, total channel interval was devided in to 12 one-meter subintervals and the water level profile was recorded with a point gage of I ±0.1 mm accuracy. In each test, to increase accuracy, the recording of water level profiles was done 2 times and their average was considered to be the flow profile depth.The results of water level profiles for each discharge are presented in Fig. 4.

In this research, 9 discharges were used for optimization and estimation of Manning roughness coefficient by Simulated Annealing algorithm. To do optimization, 6 discharges for calibration and 3 discharges for validation of optimization results were used. To compare the profiles of observation and calculation, with Manning roughness coefficient, hec-ras software was used to depict the diagram.

3. Results and discussion

In Tables 2 and 3, the values of the objective function for the simultaneous variation of the values are expressed in Table 1. The objective function didn't show any sensitivity to the change of Initial temperature. Also, there was no solution about the logarithmic and linear update function, so we didn't mention them. If the Max iterations of Reannealing interval in each epoch is taken a variable, the optimization program was run and the results are presented in Table 2. As seen in Table 2, the value of objective function is 1.281×10^{-5} and that of Manning roughness coefficient is 0.011 and in the least Max iterations 300 and 30 Reannealing interval, they were obtained with fast annealing function and exponential temperature function. The best function value and the final point are shown in Fig. 5 graphically.

Table 1. Parameters assessed in Simulated Annealing algorithm for sensitivity analysis.

Temperature update function	Annealing function	Max iterations	Initial temperature	Function tolerance	Reannealing interval
Exponential		500	100	0.000001	100
	Fast	400	50	0.00001	50
Linear		300	20	0.0001	40
	Baltzman	200	10	0.001	30
Logarithmic		100	5	0.01	20
				0.1	10

Fig. 3. Lab flume.

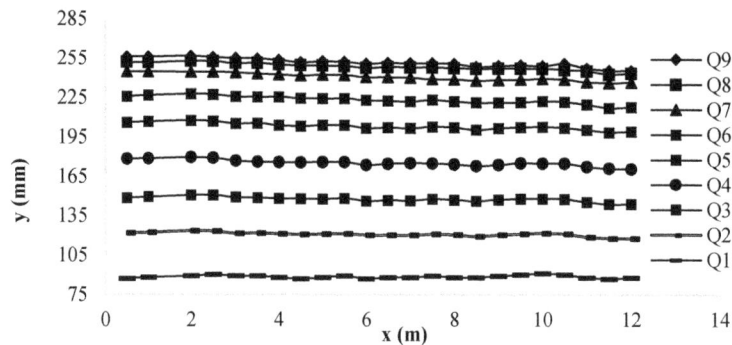

Fig. 4. Water level profile recorded for different discharge.

Table 2. The values of objective function in the sensitivity analysis test simultaneous with Simulated Annealing parameters based on the reannealing interval for max iterations.

Temperature update function	Annealing function	Function tolerance	Max iterations				
			500	400	300	200	100
Exponential	Fast	100	1.218	1.218	1.218	1.4236	3.1471
		50	1.218	1.218	1.218	1.3725	3.1471
		40	1.218	1.2181	1.218	1.3829	3.1471
		30	1.2182	1.218	1.218	1.3474	3.1471
		20	1.218	1.221	1.2189	1.3398	3.1471
		10	1.2184	1.2263	1.2831	1.409	3.1471
	Boltzman	100	1.218	1.2199	1.5145	3.1471	3.1471
		50	1.218	1.2198	1.3722	3.1471	3.1471
		40	1.220	1.2236	1.5356	3.1471	3.1471
		30	1.2245	1.2208	1.492	3.1471	3.1471
		20	1.2596	1.2798	1.5351	3.1471	3.1471
		10	1.8656	1.5173	1.3722	3.1471	3.1471

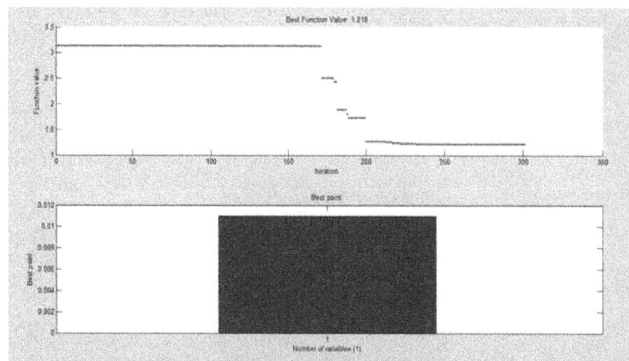

Fig. 5. The best value of objective function for best final point.

Regarding the total number of repetitions with Function tolerance to be variable, the optimization program was run as shown in Table 3. The minimum value of objective function is 1.218×10^{-5} for Manning roughness coefficient value of 0.011 at least Max iterations of 300 and Function tolerance of 10^{-6} in fast annealing function and exponential temperature update function. In Fig. 6, the best point of objective function and the best function value are shown.

Finally, the results of implementation and sensitivity analysis of Simulated Annealing algorithm and Manning roughness coefficient are shown in the table.

Table 3. The values of objective function in simultaneous sensitivity analysis test of Simulated Annealing parameters based on the function tolerance for max iterations.

temperature update function	Annealing function	Function tolerance	Max iterations				
			500	400	300	200	100
Exponential	Fast	0.000001	1.218	1.218	1.218	1.4153	3.1417
		0.00001	1.218	1.218	1.218	1.4093	3.1417
		0.0001	1.218	1.218	1.218	1.3869	3.1417
		0.001	1.218	1.218	1.218	1.3517	3.1417
		0.01	1.218	1.218	1.218	1.3822	3.1417
		0.1	1.218	1.218	1.218	1.3878	3.1417
	Boltzman	0.000001	1.218	1.2199	1.3722	3.1417	3.1417
		0.00001	1.218	1.2198	1.3722	3.1417	3.1417
		0.0001	1.218	1.2197	1.4961	3.1417	3.1417
		0.001	1.218	1.2198	1.5346	3.1417	3.1417
		0.01	1.218	1.2196	1.5142	3.1417	3.1417
		0.1	1.218	1.2197	1.4901	3.1417	3.1417

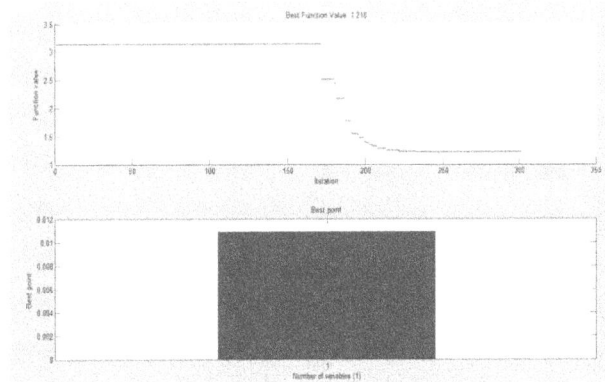

Fig. 6. The best point of objective function and best final point.

Table 4. The final results of Simulated Annealing algorithm.

Manning roughness coefficient	The best objective function value $\times 10^{-5}$
0.011	1.218

Also, for validation of Manning roughness coefficient from optimization the values of $\Delta Es/\Delta x$ were calculated and compared with observed values for 6 discharges 17.99, 28.19, 38.14, 61.74, 70.86, 83.55 (lit/sec) as in Fig. 7. The values of $\Delta Es/\Delta x$ for 3 discharges 49.26, 79.37 and 84 (lit/sec) using profile data were obtained to validate the coefficient as compared with the calculated values of $\Delta E/\Delta x = S_0 - S_f$ in Fig (8).

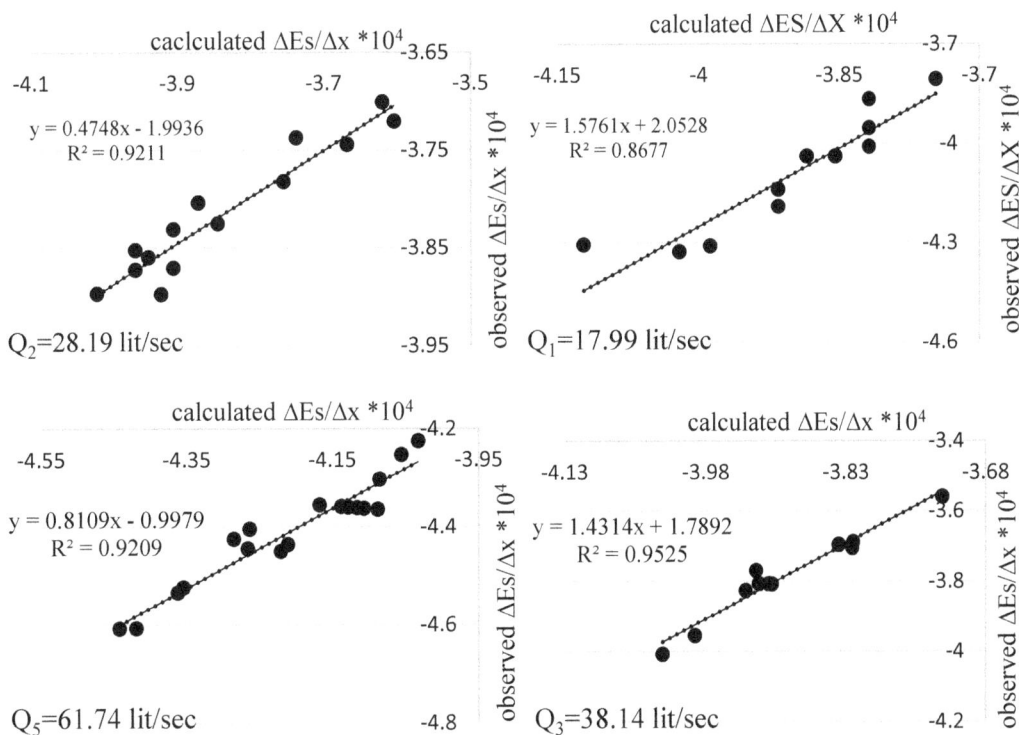

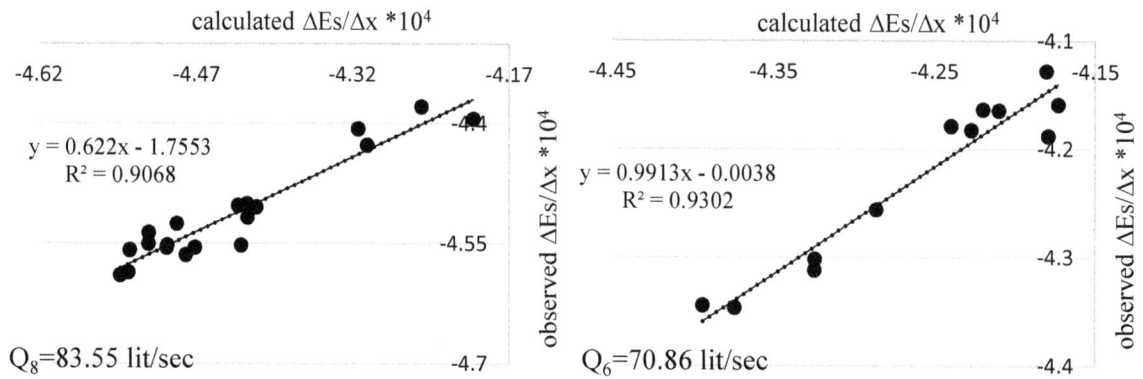

Fig. 7. The correlation diagram of calibration discharges.

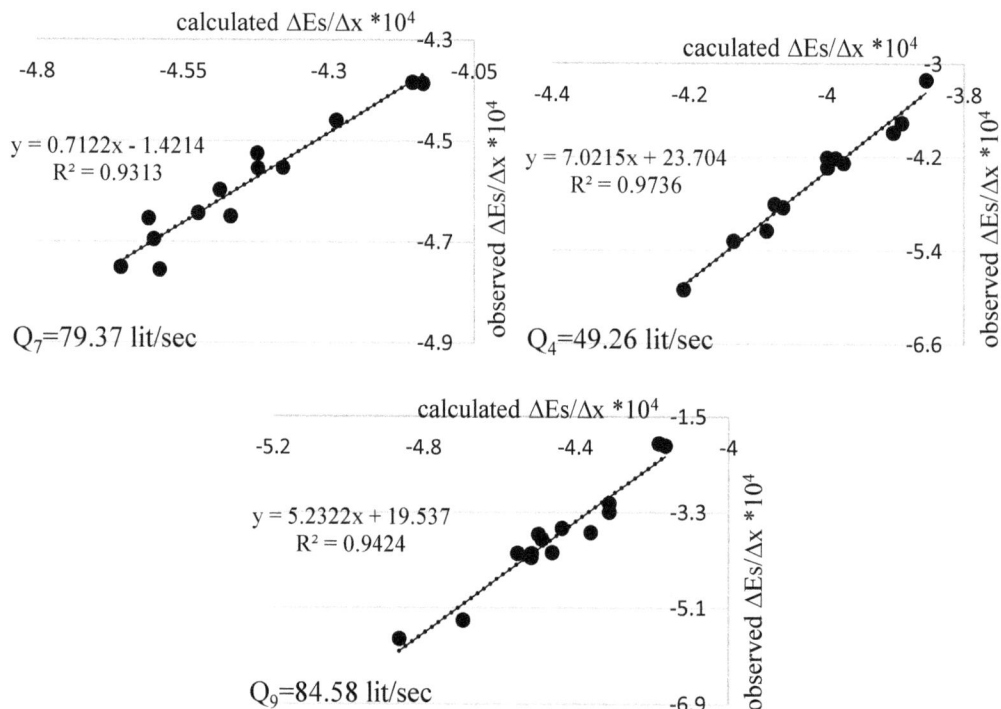

Fig. 8. The correlation diagram of validation discharges.

The statistical indices of Root mean square error (RMSE), Mean bias error (MBE) and Wilmuth or Adaptation index (d) were calculated for calibration and validation data from optimization as seen in Tables 5 and 6. As all the statistical indices are in good domain, the optimization method has been successful in estimation of Manning roughness coefficient.

With Manning roughness coefficient from optimization (n=0.011) and HEC-RAS software, calculated profiles of water level were drawn and compared with observed profiles, as in Fig. 9 and 10.

Table 5. The values of statistical indices for calibration data.

Number	Discharge (lit/sec)	RMSE	MBE	d
1	17.99	0.19978	0.19978	0.99997
2	28.19	0.06979	-0.03631	0.99999
3	38.14	0.10393	-0.10935	0.99991
4	61.74	0.20217	0.20217	0.99995
5	70.86	0.3492	-0.03336	0.99992
6	83.55	0.07801	-0.07809	0.99999

Table 6. The values of statistical indices for validation data.

number	Discharge (lit/sec)	RMSE	MBE	D
1	49.26	0.60691	0.47340	0.99969
2	79.37	0.14311	0.14311	0.99999
3	84.58	0.89873	-0.69054	0.99995

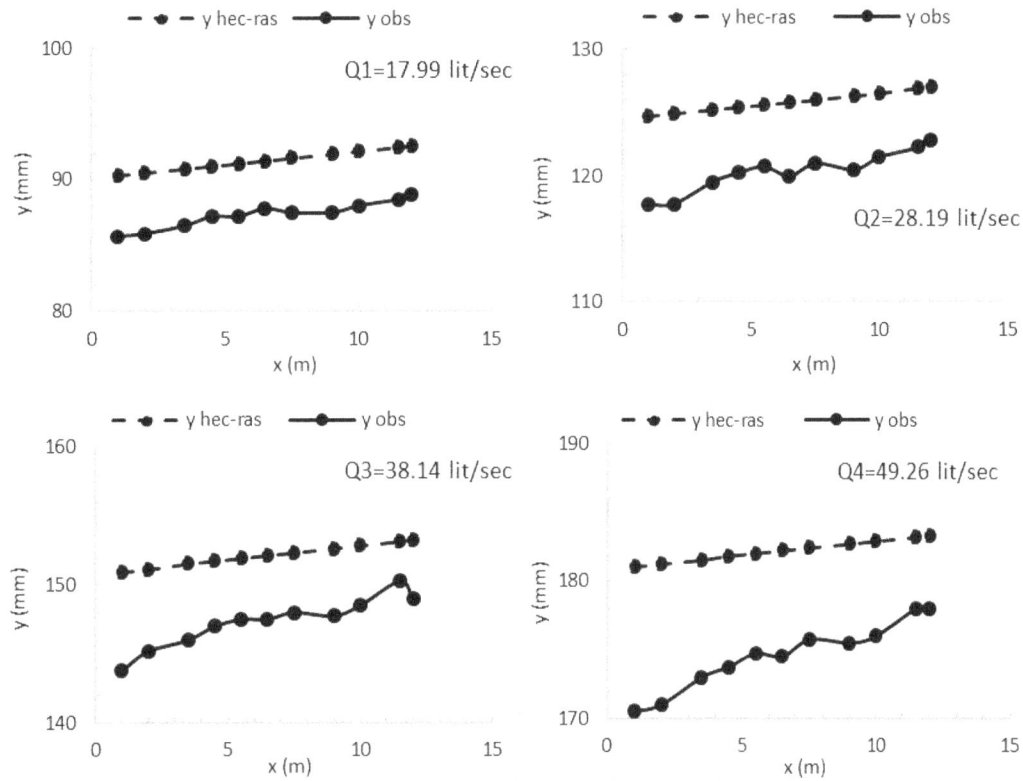

Fig. 9. Observed and calculated profiles of water level.

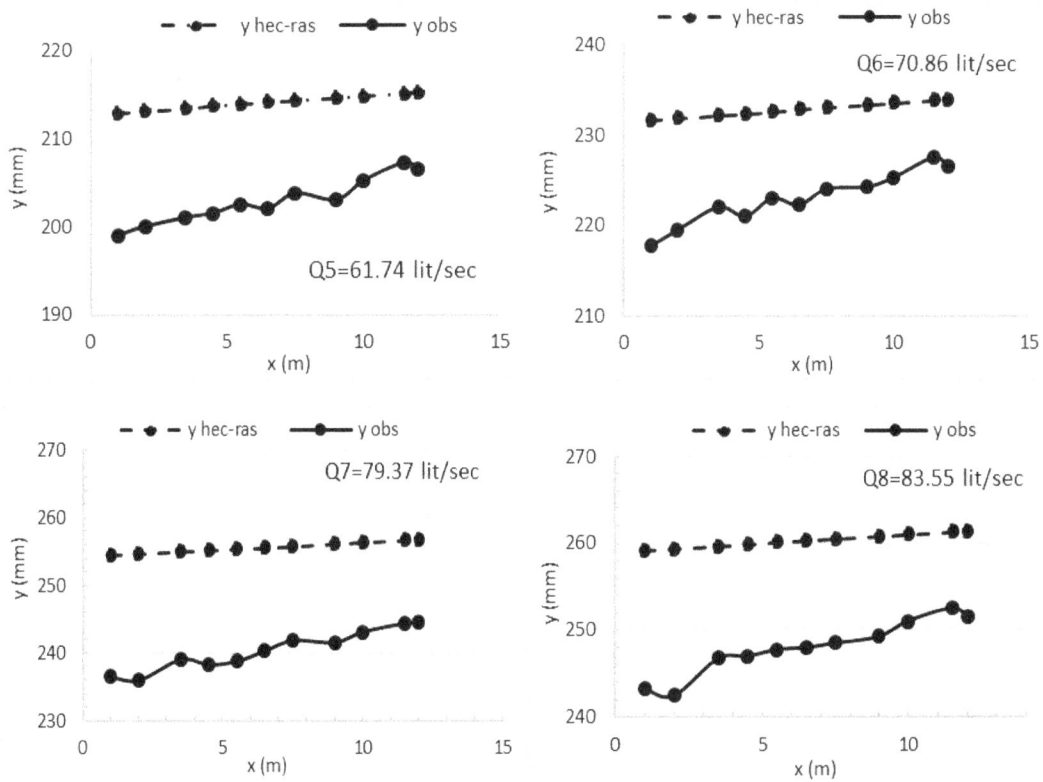

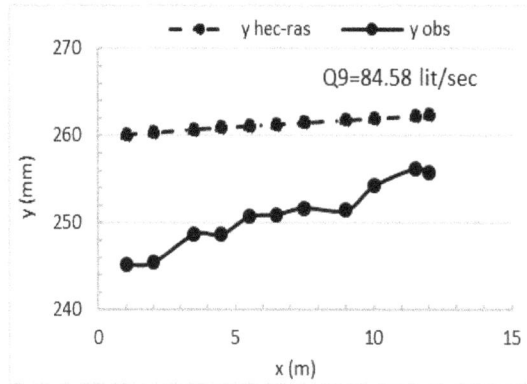

Fig. 10. Observed and calculated profiles of water level.

As seen in Fig. 9 and 10, observed profiles are near the calculated profiles showing the accurate estimation of Manning roughness coefficient using the combination of Simulated Annealing algorithm and gradually varied flow theory. The values of correlation diagram are shown in Figs. 10 and 11 for 9 discharges.

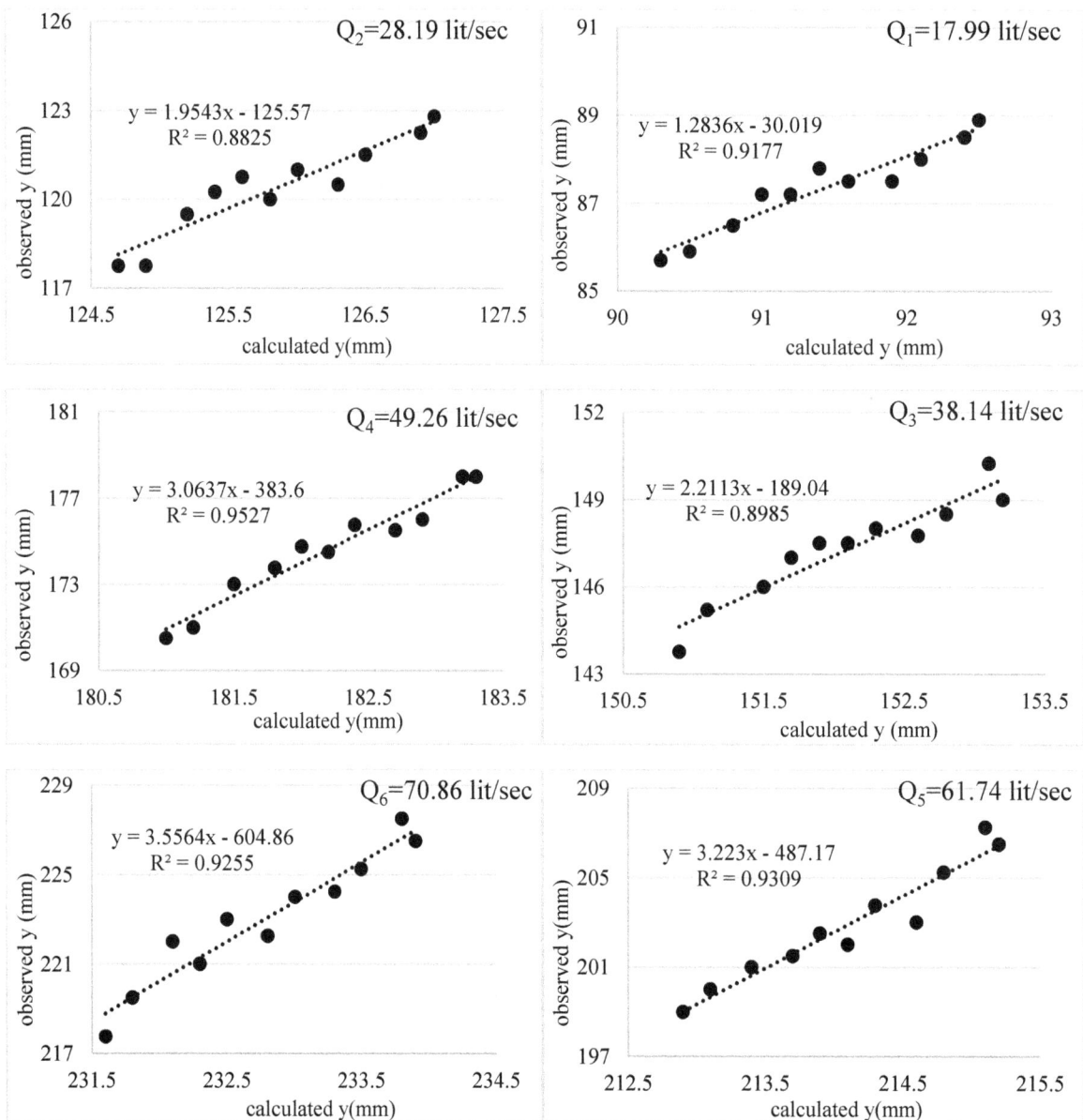

Fig. 10. The correlation for observed and calculated profiles.

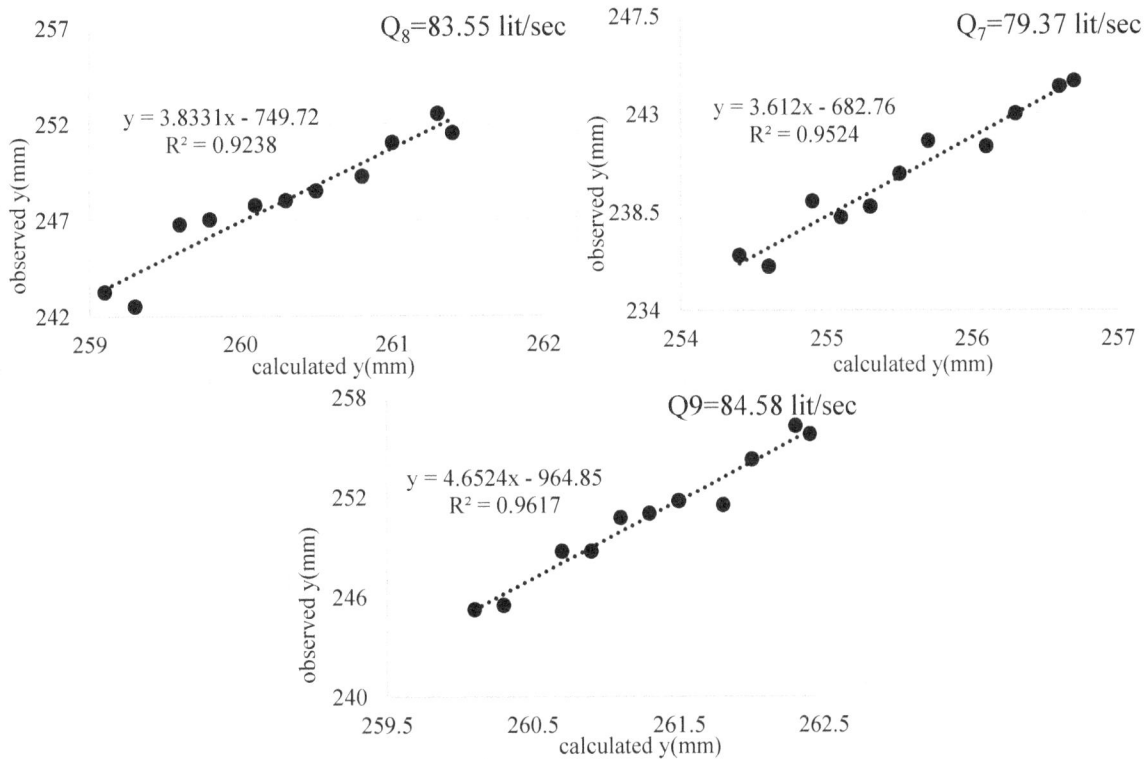

Fig. 11. The correlation for observed and calculated profiles.

Comparing the manning coefficient roughness results obtained in this study for that is equal 0.011, with the coefficient recommended by Chow [min 0.009, max 0.013] for the walls and bottom of the glass, the optimization method has been successful in estimation of manning roughness coefficient.

Also, using the equation Chen et al, Sauer and Manning, we attempted to estimate Manning roughness coefficient, the results are shown in Table 7.

Table 7. Manning roughness coefficient is calculated according to the equation.

Equation	Manning coefficient roughness calculated
Chen, et al:	0.0113
Sauer:	0.0109
Manning:	0.0112

4. Conclusions

There was presented a new method to estimate Manning roughness coefficient using Simulated Annealing (SA) algorithm and gradually varied flow equations with good results. Simulated Annealing algorithm has a good convergence to gain the optimum solution. Using this method leads to the increase of accuracy in estimating this coefficient and reduction of human error, resulting in good design and better performance of utilizing water distribution networks. As in hydraulic labs, a combination of glass, plastic and metal is used in the design of walls, the presented method can give the value of Manning roughness coefficient accurately.

References

Abrishami j., Hosseini S.M., Open-Channel Hydraulics, University of EmamReza press, Mashhad, (2011).

Cereny V., Thermodynamical approach to the traveling salesman problem: an efficient simulation algorithm, J. optimization theory and applications, Springer 45 (1985) 41-51.

Chow V.T., Open Channel Hydraulics, McGraw-Hill, New York, N.Y, (1959).

Ding Y., Jia Y., Wang S.S.Y., Indication of manning's roughness coefficient in shallow water flows, Journal of Hydraulic Engineering 130 (2004) 501-510.

Gazer Zadeh A., Manning roughness coefficient nonlinear optimization model to estimate, Master Thesis, University of Chamran, Ahvaz, Iran, (2010).

Kirkpatrick S., Gelatt C.D., Vecchi M.P., Optimization by simulated annealing, Science, 220 (1983) 671-680.

Kochakzadeh S., Maghsoudi N., Free surface flow Hydraulics, Steady one dimensional flows, 7th edition, University of Tehran press, Tehran, 1 (2011).

Metropolise N., Rosenbluth A., Teller A., Teller E., Equation of state calculations by fast computing machines, J. Chem. Phys 21 (1953), 1087-1092.

Neguyan H.T., Fenton J.D., Identification of roughness for flood routing in compound channels, Proc.31st congress., Int. Assoc., Hydraulic Eng. and Res., Seoul, Korea, published on CD, IAHR 11-16 September (2005) 847-854.

Ramesh R., Datta B., Bhallamudi S.M., Narayana A., Optimal estimation of roughness in open channel flows, Journal of Hydraulic Engineering, ASCE 126 (2000) 299-303.

Tadayonfar H.R., Manning roughness coefficient effect on the distribution of roughness elements, Master Thesis, University of Ferdosi, Mashhad, Iran, (2009).

19

Mean flow characteristics, vertical structures and bed shear stress at open channel bifurcation

Akbar Safarzadeh[1,*], Babak Khaiatrostami[2]

[1]Department of Engineering, University of Mohaghegh Ardabili, Iran.
[2]Research Department, Ardabil Regional Water Co., Iran.

ARTICLE INFO

ABSTRACT

Keywords:
Dividing flow
CFD
Secondary flow
Turbulence model
Sediment transport

Water supply from rivers is accomplished with flow diversion through an intake structure. A lateral intake like bifurcation is the simplest method to withdraw water. However, flow at a channel bifurcation is turbulent, highly three-dimensional (3D) and so has many complex features. This paper reports a 3D numerical investigation of these features in an open channel flow. Simulations have been done on rectangular channel geometry, with smooth bed and sidewalls. The standard k-ε, k-ω model of the Wilcox, and RSM turbulence models are compared using the commercial code FLUENT. The simulation results have been compared with available experimental data. It was found that all of the turbulence models tested here accurately predicted velocity profiles in the main channel but in the branch channel, the RSM model with the k-ω model performing better than the k-ε model. Predicted flow physics are in close agreement with previously reported experimental results.

1. Introduction

Rivers are a major source of water for meeting various demands. Usually, water supply from rivers is accomplished with flow diversion through an intake structure. River flow often transports sediment and designers are faced with the problem of sediment entering the canals and water conveyance systems. The art of the designer is to keep the amount of sediment entering the diversion system to a minimum.

A lateral intake is the simplest method of flow withdrawal. In spite of its simple layout, using this system leads to complex flow patterns and sedimentation problems at the junction region. Flow through lateral intakes is turbulent and highly three-dimensional consisting of secondary vortices and flow separation (Neary et al. 1996; Neary and Odgaard 1993).

The complex flow patterns can lead to sediment deposition in the intake channel (Barkdoll 1997 and Abbasi 2003). The past numerical studies of diversion flows have been mostly two-dimensional. Liepsch et al. (1982); Hayes et al. (1989); and Lee and Chiu (1992) have reported laminar flow calculations. 3D numerical investigations of laminar flow through lateral intakes have been reported by Neary and Sotiropoulos (1996). They employed a finite-volume method and used a non-staggered computational grid and demonstrated the relationship between singular points in the wall shear stress field and patterns of bed-load movement observed in the laboratory. Two-dimensional turbulent flow simulations have been reported by Shettar and Murthy (1996). They used depth-averaged mean flow equations closed with the k-ε model with standard wall functions. They are the only researchers that considered the effect of water surface variations on the flow characteristics in the junction region. Issa and Oliveira (1994) are the first researchers that have reported a 3D turbulent flow simulation for T-junction flow. They employed the Reynolds-averaged Navier-Stokes equations in conjunction with the k-ε turbulence model. Neary et al. (1999) conducted a 3D turbulent flow simulation for this problem. They employed 3D Reynolds-averaged equations closed with the k-ω model. They used the experimental measurements of Barkdoll (1997) for validation. None of the reviewed studies have attempted to compare the results of different turbulence models to identify the proper model for this problem. In this paper FLUENT, a commercially-available CFD software, has been used for simulating the turbulent flow structure through a lateral intake. The standard k-ε, k-ω model of Wilcox, and the Reynolds stress model (RSM) turbulence closure schemes are used to simulate the turbulent flow through lateral intakes in open channels and the results are compared to published experimental data.

1.1. Objectives

In this study, a 3D numerical investigation is carried out for turbulent incompressible flows through a 90-degree rectangular diversion. The objective of this work is twofold: (i) to identify the proper turbulence model for this problem; and (ii) to analyze the numerical solution in order to know the complex physics of diversion flows with emphasis on velocity profile variation along the main channel and branch channel, flow topology patterns, and shear stress variations on the solid boundaries.

2. Test case

As mentioned before, the experimental measurements of Barkdoll,1997 are used to validate the numerical results of the present study. The layout of the experimental flume is shown in Fig. 1. The experiments were conducted in an open-channel flume consisting of a T-junction of two straight rectangular channels with Ar = 2. Flow depth was determined by the volume of water in the flume and otherwise not regulated. Discharge was determined by Venturi meters on flume piping for both the branch and main channels.

The inlet discharge was 0.011 m^3/sec. Velocity measurements were obtained with a Sontek Acoustic Doppler Velocimeter (ADV). A discharge ratio of 0.32 was used to comply with the experimental results. The origin of the coordinate axis is located at the outer wall of the main channel, in front of the intake inlet. These axes are normalized with respect to the width of the channel ($X^* = X/b$).

*Corresponding author E-mail: safarzadeh@uma.ac.ir

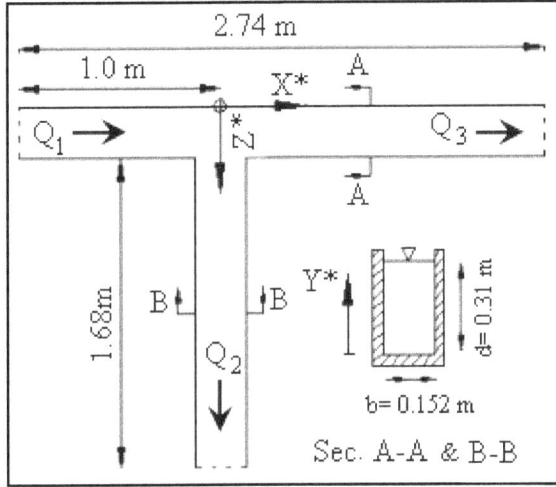

Fig. 1. Geometrical properties of the test case

3. Governing equations
3.1. Mean flow equations

For an incompressible fluid flow, the equation of continuity and balance of momentum for the mean motion, in Cartesian coordinates are given as (FLUENT, Inc., 1993):

$$\frac{\partial U_i}{\partial x_i} = 0 \tag{1}$$

$$U_j \frac{\partial U_i}{\partial x_j} = -\frac{\partial P}{\partial x_i} + g_{x_i} + \mu \frac{\partial^2 U_i}{\partial x_j \partial x_j} + \frac{\partial R_{ij}}{\partial x_j} \tag{2}$$

where U_i is the mean velocity, X_i is the position, P is the mean pressure, g is the gravity accelartion and μ is the dynamic viscosity. In Eq. 2, $R_{ij} = -\rho \overline{u_i' u_j'}$ denotes the Reynolds stress tensor. Here $u_i' = u_i - U_i$, is the ith fluid fluctuation velocity component. This parameter is modeled using the Boussinesq's assumption:

$$-\rho \overline{u_i' u_j'} = 2\mu_t S_{ij} - \frac{2}{3}\rho k \delta_{ij} \tag{3}$$

where, μ_t is the eddy viscosity. S_{ij} and k are mean rate of strain tensor and turbulent kinetic energy respectively and are defined as follows:

$$S_{ij} = \frac{1}{2}(\frac{\partial U_i}{\partial x_j} + \frac{\partial U_j}{\partial x_i}) \tag{4}$$

$$k = \frac{1}{2}(\overline{u_i' u_i'}) \tag{5}$$

3.2. Turbulence closure equations

The standard k-ε and the k-ω model of Wilcox and Reynolds stress model (RSM) turbulence closure schemes are used for turbulence modeling. In this research, these models are employed and the proper model is selected for further investigation of flow structure at this field.

3.2.1. Standard k-ε turbulence model

According to this model, the eddy viscosity is related to the turbulence kinetic energy (k) and its rate of dissipation (ε) (Celik, 1999):

$$\mu_t = \rho \, c_\mu \frac{k^2}{\varepsilon} \tag{6}$$

The turbulence quantities k and ε are calculated by the following transport equations:

$$U_i \frac{\partial k}{\partial x_i} = \frac{\partial}{\partial x_i}\left(\frac{v_t}{\delta_k}\frac{\partial k}{\partial x_i}\right) + v_t\left(\frac{\partial U_i}{\partial x_i} + \frac{\partial U_j}{\partial x_i}\right)\frac{\partial U_i}{\partial xj} - \varepsilon \tag{7}$$

$$U_i \frac{\partial \varepsilon}{\partial x_i} = \frac{\partial}{\partial x_i}\left(\frac{v_t}{\delta_\varepsilon}\frac{\partial \varepsilon}{\partial x_i}\right) + c_{\varepsilon 1}\frac{\varepsilon}{k}P - c_{\varepsilon 2}\frac{\varepsilon^2}{k} \tag{8}$$

G is the turbulence production by mean shear modeled as follows:

$$G = v_t \, (\frac{\partial U_i}{\partial x_j} + \frac{\partial U_j}{\partial x_i}) \, \frac{\partial U_i}{\partial x_j} \tag{9}$$

Closure coefficients used at this model are summarized in Table 1 (Celik, 1999).

Table 1. Closure coefficients used in k-ε model.

C_μ	$C_{\varepsilon 1}$	$C_{\varepsilon 2}$	δ_k	δ_ε
0.09	1.44	1.92	1.00	1.30

3.2.2. k-ω turbulence model

This model has been given by Wilcox (1988, 1994). In contrast to the k-ε model, which solves for the dissipation (ε) or rate of destruction of turbulent kinetic energy, the k-ω model solves for only the rate at which the dissipation occurs (the turbulent frequency, ω). Dimensionally ω can be related to ε by ω = ε / k (Celik, 1999):

$$\mu_t = \rho \frac{k}{\omega} \tag{10}$$

The turbulence quantities k and ω are calculated by the following transport equations:

$$U_i \frac{\partial k}{\partial x_i} = \frac{\partial}{\partial x_i}\left[(\frac{1}{R} + \sigma^* v_t)\frac{\partial k}{\partial x_i}\right] + G - \beta^* \omega k \tag{11}$$

$$U_i \frac{\partial \omega}{\partial x_i} = \frac{\partial}{\partial x_i}\left[(\frac{1}{R} + \sigma v_t)\frac{\partial \omega}{\partial x_i}\right] + \alpha \frac{\omega}{k}G - \beta \omega^2 \tag{12}$$

Closure coefficients used at this model are summarized in Table 2 (Celik, 1999).

Table 2. Closure coefficients used in k-ω model.

α	B^*	B	σ^*	σ
5/9	9/100	3/40	½	½

3.2.3. RSM turbulence model

The Reynolds stress model (RSM) solves the Reynolds-averaged Navier-Stokes equations by using the Reynolds stresses transport equations (seven-equations for 3D flow) and an equation for the dissipation rate, ε. The RSM accounts for the effects of the streamline curvature, vorticity, circulation, and rapid changes in the strain rate in a more efficient way than the two-equation models; however, it requires more computational effort and time. The transport equation in this model is as Eq. 9 (Launder, 1989, a and b).

Left hand side terms of Eq. 9 are the local time derivatives and convection term (C_{ij}) respectively. Terms of the right hand side are turbulent diffusion ($D_{T,ij}$), molecular diffusion ($D_{L,ij}$), stress production (P_{ij}), pressure strain (Φ_{ij}), dissipation (ε_{ij}) and production by system rotation (F_{ij}), respectively. Most of the terms in this transport equation, including C_{ij}, $D_{L,ij}$, P_{ij} do not require any modeling and are directly solved. However, $D_{T,ij}$ (Lien and Leschziner, 1994), Φ_{ij} and ε_{ij} (Gibson and Launder, 1978; Launder, 1989a and b) need to be modeled to close the transport equation. To simulation the pressure strain the linear pressure-strain method is used.

$$\frac{\partial}{\partial t}(\rho \overline{u_i' u_j'}) + \frac{\partial}{\partial x_k}(\rho u_k \overline{u_i' u_j'}) =$$

$$\frac{\partial}{\partial x_k}\left[(\rho \overline{u_i' u_j' u_k'}) + \overline{p(\delta_{kj} u_i' + \delta_{ik} u_j')}\right] +$$

$$\frac{\partial}{\partial x_k}\left[\mu \frac{\partial}{\partial x_k}(\overline{u_i' u_j'})\right]$$

$$-\rho\left(\overline{u_i' u_k'}\frac{\partial u_j}{\partial x_k} + \overline{u_j' u_k'}\frac{\partial u_i}{\partial x_k}\right) + \overline{p(\frac{\partial u_i'}{\partial x_j} + \frac{\partial u_j'}{\partial x_i})}$$

$$-2\mu(\overline{\frac{\partial u_i'}{\partial x_k}\frac{\partial u_j'}{\partial x_k}})$$

$$-2\rho\ \Omega_k(\overline{u_j' u_m'}\varepsilon_{ikm} + \overline{u_i' u_m'}\varepsilon_{jkm}) \tag{13}$$

4. Numerical solution

The CFD code used in this work is version 6.0.12 of Fluent. This software allows the solution of the three-dimensional Navier-Stokes equations in order to calculate the flow field. This code uses a finite-volume discretization method in conjunction with different turbulence models. Different schemes such as Second Order Upwind, SOU, Power Law and Quick may be used to discretize the convection terms of the transport equations. Pressure and velocity field coupling may be done by SIMPLE, SIMPLEC and PISO algorithms. Flow geometries are constructed using Gambit Software. Flow field boundary conditions and mesh generations were done by this software [8]. For this study, due to existence of circulation and separation zones in the flow field, the convection term has been discretized using the SOU scheme. Staggered meshes in conjunction with SIMPLE algorithm have been used for flow field solution. The convergence criterion is set to 10e-5.

4.1. Boundary condition

At the main channel inlet, the "Velocity Inlet" boundary condition has been used. The flow properties at the channel inlet are shown at Table 3 (Barkdoll, 1996). The flow is subcritical and has a turbulent regime. The velocity field and turbulence parameters (k,ε and ω) are imposed from a separate simulation of fully developed turbulent flow through a straight channel.

At the exits of both the main and branch channels, the "outflow" boundary condition has been used. This condition states that the gradients of all variables (except pressure) are zero in the flow direction. At these boundaries, the flow often reaches a fully developed state. To ensure that this condition was satisfied at the exit of the diversion (a fully developed separation eddy); this channel was lengthened to 2.2 meters. The experimental length of the main channel was recognized to be adequate.

Table 3. Flow properties at inlet section.

Discharge(Q_1) (lit/sec)	Froude Number	Reynolds Number
11	0.13	49600

The inlet discharge has been apportioned between channels. The Discharge ratio ($r = Q_2/Q_1 = 0.32$) was imposed to the exit of the diversion. Taylor showed that for discharge ratios between 0 and 0.45 and Froude numbers in the main channel between 0 and 0.4, there is less than 2% variation in flow depth in the vicinity of the diversion (Taylor, 1944). Using these points, the "Symmetry" boundary condition has been used to model the free surface. At the solid boundaries the "Wall" boundary condition was used. The walls were hydraulically smooth and the no-slip and no-flux conditions dictated to them. The implementation of wall boundary conditions in turbulent flows starts with the evaluation of:

$$y^+ = \frac{\Delta y\,p}{\upsilon}\sqrt{\frac{\tau_w}{\rho}} \tag{14}$$

where, Δy_p is the distance of the near-wall node to the solid surface and τ_w is the wall shear stress. The distance of the first grid surface off the walls is important and depends on the flow conditions, wall roughness and the turbulence model that is used. The k-ε model uses the wall function to "bridge" the solution variables at the near-wall cells and the corresponding quantities on the wall but the k-ω model resolves the near wall region (laminar sub layer region).

4.2. Materials and methods

A three dimensional view of a typical computational mesh for a rectangular diversion configuration is shown in Fig. 2. The grid lines were clustered near the solid walls and in the junction region. To ensure the validity of the numerical solution for each turbulence model, different mesh sizes were applied to each of the models, according to the recommended selection of the near-wall cells (FLUENT 1993). Fig. 3 illustrates comparisons between measured and predicted velocity profiles using the RSM turbulence model with different mesh sizes.

A variety of mesh sizes were employed. The coarsest of the mesh sizes with similar results are shown. For the coarser of these two grid cases there were 198,620 total nodes and 220,425 for the finer grid case. The difference between the coarse and the fine grid predictions at the main channel are small enough to conclude that the coarser of the meshes be selected for the remainder of the analysis. For the k-ε and RSM models the total nodes were thus 198,620 while for k-ω they were 252,412. Additionally, it was found that using 10 nodes in the boundary layer for the k-ε and RSM turbulence models and 2 nodes in the laminar sub-layer for the k-ω model is adequate. These were determined by the guidelines in the FLUENT User's Manual (FLUENT Inc., 1993).

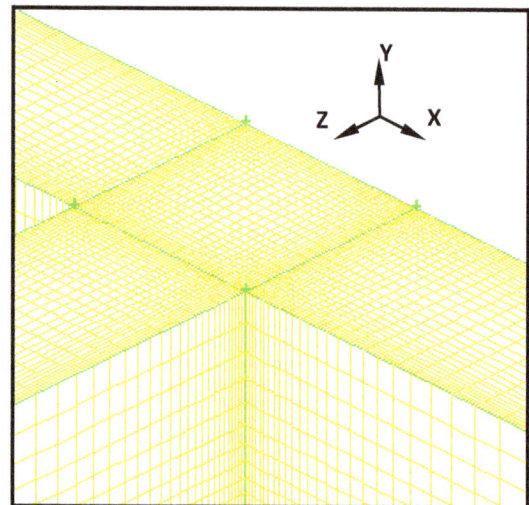

Fig. 2. Computational mesh at dividing zone.

5. Results and discussion
5.1. Appropriate turbulence model

Comparisons between measured and predicted streamwise velocity profiles near the free surface plane in both the main and branch channels for the two turbulence models employed are presented in Fig. 4. This Figure shows that all of the turbulence models used accurately predict velocity profiles in the main channel but in the branch channel, the RSM model performs very well and the k-ω model performed better than the k-ε turbulence model.

Comparison between measured and predicted results in Fig. 6 show that the k-ε model under-predicts the length of the separated flow in the intake channel but the k-ω and RSM model predictions show good agreement with the experimental results. The prediction by the k-ω model differs slightly from the experimental results, but the RSM model predicts the velocity profiles very well.

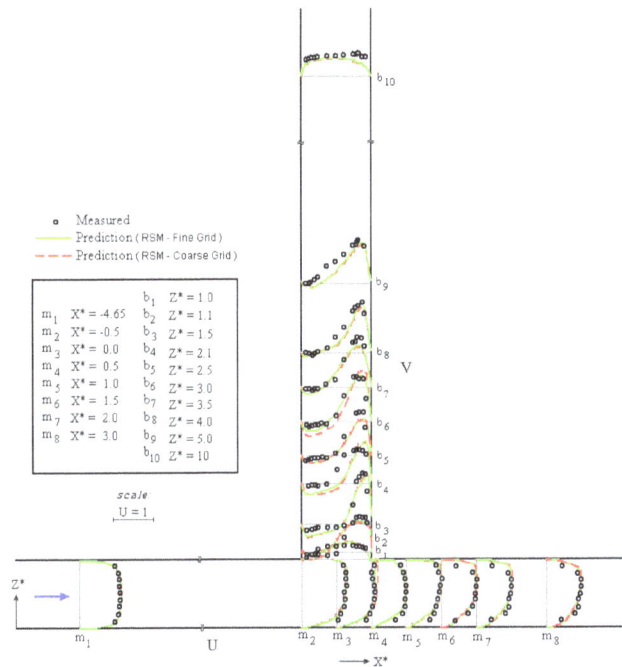

Fig. 3. Grid sensitivity results (only the coarsest two acceptable grid size results shown)."Fine grid" had 220,425 computational nodes and "Coarse grid" had 198,620. Re=49600.

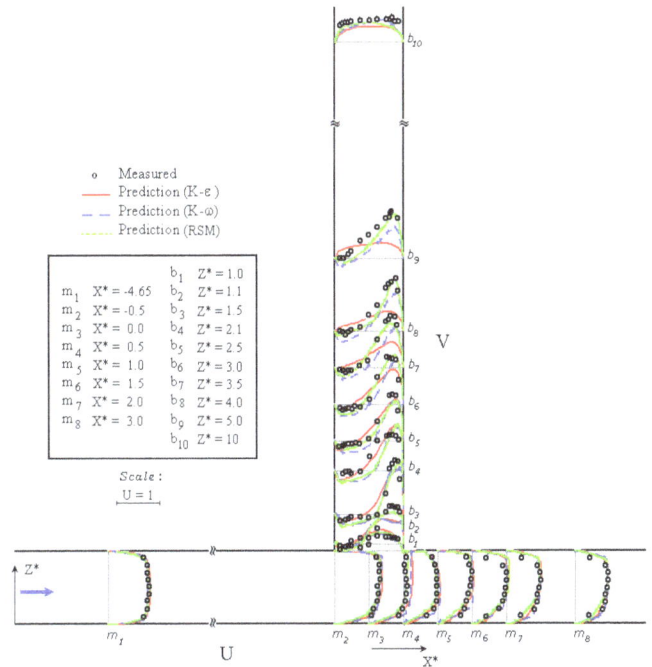

Fig. 4. Comparison of turbulence model results (Near surface 2D velocity profiles).

Forward velocity maximum shifts toward the inner bank as the flow enter the inlet region (Sec. m_2). As the flow enters into the branch, the resultant velocity along the inlet reduces (Sec. m_3) and hence, at the downstream edge of the inlet, forward velocity maximum shifts away from the inner wall (Sec. m_4). The small discrepancies at Sec. m_3 may be caused by the so-called "velocity-dip" phenomenon that cannot be simulated by isotropic models, while RSM models predicts this phenomenon well.

The failure of the two-equation models to accurately model the regions with anisotropic turbulence, such as curved-surface secondary motions and separation are the weakness of these models. In addition, the two equation turbulence models are unable to predict certain flow features because of the assumption that the flow does not depart far from local equilibrium, and that the Reynolds number is high enough that local isotropy of eddy viscosity is approximately satisfied.

5.2. Physics of diving flow

Fig. 5 illustrates the 2D streamline plots at three horizontal planes: near bed (Y/H=0.01), Mid-depth (Y/H=0.5) and near free surface planes (Y/H=1). The most distinct features in these plots are:
1- The dividing streamline. This line denoted as "SL" in these figures. Its location in the main channel changes over depth extending out further near the bed than near the surface.
2- Fig. 5 (a) indicates the separation zone along the left wall of the intake. The normalized length of this zone ($L_r=L_s/b$) is about 5.2. The length of the separation zone decreases by going downwards.
3- In Fig. 5(c) there are streamlines of flow that appear to emanate on the right wall of the branch. These are streamlines of flow that enter the branch at higher elevation, then impinges on the right wall and are deflected down ward toward the bed where they spread out in the direction of the separation zone. Interaction of deflected downward flow and cross sectional secondary flows along the lateral intake results in a tornado-like 3D motion (Fig. 7).
4- Fig. 5(c) contains many important features that may be used for sediment transport implications. This figure shows that the most of the bed flow reaching the junction region is dragged into the intake channel. There are some special points that have implications for sedimentations. These are the points where the magnitude of the shear stress vector goes to zero and its direction is indeterminate. These are known as "singular-points" and the most significant among

them are a focus of separation (F_s), located in the branch channel and a saddle point (S), located just off the downstream corner of the intake.

The focus of separation in the branch channel is a point where near-bottom particles would tend to accumulate and corresponds with the region where sandbars are commonly known to form at lateral intakes. The saddle point off the downstream corner of the intake is the origin of streamlines that sweep the bed of the junction region and divert them to the focus of the separation region (Neary et al., 1993). Fig.7 illustrates the bed-shear stress distributions. This figure can be used to identify likely regions of scour and deposition. The following three regions are observed:

1- A circular region of low bed-shear stress, which coincides with focus point (zone m), discussed earlier.

2- A high bed-shear stress zone that looks like a dagger (zone **n**) where bed-load transport is likely to occur.

3-A high shear stress zone off the downstream corner of the intake (zone **p**). Downward flow and secondary motion at this region are the causes of this scouring.

4- A low shear stress zone coinciding with the saddle point (zone **q**). At this region, the sedimentation may occur.

5- A high shear stress zone at the inlet of the diversion channel (zone l). Laterally accelerated flow due to transverse suction pressure and secondary flow due to role of the dividing surface as the outer bank of a virtual bend and outer boundary of the separation zone as the inner wall of a virtual bend are the main causes of the scorring process at this region.

Common sediment transport models use the so-called mean bed shear stress approach in which bed-load transport occurs when the bed-shear stress exceeds a threshold level. Although the present simulations were conducted with a fixed bed, the bed shear stress approach can be used to interpret the computed shear stress distributions to gain qualitative insights about the effects of the flow structure on sedimentation processes in lateral intakes.

Comparison of Figs. 7 and 8 shows that high and low shear stress zones predicted by current numerical simulations perfectly coincide with scouring and deposition zones at dividing zone of the channel which are previously reported in a movable bed experimental work.

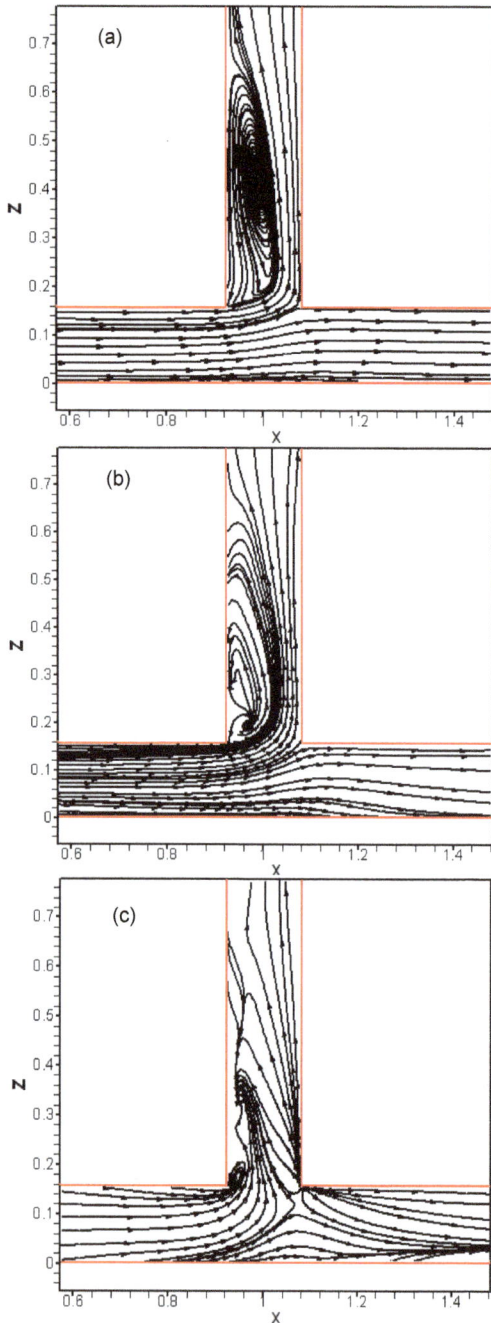

Fig. 5. 2D streamlines at horizontal planes: (a): Y/H=1, (b): Y/H=0.5 and (c): Y/H=0.01.

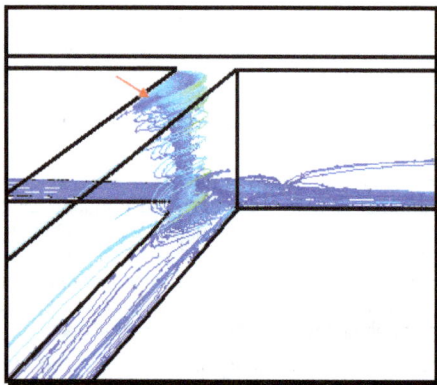

Fig. 6. 3D flow structure inside of the branch channel.

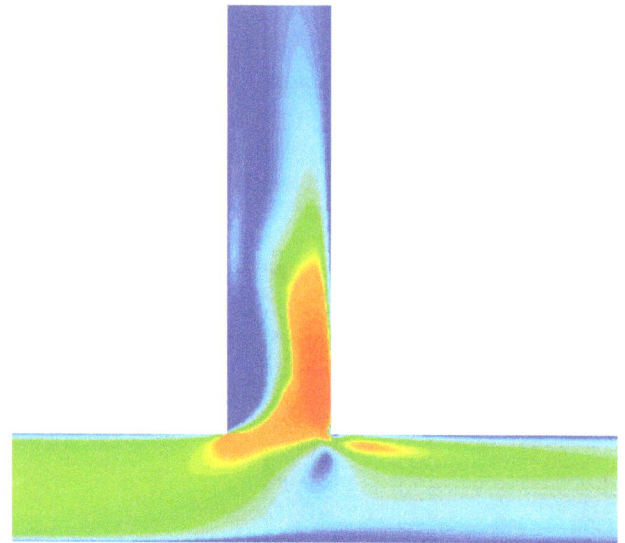

Fig. 7. Bed shear stress contour at dividing flow zone.

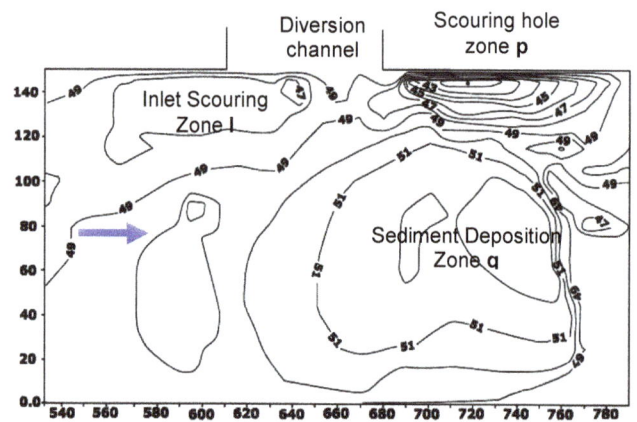

Fig. 8. Bed topography in a movable bed experimental dividing flow test (Abbasi, 2003).

6. Conclusions

A commercially available CFD code for the prediction of flow in the open channel division was used. Three turbulence closure schemes were employed and the performance of each model was evaluated using experimental data. It was found that all of the turbulence models tested here accurately predicted velocity profiles in the main channel but in the branch channel, the RSM model with the k-ω model performing better than the k-ε model. 2D streamline plot at near-bed surface exhibits a complex nature of flow at this plane. At this plane the singular points was seen. The most significant among these points are a focus of separation (Fs), located in the branch channel and a saddle point (S), located just off the downstream corner of the intake. These points have an important role in sedimentation problem. The calculated bed-shear stress distributions can be used to identify likely regions of scour and deposition.

Due to presence of highly turbulent dividing surface along the main channel and the flow separation zone along the diversion channel, instability of shear layers generated by these two flow mechanisms induce strong instantaneous vertical motions and consequently bed shear stresses. It is necessary to use sophisticated modeling approaches such as large eddy simulation (LES) in order to investigate the unsteady nature of the turbulent flow at dividing zone, especially at the near bed region to improve our knowledge about the sediment control at river diversion projects.

References

Abbasi A., Experimental study of sediment control at lateral intake in straight channel, PhD Thesis, Tarbiat Modares University, (2003).

Barkdoll B., Sediment control at lateral diversion, PhD dissertation, University of Iowa, (1997).

Celik, I.B., Introductory turbulence modeling, Western Virginia University, (1999).

Fluent Inc., FLUENT user's guide; Fluent, New Hampshire, (1993).

Gibson M.M., Launder B.E., Ground effects on pressure fluctuations in the atmospheric boundary layer, Journal of Fluid Mechanics 86 (1978) 491-511.

Launder B.E., Spalding D.B., Lectures in Mathematical Models of Turbulence, Academia press, London, England, (1972).

Launder B.E., Second-moment closure and its use in modeling turbulent industrial flows, International Journal of Numerical Methods in Fluids 9 (1989a) 963-985.

Launder B.E., Second-moment closure: present and future?, International Journal of Heat Fluid Flow 10 (1989b) 282-300.

Lien F.S., Leschziner M.A., Assessment of turbulent transport models including non-linear RNG eddy-viscosity formulation and second-moment closure, Computers and Fluids 23 (1994) 983-1004.

Neary V., Odgaard A.J., Three-dimensional flow structure at open channel diversions", Journal of Hydraulic Engineering 119 (1993) 1224–1230.

Neary V., Sotiropoulos F., Numerical investigation of laminar flows through 90-degree diversion of rectangular cross-section, Computer and Fluids 25 (1996) 95-118.

Shettar A., and Murthy K., A numerical study of division of flow in open channels, Journal of Hydraulic Research 34 (1996) 651-675.

Taylor E.H., Flow characteristics at rectangular open-channel junctions, Transactions of the American Society of Civil Engineers 109 (1944) 893–912.

20

Analyzing Tabriz metropolitan drinking water utilities by using performance benchmarking

Hamid Najaf Zadeh[1], Karim Hosseinzadeh Dalir[1, *], Mohammad Reza Pourmohammadi[2]

[1]Geography and Urban Planning Department, Marand Branch, Islamic Azad University, Marand, Iran.
[2]Geography and Urban Planning Department, Faculty of Planning and Environment Sciences, University of Tabriz, Iran.

ARTICLE INFO

Keywords:
Assessment
Existing conditions
Performance benchmarking
Tabriz Metropolitan
Water utility

ABSTRACT

Supply and maintenance of urban drinking water utilities are the most important priorities of people in the world especially in urban areas and it is very clear for urban planners or decision makers to evaluate the costs of action or weigh them against the problems of inaction. Also, specific annual budget is essential for ensuring people welfare and using water utilities with good quality. There are different issues in relation to managing of urban water utilities in terms of cultural, social, physical, environmental and even political and it is necessary to assess the existing conditions of utility by authorities and experts for making decision about those applications. So, we introduce Performance Benchmarking method for reaching this aim. This method is one of the best and update solutions in analyzing drinking water utility in developed countries especially in United State of America. So, in this paper, seven drinking water utilities of United State have been compared with Tabriz metropolitan drinking water utility that is located in North West of Iran and results of indicators' performance have been comparatively explained. Also, Results show that Tabriz metropolitan drinking water utilities are low advanced in terms of many indicators' performance than seven United States drinking water utilities. But, in some indexes almost equal to and in certain other cases are advanced than it. However, this methodology is very effective for decision makers, responsible and other experts in all regions and this model can be applied for other cities and urban areas.

1. Introduction

According to United Nation's projections, by 2050 almost half of the world's population will be experiencing either water scarcity (<1,000m3 of renewable water per capita per year) or water stress (between 1,000m3 and 1,700m3 per capita per year). It is estimated that 1 billion people in developing countries do not have access to portable water and unsafe water is implicated in the deaths of more than 3 million people annually and causes 2.4 billion episodes of illness from water-borne diseases each year (Oyegoke et al.2012). The world urban population was projected to increase from 6.7 billion in 2007 to 9.2billion in 2050(United Nations. 2008). 90% of this global entire population growth will take place in urban areas of developing economies (United Nations. 2004). Lacks of fresh water with good quality are the greatest challenges of civilization in the 21 century that threats social welfare, public health and ecosystems. So, reduction of water resources has been daily done in many countries in two ways including: historical evaluation and predicting of future in 1950 to 2010 that indicates these problems is very important and critical. In general, managing the needs to drinking water have been resolved the gap between supply and demand. reducing water harvesting in America properly confirm the Third World Water statement in Kyoto 2003 that the water crisis is not the lack of water but the problem is water management. At now, Tabriz metropolitan are faced with increasing demand for drinking water and limited water resources because of growing population trend. Because of the geographical location of Tabriz and the uncontrolled growth of water demand for new applications, water management organization of

Tabriz city has faced with difficult situations, despite the use of adjacent water resources and construction of a new urban water infrastructure. So, it can be said that Tabriz urban water decision makers are faced with three major challenges in terms of water supplying for citizen. The first challenge is population growth and increasing water demand. The second challenge is funding for the implementation of water supply, transmission and demand management in inter-basin. And the third challenge is the lack of integrated management system in the metropolitan city of Tabriz and also in Iran. For example, due to lack of water infrastructure in some of the surrounding towns of Tabriz and especially in the Khavaran town in east of this city, the development and construction of these settlements are faced with many problems. Also, surplus Density selling by municipality of Tabriz causes other urban difficulties in this city. Nowadays, Managing and utilization of water resources and the creation, operation and maintenance of water and wastewater installations and structures, policy management, watershed and water resource development and management studies and planning water supply are important issues in all human societies. Therefore, research in this subjects and presenting techniques to improve this situation and conditions have particular importance in terms of training and research centers. The studies about water infrastructure show that these infrastructures are destroyed over time. These losses occur as a result of lack of proper maintenance and this issue itself primarily related to the finance budget and facilities. In general, experts suggest that the experience gained in the planning and development to maintain and enhancing the efficiency of the current situation and future development of drinking water infrastructure is

*Corresponding author E-mail: prodalir@yahoo.com

urgent essential. But, for the maintenance and development of basic infrastructure, it is primarily needed to evaluate them. Such an assessment is possible only with performance benchmarking. Benchmarking is a tool for infrastructure managers and supervisors that it can be applied for all drinking water infrastructures in all states and cities to compare performance of them. However, learning and knowledge about indices to assessing ultimate performance of infrastructure and to better managing of them is necessary and very important. Benchmarking allows people who are not part of the utility to develop confidence that it is efficient and able to continuously improve. It also provides value in terms of cost and service by identifying factors that could delay potential improvement opportunities, prioritizing improvement opportunities, developing realistic timelines, and understanding the costs involved in completing any potential improvement. Ultimately, one of the outcomes of benchmarking is to ensure that the prices a utility charges its customers reflect efficient production costs. In general, the utility with the lowest price is not always the best performer (Berg et al. 2010).

2. Lesson from lectures

Benchmarking is popular and effective for performance evaluation not only in the water industry; there are many critical parameters for performance evaluation, and the types of parameters vary across industries, countries, and locations. A literature review reveals that if benchmarking practice is used efficiently, it can help water utilities improve overall performance. Many companies have experienced significant success in upgrading their organizational capabilities through benchmarking (Barber.2004) Benchmarking tools are important for documenting past performance, establishing baselines for gauging productivity improvements, and making comparisons across service providers (Berg et al. 2007). Cognitive, interest, values, and authority conflicts can be resolved when designing and implementing policies by using benchmarking for water utility performance, according to Berg (2006). Chen (2005) argues that service quality is an important factor in the water and sewer industries. Lin (2005), using data from the Peruvian water sector (1996–2001), examined how introducing quality variables affected performance comparisons across utilities. Corton (2009) conducted a comprehensive efficiency analysis of water utilities in six countries in Central America. The aim of that study was to provide policymakers and investment fund institutions with quantitative evidence of the effectiveness of regional water sectors and utilities from different perspectives. One conclusion of that analysis pointed toward additional efforts for improving data collection procedures in the region. According to an article by Dassler et al (2006), regulations are subject to available information, and lack of information may lead to inefficient allocation. A study by Shleifer (1985) considered the benchmarking approach by reporting on its actual use in UK regulatory bodies in telecommunications, water, and energy. There are very few studies related to Indian water utilities. None of the studies has evaluated a utility's performance using sustainability-related parameters. Singh (2010) attempted to fill this gap and suggested a sustainability-based benchmarking framework to assess the efficiency of 18 Indian urban water utilities using a data envelopment analysis approach. A few initiatives use subjective indicators; these can be eliminated by quantifying performance indicators. Most of the initiatives focus on one or a few areas of performance. A holistic evaluation of overall performance can be done using a comprehensive set of indicators that cover all major areas of a utility's performance. Very few benchmarking initiatives are web-based; this can be changed by creating a web-based benchmarking platform. Web-based benchmarking will provide a platform not only for data gathering, but also for providing utilities a platform for result visualization. (Rathor et al.2013)

3. Lesson from lectures

In this study, urban water drinking of infrastructure in Tabriz metropolitan in order to better managing have been investigated and analyzed by using benchmarking method. Benchmarking utility performance indicators is an essential element of continuous improvement, allowing utilities to track their own performance and to compare their results to peers to identify areas that could be strengthened (Rathor et al. 2013). This methodology has been applied in Tabriz metropolis in North West of Iran for the first time. But, Benchmarking methodology have been continuously applied by the AWWA. Also, performance indicators that involved in the water industry and organization are using to provide a suitable framework for the

development of infrastructure in order to providing quality and effective management for drinking water. According to specified schedule, application process has been used to identified and executive defects (AWWA. 2008). In general, Benchmarking is a multistep process that needs to be carefully defined alongside a timeline to achieve the final goal. The process starts with selecting the subject and the practice. Data collection is performed using indicators that cover the major areas that contribute to overall performance. After defining the indicators, the next step is to define the data source and to develop the data collection process. The collected data are transferred to a common platform for analysis. The data are verified and analyzed, and results are produced. These results are validated with assistance from data sources and experts. The analysis results are used to determine potential gaps in performance and areas of underperformance. Goals are adjusted according to the analysis results, and improvement in future performance is targeted. The benchmarking process has been broken down into seven basic steps, as shown in Fig 1. (Rathor et al. 2013).

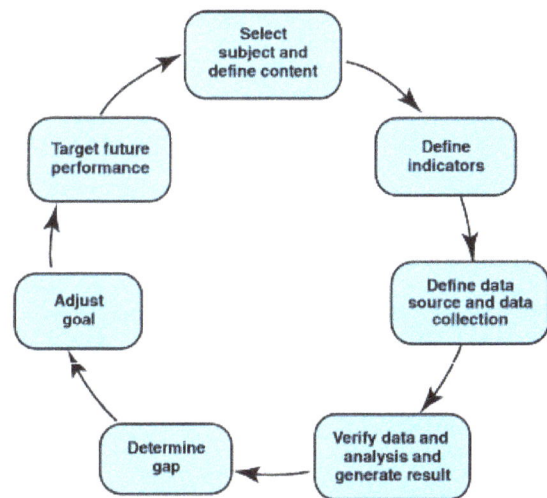

Fig.1. Basic Steps for Benchmarking (Rathor et al. 2013).

Also, the online platform known as the Water Infrastructure Database (WATERiD) was created so that utilities can compare their performance with similar-sized utilities and compare self-performance with that of previous years. The data were collected from utilities using benchmarking data collection sheets. This data collection was done using the web interface on WATERiD.

4. Results

Results of drinking water utilities of United State of America have been gathered from WateriD website. Also Tabriz Water Utility Data was gathered from Tabriz water and wastewater company. The performance indicators, methodology, and process of benchmarking were developed using research papers, books and reports on benchmarking done by various sectors. Total 89 indicators were used for data collection. Analysis results for few indicators have not been included because of the unavailability of enough data to make conclusions. The indicators were modified according to the aims and the needs pf this research. In general, for drinking water utility performance benchmarking the performance indicators are grouped under the following sections:

4.1. Eater resource utilization

Some utilities have a significant percentage of water loss (nonrevenue water) that is calculated as the percentage of treated water lost because of leakage and overflow. Most of the utilities do not reuse or recycle supplied water, even though reuse or recycling can help in conserving natural resources if done properly. In some areas, such as Florida, all wastewater is called recycled water in that it is pumped into groundwater aquifers to be used later. Availability of raw water resources varies for different utilities and depends on the geographic location of the utility. Some utilities are located in regions where extracting raw water does not require any permit and extraction is dependent on the capacity of intake structures. Other utilities are located at critical locations where there is a limitation on water extraction.   The results of water resources in Tabriz and seven water utilities have been showed in Fig 2. However, Water resources generally are divided to three section including:1- water resources availability 2- reused supplied water 3- water lost. In Tabriz city water resources availability, almost is equal to average but reused supplied water indexes amount in this city is zero and this is a negative point for Tabriz utilities. Also, water lost indexes is greater than average and this is a negative point for that utility.

4.2. Employee information

Employee-related indicators offer insight into how the participating utilities have staffed their utility, both in terms of leadership and operations; how they are structured in terms of employee levels; how they invest in their employees (e.g., training); and how they maintain a safe working environment. Alsharif et al (2008), Berg and Lin (2007), Mugisha (2007), Lin (2005), Lonborg (2005), Tupper and Resende (2004), Aida et al. (1998), and Lambert et al (1993) among others have used the number of employees (or labor or staff) as an input in their studies. The number of employees per 1,000 connections and per million gallons of water produced per day varies with a utility's size and location. Distribution of employees by percentage in upper management; human resources; financial and commercial; customer service; planning, design, and construction; and water quality monitoring was lower compared with percentage of employees in operations and management in all of the utilities. The number of employees in functions such as human resources and finance varies significantly between municipalities and drinking water authorities. In a municipality, many of these functions are provided by the general fund; the utility then disburses payment in lieu of taxes and/or transfers funds to the general fund for indirect costs. In a water authority, human resources and finance positions on usually on-staff positions. Most of the utilities invest significant time and resources on personnel training, most of which is safety-related. The percentage of employees injured is

significant, indicating that better safety training is required. The rate of absenteeism from accidents is on the lower side for all the utilities. Results of Employee percent as per Function have been showed for Tabriz Metropolitan and seven case study in fig 3. It must necessary say that Greater management percent is more than average and in human resources indexes almost is equal to average. Also, financial, commercial and customer service indexes are greater than average. Result show, planning, design and construction indexes employee percent in Tabriz are very lower than average. Operation, maintenance and water quality monitoring indexes are very lower than average in Tabriz Metropolitan.

4.3. Physical asset

A physical asset is any tangible item of economic, commercial, and/or exchange value and usually refers to cash, equipment, inventory, and/or properties owned by a business. Managing these physical assets is important for utilities to function properly. Quantifying the asset's performance and understanding the need for maintenance and replacement are important when managing any asset. Benchmarking has become a useful tool in the public debate over infrastructure compared with the treated water storage capacity. Most of the utilities shows a high value for this indicator, which implies that utilities produce more water than the storage capacity and pump it to customers as soon as the water is treated. Valve density indicates the number of valves per mile of main. The results for this indicator show that density for valves has a significant difference in value for different utilities. Hydrant density shows the number of water hydrants per length of main; this indicator value is in a similar range for all the utilities. Meter density shows the percentage of customers with meters. Most of the utilities have customer meters for almost all customers, showing a value close to 100 % for all utilities in the study. Density of valves and hydrants have been showed for Tabriz and seven water utilities of united states of America in Fig. 4. Also, the results indicate that percent of valves and hydrants are lower than average and median in Tabriz.

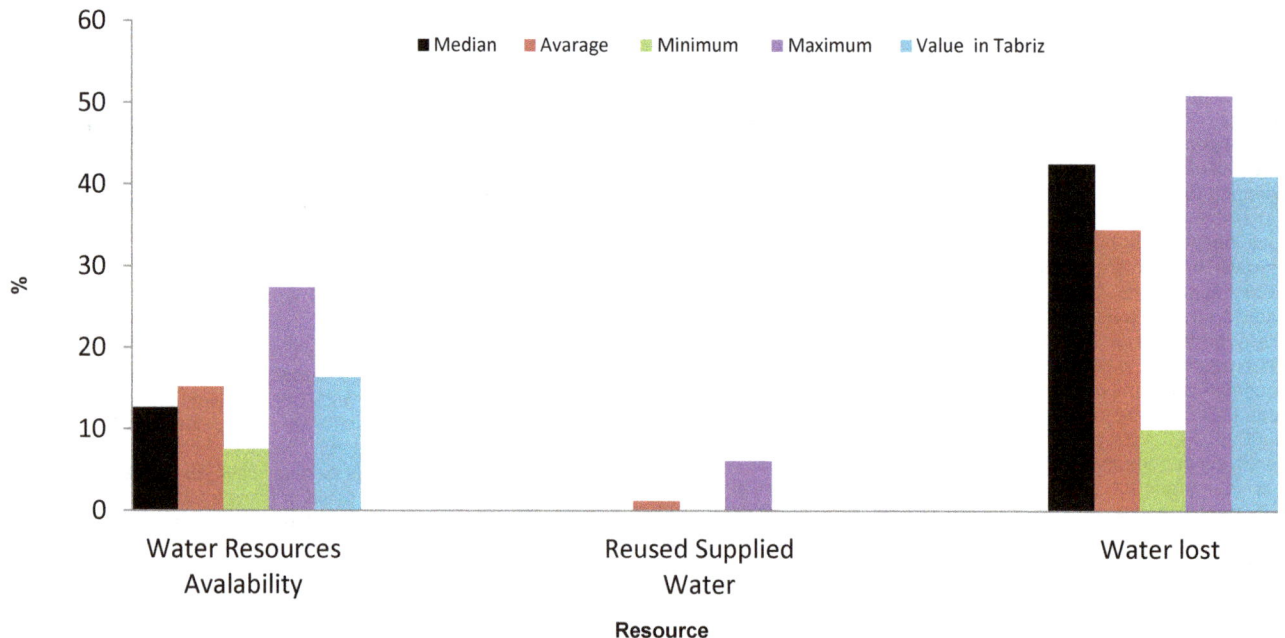

Fig. 2. Indicators Related to water resources.

4.4. Service quality

Service quality is a very important aspect of the water industry (Chen. 2005) and is a vital component in a utility's ability to maintain profitability and success. Population coverage provides the percentage of population served in the service area; for most of the utilities, this number is high. Main breaks show the number of main breaks in every 100 mi of mains in the past year and how the mains have been maintained. A few utilities have a higher number for this indicator (29

main breaks per 100 mil). Limiting water interruptions is a critical part of service because a higher number of interruptions causes higher customer dissatisfaction. Percentage connections with interruption in service show that most of the connections that experienced interruption were < 4 h, showing that most of the utilities solved the water interruption on a priority basis. Quality of supplied water is the most critical indicator in evaluating the quality of service. Before supplying the treated water, many required tests defined in water standards must be performed. The water quality indicators were divided into two categories: the total percentage of tests compliant with the standards for treated water and the total percentage of required tests done. These indicators summarize the total percentage of tests compliant with the standards and total percentage of required tests done. Most of the utilities surveyed perform more than the minimum number of required tests, and most showed an almost 100% compliance with permit conditions for the tests. Percentage of Tests complying with the standard in Tabriz water utility and seven water utility of United States of America have been showed in Fig 5. Results indicates that the amount of total, Aesthetic, Microbiological and physical- chemical tests in Tabriz city and seven water utilities of united states of America are almost equal but the percent of Radioactive test in Tabriz city is very lower than seven water utilities of united states of America.

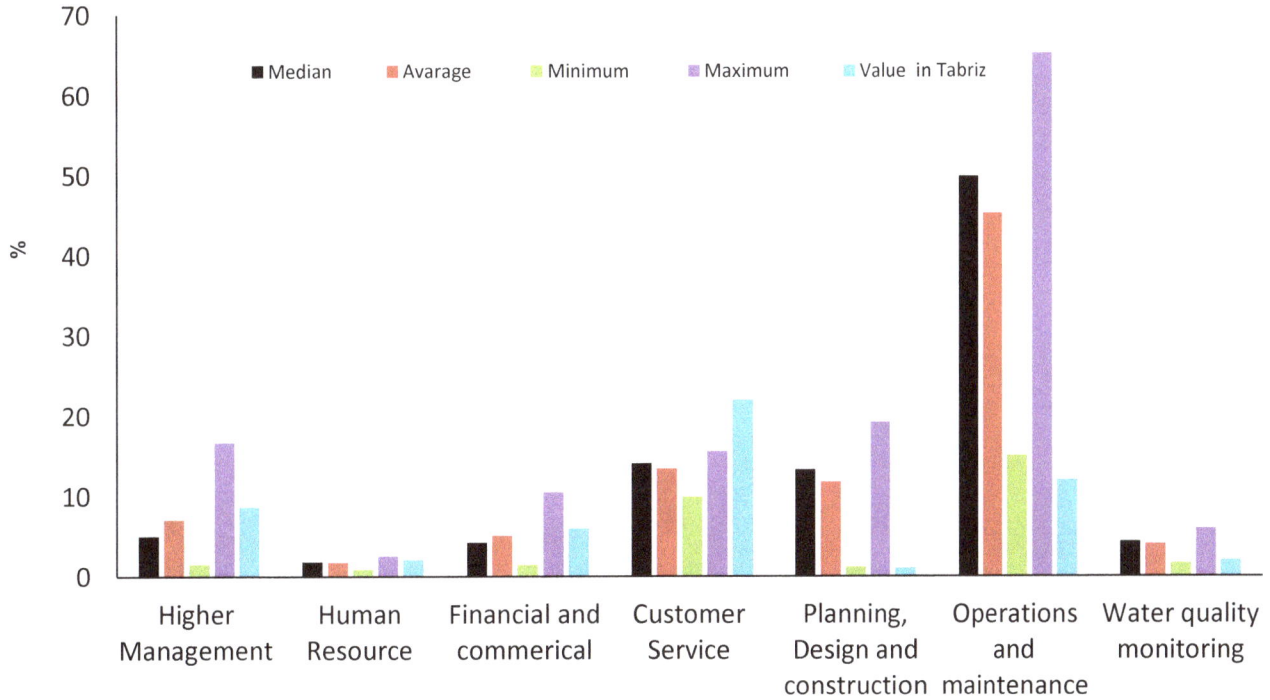

Fig. 3. Employee % as per function.

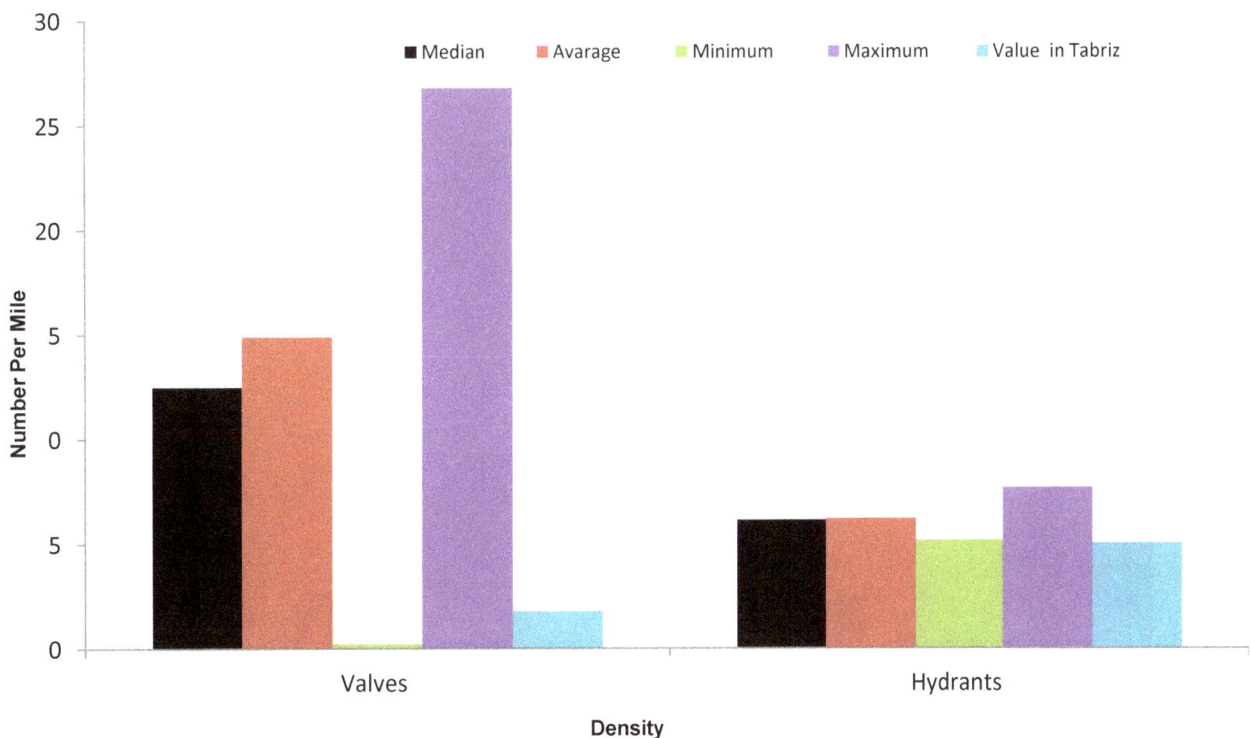

Fig. 4. Density of Valves and hydrants.

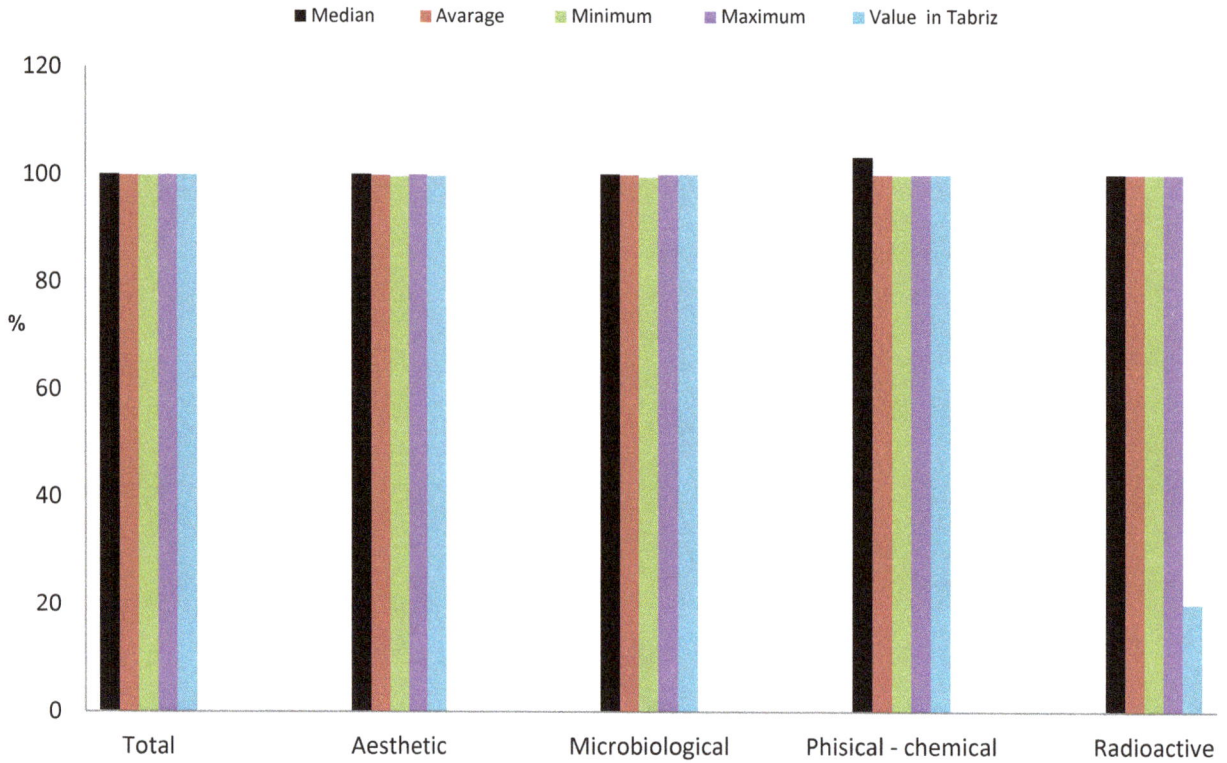

Fig. 5. Percentage of Tests complying with the standard.

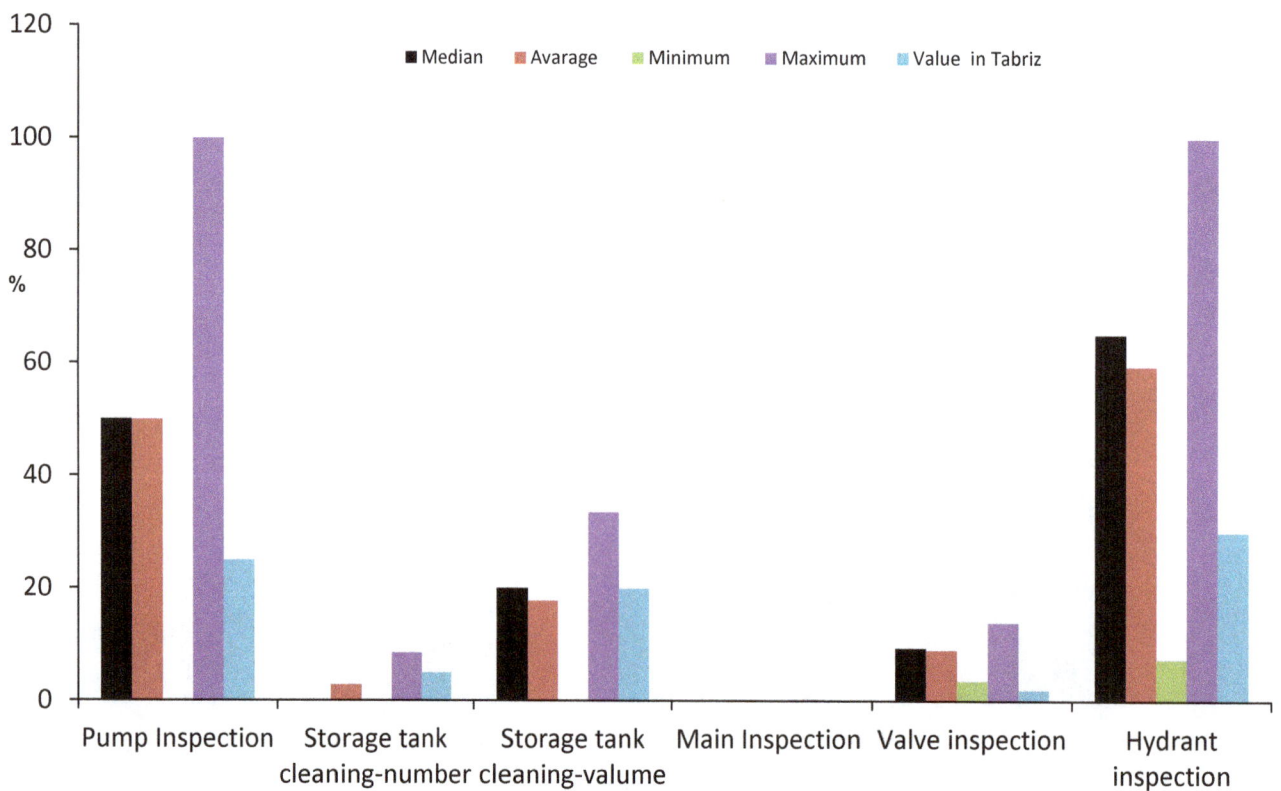

Fig. 6. Percentage of physical asset inspection and maintenance.

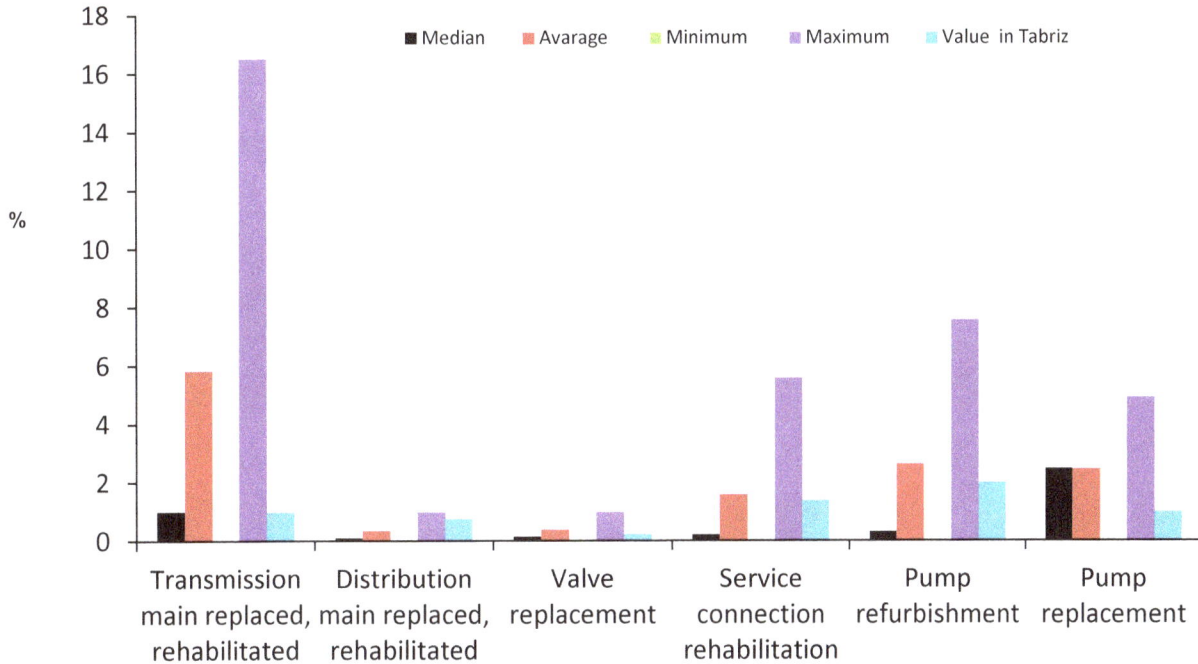

Fig. 7. Percentage of different type of operational performance.

4.5. Operational performance

Those responsible for utility operations can only manage what they measure, so having information on productivity trends and relative performance enables utility managers to direct attention to shortfalls (Berg et al. 2007). The indicators for operational performance evaluate efficient use of resources, reliability, inspection of current assets, rehabilitation of existing assets, and losses from low operational performance. Inefficient or ineffective operations lead to higher costs, which in turn lead to higher sales revenue needs. Pump inspection shows the percentage of existing pumps inspected. A few utilities inspected all their pumps, whereas some utilities did not inspect any pumps in the past year. Storage tank cleaning shows the percentage of tanks cleaned in the past year. This indicator shows a lower number: about one quarter of all storage tanks are cleaned every year, indicating that most of the utilities do not clean all their storage tanks every year. Main inspection shows the percentage length of main inspected. The values indicate that no utility focuses on inspecting the mains; the value for this indicator is close to zero for all utilities. One participating utility specified that inspection techniques for buried pressure pipe were too costly for regular use. Instead, that utility uses a criticality matrix in which pipe segments are ranked ranging from a low likelihood of failure and low consequence of failure up to a high likelihood and high consequence of failure. This matrix considers pipe age, pipe material, previous failures, and consequences of failure. For example, a main transmission pipeline that provides service to a hospital will have a much higher consequence from failure than a distribution line serving five houses on a cul-de-sac. Results for valve inspection show that most of the utilities do not inspect valves on a regular basis and merely change them whenever a problem occurs. Hydrant inspection shows the percentage inspected; the results for this indicator show that a few utilities inspect all the hydrants every year whereas others do not, instead replace hydrants when they stop working. The results for physical asset inspection and maintenance are showed in Fig. 6. Indicators for rehabilitation such as leakage control show the number of main breaks detected and repaired per 100 mi of main. Many utilities showed the value of this indicator in the percentage of meters that are working; all the utilities show a number close to 100 % for this indicator. Unmetered water shows the percentage of water that is not metered; utilities show an average value of 10% for this indicator. Results show in pump inspection, valve inspection and hydrant inspection indexes Tabriz drinking water utility is lower than average and median in seven drinking water utilities in united states of America. Also, Fig. 6 indicates

storage tank cleaning-number and storage tank cleaning-volume is greater than average and median Water loss includes water lost through leaks, breaks, backwash, flushing, and under registering meters. Water loss per connection and percentage of water lost that was treated in the past year indicates a range of 8–27 %. Water loss is a concern for every utility. Some nonrevenue water can be attributable to a poor meter replacement program because meters tend to under register as they age. The operational meter's indicator showed in the Fig. 7 indicates percentage of different type of inquiries indexes. In this figure, transmission main replaced, rehabilitated, distribution main replaced, rehabilitated, valve replacement, service connection rehabilitation and pump refurbishment indexes information have been introduced for Tabriz drinking water utilities and seven water utilities of united states of America. Also, the amount of transmission main replaced, rehabilitated, valve replacement, service connection rehabilitation and pump refurbishment indexes value in Tabriz water drinking utilities is lower than seven water drinking utilities in United States of America. Also, percentage of distribution main replaced, rehabilitated in Tabriz water drinking utilities is greater than seven water drinking utilities in United States of America.

4.6. Customer enquiries

Customer satisfaction is a measure of how services supplied by a utility meet customer expectation. Customer satisfaction is defined as "the number of customers, or percentage of total customers" whose reported experience with a firm, its products, or its services (ratings) exceeds specified satisfaction goals (Farris. 2010). In a survey of nearly 200 senior marketing managers, 71% responded that they found a customer satisfaction metric useful in managing and monitoring their businesses (Farris, 2010). Customer satisfaction is viewed as a key performance indicator within business and is often part of a balanced scorecard. In a competitive marketplace in which businesses compete for customers, customer satisfaction is a key differentiator and increasingly has become a key element of business strategy (Gitman. 2005). Service inquiries per 100 connections show total inquiries in the past year per 100 connections; the value is in the 0.5–4.1 range. Inquiries were further divided by type: percentage of pressure related reports, percentage of continuity-related reports, percentage of water quality/taste-related reports, percentage of water quality/ odor-related reports, and percentage of interruption-related reports. Pressure of water supply, continuity of water, and interruptions are the most customer-reported categories. Fig 8. indicates percent of different type of inquiries indexes including:1-presure- related, continuity related, water quality-taste related, water quality oder-related and interruption related indexes. Also, the value of pressure- related and water quality oder-related is lower than average and median. Also, the value of continuity related and interruption-related indexes are greater than average and median.

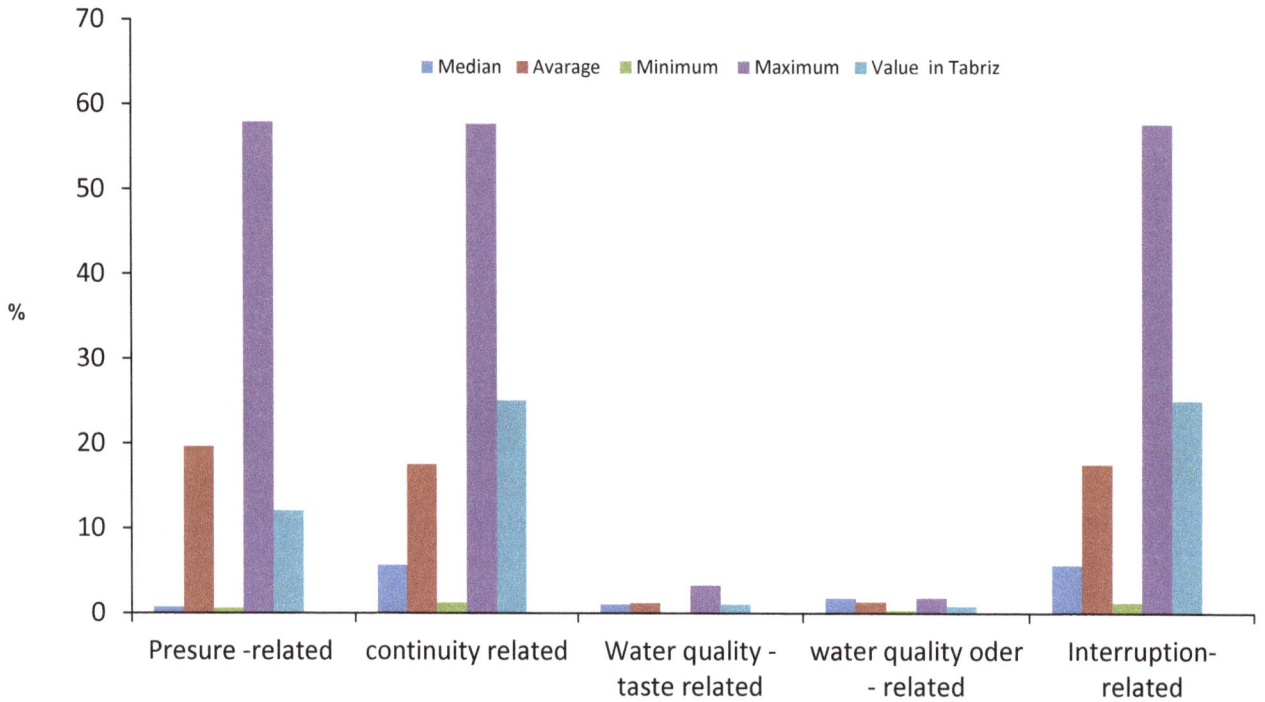

Fig. 8. Percentage of different type of customer inquiries.

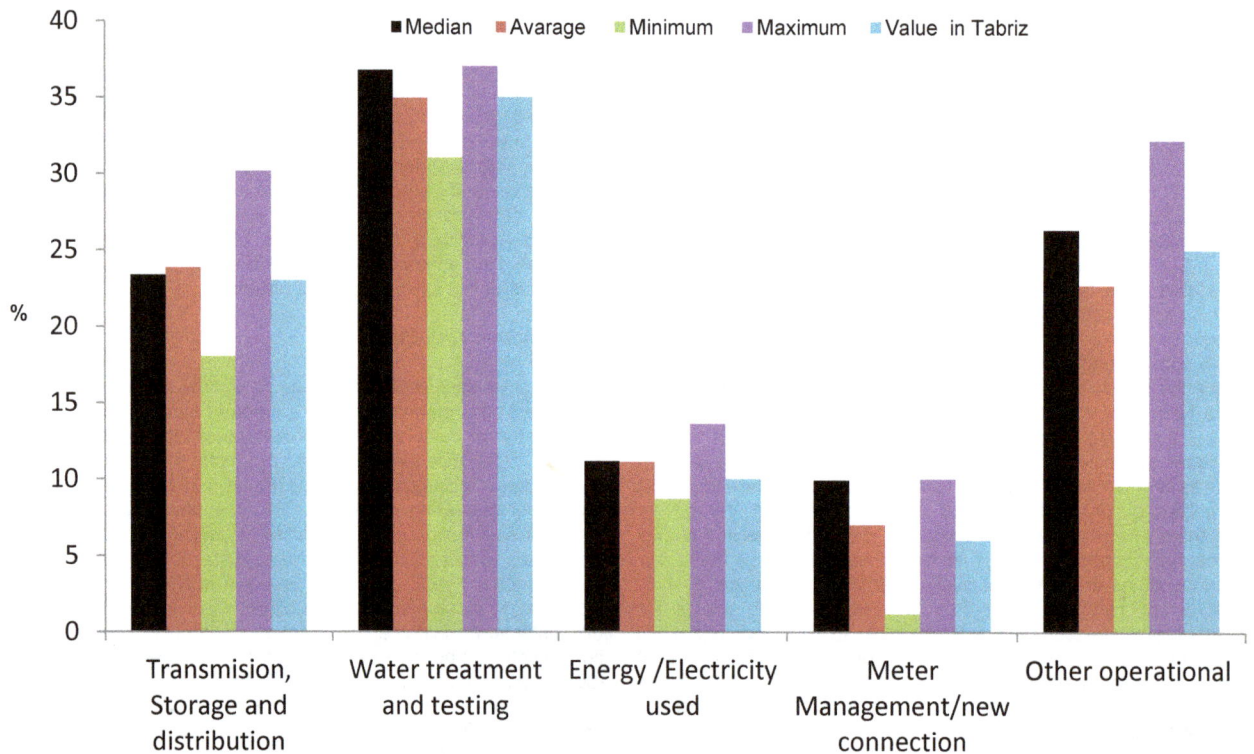

Fig. 9. Percentage of different types of operational cost.

4.6. Customer enquiries

Customer satisfaction is a measure of how services supplied by a utility meet customer expectation. Customer satisfaction is defined as "the number of customers, or percentage of total customers" whose reported experience with a firm, its products, or its services (ratings) exceeds specified satisfaction goals (Farris. 2010). In a survey of nearly 200 senior marketing managers, 71% responded that they found a customer satisfaction metric useful in managing and monitoring their businesses (Farris, 2010). Customer satisfaction is viewed as a key performance indicator within business and is often part of a balanced scorecard. In a competitive marketplace in which businesses compete for customers, customer satisfaction is a key differentiator and increasingly has become a key element of business strategy (Gitman. 2005). Service inquiries per 100 connections show total inquiries in the past year per 100 connections; the value is in the 0.5–4.1 range. Inquiries were further divided by type: percentage of pressure related reports, percentage of continuity-related reports, percentage of water quality/taste-related reports, percentage of water quality/ odor-related reports, and percentage of interruption-related reports. Pressure of water supply, continuity of water, and interruptions are the most customer-reported categories. Fig 8. indicates percent of different type of inquiries indexes including: 1-presure- related, continuity related, water quality-taste related, water quality oder-related and interruption related indexes. Also, the value of pressure- related and water quality oder-related is lower than average and median. Also, the value of continuity related and interruption-related indexes are greater than average and median.

4.7. Financial performance

Financial results are reflected in the utility's return on investment, return on assets, value added, total cost, revenue generated, cost coverage, and profit. Financial performances a measure of how well a utility can use assets from its primary mode of business and generates revenues; it is also used as a general measure of a utility's overall financial health over a given period and can be used to compare similar utilities. Consideration of financial sustainability includes examining how the role of collections, revenues, and operating expenses affect overall performance. Key financial ratios should serve as indicators of long-term performance because revenues used to facilitate future capacity investments for both network expansion and external funding can be contingent on current cash flows more than covering operating expenses (Berg et al. 2007). The indicators for the revenue section summarize revenue per million gallons of treated water produced. Revenue is further divided into percentage of sales revenue and percentage of other revenue. The largest percent of revenue comes from sales for all the utilities. The cost indicators summarize total cost per million gallons of treated water produced, capital cost per million gallons of water produced, and operating cost per million gallons of water produced. The summary of percentage of operation cost is shown by the type of operation. The percentage of operational cost for transmission, storage, distribution, and water treatment and testing

shows higher values; this result was expected because these are utilities' main functions. Investment indicators summarize total investment per million gallons of water produced, percentage of investment on new assets, and percentage of total investment on replacement and renovation. For all the utilities, the majority of investment is for new assets. The rate a utility charges the consumer changes depending on the cost of treatment and operation. Total cost coverage, operational cost coverage, liquidity ratio, asset turnover ratio, and water loss cost (nonrevenue water) are efficiency indicators. some utilities are experiencing reduction of demand; this is expected to continue, although population is expected to increase. The various factors contributing to lower demand includes conservation programs, water and sewer rates, and improved system operations. One utility reported that in the past five years the average household consumption decreased from 6.53 ccf (4,880 gal) per month to 5.62 ccf (4,200 gal) per month. Although from a conservation standpoint, this is perceived as a good thing, it is creating significant challenges for utilities. Revenues in both water and sewer funds decreased by approximately $1 million in the past year. The utility's city council was reluctant to raise water and sewer rates in this economic climate, and any increases they approved could only cover rising operational costs. As a result, the utility had to significantly cut back its capital improvement programs— from $20–25 million to $2–3 million in the sewer fund at a time when infrastructure investment is critical for long-term sustainability. All the results for the indicators can be accessed using the WATERiD website (www.waterid.org). Also, percentage of different types of operational cost indexes such as: transmission, storage and distribution, water treatment and testing, energy/ electricity used, meter management/new connection and other operational have been showed in fig 9. Results indicates values of transmission, storage and distribution, water treatment and testing, energy/electricity used and it is clear the amount of meter management/new connection indexes in Tabriz drinking water utilities is lower than median and average of seven water utilities in united states of America and the amount of other operational indexes are greater than average and median.

5. Conclusions

Results show, planning, design and construction indexes employee percent in Tabriz are very lower than average. Operation, maintenance and water quality monitoring indexes are very lower than average in Tabriz Metropolitan. Also the results of this paper indicate that performance assessment of Tabriz utility water system are very effective for managing of water network of Tabriz and planning for future needs of citizens. Also, this method introduces techniques for experts and to have a better and efficient management for welfare and security of our people in the future. Also, this method introduces techniques for experts and decision makers to create developmental plans with the correct analysis. In addition to, this technique has been applied for identifying weakness and strength points of water drinking water utilities of Tabriz city in North West of Iran for the first time.

References

Aida K., William W.C., Jesus T.P., Toshiyuki S, Evaluating Water Supply Services in Japan with RAM: A Range-Adjusted Measure of Inefficiency. Omega, International Journal of Management Science 26 (1998) 207-208.

Alsharif K., Feroz E.H., Klemer A., Raab R., Governance of Water Supply Systems in the Palestinian Territories: A Data Envelopment Analysis Approach to the Management of Water Resources, Journal of Environmental Management 87 (2008) 80-81.

AWWA, Effective utility management, A primer for water and wastewater utilities, (2008) 12-14.

Barber E., Benchmarking the Management of Projects: A Review of Current Thinking, International Journal of Project Management, 4 (2004) 301-302.

Berg S.V., Conflict Resolution: Benchmarking Water Utility Performance, Public administration and Development (2007).

Berg S.V., Padowski J., Overview of Water Utility Benchmarking Methodologies: From Indicators to Incentives Sanford. Public Utility Research Center, University of Florida, USA, 102 (2010).

Cagle R.F., "Infrastructure Asset Management: An Emerging Direction", Proceedings of the 47th Annual Meeting of the Association for the Advancement of Cost Engineering International, Orlando, Fla. (2003) 28-30.

Chen L., Service Quality and Prospects for Benchmarking: Evidence from the Peru Water Sector, Utilities Policy Journal (2005) 34-36.

Corton M.L., Berg S.V., Benchmarking Central American Water Utilities, Utilities Policy Journal 17 (2009) 267-268.

Dassler T., Parker D., Saal D.S., Methods and Trends of Performance Benchmarking in UK Utility Regulation, Utilities Policy Journal 14 (2006) 166-167.

Farris P., Marketing Metrics: The Definitive Guide to Measuring Marketing Performance. FT Press, Upper Saddle River (2010).

Gitman L.J., McDaniel C.D., The Future of Business: The Essentials. South-Western, Mason, Ohio.15 (2005) 256-257.

Lin C., Incorporating Service Quality & Prospects of Benchmarking: Evidence from the Peru Water Sector, Utilities Policy Journal 13 (2005) 230-231.

Mugisha S., Effects of Incentive Applications on Technical Efficiencies: Empirical Evidence from Ugandan Water Utilities, Utilities Policy 15 (2007) 225-226.

Oyegoke S.O., Adeyemi A.O., Sojobi A.O., The Challenges of Water Supply for a Megacity: A Case Study of Lagos Metropolis, International Journal of Scientific & Engineering Research 2 (2012) 105-106.

Shleifer A., A Theory of Yardstick Competition, Rand Journal of Economics 16 (1985) 319-320.

Singh M.R., Upadhyay V., Mittal A.K., Addressing Sustainability in Benchmarking Framework for Indian Urban Water Utilities, Journal of Infrastructure Systems 16 (2010) 81-92.

Tupper H.C., Resende M., Efficiency and Regulatory Issues in the Brazilian Water and Sewage Sector: An Empirical Study, Utilities Policy Journal 12 (2004) 29-3.

21

Numerical simulation of prevention of saltwater intrusion in panama channel by using of bubble curtain system

Mehdi Nezhad Naderi[*], Omid Zolfaghari

[1]*Department of Civil Engineering, Tonekabon Branch, Islamic Azad University, Tonekabon, Iran.*

ARTICLE INFO

ABSTRACT

Keywords:
Bubble curtain
Computational fluid dynamic method
Fluent
Multiphase flow
Prevention of Saltwater intrusion

A bubble curtain is a system that produces bubbles in a deliberate arrangement in water. The technique is based on bubbles of air (gas) being let out under the water surface, commonly on the bottom. When the bubbles rise they act as a barrier, a curtain for prevention of the spreading of particles and other contaminants. In this paper is paid to applications of bubble curtain in protection of environment of offshore. Due to the salt water intrusion in Panama navigable channel is causing environmental damage. Construction of bubble curtains along the channel can be studied as a playbook. In this study, two-phase flow is simulated with simulation software Fluent6.3 for freshwater input from the left, saltwater input from the right, air from several vertical bubbles and water injection. The model is solved by using of two-phase Mixture pattern. For problem solving is used the k-ε turbulence model. The air inlet velocity is considered 0.6 meters per second and again 0.2 meters per second. By using air curtains (bubbles) can be prevented salt water intrusion and the density also be reduced. In this paper the multiphase flow is simulated by computational fluid dynamics method in Panama channel.

1. Introduction

A bubble curtain is a system that produces bubbles in a deliberate arrangement in water. This technique is based on bubbles of air (gas) under the water surface that act commonly as a barrier. When the bubbles rise they act as a barrier or a curtain. This paper paid to applications of this system for prevention of the advance of seawater and protection of environment of offshore. The multiphase flow is simulated by computational fluid dynamics method. During high level of tide bubble curtain system can be used simultaneous in several parallel rows of air injection in across the offshore. During low level of tide air injection rate and use of the bubble curtain system is reduced. The results of the numerical models show that increasing air injection rate is caused to reduce seawater intrusion. Bubble curtain, in its simplest form, is a circle or square tube with holes in it that air is injected under pressure into the tube and the bubbles will create a curtain of bubbles. Components are needed to create a barrier bubbles: 1- compressed air from a compressor station, 2- pipe with special nozzles incorporated and anchor blocks, 3 - levels produced by bubble curtains and 4- drain valve at the end of the nozzle tubes. The subjects expressed about the use of bubble curtains to protect the marine environment.

Two researchers named Ghyben and Herzberg separately studied fresh underground water flow to the oceans along the coasts of Europe. They found that anywhere from a coastal aquifer, If depth of interface between fresh and saltwater is measured from sea level, $(h_s h_s)$, then level of fresh ground water from sea level, $(h_f h_f)$, will be 1/40 $(h_s h_s)$ in that point (Ghyben 1889; Herzberg 1901). Since these studies were started by two scientists this phenomenon is mentioned with regard to "Ghyben - Herzberg" that will be explained. Many reviews on the types of groundwater management models and their applications are made by Gorelick (1983), and Yeh (1986). The management models applications in saltwater intrusion, are relatively recent, (Cheng et al. 1999; Fatemi and Ataie-Ashtiani 2008; Bear and Cheng 1999; Cheng and Ouazar 1999; Cummings 1971; Cummings and McFarland 1974; Dagan and Bear 1968; Das Gupta et al. 1996; Naji et al. 1999; Bear and Verruijt 1987; Shahmoradi and Qavami 2008; Siddiqui et al. 2011; Reddy and et al. 2009; Mertzanides et al. 2010; David et al. 2008; Salamasi and Azamathulla; 2013; Kouzana et al. 2009; Jorreto et al. 2014; Van Camp et al. 2013; Zghibi et al. 2010; Werner et al. 2013; Sanz and Voss 2006; Rajabi and Ataie-Ashtiani 2014). In this study, flow is unsteady with two-dimensional turbulence form. Velocity and pressure are a function of time and space. For model of the velocity and pressure fluctuations is the integrated from the Navier Stokes equation at time. Integration of Navier Stokes equations at time is known Reynolds equations (Reynolds 1984).

Turbulence model equations are two equation models k-ε (Standard) that have been averaged in depth (Rastogi and Reddy 1978). ε equation is as one of the main sources of the limitations of accuracy of the standard version of the k-ε model and the Reynolds stress model. It is interesting that k-ε model includes a correction term that is dependent to strain with c13 constant in the ε equation of RNG model (Yakhot et al. 1992). WillCox provided turbulence equations of k-ω (standard) model (WillCox 1988).

$$\frac{\partial u}{\partial x} + \frac{\partial v}{\partial y} + \frac{\partial w}{\partial z} = 0 \tag{1}$$

$$\frac{\partial \rho u}{\partial t} + \frac{\partial \rho uu}{\partial x} + \frac{\partial \rho uv}{\partial y} + \frac{\partial \rho uw}{\partial z} - \rho f_c v = -\frac{\partial P}{\partial x} + \frac{\partial \tau_{xx}}{\partial x} + \frac{\partial \tau_{xy}}{\partial y} + \frac{\partial \tau_{xz}}{\partial z} \tag{2}$$

$$\frac{\partial \rho v}{\partial t} + \frac{\partial \rho uv}{\partial x} + \frac{\partial \rho vv}{\partial y} + \frac{\partial \rho vw}{\partial z} + \rho f_c u = -\frac{\partial P}{\partial y} + \frac{\partial \tau_{yx}}{\partial x} + \frac{\partial \tau_{yy}}{\partial y} + \frac{\partial \tau_{yz}}{\partial z} \tag{3}$$

$$\frac{\partial \rho w}{\partial t} + \frac{\partial \rho uw}{\partial x} + \frac{\partial \rho vw}{\partial y} + \frac{\partial \rho ww}{\partial z} = -\frac{\partial P}{\partial z} + \frac{\partial \tau_{zx}}{\partial x} + \frac{\partial \tau_{zy}}{\partial y} + \frac{\partial \tau_{zz}}{\partial z} - \rho g \tag{4}$$

*Corresponding author E-mail: Mehdi2930@yahoo.com

Fig. 1. View of a bubble curtain in place of berthing the ship in Vancouver of Canada to reduce noise pollution (Swanson 2004).

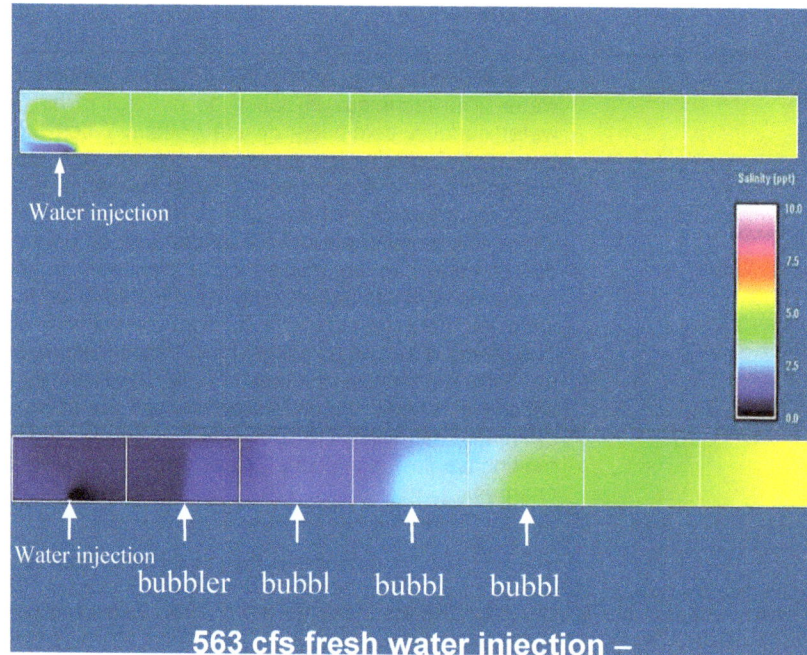

Fig. 2. Simulation of a salinity intrusion barrier, Panama Canal study with one water injection at started of canal and four bubblers in canal (Luong and et al. 2007).

2. Turbulence model equation

Known two-equation model of k-ε (Standard) are presented for averaged form in depth as follows (Rastogi and Reddy 1978):

$$\frac{\partial hk}{\partial t} + \frac{\partial U_j hk}{\partial x_j} = \frac{\partial}{\partial x_j}[(\nu + \frac{\nu_t}{\sigma_k})h\frac{\partial k}{\partial x}] + hP_k + hP_{kv} - h\varepsilon \tag{5}$$

$$\frac{\partial h\varepsilon}{\partial t} + \frac{\partial U_j h\varepsilon}{\partial x_j} = \frac{\partial}{\partial x_j}[(\nu + \frac{\nu_t}{\sigma\varepsilon})h\frac{\partial \varepsilon}{\partial x}] + hc_{1\varepsilon}\frac{\varepsilon}{k}P_k + hP_{\varepsilon v} - hc_{2\varepsilon}\frac{\varepsilon^2}{k} \tag{6}$$

$$\nu_t = c_\mu \frac{k^2}{\varepsilon}, P_k = 2\nu_t S_{ij}.S_{ij} \tag{7}$$

$$P_{kv} = c_k \frac{k^2}{\varepsilon}, c_k = \frac{1}{c_f^{1/2}}, P_{\varepsilon v} = c_\varepsilon \frac{u_f^4}{h^2}, c_\varepsilon = \frac{1}{\sqrt{e_*\sigma_t}} \frac{c_{2\varepsilon}c_\mu^{1/2}}{c_f^{3/4}}, c_f = \frac{u_f^2}{u^2 + v^2 + w^2} = \frac{n^2 g}{h^{1/3}} \tag{8}$$

where c_μ=0.09, $c_{\varepsilon1}$=1.44, σ_k=1 and σ_ε=1.31. P_{kv} and $P_{\varepsilon v}$ are production terms as result of non-uniform distribution velocity in depth that is stronger near-bed. P_k is production term of turbulent kinetic energy averaged in depth as result of velocity gradients in the plan. ν_t is the vortex viscosity. Turbulence model is used for calculation of lateral flow into one channel and is achieved much better results in comparison with ν_t for fixed parameters of rotational flow (MCGurik and Rodi 1978). C_f is the bed friction coefficient. σ_t is Schmidt number that shows relationship between turbulence viscosity and turbulent diffusion coefficient according to the following equation:

$$\varepsilon_d = \frac{\nu_t}{\sigma_t} \tag{9}$$

Amount of σ_t is considered 0.5 (Keller and Rodi 1988). Although values of σ_t are 0.5 to 2 in variable references (Gibson and launder 1978). e_* is coefficient that gives turbulence diffusion coefficient in depth by following equation (Keller and Rodi 1988).

$$\varepsilon_d = e_* h u_f \tag{10}$$

Direct measurement of color broadcasting in the fixed-width channels offers 0.15 for e_*. Although Keller and Rodi achieved better solutions for the velocity and stress within the composite channels (Keller and Rodi 1988). On the other hand, Biglari and Sturm have been assumed e_* equaled to 0.3 to get the better answer within the composite channels (Biglari and Sturm 1998). MCGurik and Rodi have considered $1/\sqrt{(e_*\sigma_t)}$ equaled to 3.6 (MCGurik and Rodi 1978). In ε equation of RNG model includes a correction term $c_{\varepsilon q}$ that is constant strain-dependent (Yakhot et al. 1992). For k-ε (RNG), we have:

$$\frac{\partial h\varepsilon}{\partial t} + \frac{\partial U_j h\varepsilon}{\partial x_j} = \frac{\partial}{\partial x_j}[(\nu + \frac{\nu_t}{\sigma\varepsilon})h\frac{\partial \varepsilon}{\partial x}] + hc_{1\varepsilon}^* \frac{\varepsilon}{k}P_k + hP_{\varepsilon v} - hc_{2\varepsilon}\frac{\varepsilon^2}{k} \tag{11}$$

$$c_\mu = 0.0845, c_{1\varepsilon}^* = c_{1\varepsilon} - \frac{\eta(1-\frac{\eta}{\eta_0})}{1+\beta\eta^3},$$

$$c_{1\varepsilon} = 1.68, \sigma_k = 1.39, \beta = 0.012, c_{1\varepsilon} = 1.42,$$ (12)

$$\eta = (2E_{ij}.E_{ij})^{1/2}\frac{k}{\varepsilon}, \eta_0 = 4.377$$

Only constant β is adjustable, high levels of turbulent data are obtained near-wall. All other constants are calculated explicitly as part of the RNG process.

$$\frac{\partial hk}{\partial t} + \frac{\partial U_j hk}{\partial x_j} = \frac{\partial}{\partial x_j}[(v+\frac{v_t}{\sigma_k})h\frac{\partial k}{\partial x}] + P_k + P_b - h\varepsilon$$ (13)

$$\frac{\partial h\varepsilon}{\partial t} + \frac{\partial U_j h\varepsilon}{\partial x_j} =$$ (14)

$$\frac{\partial}{\partial x_j}[(v+\frac{v_t}{\sigma\varepsilon})h\frac{\partial\varepsilon}{\partial x}] + hc_{1\varepsilon}\frac{\varepsilon}{k}P_k + hc_1 S_\varepsilon - hc_2\frac{\varepsilon^2}{k+\sqrt{v\varepsilon}} + S_\varepsilon$$

$$c_1 = Max[0.43, \frac{\eta}{\eta+s}], \eta = s\frac{k}{\varepsilon}, s = \sqrt{2s_{ij}s_{ij}},$$ (15)

$$\mu_t = hc_\mu\frac{k^2}{\varepsilon}, P_k = -\rho\overline{u_i'u_j'}\frac{\partial u_j}{\partial x_i},$$ (16)

$$P_k = \mu_t s^2, P_b = \beta g_i\frac{\mu_t}{Pr_t}\frac{\partial T}{\partial x_i}, \mu_t = \rho c_\mu\frac{k^2}{\varepsilon}, c_\mu = \frac{1}{A_0+A_s\frac{KU^*}{\varepsilon}}, U^* = \sqrt{s_{ij}s_{ij}+\overline{\Omega_{ij}\Omega_{ij}}},$$

$$\overline{\Omega_{ij}} = \Omega_{ij} - \varepsilon_{ijk}\omega_k, A_0 = 4.04, A_s = \sqrt{6}\cos\Phi, \Phi = \frac{1}{3}\cos^{-1}(\sqrt{6}\omega), \omega = \frac{s_{ij}s_{jk}s_{ki}}{\tilde{s}^3}, \tilde{s} = \sqrt{s_{ij}s_{ij}}.$$ (17)

$$s_{ij} = \frac{1}{2}(\frac{\partial u_j}{\partial x_i} + \frac{\partial u_i}{\partial x_j}), c_{1\varepsilon} = 1.44, c_2 = 1.9, \sigma_k = 1, \sigma_\varepsilon = 1.2, \beta = -\frac{1}{\rho}(\frac{\partial P}{\partial T})p, Pr_t = 0.85$$ (18)

WillCox, turbulence model k-ω (standard) equation to be provided as follows (WillCox 1988):

$$\frac{\partial k}{\partial t} + U_j\frac{\partial k}{\partial x_j} = \tau_{ij}\frac{\partial U_i}{\partial x_j} - \beta^* k\omega + \frac{\partial}{\partial x_j}[(v+\sigma^* v_T)\frac{\partial k}{\partial x_j}]$$ (19)

$$\frac{\partial\omega}{\partial t} + U_j\frac{\partial\omega}{\partial x_j} = \alpha\frac{\omega}{k}\tau_{ij}\frac{\partial U_i}{\partial x_j} - \beta\omega^2 k\omega + \frac{\partial}{\partial x_j}[(v+\sigma v_T)\frac{\partial\omega}{\partial x_j}]$$ (20)

where $v_t = k/\omega$, $\alpha = 5/9$, $\beta = 3/40$, $\beta^* = 9/100$, $\sigma = 1/2$, $\varepsilon = \beta^*\omega k$.

3. Numerical model

The values of the physical properties of water are considered as a default respectively, for density, viscosity, heat capacity and thermal conductivity. Solutions of all governing equations are subject to assignment of variables correctly in the boundary nodes. In steady state problems required only boundary condition but in unsteady state problems is required the initial conditions for all nodes in the network. Common boundary conditions in hydraulic issues include (Soltani and Rahimi Asl 2003):

A- Inlet boundary condition: numerical models can fit the model by means of the various boundary conditions such as velocity, mass flow, etc. For example, in modeling of flow inside a closed or open channel can be used velocity inlet as input boundary condition.

B- The outlet boundary condition is considered pressure outlet equals the atmospheric pressure. If the output is chosen at a far distance from geometric constraints, and no change in direction of flow then the flow state is developed full. Using this model is caused the output surface is perpendicular to the flow and gradient is zero in the perpendicular direction on the output surface (Soltani and Rahimi Asl 2003).

C - Wall boundary condition: the wall boundary condition is used to limit the area of between fluid and solid. The model is ready for simulation by Solutions set and defining the model. The following steps show the simulation process (Versteeg and Malalasekera 2007): selection methods of discretization equation: In this paper first order upstream difference method is used for discretization of momentum, k, ε and ω equations and the standard method is used to find the pressure. Selection methods of the relation velocity - Pressure: this step is only being studied segregated. In this paper is used from SIMPLE method for velocity - pressure coupling. Determine the discount factors: the discount factor values are used for control of calculated variables in each iteration. In this paper, the default values are used respectively for the pressure, density, momentum, k, ε and turbulent viscosity. In this paper, the initial values of the relative pressure is considered zero and the initial values of velocity components close to the average values presented in the input stream. By completing the steps in the numerical model, we can start the introduced process of problem by defining of repeat process. The frequency of reporting of results can be introduced before computing the numerical model. During solution process can be seen convergence of solution by the control of residues, integral of surface, statistics and values of the force. After finishing solution, the computation of the unknown quantities and the results can be calculated at any point of the field and can be displayed by vector in the form, contour and profile views (Versteeg and Malalasekera 2007). In this paper for solution of flow is usually introduced initial number repeat 1000 with report of every step of the calculation that conditions for convergence of the unknown parameters were satisfied after 300 to 350 iterations. The results of the numerical models show that increasing saltwater hydraulic gradient and times of tidal flow are caused to seawater intrusion from over underground dam to coast as figure 2-a to 2-e.

4. Meshing model

Gambit software version 2.3.16 is used to generate the channel geometry and meshing. Model of the network is used Quad element and the types of Map and Pave for pages and Hex elements and types of Map of Cooper for volumes. Inlet and outlet and wall boundary conditions and symmetry were introduced in the software.

5. Bubble curtain system for prevention of seawater intrusion in coastal aquifers

In this paper is paid to three-phase flow simulation by software Fluent6.3 that freshwater input is from the left side, saltwater on the right side of the entrance and the air from vertical duct. By using of mixture model and k-ε turbulence model in software the three-phase mixture is dissolved. At first the air inlet velocity is considered 0.6 meters per second then is reduced to 0.2 meters per second.

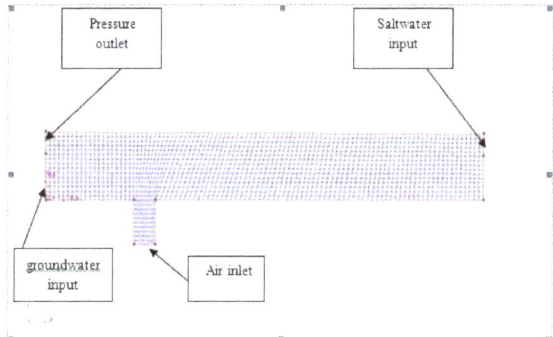

Fig. 3. Meshing of model and boundary conditions.

Fig. 4. Velocity magnitude contours for the three phase flow for seawater intrusion from right input, bubble curtain channel and freshwater from left input (air velocity is 0.6 m/s).

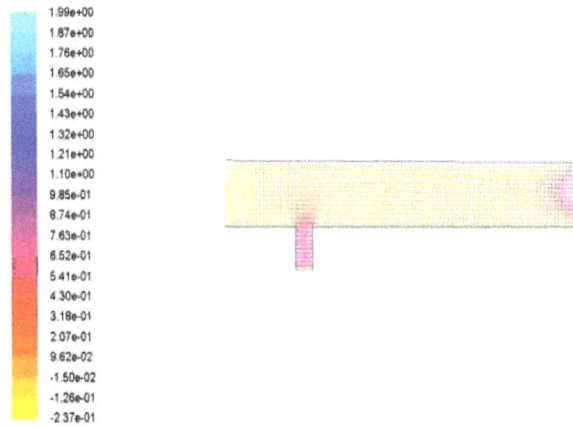

Fig. 5. the y velocity contours for the three-phase flow in the y direction (Air velocity is 0.6 m/s).

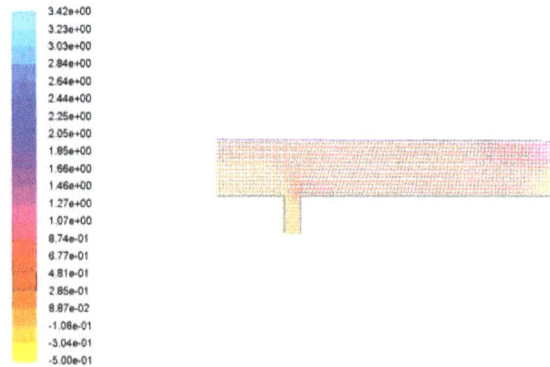

Fig. 6. the x velocity contours for the three-phase flow in the x direction (air velocity is 0.6 m/s).

Fig. 7a. The velocity magnitude contours for the three-phase flow (air velocity is 0.2 m/s).

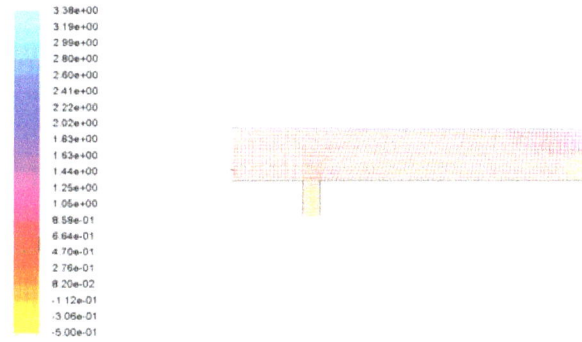

Fig. 7b. the x velocity contours for the three-phase flow in the x direction (air velocity is 0.2 m/s).

Fig. 7c. The velocity contours for the three-phase flow in the y direction, (air velocity is 0.2 m/s).

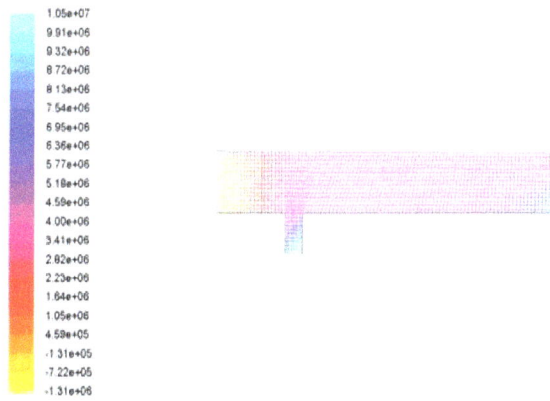

Fig. 8a. the pressure contours for the three-phase flow (air velocity is 0.2 m/s).

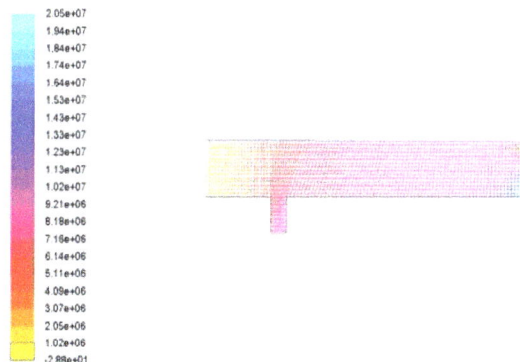

Fig. 8b. the pressure contours for the three-phase flow (air velocity is 0.6 m/s).

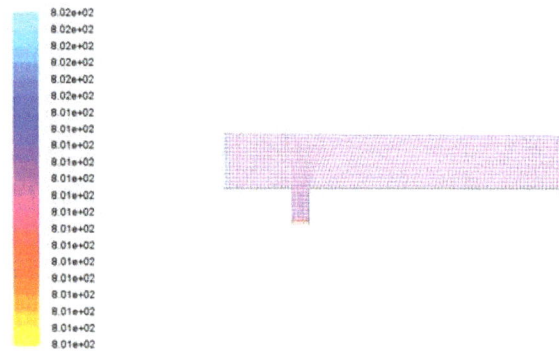

Fig. 9a. The density contours for the three-phase (air velocity is 0.2 m/s).

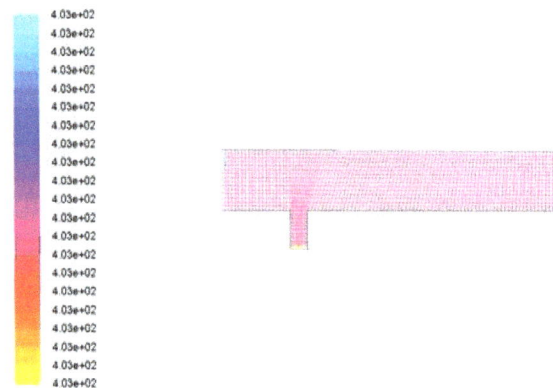

Fig. 9b. the density contours for the three-phase flow (air velocity is 0.6 m/s).

6. Conclusions

A bubble curtain is a system that produces bubbles in a deliberate arrangement in water. In this paper is paid to three-phase flow simulation by software Fluent6.3 that freshwater input is from the left side, saltwater on the right side of the entrance and the air from vertical duct. By using of mixture model and k-ε turbulence model in software the three-phase mixture is dissolved. At first the air inlet velocity is considered 0.6 meters per second then is reduced to 0.2 meters per second. The results of the numerical models show that increasing air injection rate is caused to reduce seawater intrusion. As you can see using of air (bubbles) curtain can be prevent from saltwater intrusion and also reduce density.

References

Bear J., Verruijt A., Modeling groundwater flow and pollution, Springer, Sep 30, Science, (1987) 414.

Bear J., Cheng, A.H.D., An overview, Chap. 1, in "Seawater intrusion in coastal aquifers concepts, methods, and practices", eds, (1999) 1-8.

Biglari B., Sturm T.W., Numerical modeling of flow around bridge abutments in compound channel, Journal of Hydraulic Engineering, 124 (1998) 156-163.

Cheng A.H.D., Halhal D., Naji A., Ouazar, D., Pumping optimization in saltwater – intruded coastal aquifers, J. Water Resource Research, 36 (1999) 2155-2165.

Cheng A.H.D, Ouazar D., Analytical solutions, Chap. 6, Seawater intrusion in coastal aquifers concepts, methods, and practices", eds, (1999) 163-191.

Cumming, R.G., Optimum exploitation of groundwater reserves with saltwater intrusion, Water Resources Research 7 (1971) 1415-1424.

Cummings R.G., McFarland, J.W., Groundwater management and salinity control, Water Resources Research 10 (1974) 909-915.

Dagan G., Bear J., Solving the problem of local interface upconing in a coastal aquifer by the method of small perturbations, Journal of Hydraulic Research 6 (1968) 15-44.

Das Gupta A., Nobi N., Paudyal G.N., Ground-water management model for an extensive multiaquifer system and an application, Groundwater 34 (1996) 349-357.

Fatemi E, Ataie-Ashtiani B., Simulation of seawater intrusion effect on contaminant transport in coastal aquifer of Tallar, 4th National Congress of Civil Engineering, Tehran university in Iran, (2008).

Ghyben W.B., Nota in Verband Met de Woorgenomen Putboring Nabij Amesterdam, Tijdschrift van Let Koninklijk Inst, Van Ing (1889).

Gorelick S.M., A review of distributed parameter groundwater management modeling methods, Water Resource Research 19 (1983) 305-319.

Herzberg B., Die Wasserversorgung einiger Nordseebader, Journal of Gasbeleuchtung and Wasserversorgung 44 (1992) 842-884-Mumich.

Jorreto S., Pulido-Bosch A., Gisbert J., Sánchez-Martos F., Francés, I., Investigating seawater intrusion due to groundwater pumping with schematic model simulations: The example of the Dar es Salaam coastal aquifer in Tanzania, Journal of African Earth Sciences 96 (2014) 71-78.

Keller R.J., Rodi, W., 1988. Prediction of flow characteristics in main channel/floodplain flows, Journal of Hydraulic Research 26 (1988) 425- 441.

Kouzana L., Ben Mammou A., Sfar Felfoul, M., The fresh water-seawater contact in coastal aquifers supporting intensive pumped seawater extractions: A case study, Comptes Rendus Geoscience 341 (2009) 993-1002.

Luong P.V., Sanchez J.E., Bernard R.S., Tate, C.H., Numerical Modeling and Simulation of a Salinity Intrusion Barrier, Panama Canal Study, ERDC Coastal and Hydraulics Laboratory, Sean.

Naji A, Cheng A.H.D., Ouazar D., BEM solution of stochastic seawater intrusion, Engineering Analysis Boundary Elements 23 (1999) 529-537.

Mertzanides Y., Economou N., Hamdan H., Vafidis A, Imaging seawater intrusion in coastal zone of Kavala (N. GREECE) with electrical resistivity tomography, Bulletin of the Geological Society of Greece, Proceedings of the 12th International Congress, Patras, (2010).

McGurik J.J., Rodi W., A depth-averaged mathematical model for the near fluid of side discharge into open- channel flow, Journal of Fluid Mechanics 864 (1978) 761-781.

Rajabi M.M., Ataie-Ashtiani B., Sampling efficiency in Monte Carlo based uncertainty propagation strategies: Application in seawater intrusion simulations, Advances in Water Resources 67 (2014) 46-64.

Rastogi A.K., Rodi W., 1978. Prediction of heat and mass transfer in open channels, Journal of Hydraulics Division, ASCE, 104 (1978) 397- 420.

Reddy A., Kumar K.N., David K., Varma K.S., Ground water potential and qualitative studies of Hyderabad, Andhra Pradesh (India), Journal of Environmental Research and Development 3 (2009) 1065-1074.

Reynolds O., On the dynamical theory of incompressible viscous Fluids and the determination of the criterion, Philosophical Transactions of the Royal Society of London (1984) 123-161.

Salamasi F., Azamathulla H.Md., Determination of optimum relaxation coefficient using finite difference method for ground water flow, Arabian Journal of Geosciences 6 (2013) 3409-3415.

Sanz E., Voss C.I., Inverse modeling for seawater intrusion in coastal aquifers: Insights about parameter sensitivities, variances, correlations and estimation procedures derived from the Henry problem, Advances in Water Resources 29 (2006) 439-457.

Shahmoradi B., Qavami A., A specification of the sanitary boundaries of a well, Journal of Environmental Research and Development 2 (2008) 523-529.

Siddiqui S., Mahmood A., Qari R., Determination of Fluouride concentration in seawater of different shores along the Karachi, Journal of Environmental Research and Development 5 (2011) 928-932.

Soltani M.V., Rahimi Asl R., Computational fluid dynamics by Fluent software, Tehran, Tarrah issues, (2003)

Swanson J., Fish kills drive techniques for placing bridge piles. Issue of Daily Journal of Commerce (Portland, OR), (2004).

Tarbox D.L., Hutchings W.C., Alternative approaches for water extraction in areas subject to saltwater upconing, In 20th Salt Water Intrusion Meeting, Naples, Florida, (2008) 23-27.

Van Camp M., Mtoni Y., Mjemah I.C., Bakundukize C., Walraevens K., Assessment of seawater intrusion and nitrate contamination on the groundwater quality in the Korba coastal plain of Cap-Bon (Northeast of Tunisia), Journal of African Earth Sciences 87 (2013) 1-12.

Versteeg H.K., Malalasekera W., An introduction to computational fluid dynamics: The Finite Volume Method, Prentice Hall, February 16 (2007) 503.

Werner A.D., Bakker M., Post, Vincent E.A, Vandenbohede A., Lu C., Ataie-Ashtiani B., Simmons C.T., Barry A., Seawater intrusion processes, investigation and management: Recent advances and future challenges, Advances in Water Resources 51 (2013) 3-26.

Wilcox D.C., Re-assessment of the scale-determining equation for advanced turbulence models, AIAA Journal, vol. 26 (1988) 1414-1421.

Wursig B., Greene C.R., Jefferson T.A., Development of an air bubble curtain to reduce underwater noise of percussive piling, Marine Environmental Research 49 (2000) 79-93.

Yakhot V., Orszag S.A., Thangam S., Gatski T.B., Speziale, C.G., Development of turbulence models for shear flows by a double expansion technique, Physics of Fluids A 4 (1992) 1510-1520.

Yeh W.W.G, Review of parameter identification procedures in groundwater hydrology: the inverse problem, Water Resource Research 22 (1986) 95-108.

Ziegeler NAVO MSRC Visual Analysis and Data Interpretation Center in association with Saltwater Separation, LLC.

Zghibi A., Tarhouni J., Zouhri L., Geophysical and hydrochemical study of the seawater intrusion in Mediterranean semi-arid zones. Case of the Korba coastal aquifer (Cap-Bon, Tunisia), Journal of African Earth Sciences 58 (2010) 242-254.

22

Predicting the discharge coefficient of triangular plan form weirs using radian basis function and M5' methods

Azam Akhbari[1], Amir Hossein Zaji[2],*, Hamed Azimi[2], Mohsen Vafaeifard[1]

[1]Department of Civil Engineering, Faculty of Engineering, University of Malaya, Kuala Lumpur, Malaysia.
[2]Young Researches Club, Kermanshah Islamic Azad University.

ARTICLE INFO	ABSTRACT
Keywords: Triangular plan form weir Discharge coefficient Radial basis neural networks M5' method Sensitivity analysis	Weirs are installed on open channels to adjust and measure the flow. Also, discharge coefficient is considered as the most important hydraulic parameter of a weir. In this study, using the Radial Base Neural Networks (RBNN) and M5' methods, the discharge coefficient of triangular plan form weirs is modeled. At first, the effective parameters in the prediction of the discharge coefficient are identified. Then, by combining the input parameters, for each of the RBNN and M5' methods, six different models are introduced. By analyzing the modeling results for all models, it was shown that the M5' model is capable of modeling the discharge coefficient more accurately. Also, based on the modeling results, a model that considered the impact of all input parameters was introduced as a superior model. The mean absolute percentage error (MAPE) and correlation coefficients (R^2) values for the preferred model in the test mode were calculated 2.774 and 0.831, respectively. Also, for each of the M5' models, some relationships were proposed to estimate the triangular plan form weirs. The evaluation of these relationships showed that the parameters of the ratio of head over the weir to channel width (h/B) and Froude number (Fr) were the most effective parameters in the prediction of the discharge coefficient.

1. Introduction

The weirs with various shapes are installed as a barrier, orthogonal to the flow direction, to measure and regulate the water within the open channels. Weirs typically are used as rectangular, triangular, circular, composite, sinusoidal and labyrinth shapes in irrigation channels, drainage networks, and other hydraulic targets.

Many researchers have conducted laboratory, analytical and theoretical studies on the hydraulic behavior of normal weirs. Schoder and Turner (1929) proposed an equation to calculate the discharge coefficient of a steep-rectangular weir. The discharge coefficient was presented as a function of flow head over weir crest to weir height. The discharge coefficient of the proposed equation is considered without the effects of viscous and capillary effects. Rouse (1936) also presented a discharge coefficient equation for normal weirs in the case of the ratio of head over weir to the weir height are greater than 15. Kandaswamy and Rouse (1957), using the results of a laboratory study, obtained a solution to calculate the discharge through normal weirs. Strelkoff (1964) estimated the discharge coefficient of this type of hydraulic structures for a condition that the ratio of head over weir to weir height is less than ten by assuming a two-dimensional flow through a sharp weir edges using the analytical method. Taylor (1968) examined the hydraulic behavior of the labyrinth weirs. Hay and Taylor (1969) studied the different states of labyrinth weirs. The results showed that the placement panels as the triangular shape is more efficient than the labyrinth state. Hager (1983) conducted a study to calculate the discharge coefficient of normal weirs by using the geometric specification of the channel and the depth of the brink depth. Ramamurthy, Tim et al. (1987) estimated the discharge coefficient of a sharp-crested weir using the Momentum principles. Swamee (1988) obtained an equation to calculate the discharge coefficient of normal rectangular weirs. The Swamee's equation was a function of the geometric specification of the weir and the flow head over normal weir. Tullis et al. (1995) from the trapezoidal labyrinth weirs found that the discharge capacity of this type of weirs is a function of the total head on

the weir, the effective crest length of the weir and the discharge coefficient of the weir. Wormleaton and Soufiani (1998) conducted a laboratory study concerning the hydraulic specification and aerodynamic of triangular labyrinth weir. It was found that the aeration efficiency of triangular labyrinth weir is higher than the linear weirs with more equal length. Johnson (2000) proposed a relationship to calculate discharge coefficient of the rectangular weirs in the flat-topped and sharp-crested. Emiroglu and Baylar (2006) examined the effects of the included angle and sill slope of the weir on triangular labyrinth weir aeration. Tullis, Young et al. (2007) investigated the effects of submergence on the hydraulic behavior of labyrinth weir in and obtained a relationship between the discharge through the crest and the normal weir head. Bagheri and Heidarpour (2010) obtained an equation to calculate the discharge coefficient of the normal rectangular weir using a laboratory study. The discharge coefficient equation is a function of the ratio of the width of the weir to the channel width and the ratio of the weir head to weir height. Kumar, Ahmad et al. (2011) determined the by discharge capacity of a triangular planform weirs. Ahmad et al. (2011) are considered discharge coefficient of Kumar as a function of geometric weir characteristics and hydraulic parameters of flow.

In recent decades, soft computing and artificial intelligence have been used in the prediction and modeling of nonlinear phenomena by researchers from various sciences. Also, different artificial neural network algorithms have been widely used to solve multiple problems of hydraulic science, hydrology and water resources. Savi et al. (1999) modeled and predicted the runoff of the rainfall. Giustolisi (2004) calculated the resistance coefficient of the corrugated channels using the neural network algorithm GP. Bilhan et al. (2010) predicted the discharge coefficient of the sharp-crested rectangular weir on the wall of a straight channel using various neural network techniques such as Feed Forward Neural Networks (FFNN) and Radial Basis Neural Networks (RBNN). Dursun et al. (2012), using the ANFIS model, presented a relationship to calculate the discharge coefficient of semi-elliptical side weirs over the side of the rectangular channels in sub-

*Corresponding author Email: amirzaji@gmail.com

critical flow conditions. Also, Kisi Emiroglu et al. (2012) modeled the discharge of the labyrinth triangular weir located on a rectangular channel using radial base neural networks (RBNN), generalized regression neural networks (GRNN) and gene expression programming (GEP). They obtained an equation as a function of the upstream weir of Froude number and the geometric characteristics of the weir and the main channel. Azamathulla and Ahmad (2013) predicted the discharge through a rectangular slide gate on open channels, using the gene expression programming algorithm. They proposed a relationship as a function of the flow Froude number and the ratio of the flow depth at the upstream gate to the gate opening to calculate the discharge coefficient of this type of structures. Ebtehaj et al. (2015) predicted the discharge coefficient of rectangular side orifices on the wall of rectangular channels in sub-critical conditions by the GMDH neural network model.

From the literature, it should be noted that the study of the triangular plan form weirs requires further investigation. On the other hand, the use of various artificial intelligence techniques for modeling the drainage capacity of these structures is considered as an optimal and appropriate solution that contains important and practical points. Therefore, in this study, the discharge capacity of the triangular plan form weirs which are installed in the rectangular channels is modeled using Radial Base Neural Networks (RBNN), and Method modified M5 (M5'), and the results of these two methods are compared with each other. For this purpose, the effective parameters are determined by the discharge coefficient firstly. Subsequently, by combining the input parameters, for each of the RBNN and M5' methods, six different models are introduced. Then, by analyzing the results of the modeling, the superior models to predict the discharge coefficient of the triangular plan form weirs are introduced.

2. Experimental Model

In this study, for the verification of the results of RBNN and M5' methods, Kumar et al. (2011) laboratory measurements are used. A laboratory model of Kumar et al. (2011) includes a rectangular channel with a length of 12 meters, a width of 0.28 meters and a depth of 0.41 meters. The triangular planform weir is made of steel plates, which is located in a rectangular entrance channel 11 meters. A pipe provides the input flow into the rectangular channel through a fixed-headed air reservoir. A point gage measures the head of the triangular weir crest with an accuracy. To aerate the nappe over the weir, the aeration holes are embedded on the sides of the lateral walls of the rectangular channel. In the upstream of the rectangular channel, grid walls and wave suppressors are used to reduce disturbances and eddies. In the laboratory model and are the included-angle of the weir, the height of weir, the head of the flow over the weir and the discharge flow over the weir, respectively. In Table 1, the range of laboratory measurements of Kumar et al. (2011) study is shown. Also in Fig. 1, the general scheme of the laboratory model of Kumar et al. (2011) has been depicted.

Table 1. The range of laboratory values of Kumar et al. (2011).

θ	w	H	Q
30	0.0924	0.0079–0.0346	0.0020–0.0125
60	0.1005	0.0129–0.0565	0.0021–0.0120
90	0.1029	0.0136–0.0689	0.0015–0.0121
120	0.1062	0.0197–0.0725	0.0021–0.0124
150	0.1075	0.0142–0.0710	0.0012–0.0113
180	0.1000	0.0242–0.0724	0.0022–0.0109

3. Numerical methods
3.1. M5' model tree

The decision tree is a sequence of tests that determine the appropriate test at each step. A decision tree is a tree in which branch nodes represent a choice between alternatives, and leaf nodes represent a decision of the desired class. Generally, in a decision tree, data begins to be divided by the root, and eventually, in the leaf nodes are divided into different classes. In each node, the values of a sample are tested, and then, according to some of the attributes of that sample, one of the branches of that node is selected and also move downward (Hand. 2007; Kantardzic. 2011). As it is clear, in decision trees, the response variable is a label. But to solve the problems in which target variables are numerical, regression trees presented by Breiman,

Friedman, et al. (1984) is used. It should be noted that regression trees and decision trees are not structurally different from each other, and the only difference is how to deal with target variables. The model trees introduced by Quinlan (1992) is an extension of regression trees in which each leave is a multivariable linear regression model. In fact, model trees are a decision tree that is modeled for a continuous class problem. Also, instead of class names, at each leaf, a line is fitted on each class. M5 model tree was presented for the first time by Quinlan (1992) and then examined and improved by Wang and Witten (1996) as M5 '. Similar to the situation that occurred in the decision tree, in M5 'the dataset space is divided into several parts and in the first step a regression tree is constructed.

The M5' divides the primal dataset by using the Standard Deviation (SD) of the classes' members and tries to decrease this error by testing all of the attributes of that node. The Standard Deviation Reduction (SDR) is defined as follow:

$$SDR = sd(T) - \sum \frac{|Ti|}{|T|} SD(T_i) \tag{1}$$

where T is the samples that reach to a specified node, and i is the number of the outcome of the potential set. After the tree is formed, the M5' based on the samples in each decision node and leaf node develops a linear regression for that node. Then the leaves of the tree that estimate target values with low accuracy will be pruned, and the final shape of the model is obtained.

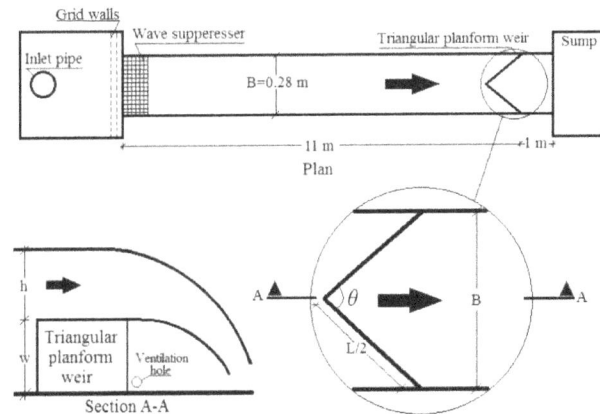

Fig. 1. The general scheme of the laboratory model of Kumar, Ahmad et al. (2011).

3.2. Radial basis neural network

The training algorithm of RBNN (Broomhead and Lowe 1988; Moody and Darken 1989; Poggio and Girosi 1990) is mostly faster than the back propagation neural networks and the probability of trapping in the local minimum is lower (Kisi et al. 2008; Ay and Kisi 2011). So that, this method is very appropriate for the complex hydraulic engineering problems. RBNN is a supervised learning algorithm that forms from three layers namely the input layer, the hidden layer, and the output layer. The input variables are introduced to the model by using the input layer. Hidden layer transfers the received information onto nonlinear future by using an activation function that is selected from a group of functions called basic kernel functions. One of the most popular basic kernel functions that are used in the RBNN method is the Gaussian function. In this function, the largest amount is obtained when the distance between the input vector and the function's centre is equal to zero. Finally, the hidden layer information is cumulated using a linear regressor, and the ultimate result is transferred to the output layer. So that, the jth output is calculated as follow:

$$x_j = f_j(u) = w_{0j} + \sum_{i=1}^{L} w_i h_i \quad j = 1, 2, ..., M \qquad (2)$$

where w is weight, h is the output of the hidden layer's neurones, L is the number of hidden layer's neurones, and M is the number of outputs that is equal to one in the present study. So that, the final equation of the RBNN model is obtained as follow:

$$x_j = f_i(u) = w_{0j} + \sum_{i=1}^{L} w_{ij} G(\|u - c_i\|) \quad j = 1, 2, ..., M \qquad (3)$$

where u is the input variable, ci is the center of the ith¬ kernel, G is the basic kernel function, and $\| . \|$ is the Euclidean distance between u and ci. The structure of a simple RBNN model is shown in the following Fig. 2.

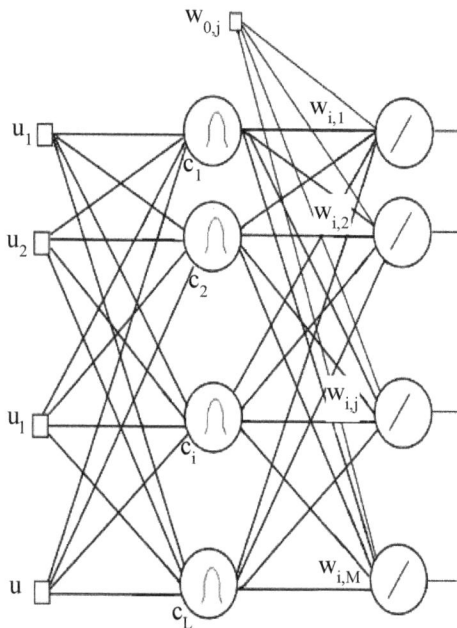

Fig. 2. RBF-NN model structure.

It is obvious that the number of input and output layers' neurons is equal to the number of input and output variables of the considered problem. However, the number of hidden layer's neurons should be determined using the trial and error method.

4. Results and discussion
4.1. Discharge coefficient

The discharge coefficient of a normal weirs is a function of weir crest height (w), total flow head over the weir (h), weir crest thickness, crest shape, vertex configuration, and vertex angle. Also, Kumar et al. (2011) presented a relationship as a function of the vertex angle (θ) and the ratio of the head over weir weir crest height (h/w) for triangular weir form plan. Therefore, in the present study, the effect of dimensionless parameters, the ratio of the weir crest length to the flow head over weir (L/h), Froude number (F_r), the ratios of head over weir to the channel width (h/B), $\sin\theta \times w/L$ and $w + h/\sin\theta \times w$ on the discharge coefficient of the triangular weir form plan (C_d) are investigated. In the following, with the use of RBNN and M5', six different models are defined as a function of those dimensionless parameters. The combination of input parameters in different models is shown in Fig. 3. To check the accuracy of the RBNN and M5' models results, statistical indices of the mean absolute percent error (MAPE) and correlation coefficients (R^2) are used:

$$MAPE = \frac{1}{n} \sum_{i=1}^{n} | \frac{C_{d\,(observed)i} - C_{d\,(predicted)i}}{C_{d(observed)i}} \times 100\% \qquad (4)$$

$$R^2 = \frac{(n \sum_{i=1}^{n} C_{d(predicted)i} C_{d(observed)i} - \sum_{i=1}^{n} C_{d(predicted)i} \sum_{i=1}^{n} C_{d(observed)i})^2}{n \sum_{i=1}^{n} (C_{d(predicted)i})^2 - \sum_{i=1}^{n} (C_{d(predicted)i})^2)(n \sum_{i=1}^{n} (C_{d(observed)i})^2 - \sum_{i=1}^{n} (C_{d(observed)i})^2)} \qquad (5)$$

where, $C_{d(Pridicted)i}$ and n is the laboratory discharge coefficient, predicted and the number of laboratory measurements, respectively. In this study, 80 % of the data for training and the remained 20 % were used to test the RBNN and M5' models.

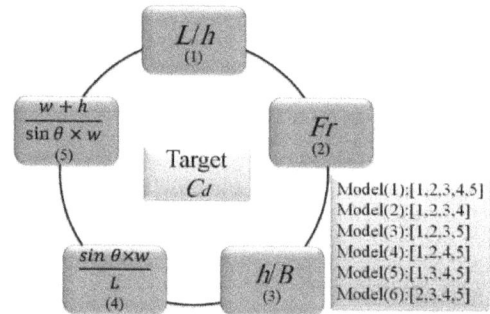

Fig. 3. Combination strategies of input parameters in different models.

4.2. Sensitivity analysis

The result of the sensitivity analysis of the training mode of RBNN method for Models 1 to 6 is shown in Fig. 4. Based on the results of the RBNN method, the MAPE value for Models 1 to 5 is almost predicted similarly. As can be seen from Figure 4a, Model (6) has the lowest error value and the highest correlation coefficients $\left(R^2 = 0.597, MAPE = 4.816 \right)$. Model (6) is a function of the Froude number (F_r), the ratio of the head over the weir to the channel width (h/B) and the ratios and $\sin\theta \times w/L$. For the M5' method, the results of the sensitivity analysis of the training mode of models 1 to 6 are shown in Fig. 4b. According to modelling results, Model (4) has the highest error $\left(MAPE = 3.957 \right)$ and the lowest correlation coefficient $\left(R^2 = 0.740 \right)$. In contrast, models 1, 3, and 6 have the least error value and the highest R^2 $\left(R^2 = 0.844, MAPE = 3.222 \right)$. As can be seen, the M5' method predicts the results of models 1, 3 and six quite similar. The mechanism of the M5' method is that calculate the most optimal results for models with lower input parameter numbers and repeated parameters in other models. Therefore, based on the M5' mechanism, Model (1) has five input parameters that L/h, F_r, h/B, $\sin\theta \times w/L$ and $w + h/\sin\theta \times w$ and in models 3 and 6 (models 3 and 6 each have four input parameters) have been repeated. Therefore, the results of all three models 1, 3 and 6 by the M5' method are predicted completely similar. In the following, the results of RBNN and M5 'for each of 1 to six are examined.

4.3. Comparison of the results of the RBNN and M5' method in prediction of the weir discharge coefficient

Fig. 5 shows the mean absolute percentage error (MAPE) between RBNN and M5' methods in prediction of the discharge coefficient of triangular weir form plan for models 1 to 6. As can be seen, the value of the Model (1) error for the M5' method in both train and test steps is less than the RBNN method. The MAPE value for the RBNN method in the train and test mode is calculated 7.967 and 6.420, while the M5' method predicts the MAPE value of 3.222 in train mode and 2.774 in the test mode. Similar to Model (2), the error value of the M5' method in both train and test situations is less than the RBNN method. The MAPE for RBNN and M5' methods in train mode is respectively 8.101 and 3.322, and in the test mode, is calculated 5.961 and 3.276, respectively.

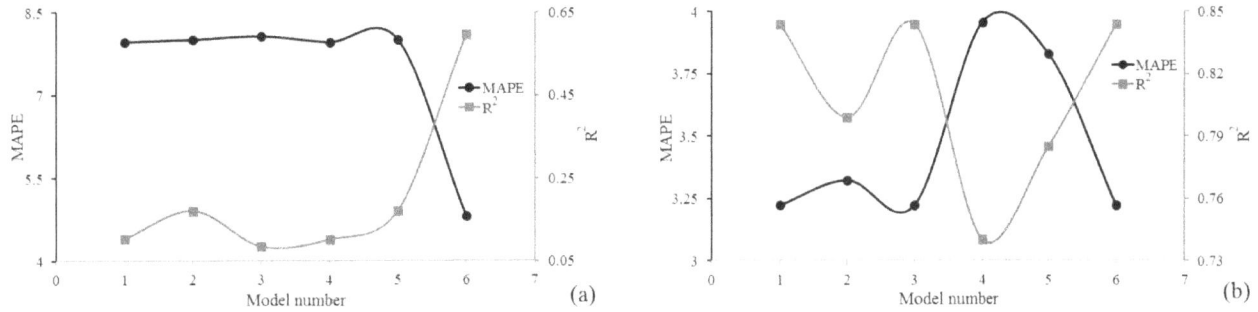

Fig. 4. Comparison of R^2 and MAPE values for training models 1 to 6 (a) RBNN (b) M5'.

The accuracy of the M5' method to predict the discharge coefficient by Model (3) is higher in the test and train mode than the RBNN method. The mean absolute percentage error calculated in the training mode for the M5' method is 3.222 and for the RBNN method is 8.072. As can be seen, for Model (3), the error of the RBNN method is 2.5 times the M5' method. In train mode, the accuracy of the M5' method is more than the RBNN method. The MAPE in the training mode for the model (4) for M5' and RBNN methods is calculated 3.957 and 7.969, respectively. As shown in Fig. 5, the accuracy of the M5 'and RBNN methods in the test mode is approximately the same. In the test mode, the MAPE value for the RBNN method is 6.421 and for the M5' method obtained 6.420.

Model (5) is a function of L/h, h/B, $\sin\theta \times w/L$ and $w + h/\sin\theta \times w$ in both modes the accuracy of the M5' method is more than the RBNN method. The mean absolute percentage error for test and train mode of M5' method is 3.828 and 4.612, respectively, and for the RBNN method is estimated 8.010 and 6.369, respectively. Similar to Models 1 to 5, in both test and train mode, the accuracy of the M5 'method is calculated more than the RBNN method. The MAPE values for test and train of the RBNN method was obtained were 4.816 and 5.018, respectively, while MAPE for the M5' method was calculated 3.222 and 2.774 in both test and train modes respectively.

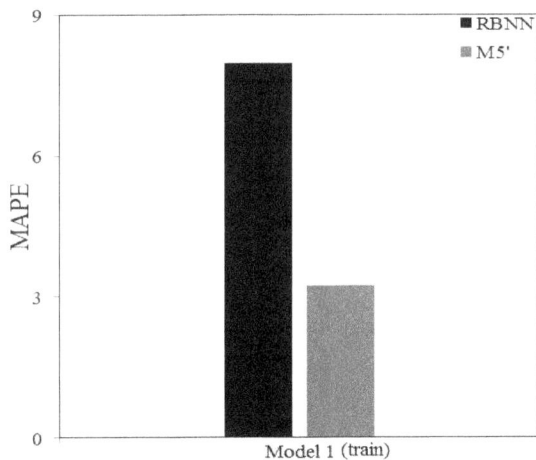

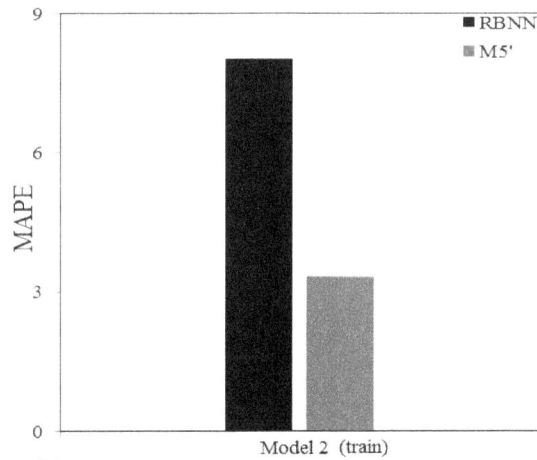

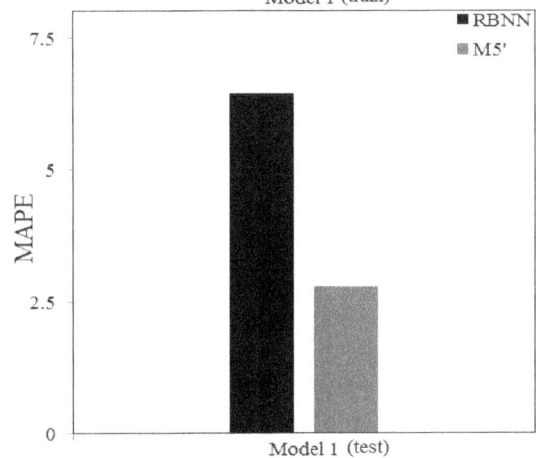

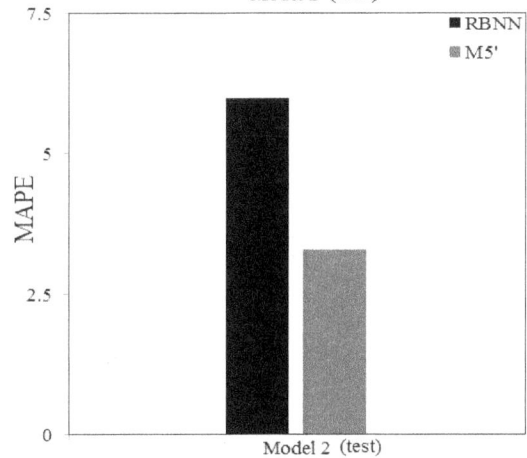

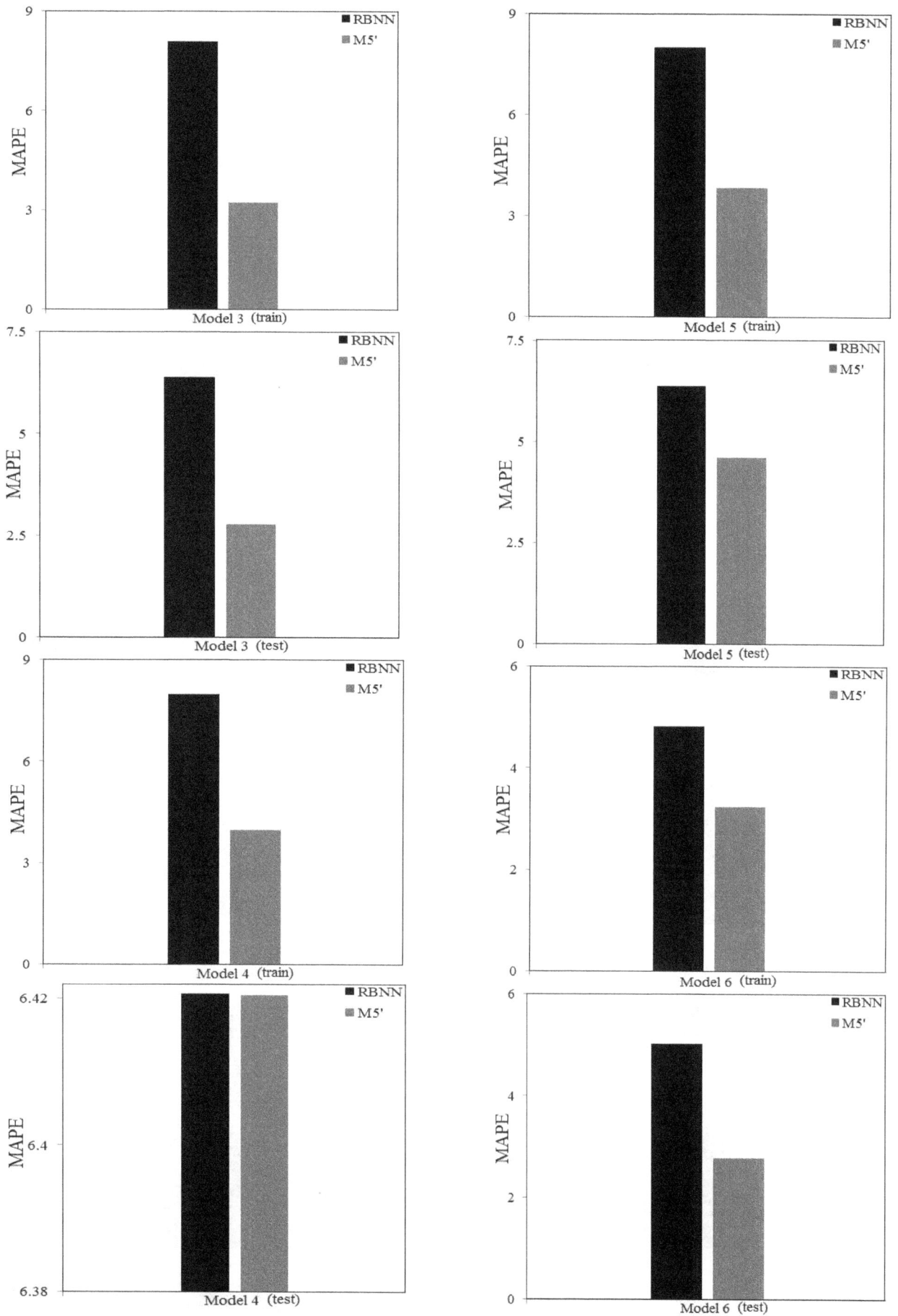

Fig. 5. Comparison of MAPE for different models of RBNN and M5'.

4.4. Derivation of discharge coefficient based on M5'

In Fig. 6, the scatter plots are illustrated for models 1 to 6 of the M5' method. As can be seen, Model (1) is a function of all input parameters. By analyzing the results of Model (1), the following relations are proposed using the M5' method. Based on the results of Model (1), if the values are h/B≤ 0.08, the discharge coefficient is a function of h/B. In other words, in this condition, the effect of the parameter h/B is perceptible. Also, if the value is $\sin\theta \times \frac{w}{L} \leq 1.499$, the discharge coefficient is predicted equally to the constant value of 0.739. However, for Froude number less than 0.824 and 0.717, the discharge coefficient is predicted regarding Froude numbers. Therefore, for Model (1) the most important parameters including h/B and Froude numbers. For Model (1):

If h/B≤ 0.08

$$C_d = 0.966 - 2.716(h/B)$$

else

if $\sin\theta \times \frac{w}{L} \leq 1.499$,

$$C_d = 0.739$$

else

if $(Fr) \leq 0.824$

if $(Fr) \leq 0.717$

$$C_d = 0.454 + 0.345(Fr)$$

else

$$C_d = 0.898 - 1.034(h/B)$$

else

if $(Fr) \leq 1.269$

$$C_d = 0.615$$

else $C_d = 0.666$

Also, by analyzing the modeling results for Model (2), the following relationships are presented. Similar to Model (1), if the value h/B is less than 0.08, the discharge coefficient is predicted only regarding the parameter h/B. For this model, if $(\sin\theta \times w/L) \leq 0.156$ discharge coefficient is a function of the $\sin\theta \times w/L$ Froude number and h/B. In this model, the effects of $w + h/\sin\theta \times w$ being negligible. For Model (2):

If $h/B \leq 0.08$, $C_d = 0.966 - 2.716(h/B)$

else

if $(\sin\theta \times w/L) \leq 0.156$

$$C_d = 0.671 - 1.149(\sin\theta \times w/L)$$

else

$$C_d = 1.156 - 0.333(Fr) - 1.079(h/B)$$

The following relationships are presented according to the modeling results. The most effective parameters of this model include h/B, $w + h/\sin\theta \times w$ and Fr. However, the effect of head over the weir to the channel width (h/B) is higher than other parameters. Also, as can be seen, the effect of the parameter L/h on Model (3) is not significant. For Model (3):

If $h/B \leq 0.08$

$$C_d = 0.966 - 2.716(h/B)$$

else

if $\sin\theta \times \frac{w}{L} \leq 1.499$,

$$C_d = 0.739$$

else

if $(Fr) \leq 0.824$

if $(Fr) \leq 0.717$

$$C_d = 0.454 + 0.354(Fr)$$

else

$$C_d = 0.898 - 1.034(h/B)$$

else

if $(Fr) \leq 1.269$

$$C_d = 0.615$$

else

$$C_d = 0.666$$

The proposed relationship for Model (4) is presented below. In this model, if the value L/h is less than 12,682 and $\sin\theta \times w/L$ smaller than 0.171, the triangular weir discharge coefficient is a function of the parameter L/h and the Froude number. In other words, in Model (4) the effect of parameters $\sin\theta \times w/L$ and $w + h/\sin\theta \times w$ is insignificant in the modeling of the discharge coefficient. For Model (4):

If $L/h \leq 12.682$

if $(\sin\theta \times w/L) \leq 0.171$

$$C_d = 0.707 - 0.008(L/h)$$

else

$$C_d = 0.947 - 0.025(L/h) - 0.533(Fr)$$

else

$$C_d = 0.813 - 0.004(L/h) - 0.122(Fr)$$

For Model (5) the effect of the Froude number has been removed. For this model, if h/B is smaller than 0.08, the discharge coefficient is a function of the h/B. However, for $w + h/\sin\theta \times w$ less than 1.499, the value of the discharge coefficient is equal to the constant value of 0.739. Also, for a $w + h/\sin\theta \times w$ lower than 1.809, the discharge coefficient is predicted based on h/B. In general, in this model, the effect of the parameter h/B is significant, and in contrast, the impact of $\sin\theta \times w/L$ and v is negligible. For Model (5):

If $h/B \leq 0.08$

$$C_d = 0.966 - 2.716(h/B)$$

else

if $(w + h/\sin\theta \times w) \leq 1.499$

$$C_d = 0.739$$

else

if $(w + h/\sin\theta \times w) \leq 1.809$

$$C_d = 0.840 - 1.112(h/B)$$

else

if $(\sin\theta \times w/L) \leq 0.114$

$$C_d = 0.669$$

else

$$C_d = 0.711$$

For Model (6) the effect of the parameter L/h is ignored. In this model, the effect of the parameters h/B and Fr is greater than the other input parameters. For example, for h/B values less than 0.08, the triangular weir discharge is considered by parameter h/B. Also, if the $w + h/\sin\theta \times w$ value is less than 1.499, the value of the discharge coefficient is equal to the constant value of 0.739. As can be seen, for Model (1) to Model (6) the parameters h/B and Fr are introduced as the most effective parameters. For Model (6):

If $h/B \le 0.08$
$C_d = 0.966 - 2.716(h/B)$
else
if $(w + h/\sin\theta \times w) \le 1.499$
$C_d = 0.739$
else
if $(Fr) \le 0.824$
if $(Fr) \le 0.717$
$C_d = 0.454 + 0.354(Fr)$
else
$C_d = 0.898 - 1.034(h/B)$
else

if $(Fr) \le 1.269$
$C_d = 0.615$
else
$C_d = 0.666$

5. Conclusions

Normal weir is installed in open channels as a simple plane to adjust and measure the flow. As the flow approaches the normal weir location, the flow from the weir crest to the downstream channel is thrown. Normal weirs are divided into two types: sharp-crested and broad-crested weirs. The sharp-crested weirs as rectangular, triangular, circular, Sutro and triangular plan form weirs are used. In this study, the triangular plan form weirs model was modeled using Radial Base Neural Networks (RBNN) and Method modified 5 (M5') methods. For this purpose, effective parameters were first identified on the discharge coefficient. Then, by combining the input parameters, for each of the RBNN and M5' methods, six different models were introduced. By comparing the results of six defined models, the M5' model is more accurate. The superior model was also introduced with sensitivity analysis. The superior model, model the discharge coefficient regarding all input parameters. In the following, for each M5' model, relationships were proposed to predict the discharge coefficient. The analysis of these relationships showed that the parameters of head over the weir to the channel width and Froude number are the most effective parameters in the prediction of the triangular plan form weirs.

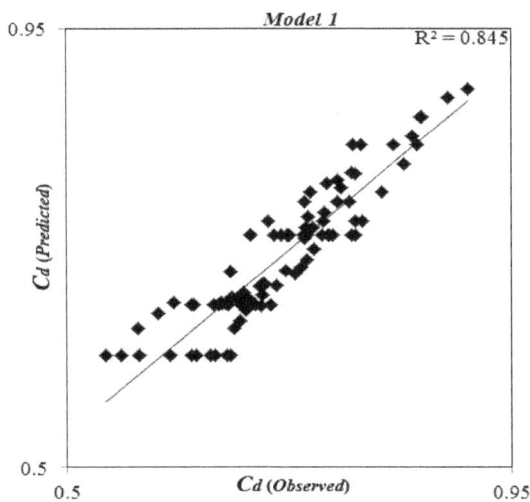

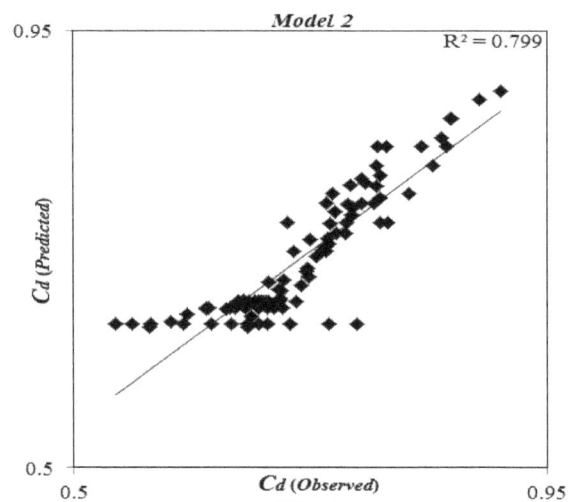

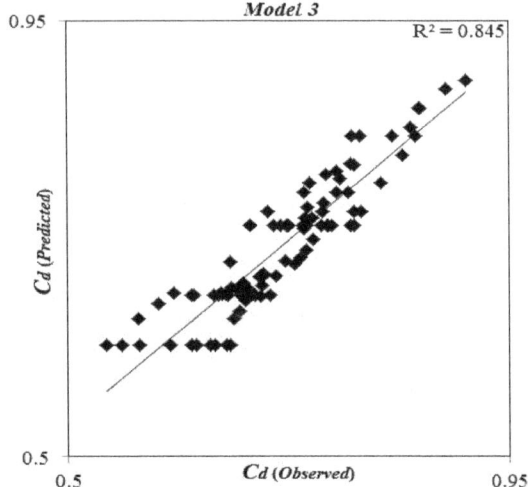

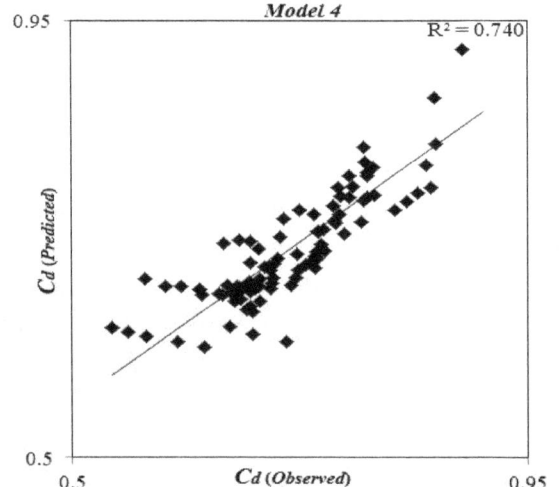

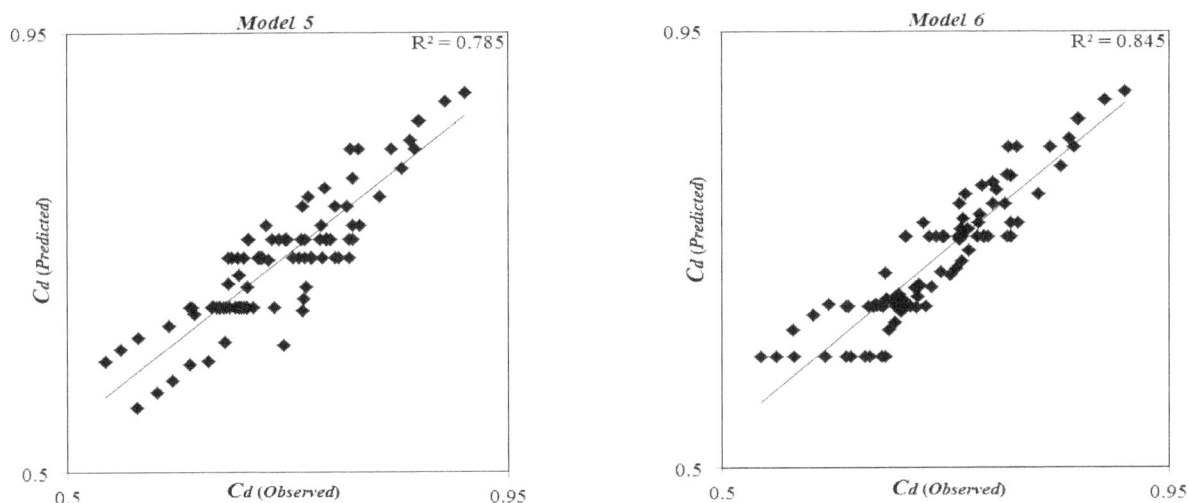

Fig. 6. Results of the predicted discharge coefficient by the M5 'method for models 1 to 6.

References

Ay M., Kisi O., Modeling of dissolved oxygen concentration using different neural network techniques in Foundation Creek, El Paso County, Colorado, Journal of Environmental Engineering 138 (2011) 654-662.

Azamathulla H.M., Ahmad Z., computation of discharge through side sluice gate using gene-expression programming, Irrigation and Drainage 62 (2013) 115-119.

Bagheri S., Heidarpour M., Flow over rectangular sharp-crested weirs, Irrigation science 28 (2010) 173.

Bilhan O., Emiroglu M.E., Kisi O., Application of two different neural network techniques to lateral outflow over rectangular side weirs located on a straight channel, Advances in Engineering Software 41 (2010) 831-837.

Breiman L., Friedman J., Stone C.J., Olshen R.A., Classification and regression trees, CRC press (1984).

Broomhead D.S., Lowe D., Radial basis functions, multi-variable functional interpolation and adaptive networks, Royal Signals and Radar Establishment Malvern (United Kingdom) (1988).

Dursun O.F., Kaya N., Firat M., Estimating discharge coefficient of semi-elliptical side weir using ANFIS, Journal of hydrology 426 (2012) 55-62.

Ebtehaj I., Bonakdari H., Khoshbin F., Azimi H., Pareto genetic design of group method of data handling type neural network for prediction discharge coefficient in rectangular side orifices, Flow Measurement and Instrumentation 41 (2015) 67-74.

Emiroglu M.E., Baylar A., Closure to Influence of Included Angle and Sill Slope on Air Entrainment of Triangular Planform Labyrinth Weirs, Journal of Hydraulic Engineering 132 (2006) 748-748.

Giustolisi O., Using genetic programming to determine Chezy resistance coefficient in corrugated channels, Journal of Hydroinformatics 6 (2004) 157-173.

Hager W.H., Hydraulics of plane free overfall, Journal of Hydraulic Engineering 109 (1983) 1683-1697.

Hand D.J., Principles of data mining, Drug safety 30 (2007) 621-622.

Hay N., Taylor G., A computer model for the determination of the performance of labyrinth weirs. 13th Congress of IAHR, Koyoto, Japan (1969).

Johnson M.C., Discharge coefficient analysis for flat-topped and sharp-crested weirs, Irrigation science 19 (2000) 133-137.

Kandaswamy P., Rouse H., Characteristics of flow over terminal weirs and sills, Journal of the Hydraulics Division 83 (1957) 1-13.

Kantardzic M., Data mining: concepts, models, methods, and algorithms, John Wiley & Sons (2011).

Kisi O., Emiroglu M. E., Bilhan O., Guven A., Prediction of lateral outflow over triangular labyrinth side weirs under subcritical conditions using soft computing approaches, Expert systems with Applications 39 (2012) 3454-3460.

Kisi O., Yuksel I., Dogan E., Modelling daily suspended sediment of rivers in Turkey using several data-driven techniques/Modélisation de la charge journalière en matières en suspension dans des rivières turques à l'aide de plusieurs techniques empiriques Hydrological Sciences Journal 53 (2008) 1270-1285.

Kumar S., Ahmad Z., Mansoor T., A new approach to improve the discharging capacity of sharp-crested triangular plan form weirs, Flow Measurement and Instrumentation 22 (2011) 175-180.

Moody J., Darken C.J., Fast learning in networks of locally-tuned processing units, Neural computation 1 (1989) 281-294.

Poggio T., Girosi T., Regularization algorithms for learning that are equivalent to multilayer networks, Science (Washington) 247 (1990) 978-982.

Quinlan J.R., Learning with continuous classes. 5th Australian joint conference on artificial intelligence, Singapore (1992).

Ramamurthy A.S., Tim U.S., Rao M., Flow over sharp-crested plate weirs, Journal of irrigation and drainage Engineering 113 (1987) 163-172.

Rouse H., Discharge characteristics of the free overfall: Use of crest section as a control provides easy means of measuring discharge, Civil Engineering 6 (1936) 257-260.

Savic D.A., Walters G.A., Davidson J.W., A genetic programming approach to rainfall-runoff modelling, Water Resources Management 13 (1999) 219-231.

Schoder E.W., Turner K.B., Precise weir measurements, Transactions of the American Society of Civil Engineers 93 (1929) 999-1110.

Strelkoff T.S., Solution of Highly Curvilinear Gravity Flow, Journal of the Engineering Mechanics Division 90 (1964) 195-222.

Swamee P.K., Generalized rectangular weir equations, Journal of Hydraulic Engineering 114 (1988) 945-949.

Taylor G., The performance of labyrinth weirs, University of Nottingham (1968).

Tullis B.P., Young J., Chandler M., Head-discharge relationships for submerged labyrinth weirs, Journal of Hydraulic Engineering 133 (2007) 248-254.

Tullis J.P., Amanian N., and Waldron D., Design of labyrinth spillways, Journal of hydraulic engineering 121 (1995) 247-255.

Wang Y., Witten I.H., Induction of model trees for predicting continuous classes (1996).

Wormleaton P.R., Soufiani E., Aeration performance of triangular planform labyrinth weirs, Journal of environmental engineering 124 (1998) 709-719.

23

Numerical simulation of the effects of downstream obstacles on malpasset dam break pattern

Hamed Azimi[1], Majeid Heydari[*,2], Saeid Shabanlou[3]

[1]Young Researchers Club, Kermanshah Branch, Islamic Azad University, Kermanshah, Iran.
[2]Department of Science and Water Engineering, Faculty of Agriculture, Bu-Ali Sina University, Hamedan, Iran.
[3]Department of Water Engineering, Kermanshah Branch, Islamic Azad University, Kermanshah, Iran.

ARTICLE INFO

ABSTRACT

Keywords:
Dam break
Malpasset dam
Numerical simulation
Flow pattern

Dam break is an important phenomenon which significantly affects the environment as well as the inhabitants of the downstream areas of the dam. In the present study, the hydraulic break of Malpasset dam as a result of sudden flooding was simulated numerically using the FLOW-3D software. The two-equation k-ε turbulence models and RNG k-ε turbulence model were used to simulate the flow field turbulence. Also, the free-surface variations of the flow were simulated using the VOF (Volume of Fluid) scheme. The results obtained from the numerical model were in good agreement with those predicted by the EDF model. Based on the simulation results, the maximum pressure occurred at the lower layers of the flow and reduced as the free surface of the flow was approached. The maximum pressure increased at each point in time. The maximum longitudinal velocity occurred at the front of the advancing wave resulting from break of the dam, and subsequently decreased due to the increasing depth at the downstream of the dam. Additionally, the effects of obstacles with different shapes on the flow pattern arising from dam break (due to sudden flooding) were also investigated. Examination of these effects revealed that the cubic obstacle placed obliquely in the flow direction produced the maximum separation region at its downstream. Conversely, this separation region was eliminated completely when a cylindrical obstacle was used. The maximum and minimum Froude numbers were obtained for the flow encountering the perpendicular cubic obstacle and the flow impacting the cylindrical obstacle, respectively.

1. Introduction

Dam break is an important consideration in the design of dams. If a dam collapses, a huge wave is containing mud, dam reservoir water and dam body materials starts moving, thus threatening the land, people, or residential areas that are located in its path. Numerous theoretical and experimental studies have been conducted regarding dam break. Bellos et al. (1992) performed a 2D laboratory experiment on dam break. Bell et al. (1992) conducted a research on one-dimensional and two-dimensional break of a dam using a channel with a 180-degree bend for the 2D break analysis, and compared the results obtained for various cases. Fraccarollo and Toro (1995) investigated the 2D break of a dam numerically and experimentally, and compared these results. Aziz Khan et al. (2000) studied dam break by taking into account the effect of suspended particles on the break wave and examined the velocity profile of the same via hexahedron particles, using mobile cameras along the channel to record their results. Soares and Zech (2002) studied dam break at bends where a 90-degree bend, a dry bed, and a rectangular channel were implemented to measure velocity and the free water surface during dam break. Using a 3D channel, Mirei and Akiyama (2003) investigated the effects of hydrodynamic forces during break on the structures located downstream of the dam. Bellos (2004) studied the 2D wave motion triggered by flooding in two different cases: dry bed and wet bed, and

subsequently compared the obtained results. Bellos (2004) measured flow depth at various hydraulic conditions and recorded the hydrographs corresponding to different flooding conditions. Eaket et al. (2005) devised a new method for recording the results obtained from dam break via imaging and converting it into photographs which were subsequently used for producing the required results to determine the surface water profile. Cagatay and Kocaman (2011) conducted experimental and numerical studies on the flow resulting from flooding of water over the body of the dam. They compared their numerical results with those obtained from the experimental study. They used the FLOW-3D software for simulating their numerical model, and implemented the Navier-Stokes (RANS) and the SWE (Shallow Water Equation) equations to solve their flow field. In their study, Cagaty and Cocamen investigated the flow field and free surface profile resulting from water flooding over the dam. Kochman and Guzel (2011) studied, numerically and experimentally, the phenomenon of dam break and the resulting waves hitting downstream obstacles. They used the FLOW-3D for simulation. For solving numerically the flow field, they implemented the Reynolds-averaged Navier–Stokes (RANS) equations and for simulating flow field turbulence, they used the two-equation turbulence model. They also studied the damage caused by the break to the structures that were situated downstream of the dam. Using the equations for shallow waters (SWE), Singh et al. (2011) developed a 2D model for simulating Malpasset Dam break. Caboussat et al. (2012)

developed a 3D model for simulating dam break flow and the consequent flooding by using the Reynolds averaged Navier-Stokes equations and the VOF (Volume of Fluid) scheme. Their 3D model was capable of taking into account the topographic effects at the break location. Caboussat et al. (2012) developed a numerical model for simulating the flooding that resulted from Malpasset Dam break. Using their own results, they also numerically simulated a hypothetical flooding situation that might have resulted from the break of Dixence Dam by considering the topography of the dam site. Developments in recent decades of computer software for numerically simulating of various phenomena has led to a considerable reduction in the expenses and the time spent for conducting experiments. In this study, the FLOW-3D software was used to simulate the break of Malpasset dam and its resulting break waves numerically. The numerical simulation of flow field turbulence was conducted via the two-equation and the RNG turbulence models. The VOF (Volume of Fluid) scheme was implemented for simulating the free surface variations of the flow.

Table 1. Specifications of the meshing used for measuring the sensitivity of the solution field.

Meshing	Cells along the x, y, and z axis
1	15×100×50
2	35×200×100
3	55×300×150
4	70×400×400

2. Materials and methods
2.1. Materials and instruments
Governing Equations

The continuity and the Reynolds-averaged Navier–Stokes equations were used for solving the incompressible fluid flow field:

$$V_F \frac{\partial \rho}{\partial t} + \frac{\partial (\rho u A_x)}{\partial x} + \frac{\partial (\rho v A_y)}{\partial y} + \frac{\partial (\rho w A_z)}{\partial z} = R_{SOR} \tag{1}$$

$$\frac{\partial u}{\partial t} + \frac{1}{V_F}\left(u A_x \frac{\partial u}{\partial x} + v A_y \frac{\partial u}{\partial y} + w A_z \frac{\partial u}{\partial z} \right) = -\frac{1}{\rho}\frac{\partial p}{\partial x} + G_x + f_x \tag{2}$$

$$\frac{\partial v}{\partial t} + \frac{1}{V_F}\left(u A_x \frac{\partial v}{\partial x} + v A_y \frac{\partial v}{\partial y} + w A_z \frac{\partial v}{\partial z} \right) = -\frac{1}{\rho}\frac{\partial p}{\partial y} + G_y + f_y \tag{3}$$

$$\frac{\partial w}{\partial t} + \frac{1}{V_F}\left(u A_x \frac{\partial w}{\partial x} + v A_y \frac{\partial w}{\partial y} + w A_z \frac{\partial w}{\partial z} \right) = -\frac{1}{\rho}\frac{\partial p}{\partial z} + G_z + f_z \tag{4}$$

where (u, v, w) , (A_x, A_y, A_z) , (G_x, G_y, G_z) and (f_x, f_y, f_z) respectively are speed component, fractional areas open to flow, gravitational force and accelerations resulted from viscosity along . Also, t, ρ, , , are time, fluid density, opening term, pressure and fractional volume open to flow respectively. The RNG turbulence model was used to simulate flow field turbulence. This model can simulate with high accuracy the turbulence of high-shear regions and low-intensity turbulent flows. Also, compared to the model, this two-equation turbulence model requires fewer empirical constants and shows a better performance when simulating separation regions. The VOF scheme was used for predicting the free surface variation of the flow. In this scheme, the following transfer equation was solved for calculating the partial-volume fluid:

$$\frac{\partial F}{\partial t} + \frac{1}{V_F}\left(\frac{\partial}{\partial x}(F u A_x) + \frac{\partial}{\partial y}(F v A_y) + \frac{\partial}{\partial z}(F w A_z) \right) = 0.0 \tag{5}$$

where, F is the partial volume of the fluid within a computational unit. If a certain computational unit is full of water, then F=1. F=0 represents the case where the cell is empty. If 0<F<1, then the cell contains both water and air [13].

Boundary Conditions

The boundary conditions for the numerical model were selected in accordance with the physical conditions of the Malpasset Dam. Since the dam collapsed as a result of flooding caused by rain and the consequent filling of the dam reservoir, forces arising from waves must be applied to the dam body at the upstream. Thus, with a given input wave, the wave amplitude and velocity, as well as the flow depth were determined at the inlet. These boundary conditions are equivalent to the wave boundary conditions in FLOW-3D. The solitary wave type was selected for the input wave to the Malpasset dam reservoir. All solid boundaries were defined as the Wall boundary conditions. The "No-slip" boundary condition was imposed on the wall and friction was neglected. Thus, no roughness is imposed at the wall boundary. The whole upper surface of the flow field was defined by the symmetry boundary condition. Under this type of boundary condition, frictions as well as temporal and spatial changes are zero for all parameters. At the downstream of the Malpasset dam, the outflow boundary condition was applied.

Flow Field Meshing

In simulating the Malpasset Dam break, the meshing within the whole computational range was implemented via a uniform block mesh consisting of rectangular elements. The sensitivity of numerical models to meshing has always been an important problem in numerical studies. Table 1 presents the specifications of the meshing implemented for break simulation of Malpasset Dam. The maximum water level in the EDF physical model measured by Electricit´e de France was used to verify the results obtained from the numerical model. Fig. 1 shows the maximum water level error obtained for various meshing. The maximum water level error predicted by the numerical model was considerably reduced with increasing number of computational cells. Fig. 2 shows the computational field meshing obtained from the FLOW-3D as well as various views of the meshing.

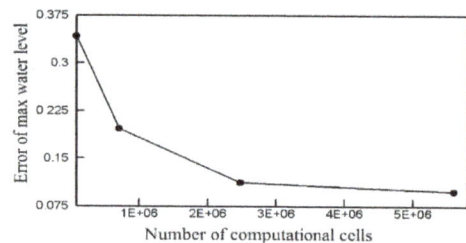

Fig. 1. Maximum water level error for different meshes.

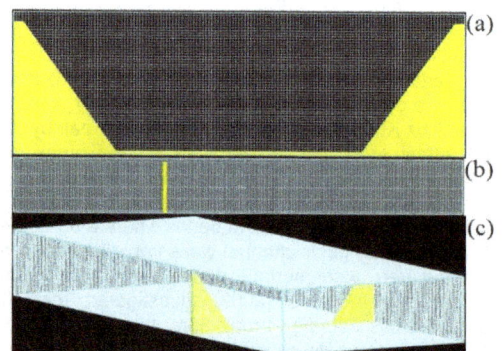

Fig. 2. Computational field meshing obtained from the FLOW-3D: a) transverse section, b) longitudinal section, c) 3D perspective.

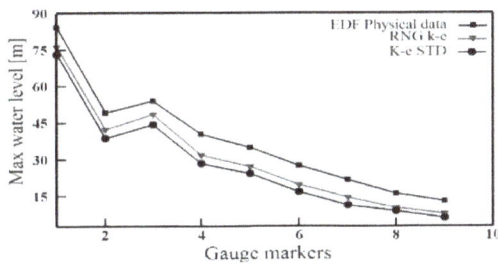

Fig. 3. Effect of standard and RNG turbulence models on maximum water level.

Effect of Turbulence Model on the Numerical Results

In this section, the effects of the two-equation turbulence model and the RNG model on hydraulic break patterns of Malpasset dam are investigated. The meshing and boundary conditions used here are set in accordance with the results obtained from the previous sections. In the standard turbulence model, two differential equations must be solved: the kinetic energy equation and the kinetic energy dissipation rate equation. The equations used in the RNG turbulence model, however, are similar to those used in the two-equation turbulence models (such as the standard model). These turbulence models are based on the Reynolds normalized groups which use statistical methods for obtaining an averaged equation for turbulence quantities. Models based on stress equations emphasize less on empirical parameters. The RNG turbulence model is an example of a turbulent model with a stress equation. However, the constants in this model are

calculated explicitly, whereas those in the turbulence model are obtained empirically. Fig. 3 shows the effects of the standard k-ε turbulence model and the RNG model on the break pattern of Malpasset Dam. As can be seen, the RNG turbulence model provides a more exact simulation for maximum water height variations.

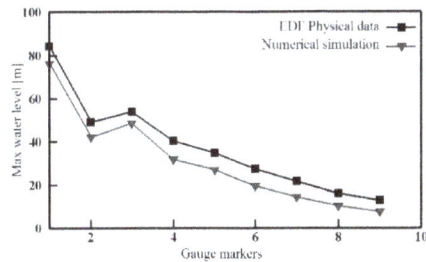

Fig. 4. Comparison of the results obtained for max water level form the numerical model and the EDF model.

Comparison of Simulation and EDF (Electricit´e de France) Results

The maximum water level values were obtained via the EDF (Electricit´e de France) physical model for verifying the numerical model and the simulation results obtained for the Malpasset dam break. Fig. 4 compares the results obtained from numerical simulation and those obtained from the physical model. Based on the simulation results, the numerical model predicted the max water level values with acceptable accuracy.

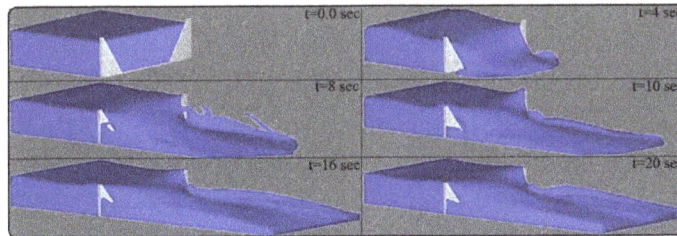

Fig. 5. 3D free surface variations of the flow at the instant of break.

3. Results and discussion

Fig. 5 shows the 3D flow free surface variation pattern at the instant the break occurred. As can be seen, at t=0.0, the water behind the dam is stable. Upon the breakage (break), a great volume of water is

displaced from the reservoir towards the downstream areas of the dam. At t=4 sec, the wave resulting from the flooding enters the computational field and, in time, water engulfs the whole downstream of the dam structure.

Table 2. Geometrical properties of the obstacles used in numerical modeling.

Obstacle shape	Obstacle orientation	Obstacle Dimensions (m)
Cubic	Perpendicular to the flow direction	$40 \times 40 \times 40$ $(\text{length} \times \text{width} \times \text{height})$
Cubic	Oblique	$40 \times 40 \times 40$ $(\text{length} \times \text{width} \times \text{height})$
Cylindrical	–	40×40 $(\text{diameter} \times \text{height})$

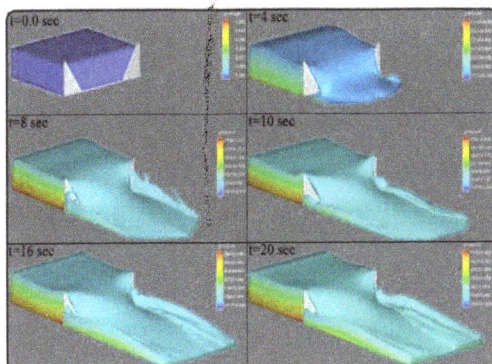

Fig. 6. Flow field pressure variations during break of the dam.

Pressure pattern as well as the minimum and maximum pressure values is important hydraulic parameters in the break pattern of a dam. In this section, pressure variations of Malpasset dam during its break are evaluated. Fig. 6 shows the pressure variations in the flow field during break of the dam. The simulation results show that the maximum pressure occurs at the lower layers of the flow and the pressure decreases towards the free surface. As time passes, the maximum pressure at each section in time increases.

During collapse of a dam, it is possible that supercritical and subcritical flows occur. Fig. 7 shows the Froude number variations during Malpasset dam collapse as well as the flow regime divisions. As can be seen in this figure, the flow undergoes supercritical conditions downstream of the dam body. A supercritical flow, due to its high energy, can cause extensive scouring and erosion. As can be seen, due to the great depth of the flow, the flow regime inside the reservoir is subcritical.

Fig. 8 shows the longitudinal velocity component variations of the flow. It can be seen that the maximum longitudinal velocity occurs at the front of the advancing wave which results from the collapse of the dam, and subsequently, this velocity component decreases due to the increased flow depth downstream of the dam.

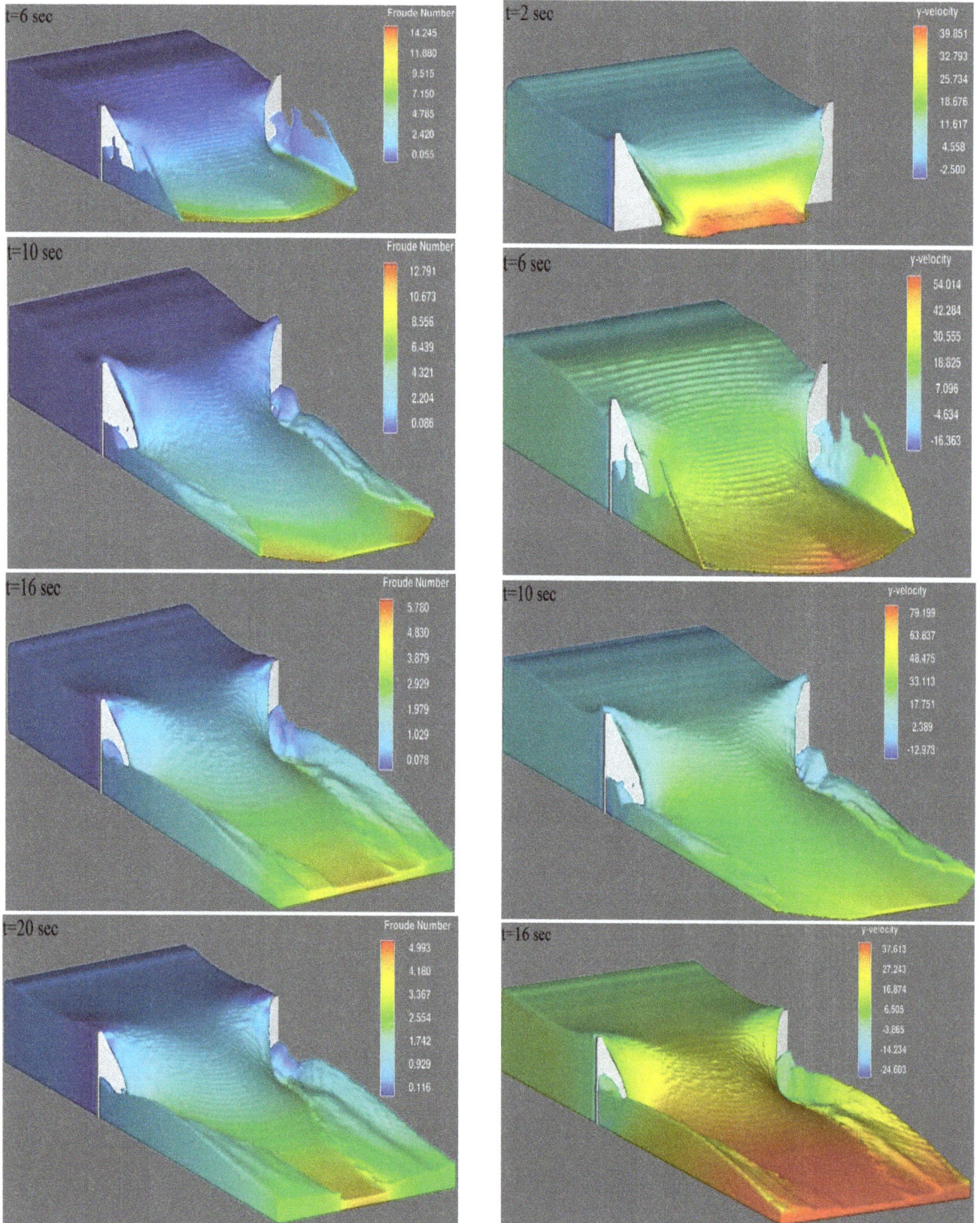

Fig. 7. Froude number variation pattern during the collapse of Malpasset dam.

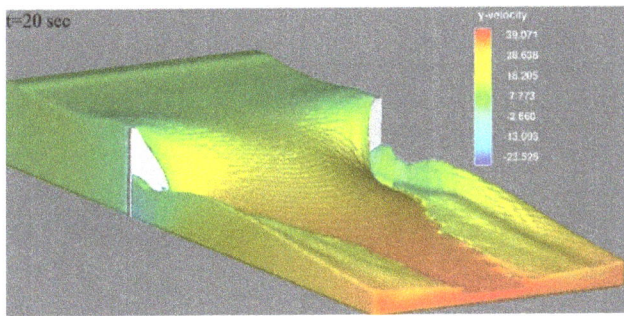

Fig. 8. Longitudinal velocity variations of the flow during collapse of Malpasset dam.

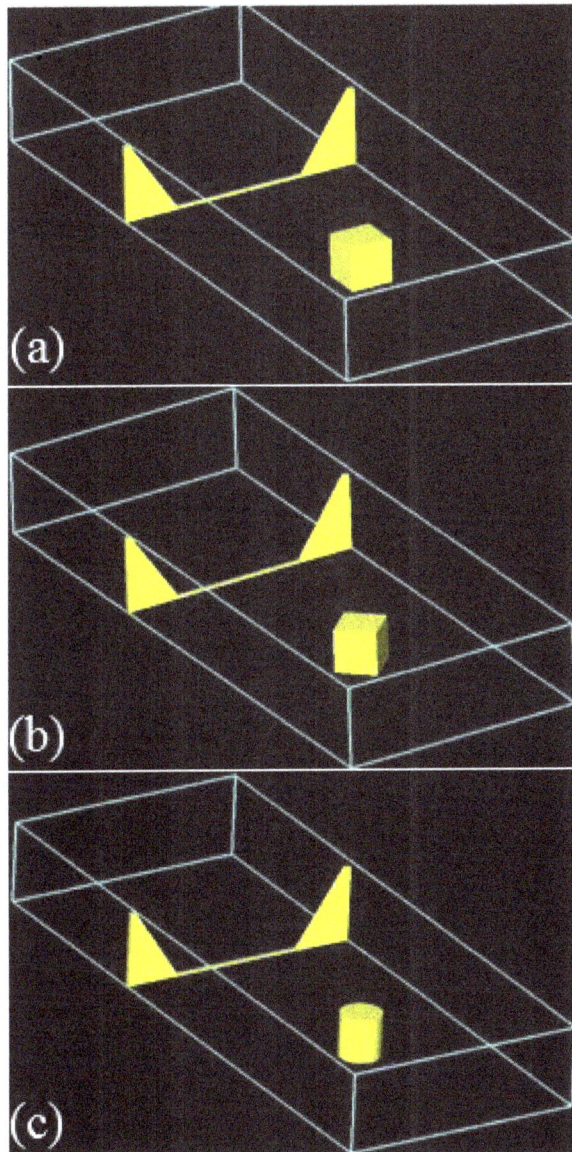

Fig. 9. Various obstacles with different shapes positioned downstream of the dam: (a) cubic obstacle perpendicular to the flow direction (b) oblique cubic obstacle (c) cylindrical obstacle.

3.2. Effect of Downstream Obstacles on the Flow Pattern during Collapse of Malpasset Dam

Dam breakage due to concurrent forces exerted on its reservoir as a result of flooding causes severe and irreparable damage to the lands and inhabitants of the areas downstream of the dam. In this study, the flow patterns during the collapse of Malpasset dam as various obstacles were encountered downstream of the dam structure were investigated.

Table 2 shows the geometrical properties of the obstacles used in the numerical modeling. Fig. 9 shows the schematic of the flow field along with variously-shaped obstacles.

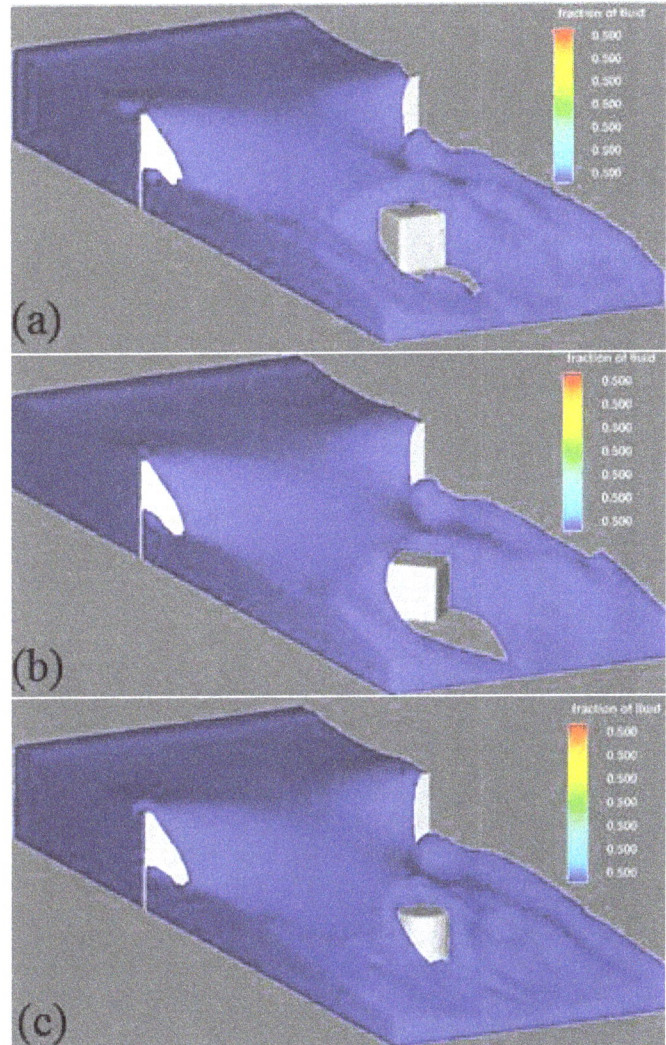

Fig. 10. 3D free surface variation patterns obtained as the flow encountered obstacles with different shapes: (a) cubic obstacle perpendicular to the flow (b) oblique cubic obstacle (c) cylindrical obstacle.

Fig. 10 shows the 3D free surface pattern of the flow as obstacles with different shapes are encountered. As can be seen, the greatest separation region occurs after the flow encounters Obstacle (b), i.e., the cubic oblique obstacle. Upon encountering Obstacle (c), however, no separation is observed due to the cylindrical shape of this obstacle.

The effects of obstacles shape on the Froude number after the dam break are shown in Fig. 11. As can be observed, the maximum Froude number at the downstream is obtained for cubic obstacle perpendicular to the flow direction. The minimum Froude number corresponds to the flow's impact with the cylindrical obstacle.

The longitudinal velocity component variations upon impact with obstacles having different shapes are shown in Fig. 12. Based on the simulation results, the maximum longitudinal velocity component is obtained for the perpendicular cubic obstacle. The variations of the longitudinal velocity component for the other two obstacles (i.e., oblique cubic and cylindrical) are similar.

4. Conclusions

Sudden unpredictable flooding of a dam reservoir is an important cause of dam break. Dam break is among the most significant hydrological and hydraulic phenomena which cause considerable damage to lands and inhabitants at the downstream of the dam. In this study, the break of Malpasset Dam and its resulting waves were numerically modeled by using the FLOW-3D. The flow field turbulence was modeled using the two-equation $k-\varepsilon$ and the RNG $k-\varepsilon$ turbulence models. Also, the variations of the free flow surface were simulated through the VOF scheme. As compared with the two-equation turbulence model, the RNG $k-\varepsilon$ model predicted the dam break pattern resulting from sudden flooding more accurately (based on the comparisons made with the results from the physical EDF model related to the break of Malpasset Dam). The 3D simulation by the RNG $k-\varepsilon$ model of the EDF physical model produced acceptable results. The simulation results showed that:

- Maximum pressure had occurred at the lower layers of the flow and that as the free surface of the flow was reached, the pressure decreased.

- The maximum pressure at each specific point in time would increase with time.

- The flow after the dam structure was supercritical which has high velocity and energy levels as well as high erosive and undermining power.

- Conversely, due to the great depth of the flow inside the reservoir, the flow regime inside the reservoir was subcritical.

- The maximum longitudinal velocity occurred at the front of the advancing wave resulting from dam break. This velocity component subsequently decreased as the flow depth increased downstream the dam.

- The greatest separation region occurred after the flow impacted the cubic oblique obstacle. The separation region was completely eliminated when the flow encountered the cylindrical obstacle.

- The maximum Froude number at the downstream of the dam was obtained for the perpendicular cubic obstacle and the minimum Froude number for the cylindrical obstacle.

- Based on the simulation results, the maximum longitudinal velocity occurred when the cubic oblique obstacle was placed in the path of the flow. The longitudinal velocity component variations were similar for the cubic oblique and the cylindrical obstacles.

Nomenclature

A_x, A_y, A_z	Fractional areas open to flow $\left(\mathrm{m}^2\right)$
F	Fluid volume fraction in a cell $(-)$
f_x, f_y, f_z	Viscous accelerations (m s^{-2})
G_x, G_y, G_z	Body accelerations (m s^{-2})
P	Pressure (N m^{-2})
R_{SOR}	Mass source $(-)$
T	Time (s)
u, v, w	Velocity components (m s^{-1})
V_F	Fractional volume open to flow $(-)$
x, y, z	Cartesian coordinate directions (m)
ρ	Fluid density (kg m^{-3})

References

Aziz khan A., Steffler P.M., Gerard R., Dam break with surges with floating debris, Journal of Hydraulic Engineering ASCE, 126 (2000) 375-379.

Bellos C.V., Experimental measurement of flood wave created by a dam break, European Water 7 (2004) 3-15.

Bellos C.V., Soulis J.V. and Sakkas J.G., Experimental investigation of two dimensional dam-break induced flows, Journal of Hydraulic Research 33 (1992) 843-864.

Bell S.W., Elliot R.C. and Chaudhry M.h., Experimental results of two dimensional dam-break flows, Journal of Hydraulic Research 30 (1992) 47-63.

Caboussat A., Boyaval S., Masserey A., Three-dimensional simulation of dam break flows, Mathicse Technical Report (2012)1-24.

Cagatay H.O, Kocaman S., Dam-break flow in the presence of obstacle: experiment and CFD simulation, Engineering Applications of Computational Fluid Mechanics 5 (2011) 41–552.

Eaket J., Hicks F.E., Peterson A.E., Use of stereoscopy for dam break flow measurement, Journal of Hydraulic Engineering ASCE, 131 (2005) 24-29.

FLOW 3D User's Manual. 2011. Version 10.0. Flow Science Inc.

Fraccarollo L., and Toro E.F., Experimental and numerical assessment of the shallow water model for two dimensional dam-break type problems, Journal of Hydraulic Research 33 (1995) 843-864.

Kocaman S., and Güzel H., Numerical and experimental investigation of dam break wave on a single building situated downstream. International Balkans Conference on Challenges of Civil Engineering, bcce, EPOKA University, Tirana, Albania, 2011.

Mirei Shige E., and Akiyama J., Numerical and experimental study of two-dimensional flood flows with and without structures, Journal of Hydraulic Engineering 129 (2003) 817-821.

Soares Frazao S., and Zech Y., Dam break in channels with 90 Bend, Journal of Hydraulic Engineering 128 (2002) 956-968.

Singh J., Altinakar M.S., Ding Y., Two-dimensional numerical modeling of dam break flows over natural terrain using a central explicit scheme, Advances in Water Resources 34 (2011) 1366-1375.

24

Virtual water strategy and its application in optimal operation of water resources

Mohammad Hossein Karimi Pashaki[*], Amir Khosrojerdi, Hossein Sedghi

Department of Water Science and Engineering, Islamic Azad University, Science and Research Branch, Tehran, Iran.

ARTICLE INFO

Keywords:
Virtual Water
Water resources Management
Virtual water trade
Sustainable development
Genetic Algorithm

ABSTRACT

The water used in the production process of an agricultural or industrial product is called "virtual water". In Iran with low average annual precipitation also lack of available water resources, concept of the virtual water and its trade is used as a strategy for optimal operation of water resources in many fields such as water scarcity, drought and so on. This concept, also, could hold some interesting new opportunities for the field of sustainable consumption. Recently, in Iran, net virtual water import reached to $(15\text{-}20)*10^9\,m^3$ per year and is one out of the top ten virtual importing countries. In this research, after virtual water applicable concepts expressing, virtual water content in some of the agricultural products in the world have been compared with products existence in Iran. Additionally, we selected some strategic agricultural products, which export and import to the country, and used an algorithm called "Genetic Algorithm", to optimize virtual water usage and trade according to demands, agricultural situation, production cost and environmental condition. Results showed which products how could help optimal water resources operation and effect of virtual water usage in economic growth.

1. Introduction

Iran has arid and semi-arid climate with 250 mm mean precipitation that it is one out of three global mean precipitation, although, it has the mean evaporation equal to three times of world mean. 94 percent of whole available water in country belongs to agriculture and 5 percent to healthy and urban usage. Just remained one percent is related to industrial usage. So, water scarcity is considered as a serious challenge in agriculture part, which is the biggest consumer of water. According to water resources scarcity and their high values as the "economical commodity", countries with crucial water conditions have introduced a new concept as "virtual water" to their strategic agricultural and industrial products.

Producing goods and services generally require water. The water used in the production process of an agricultural or industrial product is called the "virtual water" contained in the product. The concept of 'virtual water' has been introduced by Tony Allan in the early nineties (Allan 1993; 1994).

For producing 1 kg of grain we need for instance 1000-2000 kg of water, equivalent to 1-2 m^3. Producing livestock products generally requires even more water per kg of product. For producing 1 kg of cheese we need for instance 5000- 5500 kg of water and for 1 kg of beef we need in average 16000 kg of water (Chapagain and Hoekstra 2003). According to a recent study by Williams et al. (2002), the production of a 32-megabyte computer chip of 2 grams requires 32 kg of water.

If one country exports a water-intensive product to another country, it exports water in virtual form. In this way some countries support other countries in their water needs. Trade of real water between water-rich and water-poor regions is generally impossible due to the large distances and associated costs, but trade in water-intensive products (virtual water trade) is realistic. For water-scarce countries it can therefore be attractive to achieve water security by importing water-intensive products instead of producing all water-demanding products domestically. Reversibly, water-rich countries can profit from the abundance of water resources by producing water-intensive products for the export.

2. Core concepts: virtual water

There are some important and main introductions and equations which are clues in virtual water comprehensive concept. Some of the most intrinsic of them are as follows:

2.1. Virtual water conceptual parameters

In the hydrology cycle, water resources are divided into two categories: "blue" and "green" water. "Blue water" is the component of the rainfall that moves through the hydrological cycle and ends up in rivers, lakes and groundwater. This is the water that we primarily manage and use. "Green" water in the hydrological cycle, is the rainfall that is intercepted by vegetation and by the soil, and is taken up by plants to create biomass, and then evaporated back into the atmosphere. This part of the hydrological cycle has not been given much attention and is poorly managed. The blue virtual water content (BVW) was calculated as follows:

$$BVW = \frac{10 \times CIR \times CA_{irr}}{CP_{total}} \tag{1}$$

where CIR is the crop irrigation requirement (mm), CA_{irr} is the area (ha) of crop under irrigation and CP_{total} is the total amount of maize (tonnes) produced. Estimates of the area of maize under irrigation for each SADC country were obtained from the FAO Aquastat survey (FAO, 2005). Green virtual water content (GVW) was calculated as follows:

$$GVW = \frac{10 \times (CWR - CIR) \times CA_{total}}{CP_{total}} \tag{2}$$

*Corresponding author Email: m20karimi@yahoo.com

where CWR is the crop water requirement (mm) and CA_{total} is the total area under maize (ha). So, the total virtual water content is equal to the sum of the green and blue virtual water content for maize in the country.

2.2. Specific water demand

The virtual water content [VWC] of a crop, c, in a country (m^3/ton) is calculated as the ratio of total water used for the production [CWU] to the total volume of production by that country (Chapagain and Hoekstra 2004).

$$SWD_c = \frac{CWR_c}{CY_c} \qquad (3)$$

where, CWR is the crop water requirement measured at field level (m^3/ha), and Y_c the total volume of crop c produced per hectare in the country (ton/ha).

Crop water requirement is defined as the total water needed for evapotranspiration from planting to harvest for a given crop in a specific region, when adequate soil water is maintained by rainfall and/or irrigation, so that it does not limit plant growth and crop yield. Under standard conditions when a crop grows without any shortage of water, the crop evapotranspiration is equal to the CWR of a crop (Allen et al. 1998). The crop water requirement is calculated by accumulation of data on daily crop evapotranspiration, ETc (mm/day), over the complete growth period as following:

$$CWR_c = 10 \times \sum_{d=1}^{lp} ET_{c,d} \qquad (4)$$

where the factor 10 is meant to convert mm into m^3/ha and where the summation is done over the period from day 1 to the final day at the end of the growth period (lp stands for length of growth period in days). The crop evapotranspiration per day follows from multiplying the reference crop evapotranspiration, ET0, with the crop coefficient, Kc, as following:

$$ET_c = K_c \times ET_o \qquad (5)$$

The reference crop evapotranspiration, ET_0, is defined as the rate of evapotranspiration from a hypothetical reference crop with eight assumed crops of 12 cm, a fixed crop surface resistance of 70 sec/m and an albedo of 0.23.

2.3. Calculation of virtual water trade flows and national virtual water trade balance

Virtual water trade flows between nations have been calculated by multiplying international crop trade lows by their associated virtual water content. The latter depends on the specific water demand of the crop in the exporting country where the crop is produced. Virtual water trade is thus calculated as:

$$VWT_{(ne,ni,c,t)} = CT_{(ne,ni,c,t)} \times SWD_{(ne,c)} \qquad (6)$$

in which VWT denotes the virtual water trade (m^3/yr) from exporting country, n_e, to importing country, n_i, in year, t, as a result of trade in crop, c. CT represents the crop trade (ton/yr) from exporting country, n_e, to importing country, n_i, in year, t, for crop c.

The gross virtual water import (GVWI) to a country, n_i, is the sum of all imports:

$$GVWI_t = \sum_{ni,c} VWT_{(ni,c,t)} \qquad (7)$$

The gross virtual water export (GVWE) from a country, n_e, is the sum of all exports:

$$GVWE_t = \sum_{ne,c} VWT_{(ne,c,t)} \qquad (8)$$

The net virtual water import of a country is equal to the gross virtual water import minus the gross virtual water export. The virtual water trade balance of country, x, for year, t, can thus be written as:

$$NVWI_t = GVWI_t - GVWE_t \qquad (9)$$

where NVWI stands for the net virtual water import (m^3/yr) to the country. Net virtual water import to a country has either a positive sign or a negative sign. The latter indicates that there is net virtual water export from the country.

2.4. The water footprint of a country

The total water use within a country itself is not the right measurement of a nation's actual appropriation of the global water resources. In the case of net import of virtual water into a country, this virtual water volume should be added to the total domestic water use in order to get a picture of a nation's real call on the global water resources. Similarly, in the case of net export of virtual water from a country, this virtual water volume should be subtracted from the volume of domestic water use. The sum of domestic water use and net virtual water import can be seen as a kind of 'water footprint' of a country, on the analogy of the 'ecological footprint' of a nation. In simplified terms, the latter refers to the amount of land needed for the production of the goods and services consumed by the inhabitants of a country. The 'water footprint' of a country (expressed as a volume of water per year) is defined as:

Water Footprint = WU + NVWI   (10)

in which WU denotes the total domestic water use (m^3/yr) and NVWI the net virtual water imports of a country (m^3/yr). As noted earlier, the latter can have a negative sign as well.

2.5. National water scarcity, water dependency and water self-sufficiency

As an index of national "water scarcity" we use the ratio of total water use to water availability:

$$WS = \frac{WU}{WA} \times 100 \qquad (11)$$

In this equation, WS denotes national water scarcity (%), WU denotes the total water use in the country (m^3/yr) and WA denotes the national water availability (m^3/yr). Defined in this way, water scarcity will generally range between zero and hundred percent, but it can be above hundred percent in exceptional cases (e.g. groundwater mining).

The "water dependency", WD of a nation is in this paper calculated as the ratio of the net virtual water import into a country to the total national water appropriation:

$$WD = \begin{cases} \dfrac{NVWI}{WU + NVWI} \times 100 & if \quad NVWI \geq 0 \\[12pt] 0 & if \quad NVWI \prec 0 \end{cases} \qquad (12)$$

The value of the water dependency index will in each definition vary between zero and hundred percent. A value of zero means that gross virtual water import and export are in balance or that there is net virtual water export.

The counterpart of the water dependency index, the "water self-sufficiency" index is defined as follows:

WSS= 1- WD   (13)

The level of water self-sufficiency (WSS) denotes the national capability of supplying the water needed for the production of the domestic demand for goods and services. Self-sufficiency is hundred per cent if all the water needed is available and indeed taken from within the own territory. Water self-sufficiently approaches zero if a country extremely relies on virtual water imports.

2.4. Process of global virtual water calculation

According to mentioned contexts, commonly calculating process of virtual water is as the Fig. 1.

2.5. Trading of global virtual water

In this part we investigated virtual water trade and subsequent benefits in Iran compared with other different countries. Virtual water calculating and trade identities in these countries are shown in Table 1. In this paper, we selected two kinds of agricultural products in Iran. First, 16 agricultural products studied which allocate 33.8*10⁹ cubic meter of water export. On the other hand, virtual water importance for 9 studied agricultural products is 46.1*10⁹ cubic meter. Utilized amount of virtual water in mean per ton is 2869.6 m³ for exports and 3893 m³ for imports. Imported and exported total virtual water by selected products and its values are shown in Table 2 and 3, respectively. Table 1 indicates that Iran is one of the biggest users of imported virtual water.

3. Virtual water exchange, trade and optimization in Iran

According to Table 2 and Table 3, in order to optimization of virtual water exchange, we calculated two parameters by dividing (consumed water/Yield) with (m³/Kg) dimension and (Value/Amount) with ($/Kg) dimension which is used as cost and benefit. After that, we made object function to optimization by "Genetic Algorithm" based on mentioned cost and benefit indexes.

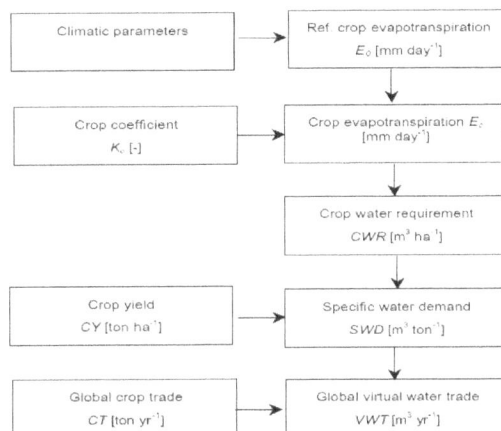

Fig. 1. Calculating process of virtual water.

Table 1. Vitual water index calculating and subsequent trade in export and import (Cubic Million Meter per Year).

Country	Water Availability	GVWE	GVWI	NVWI	Water Footprint	WS (%)	WSS (%)	WD (%)
Iran	117500	803.4	6623.1	5819.7	1457	72.9	93.6	6.4
Australia	343000	30130.3	1011	-29119.3	1085	8	100	0
Germany	171000	9671.3	13589.1	13589.1	742	27.7	77.7	22.3
Japan	547000	188.4	59443.6	59443.6	1196	16.8	60.7	39.3
Brazil	6950000	32161.8	23161.6	-9000.2	225	0.7	100	0
Cameron	268000	187.9	175.3	-12.6	33	0.2	100	0

Table 2. Total exported virtual water, amount and value by selected products (1996-2004), Iran.

Product	Amount (Ton)	Value (1000 $)	Virtual Water Volume (m³)	Consumed Water (m³/ha)	Yield (kg/ha)
Grape	4452290	549390	2174514799	8000	4091
Nut	19610	21200	115352941	850	5000
Pistachio	1100790	3757700	20408621090	431.5	8000
Onion	658970	57230	434922685	21212	14000
Tea	90130	61140	179707055	1630	3250
Date	991600	305790	1967460317	6300	12500
Cucumber	69060	17800	7847727	22000	25000
Apple	1590020	273630	636008000	13000	5200
Potato	612550	74140	320731180	25000	13090
Garlic	30070	4730	14729881	14290	7000
Walnut	25350	35260	131688311	1155	6000
Tomato	1851550	260550	608366428	35000	11500
Beans	5390	1120	2964500	10000	5500
Citrus	36440	6060	31617058	170000	14750
Sum	11782850	5520500	33812343300	-	-

3.1. Optimization

Object function determined as:
Maximum obtained benefit ((Benefit Index: β) * Production Amount) and Minimum virtual water usage ((cost Index: α) * Production Amount).

We optimized object function by Genetic Algorithm (G.A). "G.A" is a non-linear method to optimize, which is used for evolution rules to complete the process. Ability of "G.A" in finding global optimization by acceptable iteration makes it an elite method between other algorithms.

Specified characteristics of "G.A" change it to a flexible method to use in different fields of engineering sciences. According to mentioned explanation, we run "G.A" with 1000 initial population and 0.001 accuracy. In 100 iterations, evolution parameters were 0.03 for mutation coefficient and crossover coefficient equal to 0.8.

Results showed optimum possible cultivation pattern for exported product. With this pattern, we can minimize exported virtual water content and maximize obtained benefit. On the other hand, "G.A" gives another pattern for importing products which causes reduction in cost despite reduction in virtual water import. So, by this pattern, we have

165 million dollars per year increase in benefits, and economize 4150 million cubic meter in virtual water usage for exported products. For imported product, pattern results in 2140 million dollars providence per year for imported products, even though amount of imported virtual water reduced. Optimum patterns for importing and exporting products based on "G.A" results are shown in Fig. 2 and Fig. 3, respectively. In Fig. 4, convergence trend of program has shown after 100 iterations. as it shows results start to converge after approximately 70 iterations.

Table 3. Imported total virtual water, amount and value selected by products (1996-2004), Iran.

Product	Amount (Ton)	Value (1000 $)	Virtual water volume (m³)	Consumed Water (m³/ha)	Yield (kg/ha)
Sunflower	3934100	1075550	59382641610	530	8000
Peanut	77830	27710	133752296	3375	5800
Rice	10699380	2556230	61139314290	3500	20000
Barely	5245260	783370	9618310000	3000	5500
Tea	112440	310750	224190184	1630	3250
Sorghum	13986640	215420	21221108970	7250	11000
Soya	36906730	5516570	246044866700	1500	10000
Sugar	7669600	1947241	18782693880	4900	12000
Wheat	38135230	5325450	43855514500	5000	5750
Banana	1689470	721240	1013682000	25000	15000
Sum	118520500	20400361	461414074300	-	-

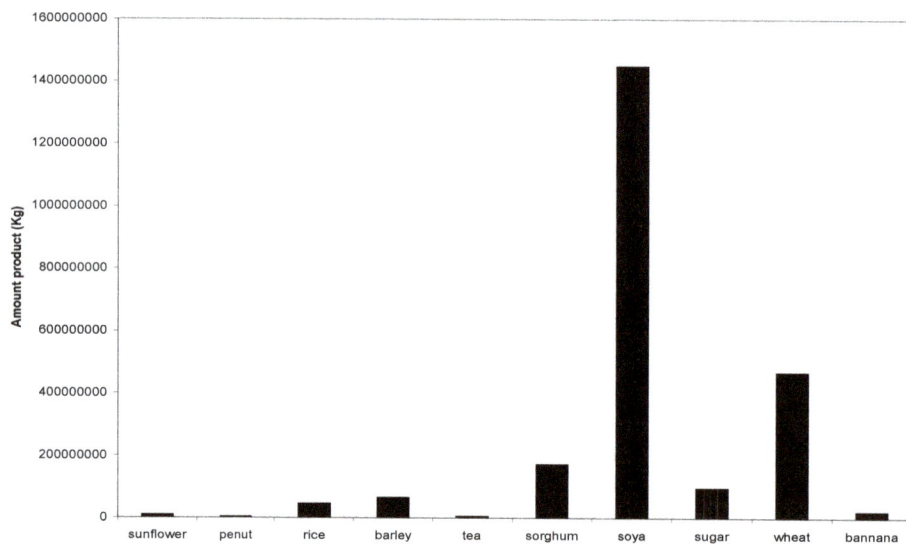

Fig. 2. Optimum pattern for importing agricultural products.

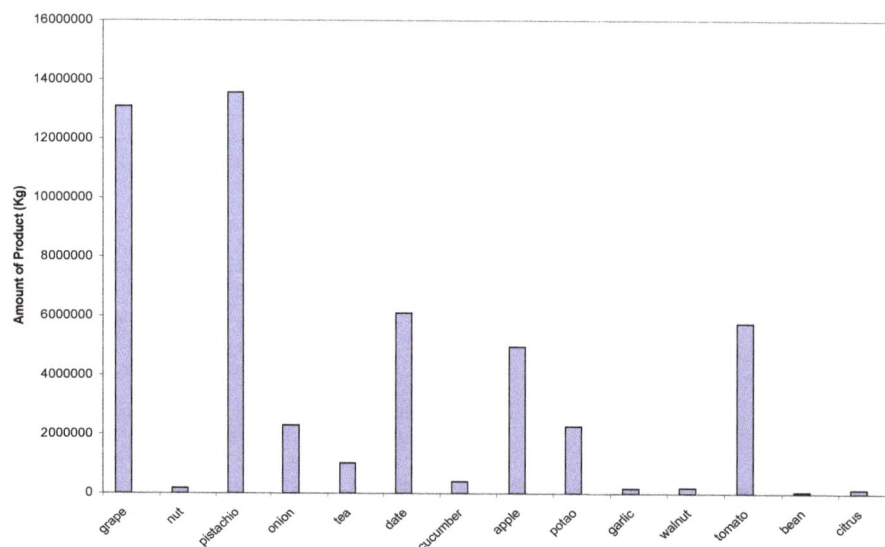

Fig. 3. Optimum pattern for exported agricultural products.

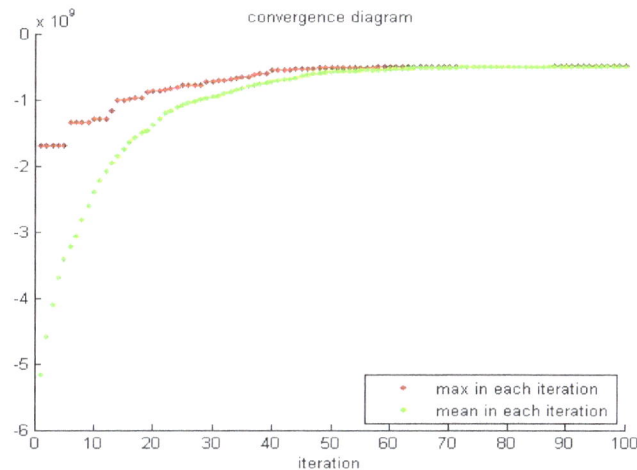

Fig. 4. Convergence diagram of "G.A" results after 100 iterations.

4. Conclusions

Finally, it seems that virtual water concept could be a useful and important factor for water scarcity in some countries or countries with arid or semi-arid climates. Additionally, its role in agricultural product trade and national economy exchange rate is considerable.

In this paper, we have studied some selected strategic agricultural products and their statistics in Iran. Virtual water parameters were also calculated and considered. Object function to optimize mentioned as a combination of cost and benefits functions. After all this, we have used "Genetic Algorithm" as an optimization algorithm and nominated a cultivation and import pattern which could optimize cost and benefits of virtual water trade in Iran. Optimization also had economical results by increasing revenues and decreasing cost of exported products. According to the optimum fitted pattern, import of "Soya" in imported product category should be increased to 32 % and "Pistachio" product for the export has been recommended to increase to 10 %. Recommended pattern for exports resulted in 4150 million cubic meter thrift each year.

References

Darowski J.M.E., Ashton P.Y., Analysis of Virtual Water Flows Associated With The Trade of Maize In The SADC Region, Hydrology and Environment System Science Jurnal 13 (2009) 1967-1977.

Hoekstra A.R., Hung P.Q., Virtual Water Trade: Proceedings of The International Expert Metting On Virtual Water trade, Value of Water Research Report Series, IHE, DELFT, (2003) 13-50.

Hoseini S.A., Arshadi M., Babai H., The First International Conference on Water Crisis, "Virtual Water: Concepts and Applications (In Persian), Zabol, IRAN, (2009).

Odularu G.O.E., Conceptual Explanat of Virtual Water Trade and Lessons For Africaion, Jurnal of Development and Agriculture Economicsl 1(2009) 162-167.

Schreier H., et al. Blue, Green and Virtual Water, for Walter and Duncan Gordon Foundation Toronto Canada, Ontario, (2008).

25

Investigating the capabilities of the NSGA-II multi-objective algorithm in automatic calibration of the WEAP model for simulating Jareh Dam and network system

Arash Azari[1,*], Milad Asadi[2]

[1]Department of Water Engineering, Razi University, Iran.
[2]Department of Hydrology and Water Resources, Shahid Chamran University, Iran.

ARTICLE INFO	ABSTRACT

Keywords:
Calibration
WEAP Model
Jareh Dam and Network
NSGA-II

In the simulation models of water resource systems, calibration processes should be performed to approximate the simulated values to the observed values due to the errors in such models. However, due to being time consuming and the difficulties associated with manual calibration, an automatic calibration model can be a resolver. In this research, the simulation of Jareh Dam and network system was conducted using the WEAP model. Then, by linking this model to the NSGA-II algorithm, its automatic calibration was performed by this algorithm. Nach statistical parameter was used to check the calibration accuracy of the model. The whole system was in the form of a multi-objective NSGA-II algorithm, in which the first objective function, which was to minimize the difference between the observed and the calculated reservoir storage volumes, was assessed versus the second objective function, which was to minimize the difference in the calculated and the simulated discharges, at two Mashin and Jokank stations. The results showed the remarkable ability of NSGA-II algorithm for automatic model calibration, so that the operation status of the dam and river was of the greatest consistency with reality.

1. Introduction

The complexity of water resources systems and the existence of different and sometimes contradictory goals together make it difficult to manage these resources properly. In this regard, simulation is a flexible tool, which is widely used for complex analysis of water resources systems. The system of interest should be introduced and described in both aspects of design variables and withdrawal policies, and then can be simulated to determine its performance quality (Bozorg Haddad and Aghmuni 2013). Various simulation models have been developed for this task and have been moved toward decision support systems (DSS) along with the advancement of software capabilities. In basin simulation models, most water resources systems are represented in a network of arms and nodes and the algorithms for solving flow network models are used for determining the spatial and temporal distribution of allocable water resources (Karimi and Mousavi 2011).

Among these models, the WEAP model can be named, which has attracted researchers' attention in recent years due to its user-friendly structure. For example, Alfarra (2004) employed the WEAP model to examine the status and problems that would be created in the future in a basin in Kenya. The results indicated that in the study area, the water allocation for agricultural sector is more than the demand in some areas, and in some other areas, this demand is not fully met. Abrishamchi et al. (2007) used the WEAP model in the Karkheh Olia River basin to study the effects of water and land resources developments on urban, industrial, and agricultural uses as well as the inflow into Karkheh Dam reservoir. The results showed that the WEAP model has a remarkable ability for studying water resources management scenarios at a river basin scale.

Purkey et al. (2008) used the WEAP model to assess the impact of water resources allocation management to the agricultural sector under the climate change conditions in California. The results showed that applying managerial constraints would improve the condition of the water resources of the region. Mutiga et al. (2010) used the WEAP model in order to balance the water demands on water resources of the study area to achieve an appropriate economic and biological sustainability. Alfarra et al. (2012) used the WEAP model to assess the water resources of their study basin. To this purpose, they investigated 5 scenarios by 2050. Hamlat et al. (2013) studied the application of the WEAP model in the western Algeria basins. The results showed that the WEAP model provides planners with reliable results to use in the future.

Lee et al. (2015) used the WEAP model to assess the sustainability of limited water resources in their study area. Hum and Abdul-Talib (2016) utilized the WEAP model to evaluate water resources and the current uses in Selangor by 2050. Rafiee Anzab et al. (2016) presented an optimization simulation model by linking the WEAP model to the PSO algorithm in order to optimally design and operate the water transfer project from Karoon River to Zohreh River in Iran. The results showed that the water transfer project could meet the water required for the development of Dah Dasht and Cheram agricultural lands in the undeveloped areas in the north of Kohkiluyeh province. Movahedian Attar and Samadi Broujeni (2013), evaluated the performance of Zayandehrud Dam using the WEAP model in four scenarios, and finally by comparing the existing reliability indicators and the existing scenarios could choose the superior scenario.

Azari et al. (2015) presented a conjunctive withdrawal model of surface and underground water resources while aiming to observe all the constraints and quality standards along Dez River. To this end, the system was simulated by the dynamic linking between the WEAP and MODFLOW models. Ghandehari et al. (2015) used the WEAP model to study the temperature and precipitation changes in Neishabour River basin and the effect of these changes on the river by means of the outputs of general atmospheric circulation models (HADCM3) and statistically downscaling methods (SDSM). Reviewing the conducted researches showed that in the models developed to simulate water

*Corresponding author E-mail: a.azari@razi.ac.ir

resources systems like the other simulation methods, calibration processes should be performed to approximate the observed and the simulated values to each other due to errors in the model.

Wang (1991), Madsen and Torsten (2001), Zhang et al. (2008), and Huang and Lei (2010) investigated various calibration models using evolutionary algorithms. Also, Arkan and Godal (2014) used a non-dominated sorting genetic algorithm to conduct a multi-objective calibration of the SWAT model, where the results showed the multi-objective calibration of model improved its capabilities. In order to calibrate the CE-QUAL-W2 qualitative model, Afshar et al. (2011) used automatic calibration in their research. The results of automatic calibration showed that the observed and the calculated values approached each other more. In order to calibrate the SWAT model, Bekele and Nicklow (2007) used automatic calibration by the NSGA-II algorithm. The results showed a better accuracy of the model due to the automatic calibration of parameters. Muleta and Nicklow (2005) used automatic calibration to calibrate the SWAT model in their study basin.

In this research, the simulation of water resources systems of Jareh Basin located on Allah River was conducted with the aim of integrated water resources management in this basin. But, due to the complexity of the system caused by the lack of proper information about the significant discharge of the medial basin resulting from the confluence of A'la River with Allah River between the Jareh Dam reservoir and the downstream diversion dam, it was time consuming and difficult to correctly simulate the system. Therefore, the main goal of this study is to investigate the capability of the NSGA-II multi-objective evolutionary algorithm in calibrating this system automatically and to evaluate the results considering the current status of the dam and river operations.

2. Materials and methods
2.1. Study area

The study area is located in the southwest of Iran in Khuzestan province, within the 35-km distance from the northeast of Ramhormoz city near Jareh village, which is in the 90-km distance from the east of Ahwaz and in the southwestern slopes of Zagros and consists of two basins including A'la River and Zard River basins. Zard River is one of the main branches forming Jarrahi River, which is one of the fullest flow rivers in the province of Khuzestan.

Zard River basin is in the northeast and east of Bagmalek city highlands (Mangasht Mountain). In the confluence of Abolabbas River and Ab-Takht River, Zard River is formed. By joining A'la River to Zard River, Allah River is formed at Zard River village, and after this river joins Maroon River, Jarrahi River is formed. The Jareh Dam reservoir is built across Zard River.

The dam is an earthy type with a clay core with a crown length of 740 meters and a crown width of 12 meters. The dam was built with the aim of providing water for the right and left banks of Ramhormoz irrigation network, and also producing hydroelectric power, which has not been operated yet. Fig. 1 shows the satellite image of the study area.

In the present study, two Mashin and Jokank hydrometric stations at the downstream of Jareh Dam were employed as well as the recorded data of the reservoir storage volume in different months in order to calibrate the model.

2.2. WEAP simulator model

The WEAP model was first developed by the Stockholm Environmental Institute. WEAP is based on the basic equations of water balance and can be used in urban and agricultural systems, independent basins, and complex boundary river systems. In addition, WEAP can cover a wide range of issues, such as analyzing the demand of each sector, water protection, rights and allocation priorities, simulating surface and underground water resources, reservoir withdrawal, hydroelectric power generation, pollution routing, ecosystem demands, evaluating vulnerability, and cost-benefit analysis of a project (Sieber et al. 2005). The information required to apply to the WEAP model include correctly locating all sources and uses within the basin and the existing hydrometric stations, as well as general parameters such as the determination of the base year, the length of the simulation period, the units of the used parameters and, in general, everything that depends on the water resources system of the region.

In this research, in order to prepare the WEAP model, the exact location of the rivers, Jareh Dam, district hydrometric stations, and the

water demands of the left and right banks of Ramhormoz were determined using GIS software. Then, by recalling it in the WEAP model, each of the complications was embedded with the provided tools. The schematic model of the studied basin is shown in Figure 1. In the created model, the simulation was considered for 4 years of the operation period of the dam from October 2010 to September 2014 because of the data are available since the start of dam operation in 2010. Then the information about all of the resources and uses, including the data on the operation of Jareh Dam such as dead volume, volume at the minimum operation level, volume at the maximum operation level, ..., information on the recorded discharges at the hydrometric stations of the area, information on the surface and underground water withdrawals on the left and right banks of Ramhormoz and the environmental water demand at the downstream of the dam were introduced into the model as a CSV file.

The water demand of the right and left banks of Ramhormoz irrigation and drainage network was calculated using the regional cropping pattern gotten through the information obtained from Khuzestan Water and Power Organization and the agricultural statistics of recent years. Afterwards, the groundwater withdrawal amount in these two plains were calculated using the information obtained from the operation wells of Ramhormoz left and right banks and were introduced into the model. The amounts of surface and groundwater withdrawal in these plains are shown in Fig. 2.

Also, the information about the discharges recorded at the hydrometric stations of the district from the Khuzestan Water and Power Organization was received daily and their monthly mean was calculated and introduced into the model. Fig. 3 shows the long-term mean monthly discharges recorded at Mashin and Jokank stations.

In order to correctly establish the water balance in the study basin and to simulate it properly, and because there is no hydrometric station for measuring the flow in A'al River, as well due to the massive flow of A'la River entering into the study basin in the rainy seasons, the inflow from A'la River to the basin was simulated. To this purpose, having data recorded at the Mashin and Jokank stations, due to the location of these stations before and after the intersection of A'la and Allah Rivers, the inflow was simulated through this branch and was introduced into the model.

Actions such as dam construction, increasing water withdrawal along rivers, water transfer, etc. can affect the natural regime of a river. This will reduce the natural flow of the river and will affect the ecosystem of the region, hence, the minimum ecological requirements of the region should be met. Therefore, in this research, in order to meet the minimum ecological demands required by the ecosystem at the downstream of Jareh Dam, an environmental node was defined. To calculate the environmental flow, the fair state of Tenant method (Montana) was applied. This method is one of the hydrological methods for determining the minimum environmental flow that determines the monthly minimum environmental flow as a percentage of the normal river flow. The calculated values for the environmental flow was 1.78 m^3/s (for Oct to Mar) and 5.35 m^3/s (for Apr to Sep).

2.3. Calibration of model

Due to lack of sufficient information for the A'la branch, in the initial investigations the simulation of the flow in this river was done using the discharge differences recorded at Mashin and Jokank stations and the initial simulation of the system was carried out. After preparing the surface water model of the area to match the results of the model with the current state, the calibration of the model was implemented through linking the NSGA-II evolutionary algorithm to the body of the WEAP model. To this purpose, the code associated with this algorithm was developed and implemented based on the issue in the MATLAB environment. At each iteration, due to the lack of sufficient information on downstream uses, the total release rate from the dam was considered as decision variables in different months during the 4-year simulation period. The algorithm was defined multi-objective, so that, the first objective was to minimize the difference between the observed and the calculated reservoir storage volumes by the model versus the second objective, which was to minimize the difference between the calculated and the simulated discharges at Mashin and Jokank stations using the Nash function. A 500-iteration run was considered as the optimization stopping criterion of the algorithm so that the objective functions reach their lowest values.

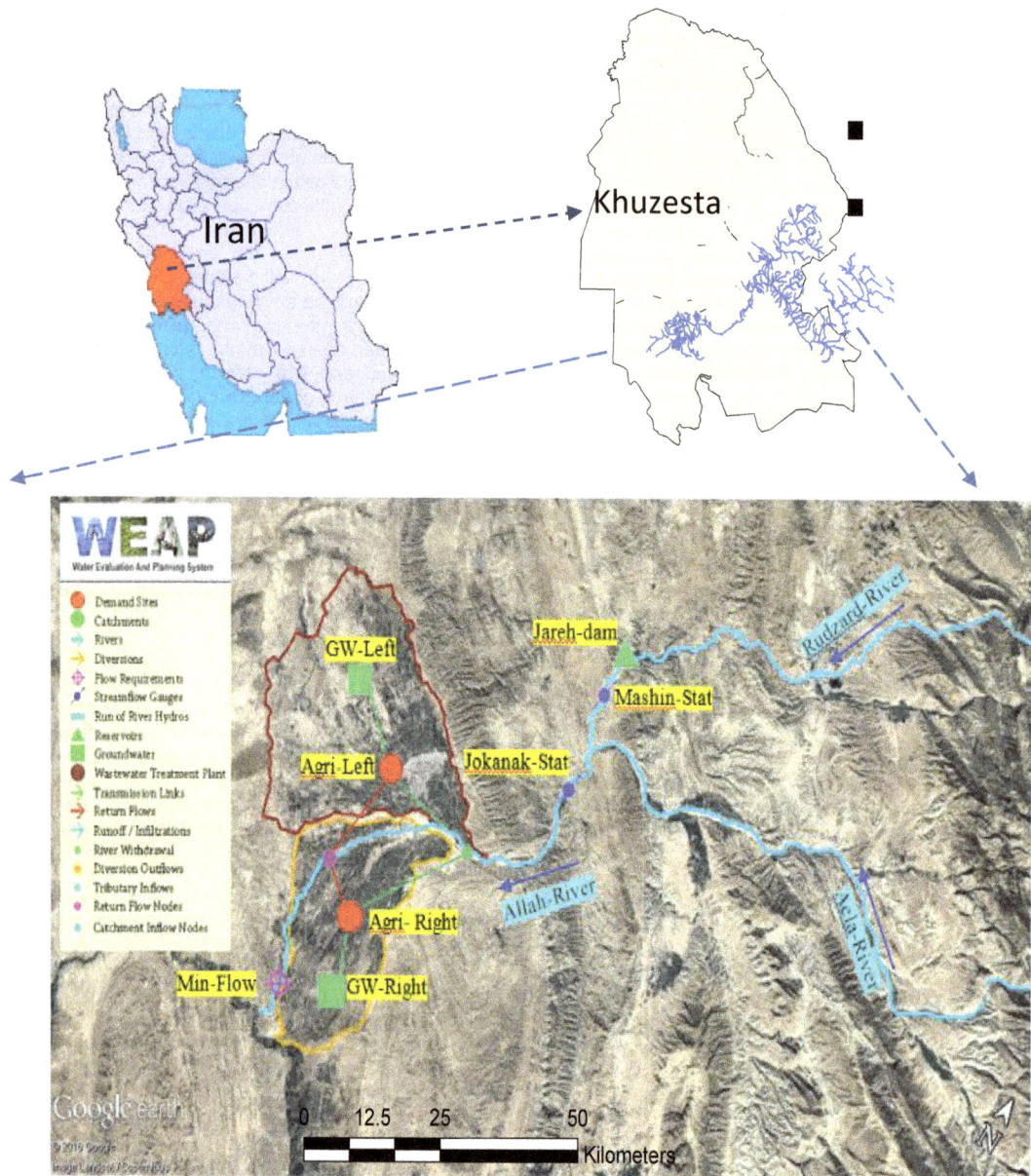

Fig. 1. Satellite image of the study area (rivers, resources, and uses).

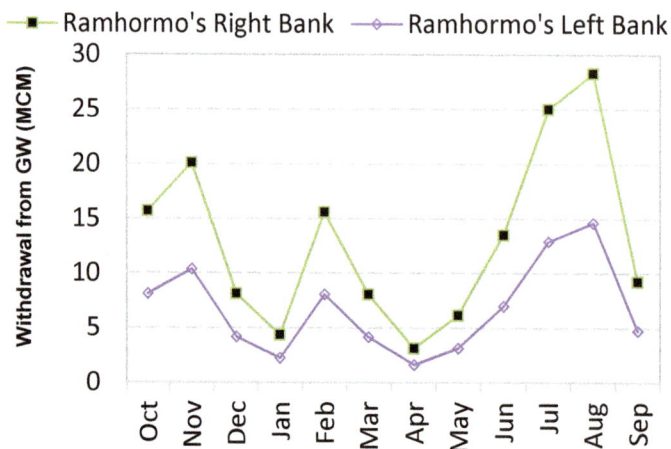

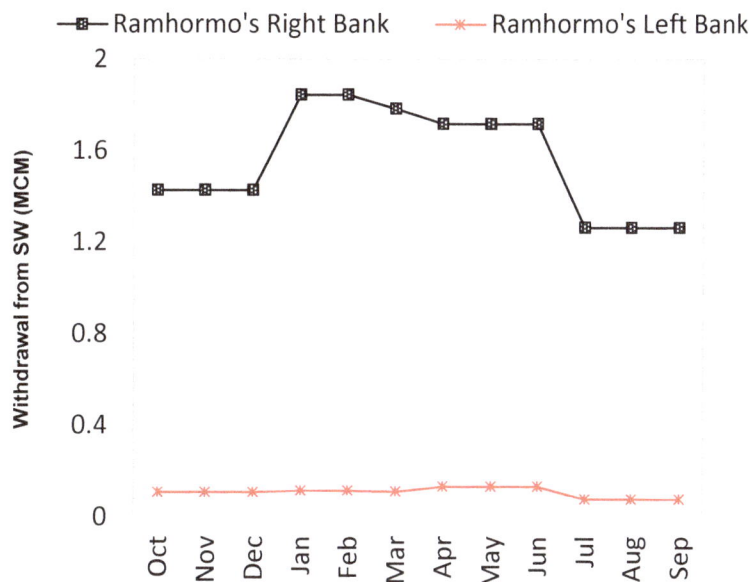

Fig. 2. Groundwater and Surface water withdrawal amounts on top and down banks of Ramhormoz.

2.4. Non-dominated Sorting Genetic Algorithm Model (NSGA-II)

Several methods are available to solve multi-objective optimization equations, including Weighted Sum Method, Epsilon-constraint Method, Goal Attainment Method, and Multi-Objective Evolutionary Algorithms (MOEAs). In this research, the non-dominated sorting genetic algorithm (NSGA-II) was used due to its ability to solve complex problems and to provide an optimal exchange curve among the goals.

This model can easily deal with problems that do not follow a particular continuous, has no integrated rational decision space, or their objective functions have random parameters. The major problems of the previous multi-objective optimization models include the enormous calculation volume at each iteration, which leads to increase the model implementation time and the inability to maintain a good number of superior values during the implementation of the model. The main structure of this model is shown in Fig. 4.

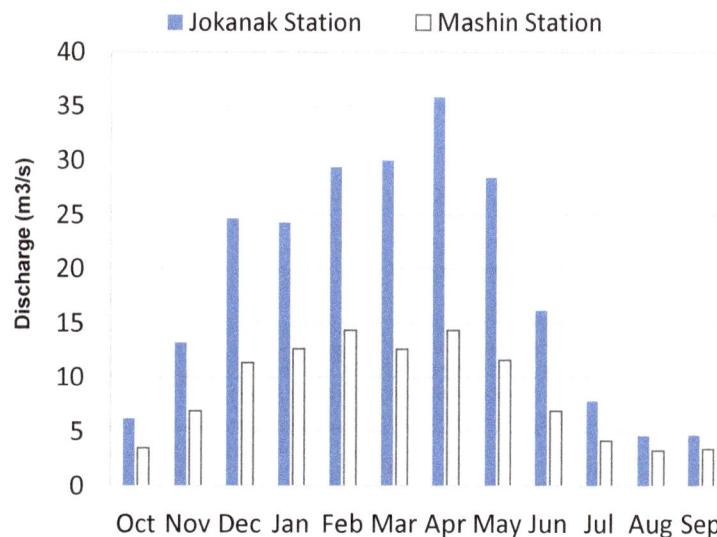

Fig. 3. Long-term mean monthly discharges recorded at Mashin and Jokank stations.

The implementation steps of this multi-objective optimization model are as follows:

1- Producing P_0 randomized parent generation equal to the number of N.

2. Sorting the initial parent generation based on non-dominated sorting method.

3- Allocating suitable ranks corresponding to the non-dominated sorting level of each non-dominated sorting, which includes rank 1 for the best balance, rank 2 for the best balance after 1, and so on.

4- Producing the offspring generation equal to N using selection, coupling and mutation operators.

5- According to the first produced generation, which includes the chromosomes of parent and offspring, a new generation is produced as follows:

*Combining parent chromosomes, P_0, with offspring chromosomes, Q_0, and producing
generation R_t equal to 2N.

* Sorting the generation R_t according to the non-dominated sorting method and identifying and categorizing non-dominated fronts (F_1, F_2..., F_L). * Producing parent generate for the next iteration (P_{t+1}) using non-dominated fronts generated equal to the number of N. At this stage, based on the number of chromosomes needed for the parent generation (N), first the chromosomes number of the first front of the parent generation is selected, and if this number does not meet the total number needed for the parent generation, then the fronts 2, 3, and etc. are taken sequentially to obtain the total amount (N). * Applying coupling and mutation operators to the newly-produced parent

generation (P_{t+1}) and producing the offspring generation (Q_{t+1}) equal to the number of N.
* Repeating step 5 to achieve the total number of iterations of interest.

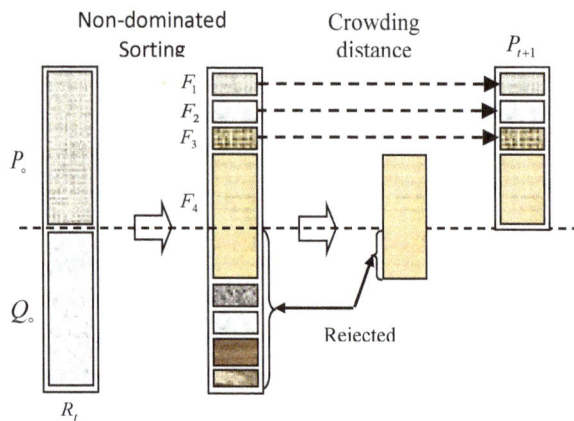

Fig. 4. Main structure of the non-dominated sorting genetic algorithm model.

2.5. Structure of the objective functions of the calibration

In this research, the Nash statistical parameter, introduced by Nash and Sutcliffe in 1970, was employed in order to calibrate the surface water model of the studied basin. This parameter is defined as follows:

$$NASH = 1 - \frac{\sum_{t=1}^{n}\left(H_t - \hat{H}_t\right)^2}{\sum_{t=1}^{n}\left(H_t - \overline{H}_t\right)^2} \tag{1}$$

where
H_t: The observed values
$\hat{H}_t$ = The simulated values
$\overline{H}_t$ = The average of the observed values
n: The number of observations.

The closer the Nash parameter to one is, the greater the correspondency of the simulated values with the observed values. The abovementioned function was utilized to calculate the calibration accuracy in comparing the monthly simulated and observed values of the storage volume of Jareh Dam reservoir, as well as assessing the calibration accuracy of the simulated and the observed discharges of Mashin and Jokank stations. In which, the objective function in the evolutionary algorithm is to minimize each of the aforementioned values using the Nash function. But because the minimization of the abovementioned function makes it closer to zero, the objective functions were defined as follows:

The first objective is to minimize the difference between the observed and the calculated reservoir storage volumes.

$$F_1 = Min\left(\frac{\sum_{t=1}^{n}\left(V_t - \hat{V}_t\right)^2}{\sum_{t=1}^{n}\left(V_t - \overline{V}\right)^2}\right) \tag{2}$$

where
V_t: The observed storage volume of Jareh Dam reservoir in month t.
$\hat{V}_t$: The simulated storage volume of Jareh Dam reservoir in month t
$\overline{V}$: The average of the observed storage volume of Jareh Dam reservoir during the period of study.
The second objective is to minimize the difference between the calculated and the simulated discharges at Mashin and Jokank stations.
The second objective is to minimize the difference between the calculated and the simulated discharges at Mashin and Jokank stations.

$$F_2 = Min\left(\frac{\sum_{i=1}^{k}\sum_{t=1}^{n}\left(Q_{it} - \hat{Q}_{it}\right)^2}{\sum_{t=1}^{n}\left(Q_{it} - \overline{Q}_i\right)^2}\right) \tag{3}$$

Q_{it}: The observed discharge at station i in month t
$\hat{Q}_i$: The simulated discharge at station i in month t
$\overline{Q}_i$: The average of the observed discharges at station i during the period of study.
By doing so, the second part of Nash relationship approaches zero, which will make Nash equation to get closer to one.

As it was mentioned earlier, in this study the evolutionary algorithm was defined in a multi-objective form. So that, the value of the objective function defined for minimizing the difference between the observed and the calculated reservoir volumes was considered versus the second objective function that was to minimize the observed and the simulated discharge differences at the stations.

3. Results and discussion

As it was mentioned earlier, the calibration process was performed using non-dominated sorting genetic algorithm. In this process, 48 decision variables (the total release rate from the dam in different months) were tested regarding the multi-objective function using objective functions. The frequent iterations of the model showed that in order to achieve better results, the initial population of chromosomes should be at least twice the number of decision variables.

Therefore, the initial population in the model was chosen equal to 96. At the first stages, choosing a larger population greatly increased the speed of convergence. Therefore, the model was run with the same population to save time. The results showed that at fewer iterations, both objective functions have significant changes, while at more iterations, the objective function of minimizing the observed and the simulated values in the reservoir volume was concerned more by the algorithm and the second objective function has a more constant value and the algorithm is approximately converged. Therefore, the number of iterating the algorithm to reach the convergence was considered equal to 500. Fig. 5 illustrates it well.

Finally, after implementing the evolutionary algorithm to calibrate the model, regarding the population size of 96 and after 500 iteration, the model was calibrated with acceptable accuracy. Moreover, the objective exchange curve (Pareto graph) between the desired objectives (the function of model calibration for the observed and the simulated reservoir volumes and the objective function of model calibration for the observed and the simulated discharges at Mashin and Jokank stations) was obtained. The Pareto graph is shown for the 500th iteration in Fig. 6.

According to the non-dominated sorting genetic algorithm, at each iteration, the best answers are selected based on valuating the objective functions and the elitism process and are stored as the optimal set to be transferred to the next generation. The points drawn in the Pareto graph are the appropriate answers for matching the observed and the calculated values of the model better. Besides, the axes of this graph are the objective functions. At the last iteration of the model, 45 optimal answers were presented, of which, according to the objective function valuation, the best answer, which was the one with the least value for the two objective functions, was selected as the superior answer and the results of its implementation were examined in the surface water model.

After calibrating the model by the algorithm, the observed and the simulated values at Mashin and Jokank stations, as well as the observed and the simulated reservoir volumes in the model, were obtained in accordance with Figures 7 to 9. As it can be seen in the figures, the calibration processes are performed with acceptable accuracy, and the observed and the simulated values are in good agreement with each other.

After calibrating the model and ensuring its accuracy, its results can be applied to interpret the current status of the region. The results of implementing the model for supplying percentage and demands reliability are presented in Table 1. As it is clear, it can be concluded that all of the demands are fully met during the entire simulation period, which indicates that there are no shortages of supplies for the studied basin. The average values of release from Jareh reservoir were calculated by the model for various uses and are presented in Fig. 10.

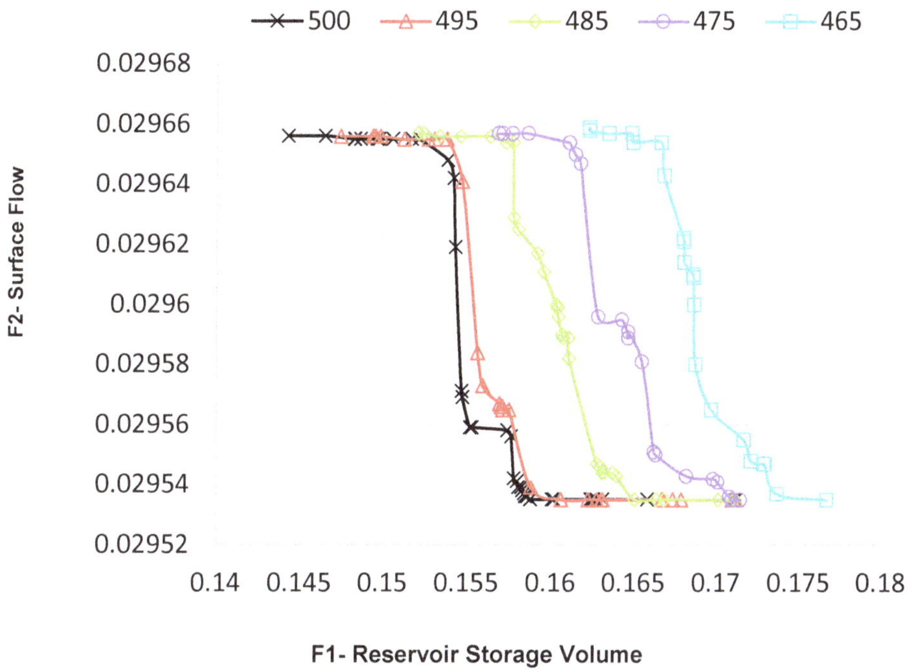

Fig. 5. Exchange curve between the objectives at different iterations.

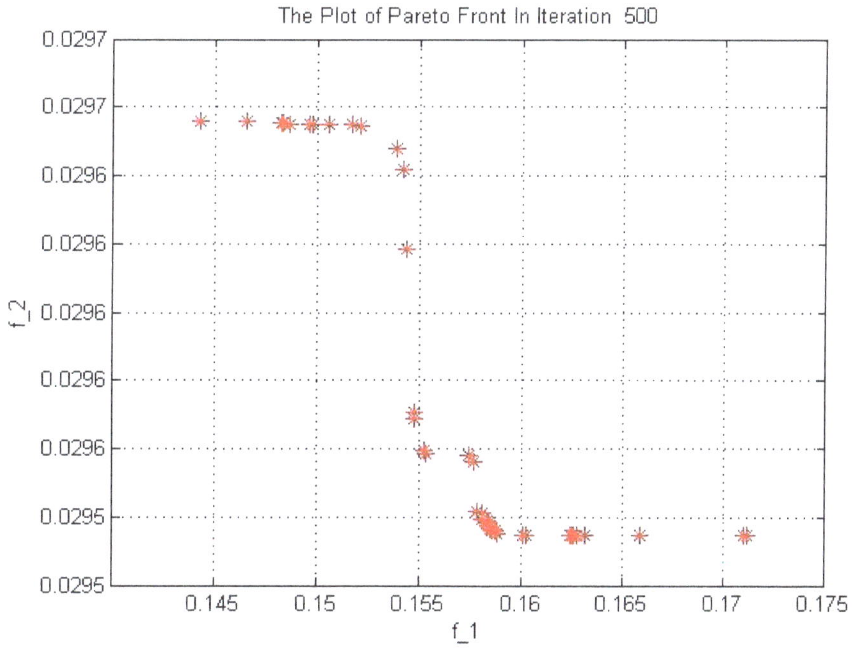

Fig. 6. Optimal exchange curve among the optimization objectives (Pareto curve) at the 500th iteration.

Table 1. Supplying percentages and demands reliability of the study area.

Demands	Supplying percentages (Coverage (%))												Reliability (%)
	Oct	Nov	Dec	Jan	Feb	Mar	Apr	May	Jun	Jul	Aug	Sep	
Ramhormo's Right Bank	99.7	100	100	100	100	100	100	100	100	100	100	100	98.9
Ramhormo's Left Bank	99.7	100	100	100	100	100	100	100	100	100	100	100	98.9
Environmental	100	100	100	100	100	100	100	100	100	100	100	100	100

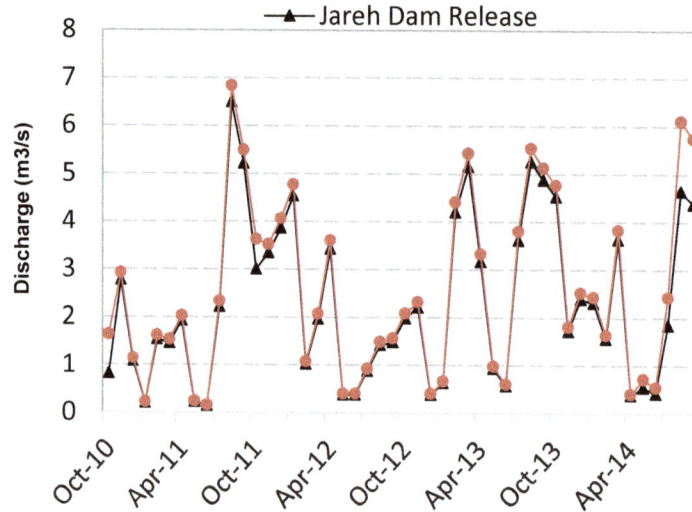

Fig. 7. Observed and simulated values at Mashin station and the outlet of Jareh Dam.

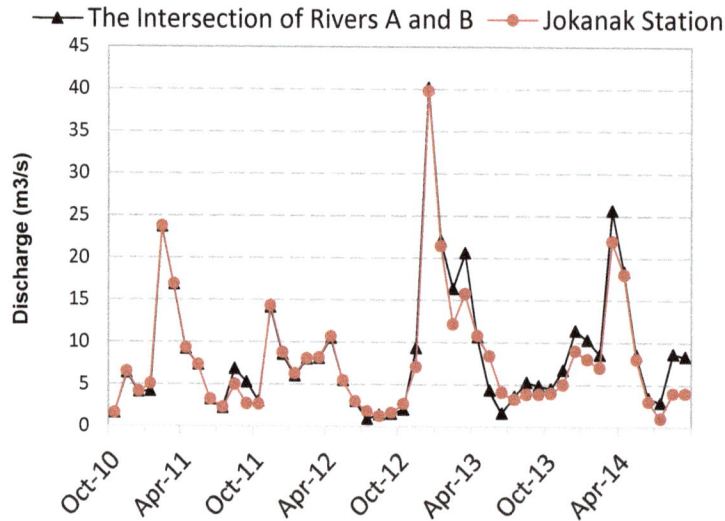

Fig. 8. Observed and simulated values at Jokank station and the entrance of A'la River into Allah river.

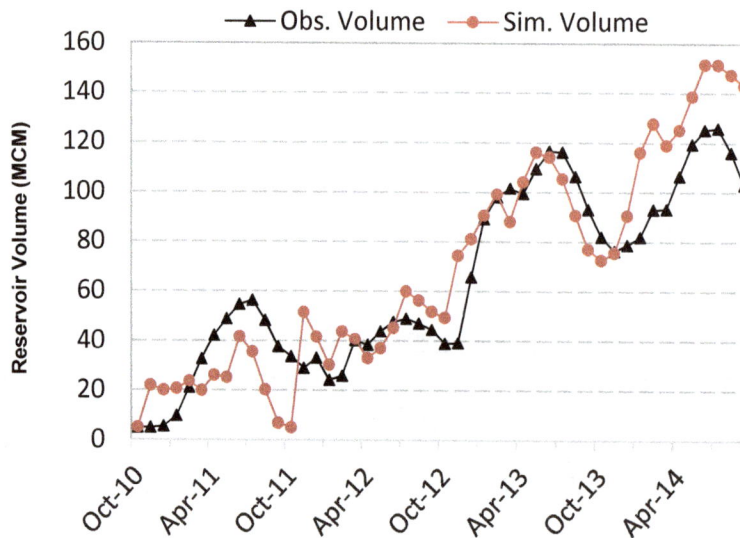

Fig. 9. Observed and simulated values of the Jareh Dam volume.

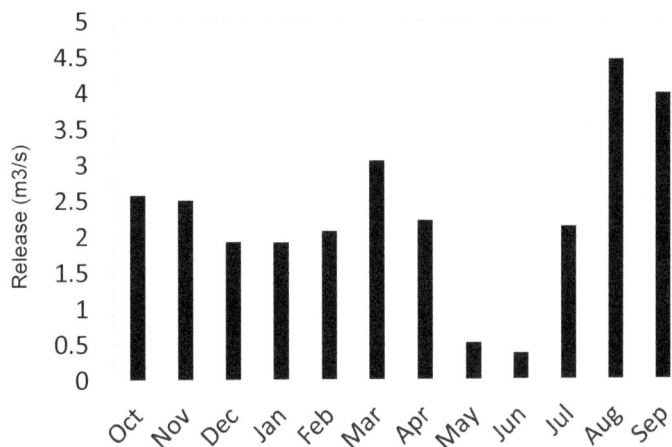

Fig. 10. Average values of flow release from the Jareh Dam reservoir.

4. Conclusions

The calibration of models is important in order to make their results closer to the observed values in modeling processes. However, due to the unique complexities of these models, especially in the absence of accurate recorded information on the river discharge of the existing medial basins, or in the absence of accurate statistics on the amount of water withdrawal at downstream of the river, manual calibration is difficult and time consuming. In this research, the ability of non-dominated sorting genetic algorithm (NSGA-II) was used by defining multi-objective functions for automatic calibration of WEAP model in the basin of Jareh Dam and network. The results showed that using this algorithm, the systematic simulation of the study area takes less time than manual calibration and is much satisfactory in terms of accuracy. The model was converged and calibrated by selecting the initial population of 96 and after 500 iterations. During the optimization process, the model focused on the reduction of the both objective functions, but at the final iteration, along with a slight decrease in the difference between the simulated and the observed discharge at the stations, it focused on reducing the simulated and the observed reservoir storage volumes more. After calibrating the model by the algorithm, it was observed that the simulated values by the model match very well to the observed values, which was analyzed using the Nash statistical parameter. At the last run of the model and after choosing the best solution among the proposed optimal solutions in the objective exchange curve, the value of this parameter for Mashin and Jokank stations as well as for the reservoir storage volume was 0.96, 0.94, and 0.82, respectively, which was completely acceptable. According to the results, it can be concluded that the algorithm has a great ability to calibrate the corresponding model, which can improve the simulated values in the model and makes them approach to reality.

References

Abrishamchi A., Alizadeh H., Tajrishy M., Abrishamchi A., Water Resources Management Scenario Analysis in Karkheh River Basin, Iran, Using WEAP Model. Hydrological Science and Technology 23 (2007) 1.

Afshar A., Kazemi H.,, Saadatpour M., Particle swarm optimization for automatic calibration of large scale water quality model (CE-QUAL-W2): application to Karkheh Reservoir, Iran. Water resources management 25 (2011) 2613-2632.

Alfarra A., Modelling water resource management in Lake Naivasha., M.Sc. thesis, International Institute for Geo-information Science and Earth Observation (2004).

Alfarra A., Kemp-Benedict E., Hötzl H., Sader N., Sonneveld, B. Modeling water supply and demand for effective water management allocation in the Jordan Valley. Journal of Agricultural and Application (JASA) 1 (2012) 1-7.

Azari A., Akhoond-Ali A.M., Radmanesh F., Haghighi A Groundwater–Surface Water Interaction Simulation in Terms of Integrated Water Resource Management (Case Study: Dez Plain). Journal of Irrigation Science and Engineering 38 (2015) 33-47. (In Persian)

Bekele E.G., Nicklow, J.W., Multi-objective automatic calibration of SWAT using NSGA-II. Journal of Hydrology 341 (2007) 165-176.

Bozorg Haddad, O., Saifollahi Aghmouni S., An Introduction to uncertainty analysis in water resources systems. Tehran university press, First edition (2013) 233 (In Persian).

Ercan M.B., Goodall, J.L., A Python tool for multi-gage calibration of SWAT models using the NSGA-II algorithm. In proceedings of the 7th International Congress on Environmental Modelling and Software, June (2014) 15-19.

Ghandhari G., Soltani J., Hamidian Pour M., Evaluation of optimal water allocation scenarios for bar river of Neishabour using WEAP model under A2 climatic changes scenario. Journal of water and soil 29 (2015) 1158-1172 (In Persian).

Hamlat, A., Errih, M., and Guidoum, A. Simulation of water resources management scenarios in western Algeria watersheds using WEAP model. Arabian Journal of Geosciences 6 (2013) 2225-2236.

Huang Y., Lei L., Multiobjective Water Quality Model Calibration Using a Hybrid Genetic Algorithm and Neural Network–Based Approach. Journal of Environmental Engineering 136 (2010) 1020-1031.

Hum N.N.M.F., Abdul-Talib, S., Modeling Optimal Water Allocation by Managing the Demands in Selangor. In ISFRAM, Springer Singapore (2016) 93-104.

Karimi, S.M., Mousavi, S.J., Priority-based allocation of water resources in the catchment area: comparison of WEAP and MODSIM models. 6th National Congress of Civil Engineering, Semnan University (2011) (In Persian).

Li X., Zhao Y., Shi C., Sha J., Wang Z.L., Wang Y., Application of Water Evaluation and Planning (WEAP) model for water resources management strategy estimation in coastal Binhai New Area, China. Ocean & Coastal Management 106 (2015) 97-109.

Madsen H, Torsten J., Automatic calibration of the MIKE SHE integrated hydrological modelling system. 4th DHI Software

Conference, 6-8 June. Helsingor, Denmark: Scanation Conference Centre (2001).

Muleta, M.K., Nicklow, J.W. Sensitivity and uncertainty analysis coupled with automatic calibration for a distributed watershed model. Journal of hydrology 306 (2005) 127-145.

Mutiga J.K., Mavengano ST., Zhongbo S., Woldai T., Becht, R., Water allocation as a planning tool to minimise water use conflicts in the Upper Ewaso Ng'iro North Basin, Kenya. Water resources management 24 (2010) 3939-3959.

Purkey, D.R., Joyce B., Vicuna S., Hanemann M.W., Dale L.L., Yates D., Dracup J.A., Robust analysis of future climate change impacts on water for agriculture and other sectors: a case study in the Sacramento Valley. Climatic Change 87 (2008) 109-122.

Rafiee Anzab N., Mousavi S.J., Rousta B.A., Kim, J.H. Simulation optimization for optimal sizing of water transfer systems. In Harmony Search Algorithm (2016) 365-375. Springer Berlin Heidelberg.

Sieber J., Swartzand C., Huber-Lee A., User guide for WEAP21. Stockholm Environment Institute Tellus Institute. (2005).

Wang Q.J. The Genetic Algorithm and Its application to Calibrating Conceptual Rainfall-Runoff Models. Water Resources Research 27 (1991) 2467-2471.

Zhang X., Srivivasan R., Van Liew, M., Multi-Site Calibration of the SWAT Model for Hydrologic Modeling. Transactions of the ASABE 51 (2008) 2039-2049.

26

Numerical modeling of flow field in prismatic compound channels with different floodplain widths

Bahram Rezaei*, Alireza Safarzade

Department of Civil Engineering, Bu-Ali Sina University, Hamedan, Iran.

ARTICLE INFO

Keywords:
Prismatic compound channel
Flow field
Numerical simulation
Turbulence model

ABSTRACT

In this paper an attempt has been made to study the effects of floodplains width and discharges on flow field in prismatic compound channels. A three-dimensional Computational Fluid Dynamic (CFD) model is used to predict the velocity distribution, secondary flow circulation and boundary shear stress in prismatic compound channels with various floodplains widths. The ANSYS-CFX software and three different turbulence models, κ–ε, κ–ε Explicit Algebraic Reynolds Stress Models (EARSM) and Eddy Viscosity Transport, are used to solve Reynolds Averaged Navier-Stokes equations. The results of the numerical modeling were then compared with experimental data on prismatic compound channels with 100 mm, 200 mm, 300 mm, and 400 mm floodplain widths. The study shows that all turbulence models are capable to predict the depth-averaged velocity in prismatic compound channels, fairly well. However, to compare with the velocity distribution, discrepancy between experimental data and boundary shear stress calculated by numerical modeling are high. Also only κ–ε EARSM model is able to predict secondary flow circulations.

1. Introduction

The prediction of the flow characteristics in compound channels with prismatic floodplains is a challenging task for engineers because of the three-dimensional nature of the flow. In compound channels flow in the main channel is faster than floodplains. This difference creates shear layer at the interface between the main channel and floodplains, leading to the generation of the vortices with vertical axes, as well as the secondary flow circulations with longitudinal axes, as shown by Sellin (1964), Tominaga and Nezu (1991), Ikeda (1999), Bousmar (2002), Rezaei (2006) and Rezaei and Knight (2011). Because of the presence of this shear layer and creation of momentum exchange between the main channel and floodplains, the conveyance capacity of the main channel decrease, while on the floodplains significantly increases. Wormleaton (1996) stated that the effects of this shear layer extend across the floodplain width and decreases to zero towards the floodplain wall. Myers (1978) also discovered that the effects of the shear layer were great at lower overbank flow depths and decrease as the flow depth increases.

There are two kinds of vortices that are generated at the interface between the main channel and the floodplain; one is the horizontal vortex due to shear layer of the stream wise flow, first observed by Sellin (1964), and the other is the secondary flow in the cross section due to anisotropy of turbulence, also called secondary flow of 2nd kind (*cf.* Nezu and Nakagawa ,1993), as shown in Fig. 1. These effects have been observed experimentally by Shiono and Knight (1991) and Tominaga and Nezu (1991) using Laser-Doppler Anemometer (LDA). Also those secondary flow cells numerically investigated by Naot et al (1993) using a non-linear κ–ε turbulence model and by Cokljat and Younis (1995), using Reynolds Stress Transport model. They have found a significant influence of secondary flows into momentum transfer and boundary shear stress. Pezzinga (1994) used a nonlinear κ–ε turbulence model to predict the uniform flow in a compound channel. He found that the proposed model is able to predict the secondary

current, created by the anisotropy of normal turbulent stress. Cokljat (1993) used a Reynolds Stress Transport model and non-lineared κ–ε turbulence model to predict flow in open channel. He found out that the Reynolds Stress Transport (RST) model is able to predict the secondary flow cells but in contrast the non-linear κ–ε model failed to reproduce this result. Both models predicted equally well the shear stress. Flow field in trapezoidal open channel was numerically investigated by Wright et al. (2004) using κ–ε and various Reynolds stress models. They revealed that while all the models generally gave similar predictions for many features of the flow, there was a clear difference in the secondary flow characteristics. The κ–ε model failed to show any recirculation and the Reynolds stress models showed some recirculation in varying degrees.

Kang and Choi (2005) used a Reynolds stress model to simulate flow field in compound channels with vegetation on the floodplains. They show that by increasing vegetation density on the floodplains the point maximum stream wise means velocity moves to the main channel also bed shear stress on the floodplains decrease while it increases in the main channel. Jing et al. (2009) modeled flow in a meandering compound channel using the Reynolds stress model (RSM). They reviled that RSM can successfully model the velocity distribution and boundary shear stress in proposed flume.

Beaman (2010) used Large Eddy Simulation (LES) to model flow field in in-bank and over-bank channels. He showed that the LES model can accurately predict the flow characteristics, specially the distribution of secondary circulations in inbank and for over-bank channels at varying depth and width ratios. The main aim of the present work is to investigate whether or not the ANSYS-CFX software is able to predict the effects of flow depth and floodplain width on flow field in prismatic compound channels. Three turbulence models including κ–ε, Eddy Viscosity Transport Equation (EDDY) and Explicit Algebraic Reynolds Stress Model (EARSM) were chosen to model the velocity distribution, depth-averaged velocity and boundary shear stress distributions.

*Corresponding author Email: b.rezaei@basu.ac.ir

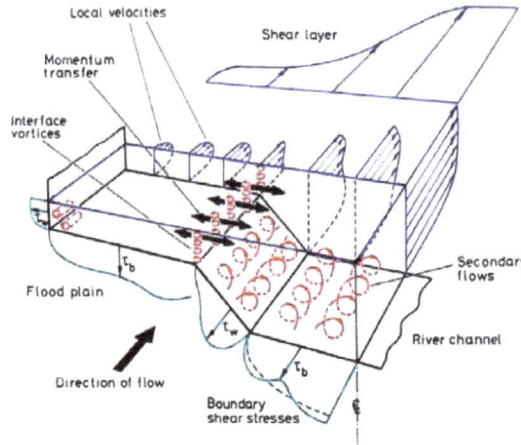

Fig. 1. Flow structure in a compound channel (Shiono and Knight. 1991).

2. Materials and methods

Experiments were carried out by Rezaei (2006), using an 18 m flume at the University of Birmingham, Department of Civil Engineering. A compound channel of simple rectangular cross-section was selected and all experiments were performed in a straight flume, almost 1200 mm in width, 400 mm in depth and with the average bottom slope of S0 = 2.003×10-3. PVC material, were used to construct rigid and smooth boundaries both for the main channel (with 398 mm width and 50 mm depth), and also for the floodplains of 400 mm wide (Rezaei. 2006).

However, for experiments in prismatic compound channels, the main channel and floodplains were isolated using L-shaped aluminum sections to make different floodplain widths, 100 mm, 200 mm, 300 mm and 400 mm (see Fig. 2).

A series of three adjustable tailgates, at the downstream flume end, controlled uniform flow in compound channel. Overbank flow in prismatic compound channel tests are denoted by OPC, the first three numbers refer to the floodplain width and two code numbers denoted the flow discharge (Rezaei. 2006).

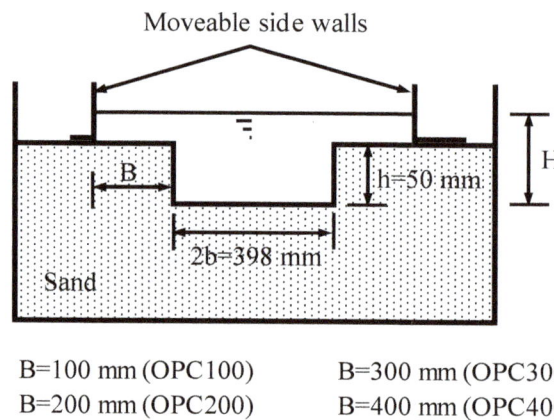

B=100 mm (OPC100)   B=300 mm (OPC300)
B=200 mm (OPC200)   B=400 mm (OPC400)

Fig. 2. Typical cross-section of prismatic compound channels with different floodplain widths.

2.1. Depth-averaged velocity measurement

The depth-averaged velocity distribution in a cross-section was measured at one section (14 m from the channel inlet) using a 13 mm diameter Novar Nixon miniature propeller current meter. Point depth-averaged velocity measurements were made laterally each 25 mm at a depth of 0.4H from the bed in the main channel and 0.4($H - h$) on the floodplains (Rezaei. 2006).

2.2. Boundary shear stress measurement

Local boundary shear stress measurements were made using a Preston tube of 4.77mm outer diameter. These measurements were performed at the same sections where velocity measurements were taken. Local boundary shear stress was measured around the wetted channel perimeter at 10 mm vertical intervals on the walls and 25 mm transverse intervals on the bed.

3. Governing equations

The conservation of mass and momentum can express the flow motion. The equation for mass is called continuity equation and expressed as follows:

$$\frac{\partial \rho}{\partial t} + \frac{\partial}{\partial x_j}(\rho U_j) = 0 \tag{1}$$

in which, ρ is flow density and U_j is time-averaged components of velocity.

$$U_j = \frac{1}{\Delta t}\int_{t_1}^{t_2} u_j\, dt \tag{2}$$

The equation of motion is an expression of the second low of Newton and can be explained in the Reynolds-averaged Navier-Stokes equation,

$$-\frac{\partial P}{\partial x_i} + \frac{\partial}{\partial x_j}(\tau_{ij} - \rho\overline{u_i'u_j'}) = \frac{\partial(\rho U_i)}{\partial t} + \frac{\partial(\rho U_i U_j)}{\partial x_j} \tag{3}$$

where P is pressure, τ is the molecular stress tensor (including both normal and shear components of the stress), $\rho\overline{u_i'u_j'}$ is called 'turbulent' or 'Reynolds' stresses and can be evaluated using Boussinesq Eddy Viscosity turbulence model.

$$-\rho\overline{u_i'u_j'} = \mu_T\left(\frac{\partial U_i}{\partial x_j} + \frac{\partial U_j}{\partial x_i}\right) - \frac{2}{3}\rho k\delta_{ij} \tag{4}$$

in which μ_t is the turbulence viscosity, k is the turbulence kinetic energy, and δ_{ij} is the Kronecker delta.

3.1. The κ–ε model

The standard κ–ε is classified as a two-equation model since it used two transport equations to describe turbulence (Launder and Spalding, 1974). These two transport equation are as follows:
Turbulent kinetic energy equation:

$$\frac{\partial(\rho k)}{\partial t} + \frac{\partial}{\partial x_j}(\rho U_j k) = \frac{\partial}{\partial x_j}\left((\mu + \frac{\mu_t}{\sigma_k})\frac{\partial k}{\partial x_j}\right) + P_k - \rho\varepsilon \tag{5}$$

Turbulent kinetic energy dissipation rate equation:

$$\frac{\partial(\rho\varepsilon)}{\partial t} + \frac{\partial}{\partial x_j}(\rho U_j\varepsilon) = \frac{\partial}{\partial x_j}\left((\mu + \frac{\mu_t}{\sigma_\varepsilon})\frac{\partial\varepsilon}{\partial x_j}\right) + \frac{\varepsilon}{k}(C_{s1}P_k - C_{s2}\rho\varepsilon) \tag{6}$$

where k is the turbulence kinetic energy and is defined as the variance of the fluctuations in velocity, ε is the turbulence eddy dissipation (the rate at which the velocity fluctuations dissipate), C_{s1}=1.44, C_{s2}=1.92, σ_k=1.00, and σ_ε=1.30 are turbulence constants. P_k is the turbulence production due to viscous forces, which is modeled using:

$$P_k = \mu_t\left(\frac{\partial U_i}{\partial x_j} + \frac{\partial U_j}{\partial x_i}\right)\frac{\partial U_i}{\partial x_j} - \frac{2}{3}\frac{\partial U_k}{\partial x_k}(\rho k + 3\mu_t\frac{\partial U_k}{\partial x_k}) \tag{7}$$

For incompressible flow,$(\partial U_k/\partial x_k)$ is small and the second term on the right side of Equation (7) does not contribute significantly to the turbulence production.

3.2. The Eddy Viscosity Transport model

A very simple one-equation model has been developed by Menter (1997). It is derived directly from the κ–ε model and is therefore named the $(\kappa$–$\varepsilon)_{hE}$ model.

$$\frac{\partial(\rho\tilde{v}_t)}{\partial t} + \frac{\partial(\rho U_j\tilde{v}_t)}{\partial x_j} = C_1\rho\tilde{v}_t S - C_2\rho(\frac{\tilde{v}_t}{L_{vk}})^2 + \left[(\mu + \frac{\rho\tilde{v}_t}{\sigma})\frac{\partial\tilde{v}}{\partial x_j}\right] \tag{8}$$

where $\tilde{v}$ is the kinematic eddy viscosity, $\tilde{v}_t$ is the turbulent kinematic eddy viscosity and C_1, C_2, and σ are model constants. The model contains a destruction term, which accounts for the structure of turbulence and is based on the Von Karman length scale:

$$(L_{vk})^2 = \left|\frac{S^2}{\frac{\partial S}{\partial x_i}\frac{\partial S}{\partial x_i}}\right| \tag{9}$$

in which S is the shear strain rate tensor. The eddy viscosity is computed from:

$$\tilde{v}_t = \frac{\mu_t}{\rho} \tag{10}$$

3.3. Explicit Algebraic Reynolds Stress model

Explicit Algebraic Reynolds Stress Models (EARSM) represents an extension of the standard two-equation models. They are developed from the Reynolds stress transport equations and give a nonlinear relation between the Reynolds stresses and the mean strain-rate and vortices tensors. Because of the higher order terms, many flow characteristics are contained within the model without the need of solving transport equations. The implementation is based on the Explicit Algebraic Reynolds Stress model of Wallin (2000) and Wallin and Johansson (2000). The Reynolds stresses are computed from the anisotropy tensor according to its definition:

$$\overline{u_i'u_j'} = k(a_{ij} + \frac{2}{3}\delta_{ij)} \tag{11}$$

where the anisotropy tensor a_{ij} is searched as a solution of the following implicit algebraic matrix equation:

$$a = \beta_1 S + \beta_2\left(S^2 - \frac{1}{3}tra(S^2)\right) + \beta_3\left(\Omega^2 - \frac{1}{3}tra(\Omega^2)\right) + \beta_4(S\Omega - \Omega S)$$
$$+ \beta_5(S^2\Omega - \Omega S^2) + \beta_6\left(S\Omega^2 + \Omega^2 S - \frac{2}{3}tra(S\Omega^2)\right) + \beta_7\left(S^2\Omega^2 + \Omega^2 S^2 - \frac{2}{3}tra(S^2\Omega^2)\right)$$
$$\beta_8(S\Omega S^2 - S^2\Omega S) + \beta_9(\Omega S\Omega^2 - \Omega^2 S\Omega) + \beta_{10}(\Omega S^2\Omega^2 - \Omega^2 S^2\Omega) \tag{12}$$

in which the β coefficients may be function of the five independent invariants of S and Ω.

$$N_a = -A_1 S + (a\Omega - \Omega a) - A_2\left(aS - Sa - \frac{2}{3}tra(aS)\right) \tag{13}$$
$$with \qquad N = A_3 + A_4(\frac{P_k}{\varepsilon})$$

The coefficients A_i in this matrix equation depend on the C_i coefficients of the pressure-strain term in the underlying Reynolds stress transport model. Their values are selected here as A_1=1.245, A_2=0, A_3=1.80, and A_4=2.25. S=S_{ij} and Ω=Ω_{ij} denote the non-dimensional strain-rate and vortices tensors, respectively. They are defined as:

$$S_{ij} = \frac{1}{2}\tau_t\left(\frac{\partial U_i}{\partial x_j} + \frac{\partial U_j}{\partial x_i}\right) \tag{14}$$

$$\Omega_{ij} = \frac{1}{2}\tau_t\left(\frac{\partial U_i}{\partial x_j} - \frac{\partial U_j}{\partial x_i}\right) \tag{15}$$

where the time-scale t_t is given by:

$$\tau_t = \frac{k}{\varepsilon} = \frac{1}{C_\mu\omega} \quad, C_\mu = 0.09 \tag{16}$$

3.4. Introducing ANSYS-CFX

The Computation Fluid Dynamic (CFD) is a capable computer-based tool for simulating the behavior of systems involving flow field, heat transfer and other physical processes. It works by numerically solving the Reynolds Averaged Navier-Stokes equations over a region of interest, with specified boundary conditions. The solution is advanced through space and time to obtain a numerical description of the flow field. The ANSYS-CFX software is a commercial CFD code; it uses the finite volume approach to solve the Reynolds Averaged Navier-Stokes equations.

The advantage of using ANSYS-CFX to other codes is that it offers multiple validated solutions as well as powerful algorithms and discrete techniques, and is also flexible in implantation of boundary condition via user defined FORTRAN subroutines (Morvan et al., 2001; Bonakdari et al., 2011).

4. Simulation of flow field

The height and width of the numerical modeling domain were exactly the same as experimental flume used by Rezaei (2006). Also, an important task was to decide which solver to use. By considering the running time of software and uniform flow condition, the Rigid Lid approach was chosen, which means that the free surface would be fixed at a certain depth by adopting a top boundary with no friction.

4.1. Mesh gridding

In numerical modeling an optimal mesh refinement was pursued, with the aim of optimal computational costs besides maintaining accuracy. Numerical simulations must have a sufficiently fine mesh to resolve the flow field near the main channel and floodplain beds and walls. To study the effect of mesh size on flow modeling, three sets of course, medium and fine mesh size for prismatic compound channel with 400 mm floodplain widths were chosen. The mesh sizes were chosen in such a way that the average refinement ratio was above the recommended minimum value of 1.3 (see Celik et al., 2008). Using $\kappa-\varepsilon$ turbulence model and discharge of 40 l/s, the depth-averaged velocity was numerically modeled (see Fig. 3). Fig. 3 indicated that by decreasing cells size accuracy of numerical modeling increase. Also the dispersion diagrams for numerical and experimental data together with an ideal line function of $y=x$ are presented in Figs. 4. As seen in figures by increasing the number of nodes, the points get close to ideal line. It should be noted that the mesh spacing in fine case was chosen in such a way that the dimensionless distance to the wall, $y^*(=yu^*/v)$, was into the range $30<y^*<500$.

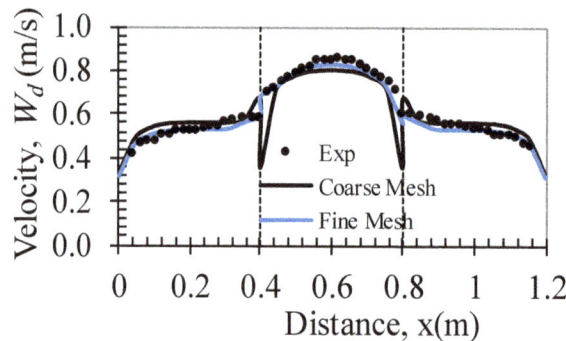

Fig. 3. A comparison between experimental and numerical modeling of depth-averaged velocity for two mesh sizes and discharge of 40 l/s.

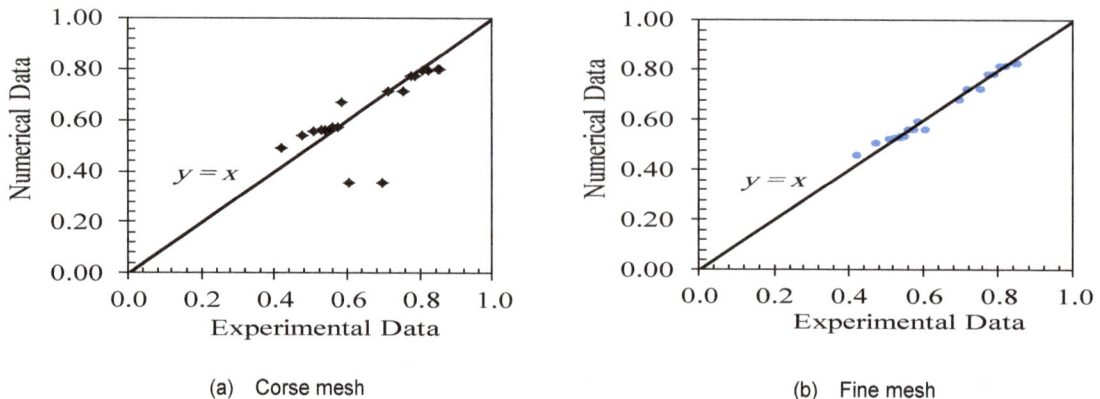

(a) Corse mesh

(b) Fine mesh

Fig. 4. Dispersion diagram of velocity for mesh size independency analysis.

The ICEM software is used to get a good grid in numerical model. Since higher accuracy is needed, grids near the water surface, the beds and the interfaces between the main channel and floodplains have been made finer than other parts of flume cross section. Along the flume, because of simple geometry, coarse grids with 0.2 m spaces have been used. Details of gridding are shown plotted in Table 1 and Figs. 5. As seen in Table 1 the maximum and minimum size of the elements in the main channel and on the floodplains are 0.006 m and 0.002 m, respectively. To make sure that the flow field in the numerical model is fully developed, 7 m has been added to the flume length.

4.2. Boundary conditions

The solution of flow field was carried out using ANSYS-CFX software with three turbulence models and an iteration procedure with accuracy of 1×10^{-8}. The boundary conditions are as follow: (a) uniform velocity distribution at the flume inlet, (b) hydrostatics pressure condition at the outlet, (c) smooth solid wall with no slip condition in the main channel and on the floodplains walls and beds, and (d) free surface condition on the water surface.

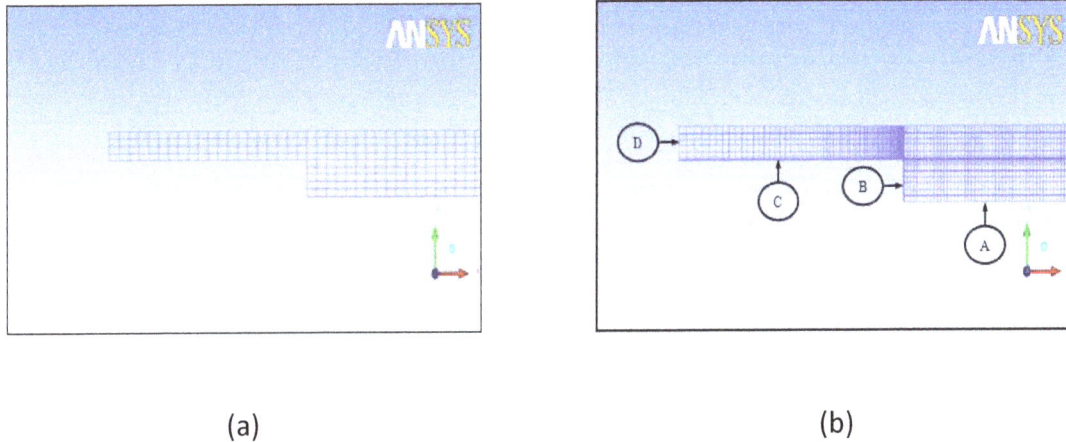

(a)   (b)

Fig. 5. Details of gridding in section; (a) coarse mesh, (b) fine mesh.

Table1. Details of gridding for fine mesh.

Element	Number of elements	Max. mesh spacing (m)	Min. mesh spacing (m)
A	70	0.006	0.002
B	10	0.006	0.002
C	Depend on width	0.006	0.002
D	Depend on water depth	0.006	0.002

$$MAPE = \frac{\sum \left| \dfrac{W_{d\,exp} - W_{dmes}}{W_{exp}} \right|}{n} \times 100 \qquad (17)$$

5. Results
5.1. Velocity distributions

To study the effects of flume geometry on flow field, the stream wise depth-averaged velocity in prismatic compound channel with four different floodplain widths and 12 discharges (Q=12 1/s, 15, 18, 21, 24, 27, 30, 35, 40, 45 and 50 1/s) were modeled using κ-ε, κ-ε EARSM, and Eddy Viscosity Transport turbulence models.

The results of depth averaged-velocity modeled by ANSYS-CFX for two discharges of 24 1/s and 45 l/s are shown in Figs. 6 and 7. As seen in figures the κ-ε and Eddy (e.g. Eddy Viscosity Transport Equation) models are able to predict the depth-averaged velocity distribution, quite well, especially in the main channel. In addition, it is clear that the κ-ε EARSM model cannot predict maximum velocity in the middle of main channel. Figures also show that, in general, by increasing discharges and floodplain widths the discrepancy between the experimental and numerical data decrease. The streamwise velocity predicted by the k-ε and Eddy Viscosity Transport turbulence models are shown in Figs 8(a), 8(b) and Figs. 8(e), 8(f), respectively. These turbulence models do not produce secondary flow, and accordingly its influence is not reflected in stream wise velocity contours. In order to study the effects of floodplain widths and discharges on flow:

The mean absolute percentage error (MAPE) of depth average velocity for two discharge of, Q=24 1/s and 45 1/s are also calculated using equation (17) and shown in Table 2. As seen in Table 2 the mean absolute percentage errors for three turbulence models are, usually, less than 8 percent. In which W_{dexp} is experimental point depth-averaged velocity, W_{dmes} is numerical depth-averaged velocity and n is the number of data The streamwise velocity distribution for experimental cases of OPC100-45 and OPC200-45 are also modeled by three turbulence models and shown plotted in Figs. 8. The figures indicate that, all turbulence models are able to predict velocity distribution fairly well. The bulging of the isovels towards the main channel from the floodplain edges in Figs. 8(c) and 8(d) are characteristic of flows where the secondary currents are present. field, the depth-averaged velocity distribution has been normalized. The average velocity in the whole cross section has been used for normalization (see Figs. 9). As seen in the Fig. 9(a), for experimental tests of OPC100-24 and OPC400-24 the normalized velocities on the floodplains are almost the same, while in the main channel the discrepancy between the results of two experimental series are apparent. This fact indicates that for discharge of 24 1/s by increasing floodplain width from 100 mm to 400mm, the interaction between the main channel and floodplains increases. As discharge increases from 24 1/s to 45 1/s, the difference between normalized velocity in the main channel and floodplain decrease which means the less interaction between those subsections (see Fig. 9(b)).

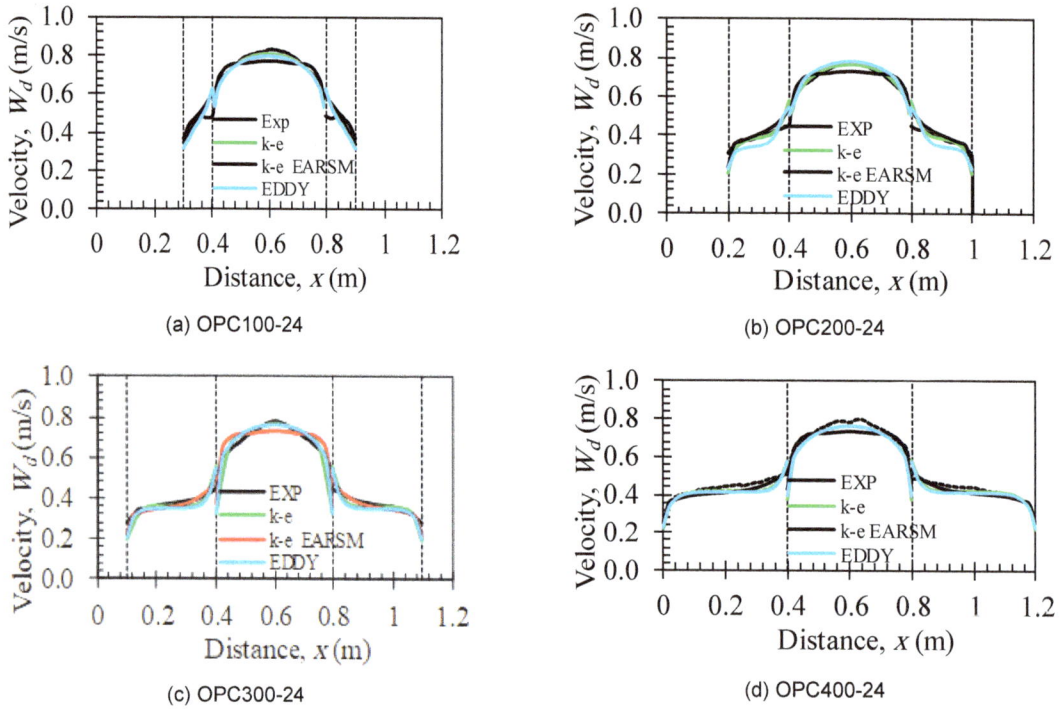

(a) OPC100-24

(b) OPC200-24

(c) OPC300-24

(d) OPC400-24

Fig. 6. Depth–averaged velocity distribution in prismatic compound channels with different floodplain widths and $Q = 24$ l/s.

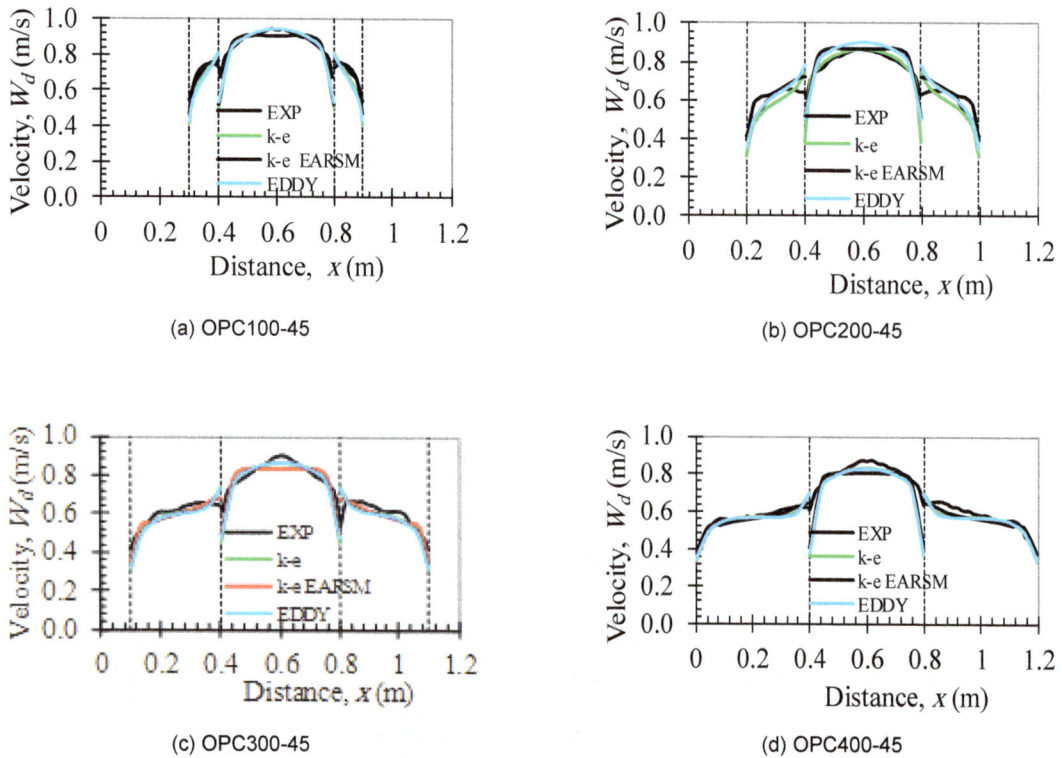

(a) OPC100-45

(b) OPC200-45

(c) OPC300-45

(d) OPC400-45

Fig. 7. Depth–averaged velocity distribution in prismatic compound channels with different floodplain widths and $Q = 45$ l/s.

Table 2. The mean absolute percentage error (MAPE) of depth average velocity for three turbulence model; a) Q=24 l/s, b) Q=45 l/s.

a) Q=24 l/s

Experimental	$\kappa-\varepsilon$	$\kappa-\varepsilon$ EARSM	Eddy
OPC100	5.18	6.65	5.91
OPC200	5.15	6.21	13.68
OPC400	7.23	6.91	7.89

b) Q=45 l/s

Experimental	$\kappa-\varepsilon$	$\kappa-\varepsilon$ EARSM	Eddy
OPC100	5.22	3.52	5.80
OPC200	5.93	4.23	6.34
OPC400	5.00	4.77	5.28

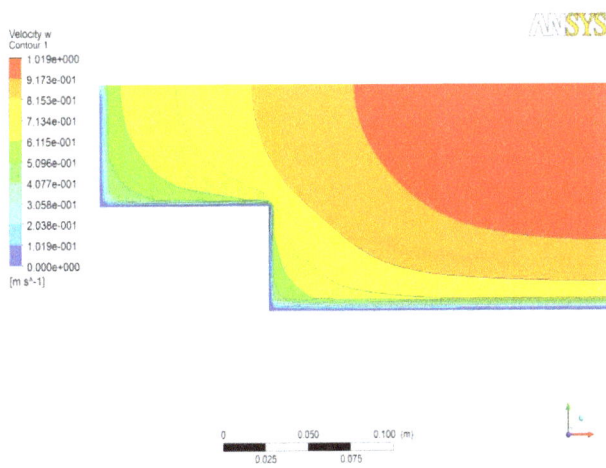

(a) OPC100-45 ($\kappa-\varepsilon$ model)

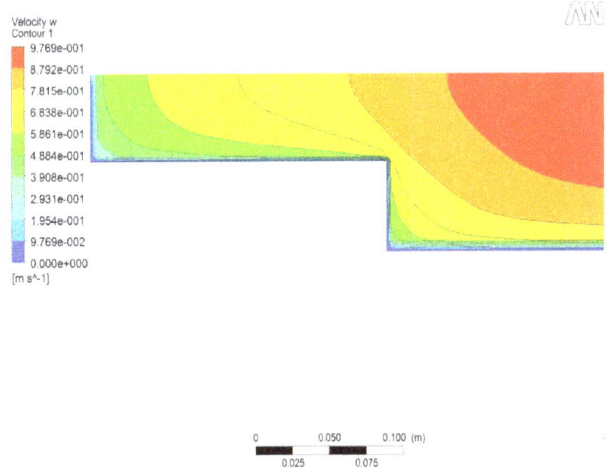

(b) OPC200-45 ($\kappa-\varepsilon$ model)

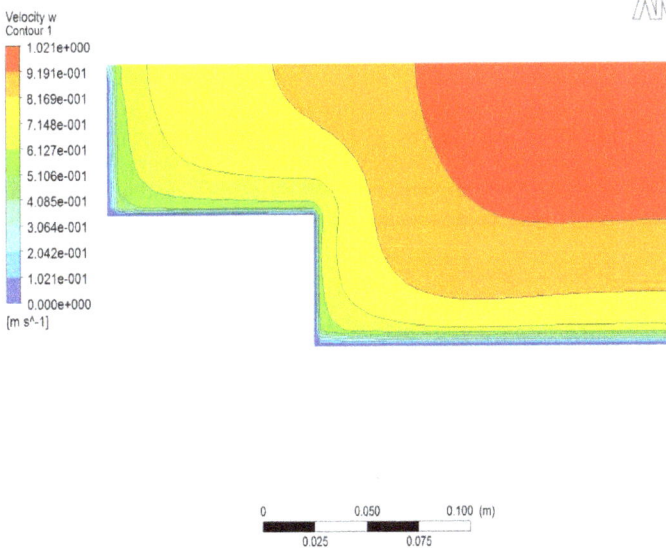

(c) OPC100-45 $\kappa-\varepsilon$ EARSM model)

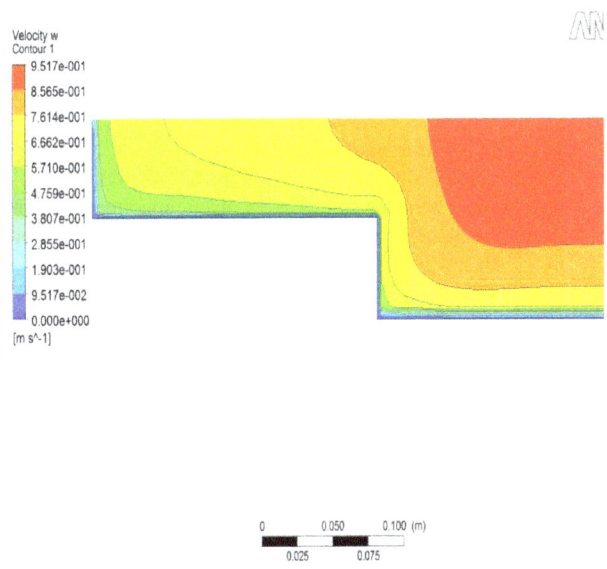

(d) OPC200-45 ($\kappa-\varepsilon$ EARSM model)

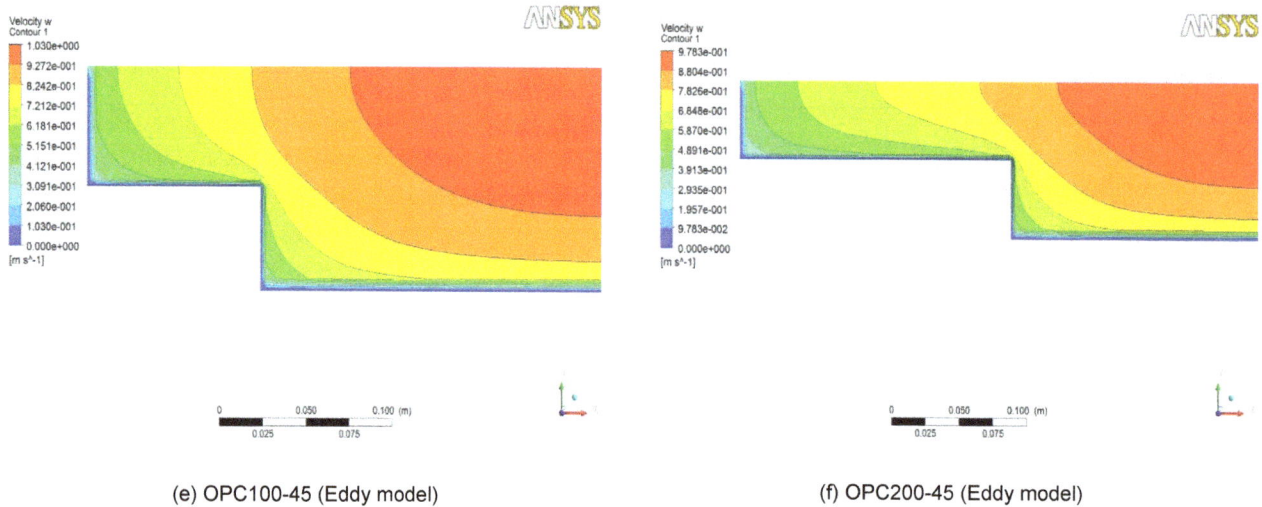

(e) OPC100-45 (Eddy model)   (f) OPC200-45 (Eddy model)

Fig. 8. Streamwise velocity Distribution simulated using turbulence models for experimental series of OPC100-45 and OPC200-45 (a, b) $K-\varepsilon$, (c, d) $K-\varepsilon$ EARSM, and (e, f) Eddy Viscosity Transport Equation.

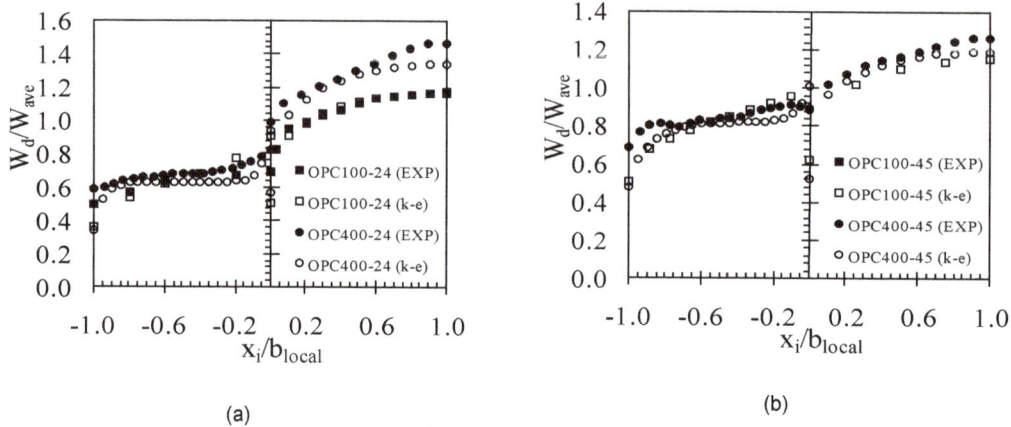

(a)   (b)

Fig. 9. A comparison between experimental and numerical normalized depth-averaged velocity for two experimental series of OPC100 and OPC400; (a) Q=24 l/s, (b) Q=45 l/s

5.2. Boundary shear stress distributions

Boundary shear stress distribution is another important parameter in river engineering when studying sediment transport and riverbank protection. It also is important for river modelers when calibrating a mathematical model, which commonly requires numerical values of resistance coefficients. The boundary shear stress distributions calculated using the three turbulence models. The results of numerical modeling were then compared with experimental data (see Figs. 10). Figs. show that; (a) all three turbulence models always underestimate boundary shear stress on the floodplain, (b) discrepancy between numerical results and experimental data near the interface of main channel and floodplain, significantly, increase, (c) the $k-\varepsilon$ turbulence model underestimates shear stress in the main channel, while the EDDY turbulence model overestimates it, (e) similar to depth-averaged velocity distributions, the $k-\varepsilon$ EARSM turbulence model is not able to predict maximum shear stress in the main channel, (f) the $k-\varepsilon$ EARSM model predicts two picks near the main channel walls which indicates the presence of strong secondary flow cells in this part of the flume. The mean absolute percentage error (MAPE) of shear stress for two discharge of, $Q=24$ l/s and 45 l/s are also calculated and shown in Table

3. As seen in the table, among those turbulence models, the $\kappa-\varepsilon$ EARSM model has the minimum MAPE also by increasing floodplain widths and flow discharges the difference between the experimental and numerical data decrease.

5.3. Secondary flow

The secondary flow patterns for two floodplain widths (300 mm and 400 mm) and discharges Q= 24 1/s and 45 l/s are simulated using $\kappa-\varepsilon$ EARSM turbulence model (see Fig. 11). The figures clearly show the effects of geometry and discharge on secondary flow pattern. For the compound channel with 300 mm and 400 mm floodplain widths and discharge of 24 1/s, presence of one strong secondary flow cell in the main channel are clear, as the discharge raise to 45 1/s the number of secondary flow circulations increase to three cells, one cell in the main channel and two strong secondary flow cells near the interface between the main channel and floodplains. This emphasizes that two secondary flow cells interacted near the interface is responsible for pushing upwards particles with smaller velocities, causing the inflection of the isovel lines and increasing depth-averaged velocity and shear stress near the main channel walls.

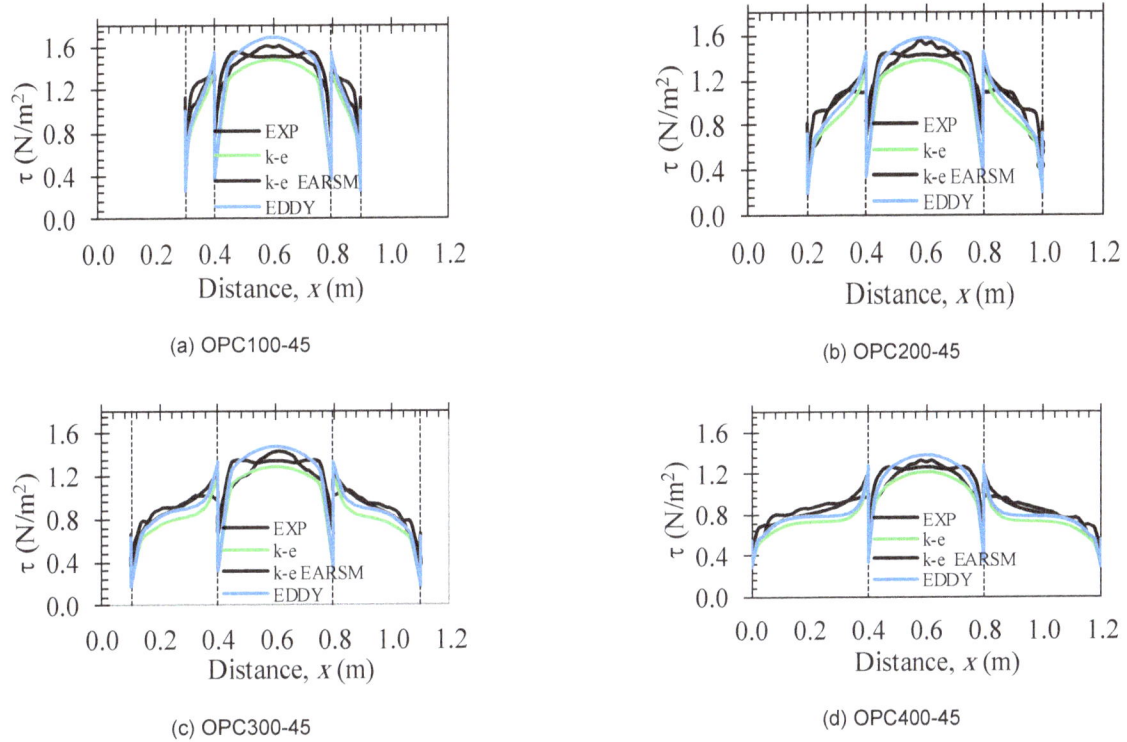

(a) OPC100-45

(b) OPC200-45

(c) OPC300-45

(d) OPC400-45

Fig. 10. Shear stress distribution in prismatic compound channel with different floodplain widths and Q = 45 l/s.

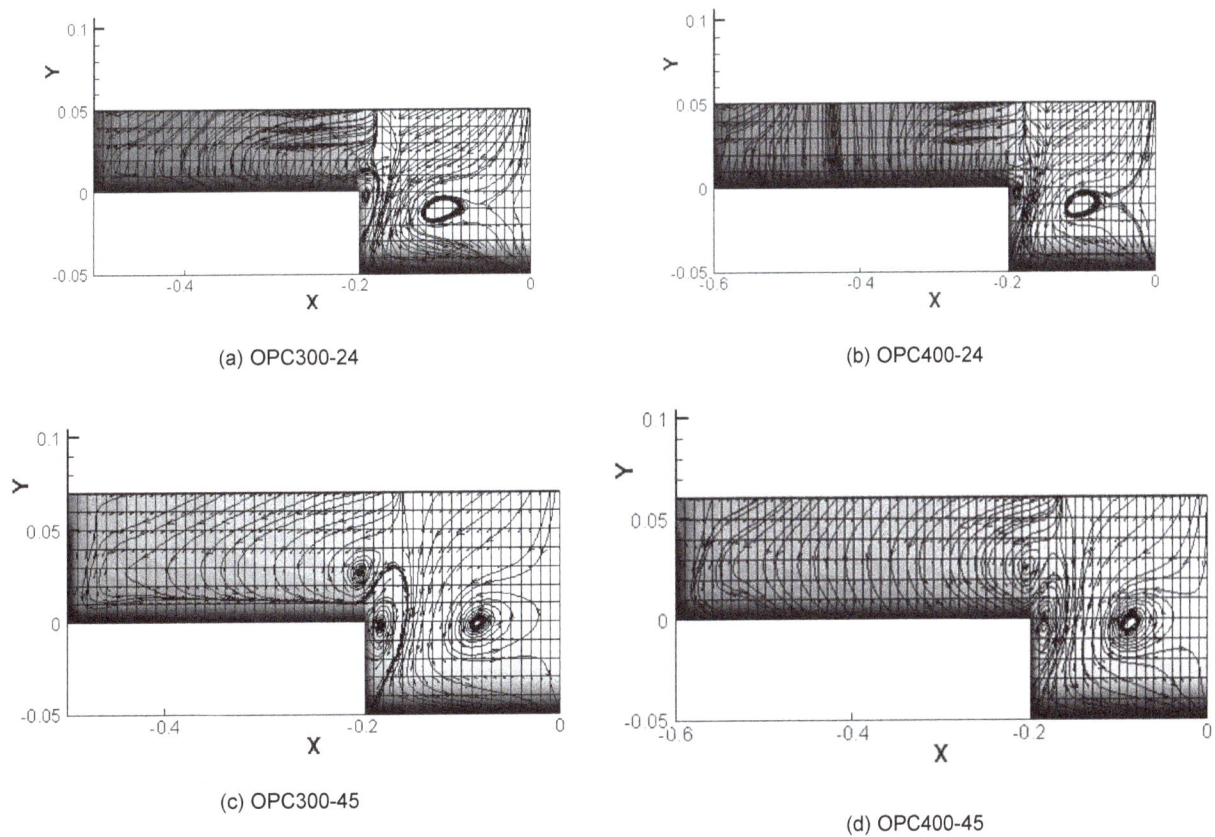

(a) OPC300-24

(b) OPC400-24

(c) OPC300-45

(d) OPC400-45

Fig. 11. Secondary current circulations predicted using $\kappa-\varepsilon$ EARSM turbulence model in prismatic compound channels for two discharges of 24 l/s and 45 l/s.

Table 3. The mean absolute percentage error (MAPE) of shear stress distribution for three turbulence model; a) Q=24 l/s, b) Q=45 l/s.

a) Q=24 l/s

Experimental	$\kappa-\varepsilon$	$\kappa-\varepsilon$ EARSM	Eddy
OPC100	17.87	12.17	22.23
OPC200	14.94	10.48	33.55
OPC400	19.50	12.51	18.77

b) Q=45 l/s

Experimental	$\kappa-\varepsilon$	$\kappa-\varepsilon$ EARSM	Eddy
OPC100	16.81	6.77	19.88
OPC200	16.09	8.88	16.50
OPC400	14.36	6.69	13.42

6. Conclusions

The velocity and boundary shear stress distributions in prismatic compound channels with different floodplain widths were numerical simulated using three turbulence models, including the $\kappa-\varepsilon$, $\kappa-\varepsilon$ EARSM and Eddy Viscosity Transport turbulence models. The results of numerical modeling were then compared with the experimental data.
1. The depth-averaged velocity distribution predicted using three turbulence models are in good agreement with the experimental data. Comparing to the $\kappa-\varepsilon$ and $\kappa-\varepsilon$ EARSM models, the Eddy Viscosity Transport Equation turbulence model can predict the depth-averaged velocity reasonably well, especially in the main channel. Also by increasing the floodplain width, the divergence between numerical modeling and the experimental data decrease.

2. To compare with depth-averaged velocity, the shear stress distribution predicted by the $\kappa-\varepsilon$, $\kappa-\varepsilon$ EARSM and Eddy Viscosity Transport turbulence models are in less agreement with the experimental data.
3. By increasing flow discharge (water depth) and floodplains width, the accuracy of three turbulence models improves.
4. Among those three turbulence models, only the k-e EARSM are able to predict secondary flow cells in the main channel and floodplains. As a result of interaction between the secondary flow cells, the k-e EARSM turbulence model shows two local peaks in shear stress distributions near the main channel walls.

References

Beaman F., Large eddy simulation of open channel flows for conveyance estimation, PhD Thesis, Nottingham University, UK (2010).

Bonakdari H., Baghalian S., Nazari F., Fazli M., Numerical analysis and prediction of the velocity field in curved channel using artificial neural network and genetic algorithm, Engineering Applications of Computation Fluid Mechanics 5(2011) 384-396.

Bousmar D., Flow modeling in compound channels – Momentum transfer between main channel and prismatic or non-prismatic floodplains, PhD Thesis, Universite Catholique de Louvain, Belgium (2002).

Celik I.B., Ghia U., Roache P.J., Procedure for Estimation and Reporting of Uncertainty Due to Discretization in CFD Applications, Journal of Fluids Engineering 130 (2008) 1-4.

Cokljat D., Turbulence models for non-circular ducts and channels, PhD Thesis, City University London, UK (1993).

Okljat D., Younis B., Second order closure study of open-channel flow, Journal of Hydraulic Engineering 121 (1995) 773-788.

Ikeda S., Role of lateral eddies in sediment transport and channel formation, River Sedimentation, Jayawardena, Lee and Wang, eds., Balkema Rotterdam, (1999) 195-203.

Jing H., Guo Y., Li C., Zhang J., Three-dimensional numerical simulation of compound meandering open channel flow by Reynolds stress model, International Journal for Numerical Method in Fluid 59 (2009) 927-943.

Kang H., Choi S.U., 3D Numerical simulation of compound open-channel flow with vegetated floodplains by Reynolds stress model, KSCE Journal of Civil Engineering 9 (2005) 7-11.

Launder B.E., Spalding D.B., The numerical computation of turbulent flows, Computer Method and Application and Engineering 3 (1974) 269-289.

Menter F.R., Eddy viscosity transport equations and their relation to the k-ε model, Journal of Fluids Engineering 119 (1997) 876-884.

Morvan N., Pender G., Ervine D.A., Three-Dimensional hydrodynamics of meandering compound channels, Journal of Hydraulic Engineering 128 (2001) 674-682.

Myers W.R.C., Momentum transfer in a compound channel, Journal of Hydraulic Research, 16 (1978) 139-150.

Naot D., Nezu I., Nakagava H., Calculation of compound open channel flow, Journal of Hydraulic Engineering 119 (1993) 1418-1426.

Nezu I., Nakagawa H., Turbulence in open-channel flows, IAHR Monograph Series, A.A., Balkema, Rotterdam, Netherlands, (1993).

Pezzinga G., Velocity distribution in compound channel flows by numerical modeling, Journal of Hydraulic Engineering 120 (1994) 1176-1197.

Rezaei B., Overbank flow in compound channels with prismatic and non-prismatic floodplains, PhD Thesis, Birmingham University, UK (2006).

Rezaei B., Knight D.W., Overbank flow in compound channels with nonprismatic floodplains, Journal of Hydraulic Engineering 137 (2011) 815-824.

Sellin R.H.J., A laboratory investigation into the interaction between the flow in the channel of a river and that over its floodplain, La Houille Blanche 7 (1964) 793-802.

Shiono K., Knight D.W., Turbulent open-channel flows with variable depth across the channel, Journal of Fluid Mechanics, 222 (1991) 617-646.

Tominaga A., Nezu I., Turbulent structure in compound open-channel flows, Journal of Hydraulic Engineering 117 (1991) 21–41.

Wallin S., Engineering turbulence modeling for CFD with a focus on explicit algebraic Reynolds stress models, PhD Thesis, Norsted Tryckeri AB, Stockholm, Sweden (2000).

Wallin S., Johansson A., A complete explicit algebraic Reynolds stress model for incompressible and compressible flows, Journal of Fluid Mechanics 403 (2000) 89-132.

Wormleaton P.R., Floodplain secondary circulation as a mechanism for flow and shear stress redistribution in straight compound channels, Chap. 28, Coherent Flow Structures in Open Channels, Editors Ashworth, Bennett, Best, McClelland, J Wiley, (1996), 581-608.

Wright N.G., Crosseley A.J., Morvan H.P., Stoesser T., Detailed validation of CFD for flows in straight channels, River Flow 2004, Naples, Italy, (2004) 1041-1048.

27

The efficiency of genetic programming model in simulating rainfall-runoff process

Hamidreza Babaali[*,1], Zohreh Ramak[2], Reza Sepahvand[3]

[1]Department of Civil Engineering, Khorramabad Branch, Islamic Azad University, Khorramabad, Iran.
[2]Department of Civil Engineering, Science and Research of Branch, Islamic Azad University, Tehran, Iran.
[3]Faculty of Civil Engineering, Isfahan University of Technology, Isfahan, Iran.

ARTICLE INFO

Keywords:
Simulation
Rainfall-runoff
Heuristic algorithm
Genetic programming

ABSTRACT

Predicting the river discharge is one of the important subjects in water resources engineering. This subject is of utmost importance in terms of planning, management, and policy of water resources with the aim of economic and environmental development, especially in a country like Iran with limited water resources. Awareness of the relation between rainfall and runoff of basins is an inseparable past of water design studies. Lack of sufficient data on rainfall-runoff due to the absence of appropriate hydrometric stations reveals the importance of using indirect methods and heuristic algorithms for estimating the basins' runoff more than before. In the present research, the genetic programming model has been employed to simulate the rainfall-runoff process of Khorramabad River basin, and in order to introduce the patterns and identify the best pattern dominating the nature of flow, all statistical data were divided into two groups of training and experiment (52 percent training and 48 percent experiment) and the program was implemented for 1000 replications using fitting functions and going through replication and developmental processes so as to find the optimal replication. Moreover, in order to evaluate the relations obtained from the simulator model, Root Mean Square Error (RMSE) and Mean Squared Error (MSE) indexes and Coefficient of Determination (R^2) have been used. The investigations demonstrate that the employed equation 3 has the greatest relevance with the observational data. Therefore, it is recommended that the said equation be used for the rainfall-runoff studies of the abovementioned basin. Based on the results, the genetic programming model is an accurate direct method for predicting the discharge of Khorramabad River basin.

1. Introduction

Accurate prediction of streamflow is an essential ingredient for both water quantity and quality management. Generally, there are two possible approaches to predict streamflow. The first approach is the process modelling that involves the study of rainfall-runoff processes in order to model the underlying physical laws (Kuchment et al. 1996). The rainfall–runoff process can be influenced by many factors such as weather conditions, land-use and vegetation cover, infiltration, and evapotranspiration. Therefore, it is subject to many simplification assumptions or excessive data requirements about the physics of the catchment. The second approach to streamflow prediction is the pattern recognition methodology which attempts to recognize streamflow patterns based on their antecedent records. In this approach, thorough understanding of the physical laws is not required and the data requirements are not as extensive as for the process model (Nourani et al. 2011). The logic behind this approach is to find out relevant spatial and temporal features of historical streamflow records and to use these to predict the evolution of prospective flows. As inputs of the models in pattern recognition method are only time-lagged streamflow observations, this approach appears more useful for the catchments with no or sparse rain gauge stations (Besaw et al. 2010).

In recent years, artificial intelligence (AI) techniques such as artificial neural network (ANN) and genetic programming (GP) have been pronounced as a branch of computer science to model wide range of hydrological processes (Whigham and Crapper,2001; Dolling and Varas, 2002). Following this, comparative studies between different AI techniques have been appeared in the relevant literature and still attempting to find out the most appropriate one (Ghorbani et al. 2010; Nourani et al. 2011; Abrahart et al. 2012). Hence we initially developed a hybrid waveletartificial neural network (WANN) model as an optimized ANN technique for monthly streamflow prediction in a particular catchment in this investigation. Then we, as a first time, compared the results of WANN with those of linear genetic programming (LGP) technique. The pattern recognition methodology is adopted as our prediction approach in this study. ANN is an effective approach to manage large amounts of dynamic, non-linear and noisy data, especially when the underlying physical relationships are not necessary to fully understanding (Nourani et al. 2011). ANNs were widely used in various fields of hydrological predictions and successful results have also been reported in streamflow prediction (Abrahart et al. 2012; Besaw et al. 2010; Can et al. 2012; Dolling and Varas 2002; Kisi and Cigizoglu 2007; Nourani et al. 2011). In the last decade, GP has been pronounced as a new robust method to solve wide range of modelling problems in water resources engineering such as rainfall-runoff modelling (Dorado et al. 2003; Nourani et al. 2012; Whigham and Crapper. 2001), unit hydrograph determination (Rabunal et al. 2007), flood routing (Sivapragasam et al. 2008), and sea level forecasting (Ghorbani et al. 2010). It was observed that a few studies existed in the literatures related to the comparison of the performance of GP and ANN in time series modelling of streamflow. Guven (2009) applied LGP, a variant of GP, and two versions of neural networks for prediction of daily flow of Schuylkill River in the USA and showed that the performance of LGP was moderately better than that of ANN. Wang et al. (2009) developed and compared several AI techniques include ANN, neural-based fuzzy inference system (ANFIS), GP and support vector machine (SVM) for monthly flow forecasting using long-term observations in China. Their results indicated that the best performance can be obtained by ANFIS, GP and SVM, in terms of different evaluation criteria. Londhe and Charhate (2010) used ANN, GP and model trees (MT) to forecast river flow one day in advance at two stations in Narmada catchment of India. The results showed the ANNs and MT techniques performed almost equally well, but GP performed better than its counterparts. All aforementioned researches show that GP models result in higher accuracy than regular ANN based modelling approaches. The fact behind this is that the ANN models are not very satisfactory in terms of precision when a time series is highly non-stationary and hydrologic process being operated under a large range of time scales (Nourani et al. 2012). To improve the results of the ANN models, input and/or output data pre-processing by wavelet decomposition technique (hybrid wavelet-ANN models) are suggested and successful results have been reported (Kisi. 2008; Labat 2005; Nourani et al. 2009a, 2011). In recent years, this hybrid model was also compared with some other classic models such as multiple linear regressions and regular ANNs. The findings deduced that the hybrid WANN model can be considered as an effective tool in modelling complex hydrological processes (Anctil and Tape. 2004; Adamowski and Sun. 2010; Partal and Kisi. 2007; Rajaee et al. 2010). Using genetic programming, Soltani et al. (2009) modeled the daily rainfall-runoff of Lighvan basin. In that research, the initial modeling of rainfall-runoff was done with 15 variables to determine meaningful variables, and the final modeling was done with meaningful variables and two sets of mathematical operators. The obtained model was recommended as the model of Lighvan basin's rainfall-runoff because of its simplicity and higher accuracy.

Gholizadeh et al. (2016) investigated the function of genetic programming as an intelligent modern method for estimating Kasilian basin's rainfall-runoff. The results revealed that the aforementioned model had a proven ability in estimating rainfall-runoff of Kasilian basin using meteorological and hydraulic parameters. While introducing the genetic programming method as a direct method for predicting the river flow, Danandeh Mehr and Majdzadeh. (2010) used this method in order to investigate the effect of daily river discharge in predicting the daily discharge of Absardeh River basin located in Lorestan Province and compared the results with those obtained from the artificial neural network. The results indicated that genetic programming had proper efficiency and high accuracy in predicting the rivers flow in comparison with the artificial neural network. Farboodnam et al. (2009) predicted the daily discharge of Lighvan River using the genetic programming method and compared the results with the artificial neural network. According to the results of their research, the genetic programming method had much higher accuracy in predicting daily river discharge in comparison with the artificial neural network.

One of the most complex hydrologic processes is the process of changing rainfall to runoff that is affected by different physical and hydrological parameters. Modeling the rainfall-runoff model and predicting the river discharge is an important measure to take in managing and controlling floodwater, designing hydraulic structures in basins and managing drought (Salajghe et al. 2009). Since when (Sherman. 1932) suggested the concept of unit hydrograph, several linear and non-linear hydrological models have been widely used for simulating the rainfall-runoff process. These models have developed over time, quality of the recommended models has constantly improved as a result of the introduction of new devices and enhancement of human knowledge. Despite the long history of statistical hydraulic and hydrological models, experience shows that regardless of their strengths, such models have lots of weaknesses as well. The fact that most of the hydrometric stations are new, the defects in the statistics of most of these stations, the need to long historical information and different meteorological parameters, the need to various parameters such as river geometry, time-consuming deduction of physical models,

and on the other hand, the non-linear feature and inherent uncertainty of the rainfall-runoff process, have led the researchers to using artificial intelligence methods and heuristic algorithms. The methods inspired by nature such as genetic programming (GP), which are introduced as soft computing, are among those models that are used in complex accurate researches (Qolizadeh et al. 2016; Fazaeili. 2010; Harun et al. 2002). Zahiri and Azamathulla (2014) used two methods of linear genetic programming and M5 tree model in order to predict the flow rate in compound cross-sections. the results showed that although both models had high accuracy in predicting the flow rate, the accuracy of the linear genetic programming was greater than the M5 tree model. Using genetic programming and artificial neural networks, (Ghorbani et al. 2010) modeled the fluctuations of the surface of water on Cocos Island in Indian Ocean with 12-hour, 24-hour, 5-day, and 10-day data. According to the results, in most of the cases, the results obtained from genetic programming were more accurate than those obtained from artificial neural networks. Whigham and Crapper (2001) modeled the daily rainfall-runoff process in Teifi and Namoi basins using genetic programming and IHACRES. The results obtained from genetic programming had higher accuracy than those obtained from the model. In a research on Orgeval basin in France, (Khu et al., 2001) took advantage of genetic programing to predict the hourly runoff and compared the results obtained from observational values and those produced through classic methods. The research results indicated the acceptable accuracy of genetic programming.

Studying the rainfall-runoff relation in different times, Liong et al. (2001) came to the conclusion that using the genetic programming method in predicting the rainfall-runoff behavior in basins had fewer errors. Jayawardena et al. (2005) modeled the rainfall-runoff process of a small basin in Hong Kong and two almost big basins in China using genetic programming. The results obtained from the modeling with daily data for the basin with steep slope and small area were not appropriate, and to solve this problem, they suggested using data with shorter time intervals (15 to 30 minutes). The results of two other basins were in line with the real data.

To the best of our knowledge, there is no research examining the performance of WANN and LGP models in monthly streamflow prediction. Thus, in this study we initially developed WANN and LGP models to predict monthly streamflow at a particular river and then compared the prediction results with the observations. Following this, based upon two successive gauging stations records we put forward six black-box ANN structures as reference models for monthly streamflow prediction on Coruh River located in eastern Black Sea region (Turkey). Then we applied wavelet transform to our ANN-based reference models. The ANN component of the models can handle the nonlinearity and non-stationary elements, while the wavelet component can deal with seasonal (cyclic) non-stationary elements of the phenomenon. In the second step, we developed monthly streamflow prediction models based on explicit LGP technique. Ultimately, we discussed the both accuracy and applicability of ANN, WANN, and LGP techniques via the comparison of their performances. Such a comparison has also been done by Wang et al. (2009) among GP and ANN for monthly discharge forecasting but they did not consider wavelet transform in their neural networks. Aytec et al. (2008) made use of neural networks and genetic programming for modeling daily rainfall-runoff of Juniata River basin in Pennsylvania in the US and found out that genetic programming models the rainfall-runoff process more accurately than neural networks.

2. Materials and methods
2.1. The region under study

Khorramabad River basin with an area of 1640 square kilometers is located in the southwest of Iran in Lorestan Province with geographical location of 33 degrees and 26 minutes and 76 seconds of northern latitude to 48 degrees and 14 minutes and 77 seconds of eastern longitude. This basin makes up about 16 percent of the Kashkan River basin. The maximum height of the basin under study is 2876 meters and its minimum height is 1112 meters. Fig. 1 the location of the basin under study.

The required data for conducting the current research include daily rainometric and hydrometric data of between 1991 and 2013, the data related to the basin physiography as well as the data related to CN value or curve number. Cham Anjir Station has been used as the rainometric and hydrometric station in the present research. The fact that the data are more complete in such stations, these stations have been selected. The following table illustrates the features of employed rainometric and hydrometric stations.

2.2. Genetic Programming

Genetic programming, which was initially innovated by (Cramer, 1985) and then developed by (Koza. 1992), is a method of evolutionary algorithms that is practically developed from Genetic Algorithm (GA). The step-by-step process of genetic programming is as follows:

1- An initial population of compound functions indicating prediction patterns are considered randomly (creation of chromosomes). 2- Introducing the initial population (chromosomes) to the computer and evaluating each person (genes) of the said population using fitting functions (identification of the most effective individuals in the nature of phenomenon). 3- Selecting the effective genes in order to replicate, mutate, have sexual intercourse, and reproduce new individuals with modified traits (children). 4- Initiating the repetitive developmental process for children in each reproduction. The fourth step will be repeated in a specified number of times until the best response is obtained.

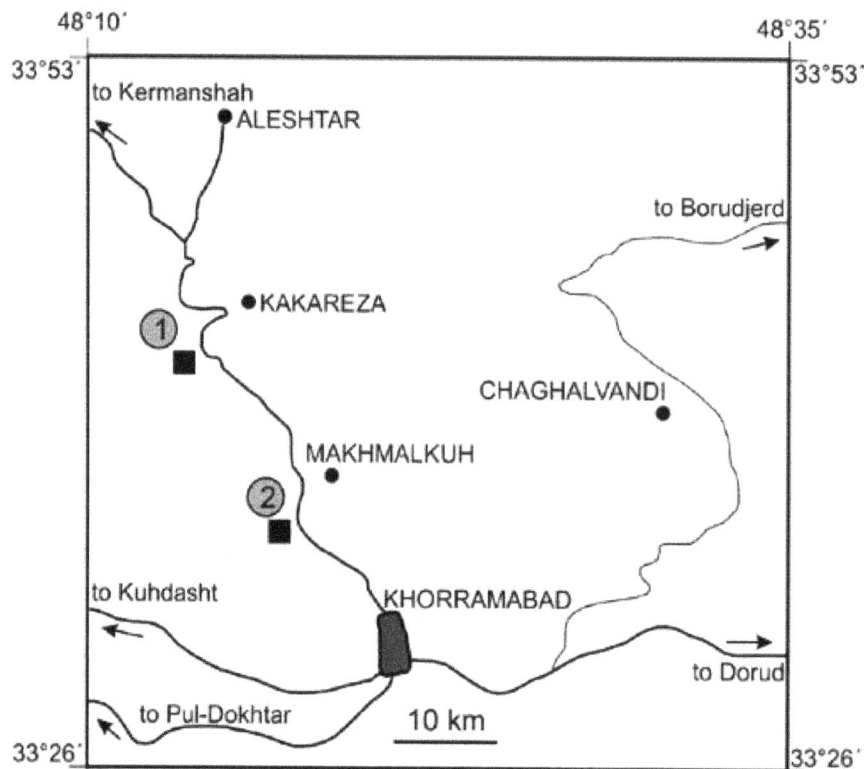

Fig. 1. Geographical location of the basin under study.

Table 1. The features of employed rainometric and hydrometric stations.

Station Name	Type of Station	Geographical Longitude	Geographical Latitude	Height
Cham Anjir	Hydrometric	48°-10'	33°-26'	1140
Cham Anjir	Rainometric	48°-35'	33°-25'	1166

Like GA, GP starts its work with the initial sets of random solutions called population. Each individual in the population is called a chromosome that represents the solution for the intended problem. Then, the chromosomes, which are illustrated as an Element Tree (ET), are examined by a fitting function so that the level of appropriateness of a solution to a particular problem can be determined. This population evolves with the help of genetic operators in many future generations. The purpose of this, exactly like genetic natural compatibility, is to create populations or generations which are more adaptable to the environment than the previous ones (Liong. 2002). Fig. 2 demonstrates the structure of the genetic programing algorithm. In the present research, the GP-LAB software has been employed to use genetic programing, which will be explained in the following.

GP-LAB software is a simple understandable application in the MATLAB programming language that is adjusted as a set of programming codes and functions. One of the most important advantages of this method is the possibility to change and adjust the application for different status or to develop its codes as desired. This software application, like other black box applications, make use of reverse engineering viewpoint as a strategy to reach the proper

solution. In addition, this software is capable of using any type of functions, phrases and parameters in the solution process and system.

GP-LAB has a parametric structure that is shown as a tree and is made up of three main patterns including Set Vare, Gen Pop, and generation. Each pattern has a substructure set that does a particular task depending on the type of problem. Furthermore, each of the patterns can use one to several parameters and functions depending on the conditions.

3. Modeling the flow using the genetic programing method

In this section of the research, the genetic programming method has been employed in order to make and produce a model that can predict and estimate Khorramabad River discharge based on the rate of precipitation. In this method, the following parameters have been used as the inputs of the simulated model so as to make close-to-reality predictions.

- Precipitation on day t (P_t)
- Precipitation on day t-1 (P_{t-1})
- Precipitation on day t-2 (P_{t-2})

- Discharge on day t-1 (Q_{t-1})

In the current research, we have tried to use the abovementioned parameters to present an equation to simulate the river discharge with the highest relevance with the observational data in each prediction time step. Table 3 represents the initial settings of the simulator model to this aim.

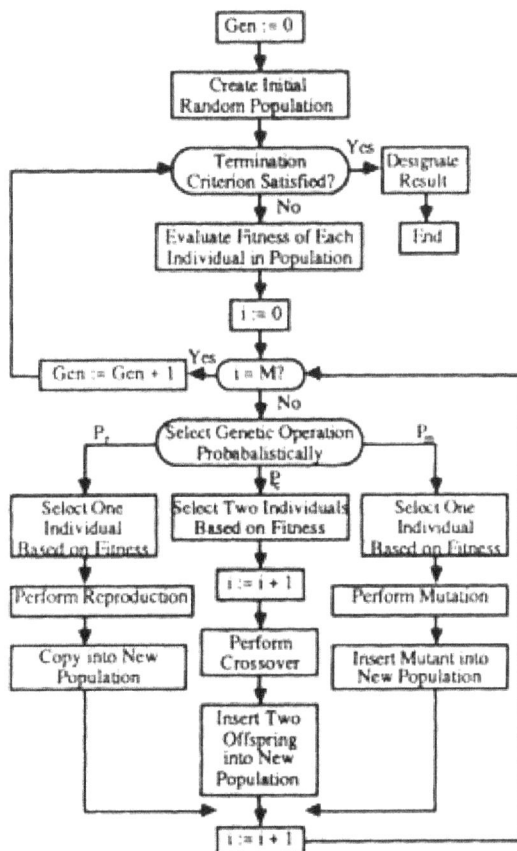

Fig. 2. The cycle and structure of genetic programming (Koza. 1992).

Table 2. Statistical characteristics of the data used by cham anjir hydrometric station.

Statistical Characteristic	Discharge (m³/s)
Number of Data	3652
Maximum Value	186.35
Minimum Value	0
Mean	11.03
Variance	141.41
Standard Deviation	11.89

Table 3. Initial settings employed in the simulator model of genetic programming.

Number of Trees of each Generation	Depth of each Tree	Number of Replications	Possibility of Mutation	Possibility of Coupling
250	3	450	0.5	0.6

In order to introduce the patterns and identify the best pattern dominating the nature of flow in the present research, we firstly used fitting functions and conducted replication and developmental processes to find the number of optimal replication and after dividing all statistical data into two sets of training and experiment (52% training and 48% experiment), we implemented the program for 1000 replications and observed that after 450 replications, RMSE value became almost constant and the tilt-shift turned to zero. This number of replications is the optimal number for replications of the simulator model. Fig. 3 illustrates the results obtained from the optimal number of replications to be used in the simulator model of genetic programming.

After specifying the initial settings of the simulator model as well as determining the number of optimal replications to be used in the genetic programming model, we began to determine the optimal equations obtained from the simulator model. the optimal equations obtained from the simulator model have been presented in Table 4.

Table 4. Three optimal equations obtained in the simulator model of genetic programming.

Equations Obtained from the Simulator Model
1) 0.36 (Q_{t-1}) Plus 0.366(P_t) Plus COS(P_{t-1}) Plus(0.8)
2) Tan(Q_{t-1}) Plus (P_t) Plus (SQRT(SQRT(P_{t-2}) Plus (P_t))) Plus SQRT(Sin(P_{t-1}))
3) Plus(Plus(Plus(Plus(Plus(P_t,cos(Mysqrt(Plus(P_t,Minus(Mysqrt(Q_{t-1},Q_{t-1})))))))

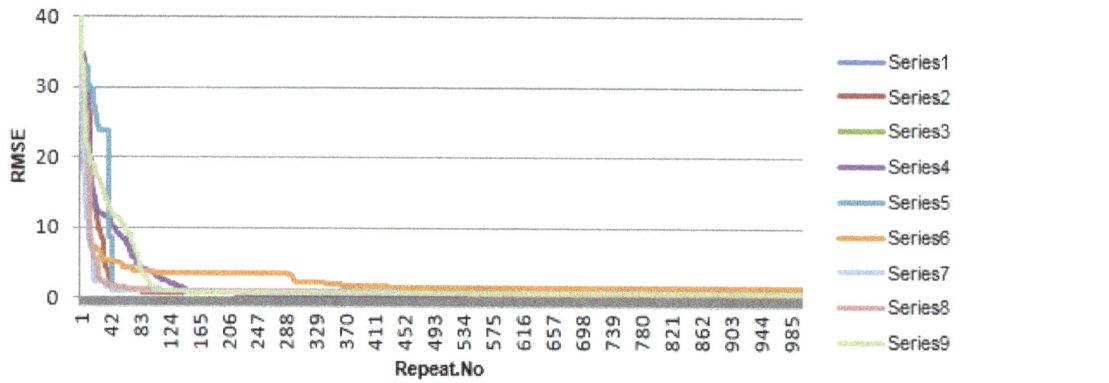

Fig. 3. Determining the number of optimal replication to be used in the simulator model of genetic programming.

In order to examine the aforementioned equations, Root Mean Square Error (RMSE) and Mean Squared Error (MSE) indexes and Coefficient of Determination (R^2) have been used, which can be calculated through equations 1, 2, and 3, respectively.

$$R^2 = \frac{(\sum_{i=1}^{n}(X_i - \bar{X})(Y_i - \bar{Y}))^2}{(\sum_{i=1}^{n}(X_i - \bar{X})^2(Y_i - \bar{Y})^2)} \tag{1}$$

$$RMSE = \sqrt{\frac{\sum_{i=1}^{n}(X_i - Y_i)^2}{n}} \tag{2}$$

$$MAE = \frac{\sum_{i=1}^{n}|X_i - Y_i|}{n} \tag{3}$$

Where X_i and Y_i are the ith real and estimated datum, $\bar{X}$ is the mean of the real data, $\bar{Y}$ is the mean of the estimated data, and n is the number of evaluation samples. Low values of RMSE and MSE and the high value of R^2 can show the accuracy of each pattern in comparison with other competitive patterns. The results of the mentioned equations have been given in Table 5.

Table 5. Results of evaluating the obtained optimal equations from the simulator model of genetic programming.

Equation	Training			Experiment		
	RMSE	MSE	R^2	RMSE	MSE	R^2
1	0.75	0.65	0.92	1.01	1.33	0.89
2	0.77	0.5	0.94	0.98	1.2	0.92
3	0.65	0.433	0.97	0.9	1.02	0.96

According to Tables 4 and 5, using equation 3 in Table 4 had the highest relevance with the observational data. Therefore, simulating the discharge corresponding to different precipitations has been done using this equation. Fig. 4 shows the simulated discharge using this equation through genetic programming in comparison with the real discharge. As seen in that Fig., high consistency is evident between the real data and the simulated ones

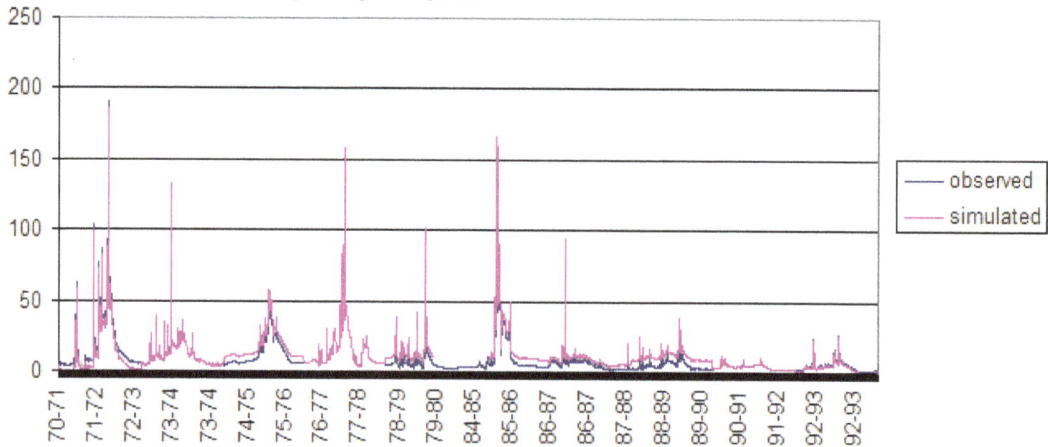

Fig. 4. Results of the simulator model of genetic programming.

4. Conclusions

In the current research, the genetic programming model was used to simulate the rainfall-runoff process. The statistical indexes related to using equations 1 to 3 in table 5 indicate that equation 3 with the lowest RMSE and MSE and the highest R^2 have the best efficiency and consistency in simulating this process. According to the results of this study, the genetic programming model is generally a direct accurate way for predicting the river discharge in Khorramabad River basin. In addition, the results of this research are in line with the results of (Guven. 2009) on predicting Schuylkill River discharge as well as the research by (Danandeh Mehr and Majdzadeh Tabatabaei. 2010) on predicting Absardeh River discharge using genetic programing. Consequently, genetic programing can be considered among the most accurate available methods in predicting river discharge.

Acknowledgement

The authors would like to thank Islamic Azad University for financial support of this study, Khorramabad, Iran.

References

Abrahart R.J., Anctil F., Coulibaly P., Dawson C.W., Mount N.J., See L.M., Shamseldin A.Y., Solomatine D.P., Toth E., Wilby R.L., Two decades of anarchy emerging themes and outstanding challenges for neural network modelling of surface hydrology, Progress in Physical Geography 36 (2012) 480–513.

Anctil F., Tape D., An exploration of artificial neural network rainfall-runoff forecasting combined with wavelet decomposition, Journal of Environmental Engineering and Science 3 (2004) 121–128.

Aytek A., and Asce M., An application of artificial intelligence for rainfall runoff modeling, Journal of Earth System Science 117 (2008) 145-155.

Aytek A.. and Kisi O.A., Genetic programming approach to suspended sediment modelling, Journal of Hydrology 351 (2007) 288-298.

Besaw L.E., Rizzo D.M., Bierman P.R., Hackett W.R., Advances in ungauged streamflow prediction using artificial neural networks, Journal of Hydrology 386 (2010) 27–37.

Can I., Tosunogulu F., Kahya E., Daily streamflow modelling using autoregressive moving average and artificial neural networks models: case study of Coruh basin, Turkey, Water and Environment Journa 26 (2012) 567–576.

Danandeh Mehr A., and Majdzadeh Tabatabaei M., Investigating the effect of daily discharge on predicting the rivers' flow using genetic programing, Water and Soil Magazine (Science and Agricultural Industries) 24 (2010) 325-333.

Dolling O.R., Varas E.A., Artificial neural networks for streamflow prediction, Journal of Hydraulic Research 40 (2002) 547–554.

Dorado J., Rabunal J.R., Pazos A., Rivero D., Santos A., Puertas J., Prediction and modeling of the rainfall–runoff transformation of a typical urban basin using ANN and GP, Engineering Applications of Artificial Intelligence 17 (2003) 329–343.

Farboodnam N., Ghorbani M.A., and A'alami M., Predicting river flow using genetic programing (CaseStudy: Lighvan River), Water and Soil Science Journal, University of Tbariz 19 (2009) 107-123.

Fazaeili H., Using the genetic programing method in modeling rainfall-runoff, ms thesis, Water Engineering Department, University of Tabriz (2010).

Ghorbani M., Khatibi R., Aytek A., Makarynskyy O., Sea water level forecasting using genetic programming and artificial neural networks, Comput. Geosci. UK 36 (2010) 620–627.

Ghorbani M.A., Kisi O., Aalinezhad M., A probe into the chaotic nature of daily streamflow time series by correlation dimension and largest lyapunov methods, Applied Mathematical Modelling 34 (2010) 4050–4057.

Guven A., Linear genetic programming for time-series modelling of daily flow rate, Journal of Earth System Science 118 (2009) 137–146.

Harun S., Ahmat Nor N.I., and Kassim A.H.M., Artificial neural network model for rainfall-runoff relationship, Journal Technology 37 (2002) 1–12.

Huo Z., Feng S., Kang S., Huang G., Wang F., and Guo P., Integrated neural networks for monthly river flow estimation in arid inland basin of northwest china, Journal of Hydrology 420-421 (2012) 159-170.

Imam Qolizadeh S., Karimi Damaneh R., and Mehdi Panah H., Estimating the runoff of kasilian basin using the gene expression programing method, natural resources, Environment, and Agriculture Studies, Vol. 3, 2nd year- No. 4 and 5 (11 and 12), (2016) 1-7.

Jayawardena A.W., Muttil N., and Fernando T.M.K.G., Rainfall-runoff modeling using genetic programming, International Congress on Modeling and Simulation Society of Australia and New Zealand (2005) 1841-1847.

Khu S.T., Liong S.Y., Babovic V., Madsen H., Muttil N., Genetic programming and its application in real-time runoff forecasting, Journal of the American Water Resources Association 37 (2001) 439-451.

Kisi O., and Cigizoglu H.K., Comparison of different ann techniques in river flow prediction, Civil Engineering and Environmental Systems 24 (2007) 211–231.

Kisi O., Stream flow forecasting using neuro-wavelet technique, Hydrological Processes 22 (2008) 4142–4152.

Koza J.R., Genetic programming: on the programming of computers by means of natural selection, Cambridge, MA: MIT Press, (1992) 35-80.

Kuchment L.S., Demidov V.N., Naden P.S., Cooper D.M., Broadhurst P., Rainfall-runoff modelling of the ouse basin, north yorkshire: an application of a physically based distributed model, Journal of Hydrology 181 (1996) 323–342.

Labat D., Recent advances in wavelet analyses: Part 1 – a review of concepts, Journal of Hydrology (Amsterdam) 314 (2005) 275–288.

Liong S.Y., Gautam T.R., Khu S.T., Babovic V., and Muttil N., Genetic programming: a new paradigm in rainfall- runoff modelling, Journal of American Water Resources Association 38 (2001) 705-718.

Londhe S., Charhate S., Comparison of data-driven modelling techniques for river flow forecasting, Hydrological Sciences Journal 55 (2010) 1163–1174.

Lopes HS., and Weinert WR., EGIPSYS: An enhanced gene expression programming approach for symbolic regression problems, International Journal of Applied Mathematics and Computer Science 14 (2004) 375-384.

Mallat S., A wavelet tour of signal processing, second ed. Academic Press, San Diego (CA), (1998).

Mohammad Salehi P., Raeini Sarjaz M., and Zia Tabar Ahmadi M., Simulating rainfall-runoff with the mathematical model based on geographical information system for emameh basin, Journal of Agricultural Sciences and Natural Resources, Natural Resources Special 15 (2007) 1-11.

Nourani V., Alami M.T., Aminfar M.H., A combined neural-wavelet model for prediction of Ligvanchai watershed precipitation, Engineering Applications of Artificial Intelligence 22 (2009a) 466–472.

Nourani V., Kisi O., Komasi M., Two hybrid artificial intelligence approaches for modeling rainfall–runoff process, Journal of Hydrology 402 (2011) 41–59.

Nourani V., Komasi M., Alami M.T., Hybrid wavelet–genetic programming approach to optimize ANN modelling of rainfall–runoff process, Journal of Hydrologic Engineering 17 (2012) 724–741.

Omidi R., Comparing the prediction of khorramabad river discharge using time series and the artificial neural network, hydrology and water resources department, ms thesis, Faculty of Water Sciences, Shahid Chamran Univeristy, Ahwaz (2014).

Ozger M., Significant wave height forecasting using wavelet fuzzy logic approach, Ocean Engineering 37 (2010) 1443–1451.

Rabunal J.R., Puertas J., Suarez J., Rivero D., Determination of the unit hydrograph of a typical urban basin using genetic programming and artificial neural networks, Hydrological Processes 21 (2007) 476–485.

Rajaee T., Mirbagheri S.A., Nourani V., Alikhani A., Prediction of daily suspended sediment load using wavelet and neuro-fuzzy combined model, International Journal of Environmental Science and Technology 7 (2010) 93–110.

Rathinasamy M.,Khosa R., Multi-scale nonlinearmodel formonthly streamflow forecasting: a wavelet-based approach, Journal of Hydroinformatics 14 (2012) 424–442.

Salajghe A., Fathabadi A., and Mahdavi M., Investigating the efficiency of neural-fuzzy methods and statistical models in simulating the rainfall-runoff process, Iran's Natural Resources Journal, Grassland and Watershed Management Journal 1 (2009) 65-80.

Sivapragasam C., Maheswaran R., Veena V., Genetic programming approach for flood routing in natural channels, Hydrological Processes 22 (2008) 623–628.

Soltani A., Modeling rainfall-runoff using genetic programing and stochastic differential equations (sde) in lighvan basin, MS Thesis in Water Resources Filed of Study, Water Engineering Department, University of Tabriz (2010).

Wang W.C., Chau K.W., Cheng C.T., Qiu L., A comparison of performance of several artificial intelligence methods for forecasting monthly discharge time series, Journal of Hydrology 374 (2009) 294–306.

Whigham P.A., and Crapper P.F., Modelling Rainfall-runoff using genetic programming, CSIRO Land and Water, P.O. Box 1666, Canberra, A.C.T. 2601, Australia, Mathematical and Computer Modellhrg 33 (2001) 707-721.

Zahiri A., and Azamathulla H., Comparison between linear genetic programming and M5 tree models to predict flow discharge in compound channels, Neural Computing and Applications 24 (2014) 413-420.

28

Reverse osmosis design with IMS design software to produce drinking water in Bandar Abbas, Iran

Nafiseh Aghababaei

Department of Chemical Engineering, Tafresh University, Tafresh, Iran.

ARTICLE INFO	ABSTRACT

Keywords:
Desalination
Process Design
Reverse osmosis
Salt rejection
Water treatment

Reverse osmosis (RO) has proven to be an efficient technique for desalination of seawater, brackish water, and reclaimed wastewater. However, the performance of RO desalination is sensitive to its design parameters and operating conditions. The purpose of this study was to model the removal of total dissolved solids (TDS) and Rejection of different ions are reported, from water to the city of Bandar Abbas. The main purpose of this research was modeling and simulation of desalination of blended water by RO. In this study, a design method based on a simulation technique has been developed for optimizing RO desalination systems. The design is made with the use of Hydranautics design software version 2011. In this paper main focus is on the design part with software. The desalinated water obtained from reverse osmosis at a pressure of 1.2 MPa showed rejections of approximately 88.49 % for SO_4^{2-}, 61.42 % for TDS, 70.34 for Cl^- and 50.85 for Na^+. It shows that software gives accurate design with least possible error and user friendly so the world while accepted. Blended water was proposed to optimize the produce drinking water with a recovery rate of 95 %. Reverse osmosis is an excellent alternative for the supply of water in Bandar Abbas.

1. Introduction

Availability of fresh potable water is a fundamental need for most aspects of life and key element for all societies. Safe drinking water, excellent sanitation and hygiene are important for good health, human survival and development (WHO. 2006), However, about 1 billion people around the world have no access to safe, clean drinking water sources; 2.5 billion people have no access to improved sanitation facilities (UNICEF and WHO. 2012). The most widely used methods for desalination include thermal and membrane processes. Membrane based desalination processes, such as Reverse osmosis (RO) and Electro dialysis (ED) are barrier technologies for producing potable water from brackish water (Kim et al. 2009). Techniques used for water desalination depend on specific requirements (TDS of feed water), in many cases more research is needed to conclude on the right technique to be applied, on the process parameters. Membrane based water desalination techniques have become reliable for providing potable drinking water (Strathman 1992). Membrane based desalination processes, such as RO and ED are barrier technologies for producing drinkable water (TDS: b500 ppm) from brackish water (Malaeb and Ayoub 2011).

Reverse Osmosis (RO) is a method which is used around the globe to produce fresh water. This technology is typically used to desalinate seawater or brackish water. Reverse osmosis (RO) is a membrane process technique, which is more popular compared to the conventional thermal process technology. One of the biggest advantages of the RO system is its low energy consumption compared to all other desalination systems (Fritzmann et al. 2007; Li 2011; Voros et al. 1998). Over the last decades, the membrane technology has experienced significant advancement that reduces the cost of filtration and enhances the quality of drinking water. As a consequence, this technology can be considered as the lowest cost technology for water desalination (Carter 2015) in comparison to others existing technologies, such as thermal desalination (multistage flash desalination, MSF; multi-effect distillation, MED) (Moonkhum et al. 2010). As a result, a number of researchers in the past decades developed several mathematical models of a RO desalination process in order to explain the separation technique and to carry out model based optimization to enhance the efficiency of the production process (Kamal et al. 2012, 2013). The global applications of RO technology for seawater desalination are projected to grow from a capacity of 40 to 100 million m^3 per day from 2008 to 2015 (Schiermeier 2008).

The fresh water production rate is dependent on the salt concentration in the feed water and on membrane properties such as the selectivity and the permeability. Several investigators studied the influence of spacers on the membrane performance for desalination processes. Karode and Kumar (2001) conducted three-dimensional flow simulations and experiments to examine flow characteristics in a feed channel containing different types of commercial spacers. Since the success of an RO system largely depends on the system design and operation condition, the availability of reliable RO models is essential for efficient design and operation (Abbas and Al-Bastaki 2005). Although the membrane makers have developed computer models to help possible customers to design an RO plant, they mainly focus on the performance analysis of some RO modules rather than the optimization of RO process in terms of energy consumption and product water quality. In optimum case initial reduction of TDS below 2000 ppm and further desalination by RO for best water recovery was considered.

Two fouling mechanisms are generally observed for membrane processes: surface fouling and fouling in pores. However, RO membranes do not have distinguishable pores; these are considered to be essentially non-porous. Thus, the main fouling mechanism for RO membranes is surface fouling. Surface fouling can occur from a variety of contaminants, including suspended particulate matter (inorganic or organic), dissolved organic matter, dissolved solids, and biogenic material (Amiri and Samiei 2007).

In addition, fouling can develop unevenly through a membrane module or element and can occur between the membrane sheets of a module, where spacers are located to create space for the concentrate stream (Tran et al. 2007). Dindarlu, Alipoor and Farshidfar studied on Chemical quality of drinking water in Bandar Abbas (Dindarlu et al. 2006). The present study was simulation model of RO systems for improved quality of drinking water in city of Bandar Abbas (Aghababaei).

This study demonstrates how the application of a simple experimental plan combined with statistical analysis can be used to

*Corresponding author E-mail: aghababaei@tafreshu.ac.ir

define the operating envelope of IMS membrane unit operations for minimizing fouling at large pilot scale.

2. Materials and methods

Groundwater and surface water are considered to be the sole natural waters resource in the study area. The main purpose of this research was modeling and simulation of desalination of blended water by RO in Bandar Abbas. Reverse Osmosis System Analysis (IMS design) (Hydranautics 2011), which is a sophisticated reverse osmosis (RO) design program that predict performances based on types of membranes. IMS design is a comprehensive software design program that allows the user to design a membrane system using Hydranautics membranes (Hydranautics 2011). It tracks RO system performance and specifically designed to be a user-friendly interface for RO system operators. To assure the highest standards of data integrity, normalization program is in compliance with ASTM Standard D 4516-85, Standard Practice for standardizing reverse osmosis performance data (Mehta and Patel 2012). The lab scale RO with a small pretreatment of RO downy membrane filter was installed, the dematerialized water was allowed to pass through it (Michael 2003). Drinking water resources of Bandar Abbas include two surface and groundwater. A pilot-scale cross-flow RO filtration system was used in this investigation. The pilot system comprises four in parts. The flow in a RO membrane system for desalination shows (Fig. 1). The feed solution was delivered from the feed reservoir to the first stage by a pump; the concentrate of the first stage was transferred to the second and third stage followed by the fourth stage. Four ESPA1 (Hydronautics, Oceanside, CA, USA) (2011) spiral wound elements were used. ESPA1 reverse osmosis elements offer the highest productivity while maintaining excellent salt rejection (99.4% (99.2% minimum)). It has the highest flow rates available to meet the water demands of desalinator. Application Data in ESPA1 was maximum applied pressure 600 psig, maximum operating temperature 113 °F (45 °C) and maximum feed water SDI (15 min) 5. Table 1 shown characterization of applied RO in this study.

Membrane rejection was calculated by relating the concentration of each compound in permeate and in feed water as where reported as a dimensionless number (Favaretto 2014).

$$\text{Salt Rejection} = [1 - \frac{C_p}{C_a}] \tag{1}$$

Flux (J) is volume of permeate (V) collected per unit membrane area (A) per unit time (t); flux is related directly to driving force and total resistance offered by the membrane as well as the interfacial region adjacent to it, which is calculated as:

$$J = \frac{V}{At} \tag{2}$$

Table 1. RO system components configuration.

Parameter	Value
HP pump flow	12505.3 gpm (Hydranautics 2011)
Feed pressure	0.98 MPa
Fouling factor	0.93
Permeate flow	11880 gpm
Raw water flow	12505.3 gpm (Hydranautics 2011)
Average flux rate	15.9 gfd

In order to evaluate the performance of reverse osmosis systems, raw and treated water experiments results were analyzed by descriptive statistics and the results in terms of mean values, standard deviation and removal efficiency were reported. The average life of RO membranes is one year.

The ability of a given membrane to withstand extended exposure to mono chloramine, other oxidants or any other non-standard chemical should be thoroughly investigated to ensure long-term membrane viability. Six to twelve months or more of pilot operation may be required for any integrity losses to become evident. The membrane warranty should reflect anticipated pretreatment methods, including addition of oxidants.

3. Results and discussion

Main aim was to design the plant for treatment for water. After application of reverse osmosis, the water was coming in the range of reusable quality. So for RO design was suggested using Hydranautics design software (Hydranautics 2011). The results of the modeling parameters measured in city of Bandar Abbas are shown in Table 2. Testing pilot reclamation plants are based on the Integrated Membrane System (IMS) (Hydranautics 2011) approach and RO membranes. Management models provide an effective means of rapidly testing and evaluating different scenarios for a given set of conditions (Bick and Oron 2000). Good resistance to fouling and scaling and high recovery operation are important in most brackish water desalination, industrial water treatment and water reuse applications. During the design and operation of RO processes, the recovery ratio is one of the consequential factors affecting the effectiveness of desalination. RO systems provide new or enhanced means for addressing these challenges. Cross flow supplied by a circulation pump washes the membranes and reduces the effects of scaling and fouling. As the salinity throughout the RO process cycles from just above the feed water salinity to that of the most concentrated brine, biofilm formation and scale precipitation can be disrupted. Greater than 99% recovery is possible by control panel setting using the same equipment, cutting brine waste production in half or more.

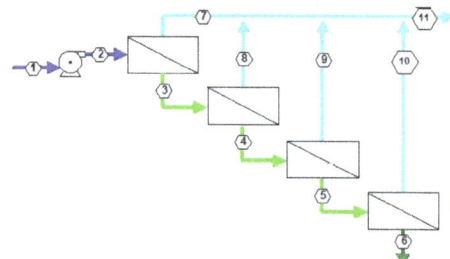

Fig.1. Schematic diagram of the RO system.

Over 95 % recovery operation has been demonstrated in the field. Studies have shown that recovery rates that produce high degrees of super-saturation of sparingly-soluble salts can be achieved in batch RO processes without the use of scale inhibitors. Specifically, recovery rates of over 95 % have been achieved and maintained from water sources with high concentrations of silica and calcium sulfate. Furthermore, scale depositions can be dissolved by batch cycling, making sustained run times at high recovery rates possible even with source waters with high levels of sparingly soluble salts (Sarkar et al. 2008). where C_p is concentration of a particular component of permeate and C_a is retentate concentration. RO membranes achieve NaCl rejections of 98–99.8 % (Bartels et al. 2005). Membrane manufacturers offer high salt rejection membranes for RO plants; the membranes do not retain the initial salt rejection throughout the membrane's lifetime (up to seven years with effective pretreatment). Normal membrane aging causes the salt passage (salt passage % = 100- R_s) to increase approximately 10 % per year (Wilf and Klinko 2001), and other factors, such as temperature, salinity, target recovery, and cleaning methods, can also affect salt passage. The desalinated water obtained from reverse osmosis at a pressure of 1.2 MPa showed rejections 88.49% for SO_4^{2-}, 61.42 % for TDS, 70.34 for Cl^- and 50.85 for Na^+.

Pressure was adjusted to a value greater than osmotic pressure by means of the restricting needle valve. The features of the resulting process configuration will be high yields, elevated removal of contaminants, and reliable operation. The higher average fluxes and pressure vessel configurations employed in RO can save substantial capital costs. For the case study systems, the conventional RO system would cost up to 16 % more than the high-flux RO system. Desalination is a general term for the process of removing salt from water to produce fresh water. Fresh water is defined as containing less than 1000 mg/L of salts or total dissolved solids (TDS) (Sandia 2003). Above 1000 mg/L, properties such as taste, color, corrosion propensity, and odor can be adversely affected. Many countries have adopted national drinking water standards for specific contaminants, as well as for total dissolved solids (TDS), but the standard limits vary from country to country or from region to region within the same country. In Table 2, for TDS, WHO (2011) has a drinking water taste threshold 250 mg TDS/L and the ISIRI (1978) has secondary standards of 200 mg chloride/L and 500 mg TDS/L. As shown in the Table 2, the most problems in terms of Physico-Chemical quality of drinking water were an increase in the rate of Chloride, sulfate, Sodium and TDS, above the maximum level recommended. However, Table 2 reveals that SO_4^{2-}, Cl^-, Na^+ and TDS does not comply with WHO guidelines and ISIRI for potability. No health-based guideline value has been developed for TDS and sulfate concentrations, but both chemical parameters can have effects on the acceptability of drinking water by consumers. The presence of sulfate in drinking water can cause a strong taste, and very high levels can have a laxative effect on unaccustomed consumers (WHO 2011). Water with high dissolved solids is of inferior palatability, an d highly mineralized water has restricted industrial applications (APHA 2005). A TDS level of less than about 500 mg·L^{-1} is generally considered to be good; drinking water becomes increasingly unpalatable at TDS levels greater than approximately 1000 mg·L^{-1} (WHO 2011). All analytical methods followed the World Health Organization (WHO 2011). The major scaling salts (Ca^{2+}, CO_3^{2-}, Mg^{2+}) showed reductions higher. Hydrated sulfate ions are very large molecules, as such, they do not exhibit preferential desalination like the other divalent ions. Another limiting parameter of an RO system design with a shorter combined element length is the concentration polarization factor (CPF). The CPF expresses an excess of dissolved ion concentration at the membrane surface. Because the concentration polarization phenomenon is inevitable in the RO process sparingly soluble salts ($CaSO_4$, $CaCO_3$, etc.) may concentrate within the membrane element beyond their solubility limit, and precipitate on the membrane leading to higher operating pressure.

Table 2. Quality of the Blended water (Feed) treated in the integrated membrane system 2011 (mg/L) (Hydranautics 2011).

Ion	Raw water mg/l	Feed water mg/l	Permeate mg/l	Concentrae mg/l	WHO recommendation	Iranian standard
Ca^{+2}	70.95	70.95	8.505	1256.40	-	75
Mg^{+2}	26.94	26.94	3.227	476.7	-	50
Na^+	270.30	270.30	132.832	3134.2	200	250
CO_3^{-2}	0.90	0.90	0.031	17.400	-	-
HCO_3^-	239.5	239.5	191.777	1146.2	-	-
SO_4^{-2}	261.13	261.13	30.048	4651.1	250	200
Cl^-	304.19	304.19	90.223	4367.800	250	200
F^-	1.03	1.03	0.673	7.900	1.5	1.5
NO_3^-	1.21	1.21	0.2	1.255	50	10
TDS	1188.50	1188.50	458.5	15057.80	1000	500
pH	7.66	7.66	7.56	8.24	$6.5 \leq pH \leq 8.5$	$6.5 \leq pH \leq 8.5$

Table 3 shows design of four-stage RO system with hydranautics version 2011 (Hydranautics 2011) in Fig. 1, the design and the practical operational conditions of the process units. This information is crucial for the conception, design, scale up and optimization of the RO unit. To explain these higher values, correlations between input and output are sought, it may be attributed to a statistical fluctuation. Removing either the calcium or the sulfate from the water supply is one way to prevent calcium sulfate scaling in the process.

Table 3. Design of four stage RO system with hydranautics version 2011 (Hydranautics 2011).

Number	Flow Gpm	Pressure Psi	TDS Ppm
1	2505.3	0	1188.5
2	12505	177.7	1188.5
3	2087.1	155.1	6936.0
4	765.2	147.2	1576.9
5	656.1	138.1	15746.2
6	625.3	124.6	15057.8
7	10418	0.0	37.1
8	1321.8	0.0	1826.5
9	109.1	0.0	15856.6
10	30.9	0.0	29695.9
11	11880	0.0	458.5

Knowing the membrane permeability, the local permeate flux can be calculated based on the local pressure. Subsequently, the overall permeate flux can also be calculated. In fact, the simulated permeate flux only deviated slightly from the observed value at the applied pressure of 1MPa (Fig. 2).

Flux rate is important in a RO system because it governs the cross flow rate and mass transport rate through the membrane. There is an optimum ratio of cross flow to permeate flow to keep the membrane surface clean, and the energy consumption at a minimum. Furthermore, if the cross flow becomes too low, the concentration polarization at the membrane surface becomes high. This causes the apparent ion and contaminant concentration at the surface of the membrane to be much higher than the incoming feed. Beta value is the measure that is used to determine the degree of concentration polarization. A beta value of 1.2, or 20 % higher concentration at the membrane surface, is considered to be the maximum beta value that is typically allowed.

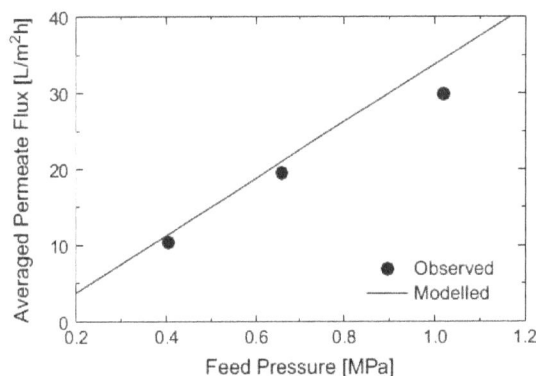

Fig. 2. Observed and modeled overall permeate flux as a function of the feed pressure at the RO system.

4. Conclusions

This study gave a brief description of the water supply in Bandar Abbas, showing that the city could implement desalination to meet drinking water. The groundwater and surface are a potential source for water harvesting, but the total dissolved solids, sulfate, chloride and sodium concentrations exceed the limits for drinking water without additional treatment. Reverse osmosis has been shown to be an efficient method for improving the quality of water from the Blended water with the quality required for drinking water. Performances of RO systems were separately evaluated for water desalination. The RO produced permeate with TDS of 458.5 mg/L at the time of sampling.

According to the results, the removal efficiencies of TDS, Sodium, Chloride and Sulfate in RO were 61.42 %, 50.85 %, 70.34 %, 88.49%. The pump efficiency is 83 %. The results indicated that the pretreatment through RO proved to be effective and reliable by removing both suspended solids and turbidity. Membrane lifetime and permeate flux, however, are primarily affected by the phenomena of concentration polarization and fouling at the membrane surface. For efficient and dependable water supply to areas like Bandar Abbas, at which potable water is scarce, RO desalination plants offer the practical option. Some variations in the quality of the feed water were taken into account in the IMS design, but generally it has had many unforeseen problems in operation and maintenance. It has been observed that this software is user-friendly and gives accurate design with possible micro details. It can save time for lab experiments. So for design of RO system in the field. The results showed the quality of feed water; and pretreatment plays an extremely important role in performance and operational problems such as fouling of RO systems.

Table 4. Projected individual element performance for RO System.

Stage	Perm. Flow gpm	Flow/Vessel Feed gpm	Conc gpm	Flux gfd	Beta	Conc.&Throt. Pressures psi	psi	Element Type	Elem. No.	Array
1-1	11590.4	52.8	3.9	29.3	1.06	135.7	0.0	ESPA1	1422	237x6
1-2	212.3	7.9	6.1	1.1	1.31	130.3	0.0	ESPA1	696	116x6
1-3	69.0	12.1	10.9	0.7	1.10	122.4	0.0	ESPA1	348	58x6
1-4	8.3	17.6	17.4	0.1	1.10	110.7	0.0	ESPA1	216	36x6

Stg	Elem no.	Feed pres psi	Pres drop psi	Perm flow gpm	Perm Flux gfd	Beta	Perm sal TDS	Conc osm pres	Concentrate saturation levels CaSO4	SrSO4	BaSO4	SiO2	Lang.
1-1	1	147.7	5.7	14.0	50.4	1.32	14.2	15.1	6	0	0	0	1.1
1-1	2	142.0	3.5	12.9	46.4	1.43	18.4	22.5	10	0	0	0	1.6
1-1	3	138.5	1.8	11.3	40.5	1.66	27.4	39.5	20	0	0	0	2.3
1-1	4	136.6	0.7	7.8	28.0	2.00	52.0	82.5	52	0	0	0	3.2
1-1	5	135.9	0.3	2.7	9.9	1.55	110.7	130.1	100	0	0	0	3.7
1-1	6	135.6	0.2	0.2	0.9	1.06	194.8	127.1	108	0	0	0	3.6
1-2	1	132.4	0.5	0.4	1.3	1.03	230.8	154.8	114	0	0	0	3.9
1-2	2	131.9	0.5	0.0	0.0	1.00	281.2	147.9	114	0	0	0	3.8
1-2	3	131.5	0.5	0.0	0.0	1.10	326.9	141.5	115	0	0	0	3.8
1-2	4	131.0	0.5	0.0	0.0	1.10	369.8	135.5	115	0	0	0	3.7
1-2	5	130.6	0.5	0.0	0.0	1.10	410.0	103.0	118	0	0	0	2.8
1-2	6	130.1	0.4	1.6	5.9	1.31	430.1	126.7	160	0	0	0	3.0
1-3	1	126.8	0.8	0.3	1.0	1.02	444.5	206.6	159	0	0	0	4.3
1-3	2	125.9	0.8	0.0	0.0	1.00	478.7	200.5	159	0	0	0	4.3
1-3	3	125.1	0.8	0.0	0.0	1.10	509.9	194.7	159	0	0	0	4.2
1-3	4	124.3	0.8	0.0	0.0	1.10	539.7	189.1	160	0	0	0	4.1
1-3	5	123.5	0.8	0.0	0.0	1.10	568.3	183.8	160	0	0	0	4.1
1-3	6	122.6	0.0	0.0	0.0	1.10	595.7	102.2	160	0	0	0	0.0
1-4	1	118.6	1.5	1.7	6.0	1.09	596.7	228.9	178	0	0	0	4.4
1-4	2	117.3	1.4	0.0	0.0	1.00	622.2	224.2	179	0	0	0	4.4
1-4	3	115.9	1.4	0.0	0.0	1.10	643.9	219.8	179	0	0	0	4.4
1-4	4	114.5	1.4	0.0	0.0	1.10	665.0	215.4	179	0	0	0	4.3
1-4	5	113.1	1.4	0.0	0.0	1.10	685.5	211.2	179	0	0	0	4.3
1-4	6	111.7	1.4	0.0	0.0	1.10	705.4	207.1	180	0	0	0	4.2

References

Abbas A., Al-Bastaki N., Modeling of an RO water desalination unit using neural networks, Chemical Engineering Journal 114 (1–3) (2005) 139-143.

Aghababaei N., Modeling improvement of drinking water by using membrane technique in Bandar Abbas, Iran, The 9th International Chemical Engineering Congress & Exhibition Shiraz, Iran.

Amiri M.C., Samiei M., Enhancing permeate flux in a RO plant by controlling membrane fouling, Desalination 207(2007) 361–369.

American Public Health Association (APHA)., Standard methods for the examination of water and wastewater, A Joint Publication of the American Public Health Association (APHA), the American Water Works Association (AWWA), and the Water Environment Federation (WEF), 20th edition (2005).

Bartels C., Franks R., Rybar S., Schierach M., Wilf M., The effect of feed ionic strength on salt passage through reverse osmosis membranes, Desalination 184 (2005)185–195.

Bick A., Oron G., Desalination technology for optimal renovation of saline groundwater in a natural reservoir, Desalination 131(2000) 97–104.

Carter N., Desalination and membrane technologies: federal research and adoption issues, Congressional Research Service 700 (2015) 7-5.

Dindarlu K., Alipoor V., Farshidfar G.R., Chemical quality of drinking water in Bandar Abbas, Bimonthly Journal of Hormozgan University of Medical Sciences (2006) 57-62.

Favaretto P.C., Reverse osmosis for desalination of water from the Guarani Aquifer System to produce drinking water in southern Brazil, Desalination 344 (2014) 402–411.

Fritzmann C., Leowenberg J., Wintgens T., Melin T., State-of-the-art of reverse osmosis desalination, Desalination 216 (2007)1-76.

Hydranautics, Membrane Solution Design Software version 2011.

Institute of Standards and Industrial Research of Iran (ISIRI)., Physical and chemical specications 2347, 5th.revision (1978).

Kamal S., Mujtaba M., Iqbal M., Effective design of reverse osmosis based desalination process considering wide range of salinity and seawater temperature, Desalination 306 (2012) 8–16.

Kamal S., Mujtaba M., Iqbal M., MINLP based superstructure optimization for boron removal during desalination by reverse osmosis, Journal of Membrane Science 440 (2013) 29–39.

Karode S.K., Kumar A., Flow visualization through spacer filled channels by computational fluid dynamics I, pressure drop and shear rate calculations for flat sheet geometry, Journal of Membrane Science 193 (2001) 69–84.

Kim Y.M., Kim S.J., Kim Y.S., Lee S., Kim S., Ha Kim J., Overview of systems engineering approaches for a large-scale seawater desalination plant with a reverse osmosis network, Desalination 238 (2009) 312–332.

Li M., Reducing specific energy consumption in Reverse Osmosis (RO) water desalination: An analysis from first principles, Desalination 276 (2011) 128-135.

Malaeb L., Ayoub G.M., Reverse osmosis technology for water treatment: state of the art review, Desalination 267 (2011)1–8.

Mehta K.P., Patel A.S., Treatment of waste water to meet disposal standards & to explore the possibilities for reuse of wastewater of common effluent treatment plant, water special volume, Environmental Pollution and Control 15 (2012) 67-70.

Michael E., A Review of wastewater treatment by reverse osmosis, EET Corporation and Williams Engineering Services Company, Inc., All Rights Reserved (2003).

Moonkhum M., Lee Y.G., Lee Y.S., Kim J.H., Review of seawater natural organic matter fouling and reverse osmosis transport modeling for seawater reverse osmosis desalination, Journal of Desalination, Water Treatment 15 (2010) 92–107.

Sandia., Desalination and Water Purification Roadmap – A Report of the Executive Committee, DWPR Program Report 95. U.S. Department of the Interior, Bureau of Reclamation and Sandia National Laboratories. Available from: http://wrri. nmsu.edu/tbndrc/roadmapreport.pdf (accessed 25.05.08.) (2003).

Sarkar P., Goswami D., Prabhakar S., Tewari P., Optimized design of a reverse osmosis system with a recycle, Desalination 230 (2008) 128-139.

Schiermeier Q., Water purification with a pinch of salt, Nature 452 (2008) 260–261.

Strathman H., Membrane Handbook., Van Nostrand Reinhold Electrodialysis, in:W.S.H.Ho, K.K. Sirkar (Eds.), New York, pp (1992) 217–262.

Tran T., Bolto B., Gray S., Hoang M., Ostarcevic E., An autopsy study of a fouled reverse osmosis membrane element used in a brackish water treatment plant, Water Research 41 (2007) 3915–3923.

UNICEF and WHO., Progress on drinking water and sanitation, 2012 update WHO/UNICEF Joint Monitoring Program for water supply and sanitation, United States of America., (2012).

Voros Kiranoudis C., Maroulis Z., Solar energy exploitation for reverse osmosis desalination plants, Desalination 115 (1998) 83-101.

WHO and UNICEF, Meeting the MDG., Drinking Water and Sanitation, Target the Urban and Rural Challenge of the Decade, www.who.int/water sanitation/monitoring/jmpfinl., (2006).

WHO, Guidelines for Drinking-Water Quality, Recommendation., World Health Organization, fourth edition, ISBN 978 92 4 1548151 (2011).

Wilf M., Klinko K., Optimization of seawater RO systems design, Desalination 138 (2001) 299–306.

29

Optimizing ANFIS for sediment transport in open channels using different evolutionary algorithms

Sultan Noman Qasem[1], Isa Ebtehaj[2,*], Hossien Riahi Madavar[3]

[1]Computer Science Department, College of Computer and Information Sciences, Al Imam Mohammad Ibn Saud, Islamic University (IMSIU), Riyadh, Saudi Arabia.
[2]Young Researchers and Elite Club, Kermanshah Branch, Islamic Azad University, Kermanshah, Iran.
[3]Department of water engineering, Vali-e-Asr University of Rafsanjan, Rafsanjan, Iran.

ARTICLE INFO	ABSTRACT
Keywords: ANFIS Differential Evolution (DE) Genetic Algorithm (GA) non-deposition sediment transport Particle Swarm Optimization (PSO)	Flow through open channels can contain solids. The deposition of solids occasionally occurs due to insufficient flow velocity to transfer the solid particles, causing many problems with transfer systems. Therefore, a method to determine the limiting velocity (i.e. Fr) is required. In this paper, three alternative, hybrid evolutionary algorithm methods, including differential evolution (DE), genetic algorithm (GA) and particle swarm optimization (PSO) based on the adaptive network-based fuzzy inference system are presented: ANFIS-GA, ANFIS-DE and ANFIS-PSO. In these methods, evolutionary algorithms optimize the membership functions, and ANFIS adjusts the premises and consequent parameters to optimize prediction performance. The performance of the proposed methods is compared with that of the general ANFIS using three different datasets comprising a wide range of data. The results show that the hybrid models (ANFIS-GA, ANFIS-DE and ANFIS-PSO) are more accurate than general ANFIS in training with a hybrid algorithm (hybrid of back propagation and least squares). Among the evolutionary algorithms, ANFIS-PSO performed the best (R^2=0.976, RMSE=0.26, MARE=0.057, BIAS=-0.004 and SI=0.059).

1. Introduction

Solid particles that deposit in entry flow through open channel systems can eventually become consolidated or cemented, especially during dry weather flow (DWF). In DWF, the lowest discharge passes through the channel. The flow velocity is below the minimum velocity required to sediment transport without deposition (limiting velocity), causing solid deposits on the channel bottom. Solid deposition on the channel bottom besides increasing bed roughness cause reduced cross-sectional area. Consequent to material deposition due to changes in velocity and shear stress distribution, reduced transmission capacity is expected.

Conventional methods of determining the limiting velocity employ constant velocity or shear stress by applying practical engineering experience obtained from project to project. Constant velocity and shear stress were presented comprehensively by Vongvisessomjai et al. (2010). These methods do not consider sediment and flow characteristics, therefore the limiting velocity, which is often presented as underestimated or overestimated, leads to sediment deposits or uneconomical plans, respectively (Bonakdari and Ebtehaj 2014a; Safarzadeh and Mohajeri 2016). Hence, many experimental and analytical studies have been conducted to investigate the factors influencing limiting velocity estimation (Nalluri and Ab Ghani 1996; Ota and Nalluri 2003; Banasiak 2008; Vongvisessomjai et al. 2010; Bonakdari and Ebtehaj 2014b) and several regression-based equations have been recommended. Since understanding the mechanism of sediment transport in open channels due to its complex three-dimensional nature is difficult, existing equations cannot provide precise estimates of limiting velocity (Ebtehaj et al. 2014). Soft Computing (SC) performs well in different engineering fields, such as

Multi-reservoir real-time operation rules (Akbari-Alashti et al. 2014); sediment transport in sewer systems (Ebtehaj and Bonakdari 2013; 2016a); wastewater treatment (Amiri et al. 2015); Scour depth (Khan and Azamathulla 2012; Najafzadeh et al. 2014); side weir discharge capacity (Parsaie and Haghiabi 2014; Ebtehaj et al. 2015); and longitudinal velocity field (Zaji and Bonakdari 2015). Fuzzy systems are one of the most widely applied methods that yield good prediction results. Azamathulla et al. (2012) predicted sediment transport in clean sewers using adaptive neuro fuzzy inference systems (ANFIS). The authors compared the results of ANFIS with a non-linear regression (NLR) equation and found that ANFIS is more accurate than NLR. Reservoir water level was estimated using ANFIS by Valizadeh and El-Shafie (2013). To improve the ANFIS ability, they used a certain membership function for each input parameter. A performance evaluation of the developed ANFIS indicated its higher accuracy over general ANFIS. Akrami et al. (2014) predicted rainfall using ANFIS. To eliminate prediction error due to noisy data, the authors combined a wavelet transform with ANFIS (Wavelet-ANFIS). The results showed that Wavelet-ANFIS performed better than ANFIS.

Combining evolutionary algorithms with other SC methods in engineering problems leads to increased accuracy. Kisi (2010) modeled suspended sediment concentration by combining differential evolution (DE) and artificial neural networks (ANN). A comparison of ANN-DE with general ANN and ANFIS showed the superior performance of ANN-DE in forecasting suspended sediment concentration. Afshar et al. (2013) evaluated the performance of multi-objective particle swarm optimization (MOPSO) in automatic calibration of water quality and hydrodynamic parameters. The results indicated that MOPSO provides a wide range of all potential calibration solutions for better decision-making. Tayfur et al. (2013)

*Corresponding author E-mail: isa.ebtehaj@gmail.com

surveyed sediment load prediction by a hybrid genetic algorithm and artificial neural network (ANN-GA) method. They found that ANN-GA is able to predict total sediment load with good accuracy. Sahay and Srivastava (2014) forecasted monsoon floods in rivers using a combination of GA, ANN and wavelet transform. To process the time series and optimize ANN's initial parameter, they used a wavelet transform and GA, respectively.

Ebtehaj and Bonakdari (2014a) evaluated the performance of ANFIS in forecasting sediment transport. The authors suggested the application of evolutionary algorithms rather than back propagation and hybrid algorithms, and compared them. Thus, in this paper, the performance of evolutionary algorithms, i.e. differential evolution (DE), genetic algorithm (GA) and particle swarm optimization (PSO) based on ANFIS (ANFIS-DE, ANFIS-GA and ANFIS-PSO) in predicting sediment transport in pipe channels using three datasets is surveyed. using dimensional analysis, the effective parameters on limiting velocity (Fr) prediction are placed in 5 groups and 6 proposed models. As a result, the performance of all models is evaluated using ANFIS, ANFIS-DE, ANFIS-GA and ANFIS-PSO.

2. The concept of ANFIS
2.1. ANFIS structure

In this paper, in order to predict the limiting velocity to prevent sediment deposition on the channel bed, a neuro fuzzy system is used. The neuro fuzzy system is a modeling framework in the form of a hybrid artificial neural network (ANN) and fuzz logic (FL). The hybrid model is presented to overcome the limitations in both FL and ANN methods. In fact, inspired by fuzzy systems, basic knowledge in a collection of constraints showed a reduction in the optimization search space. Meanwhile, a structured network is inspired by ANN back propagation. In this method, ANN tunes the membership functions (MFs) (Singh et al. 2012). There are also non-linear membership functions in the neuro fuzzy method, which lead to notable reduction in the implementation cost of a simple design based on rules and memory consumption. Thus, it is clear that a hybrid of neural networks and fuzzy systems reduces the limitations of each of two methods, hence a data mining technique is proposed to solve complex engineering problems. Adaptive neuro fuzzy inference system (ANFIS) is one of the known methods of simultaneously combining neural networks and fuzzy systems. This method is used to identify the performance of nonlinear systems with input and output datasets defined for the model. ANFIS is a structured, fuzzy inference system (FIS) model. Two of the most prominent FIS used in ANFIS are Mamdani (Mamdani and Assilan 1975) and Takagi-Sugeno-Kang (TSK) (Takagi and Sugeno 1985; Jang 1992). TSK is simpler because there is less need for rules and it uses known training methods like back propagation. Two of the algorithms by default in ANFIS training are back propagation and hybrid (hybrid of back propagation and least squares).

A schematic ANFIS structure for a network with two inputs (x, y) and one output (f) is presented in Fig. 1. The considered rules with two IF-THEN rules for FIS of TSK-type can be expressed as follows:

Rule 1
$$IF \quad x \ is \ A_1, \quad and \ y \ is \ B_1,$$
$$Then \quad f_1 = p_1x + q_1y + r_1 \tag{1}$$

Rule 2
$$IF \quad x \ is \ A_2, \quad and \ y \ is \ B_2,$$
$$Then \quad f_2 = p_2x + q_2y + r_2 \tag{2}$$

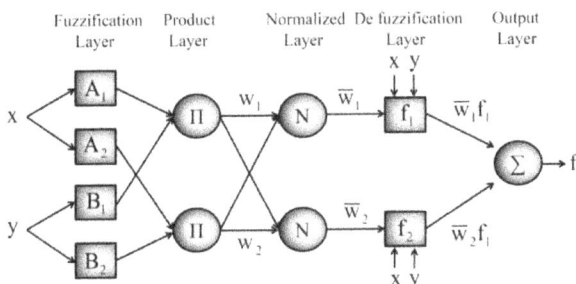

Fig. 1. ANFIS architecture for 2 inputs and 1 output

The number of nodes in the first layer with the use of entire nodes and the number of membership functions (n) for each input is determined. However, the number of nodes in other layers (layers 2 to 4) depends on the rule (R) in each fuzzy rule base. The ANFIS layer structure is as follows:

The first layer (fuzification layer): The Xi input comprising the membership degree label of fuzzy set A_{ij}, shows the membership degree of each fuzzy collection. The node function of this layer can be expressed as follows:

$$O_{ij}^1 = \mu_{ij}(X_i) \quad i = 1,2,....number\ of\ inputs, \quad j = 1,2,....n \tag{3}$$

where μ_{ij} is the jth MF for the X_i input and O_{ij}^1 is the output of node ij. Due to the satisfactory performance of Gaussian membership functions in various engineering applications (Zanganeh et al. 2009; Güneri et al. 2011; Abdi et al. 2012; Karasakal et al. 2013; Bosque et al. 2014; Premkumar and Manikandan 2015), the membership function used in this study is a Gaussian-shaped MF. This function has smoothness and concise notation as well. The Gaussian MF mathematical relationship is as follows:

$$\mu(X) = exp\left(-\left(\frac{X-a}{b}\right)^2\right) \tag{4}$$

where a and b are the parameter set.

The second layer (produce layer): the nodes in this layer (k), which are provided as circular nodes (Π), generate output using the received input.

$$O_k^2 = W_k = \mu_{e1}(X_1)\mu_{e2}(X_2)...\mu_{ek}(X_k)$$
$$k = 1,2,...,R \quad e1...e2 = 1,2,...n \tag{5}$$

The third layer (normalized layer): in this layer, the kth node determines the relative firing strength of the kth rule to the firing strengths of all rules as follows:

$$O_k^3 = \overline{W}_k = \frac{W_k}{W_1 + W_2 + ... + W_R} \quad k = 1,2,...,R \tag{6}$$

The fourth layer (de-fuzzification layer): Each node in this layer is an FIS weighted output performed as follows:

$$O_k^4 = \overline{W}_k f_k \tag{7}$$

where W_k and f_k are the output layer of de-fuzzification and k^{th} TSK-FIS, respectively. The m TSK-FIS number rules are as follows:

$$f_k = \sum_{i=1}^{m} p_{t_{ei}} + r_k \tag{8}$$

where $p_{i,ei}$ and r_k comprise the parameter set. The parameters of this layer are referred to as consequent parameters.

$$O_i^5 = Y = \sum_{k=1}^{n} \overline{w}_k f_k = \frac{\sum w_k f_k}{\sum w_k} \tag{9}$$

where Y represents all network outputs. To evaluate ANFIS performance, mean squared error (MSE) indicators are used, which are calculated as follows:

$$MSE = \frac{1}{n}\sum_{i=1}^{n}(Fr_{Actual} - Fr_{Predicted})^2 \tag{10}$$

where n is the pattern number, Fr_{Actual} is the observed Fr in the experimental tests and $Fr_{Predicted}$ is the Fr predicted by ANFIS.

Fig. 2 shows a flowchart of ANFIS. Firstly, the datasets are classified into two parts: training and testing. In this study, 70% and 30% of data were used for training and testing of the model, respectively. FIS generation is done following categorization. There are two methods, namely grid partitioning and subtractive clustering to carry out FIS generation. In this study, grid partitioning (Ebtehaj and Bonakdari 2014a) is used owing to better performance. At this stage, the numbers and types of input and output MFs should be determined. In this study, the Gaussian membership function (Eq. 4) is used as an MF. The number of MFs is determined by trial and error and considered equal to 3. The training network algorithm should be determined following FIS generation. For ANFIS training, back propagation and hybrid (back propagation algorithm with least squares) are normally used (Sobhani *et al.* 2010; Bilgehan 2011; Behera and Guruprasad 2012). Ebtehaj and Bonakdari (2014a) indicated that hybrid algorithm performance is better than back propagation. Therefore, in this study the performance of a hybrid algorithm is compared with evolutionary algorithms, namely genetic algorithm (GA), Particle Swarm Optimization (PSO) and Differential Evolution (DE), as suggested by researchers for future studies. After training ANFIS, the prediction accuracy is assessed using test datasets and if satisfactory results are achieved, the modeling process ends; otherwise the FIS generation process is repeated to reach an acceptable solution.

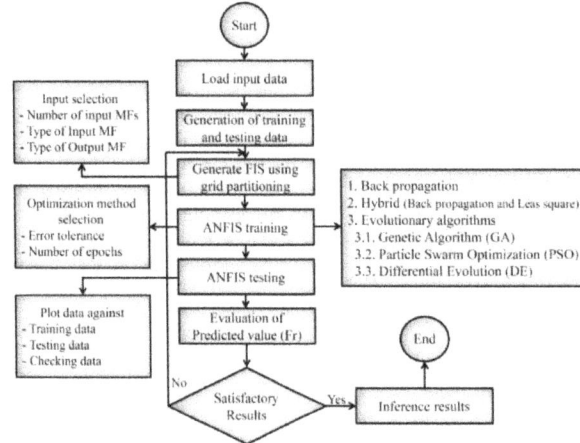

Fig. 2. General ANFIS flowchart.

2.2. Learning algorithms
2.2.1. Genetic Algorithm (GA)

Genetic algorithm (GA) is a stochastic optimization method that has performed successfully for various engineering issues. GA is capable of solving problems that gradient-based methods cannot solve well, such as nonlinear, stochastic and non-differentiable problems (Ocak. 2013). In conventional optimization approaches, in order to achieve the optimum solution, every point is generated using deterministic computations in each iteration and sequence of point approaches. In GA, the points of a population for each iteration are generated randomly and the best population point has a desire to the optimum solution that is similar to the final result (Goldberg 1989; Melanie 1996).

One of the most important steps in using GA to investigate an optimization problem is to provide an optimized potential solution to the problem as an individual or gene sequence known as chromosomes. The most common way of encoding problems as binary strings is to use zero and one strings. The basic steps in a genetic algorithm are presented in Fig. 3.

GA includes three essential components. The first part concerns creating an initial population using the mth individual that was randomly selected. The initial population produces the first generation. The second part consists of entering the mth individual and generating the output; evaluating each is based on the objective function known as a fitness function (Fig. 3). The evaluation determines the demands expected of each individual in order to achieve the ultimate goal. Finally, the third component is responsible for the new generation. A new generation is created based on the fittest individual from the previous generation.

Evaluating the process of producing generation N and generation N + 1 based on the N generation continues until the desired function is achieved. Offspring generation, which is based on the fittest individual related to the previous generation, is known as breeding. The breeding process consists of three basic steps in GA: reproduction, crossover and mutation.

Reproduction is done by two genetic operators, namely crossover and mutation. Crossover is a process in which the parent's genes change, but mutation is where genes randomly modify in the parent chromosome. Both of these operations have significant impact on the search space and a lack of appropriate values may not provide good results. Crossover and mutation represent searching the new solution area and the behavior of a random jump into unknown areas in the search space, respectively (Holland 1992). The genetic evolution algorithm process for different generations continues until a termination condition is fulfilled. The best gene call is decoded in the last generation in order to achieve the desired, optimum solution to the problem.

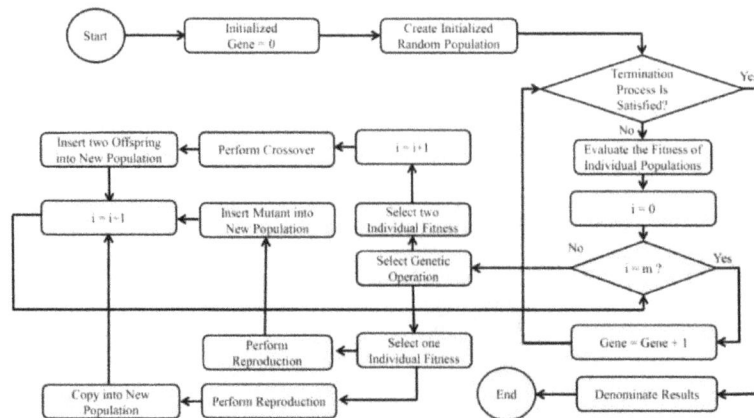

Fig. 3. GA flowchart

2.2.2. Particle Swarm Optimization (PSO)

Particle Swarm Optimization (PSO) is a smart method group (swarm) presented by Eberhart and Kennedy (1995). It is inspired by the social behavior of birds and fish. PSO is a population-base search method, in which every particle could be a candidate for the optimum solution. Particles change their positions in a multi-dimensional search space to achieve an optimal condition or limited circumstance calculation.

Empirical observations show good performance in the optimization methods field (Naka et al. 2002; Mendes et al. 2004; Yu and Li 2004). Thus, this method has been widely used in engineering optimization problems (Eberhart and Hu 1999; Yoshida et al. 2000; Ciuprina et al. 2002; Ratnaweera et al. 2004; Heo et al. 2006; Del Valle et al. 2008; Pedersen and Chipperfield 2010; Mousa et al. 2012; Ebtehaj and Bonakdari 2016b). A PSO algorithm flowchart is presented in Fig. 4. The first step is to determine the initial particle swarm, P(k), so that the $x_{is}(k)$ position of each particle ($P_i \in P(k)$) in hyperspace is equal to k = 0. The second stage is to evaluate the F function performance for each particle using the particle position ($x_i(k)$).

$$if \quad F(x_i(k)) < pbest_i \quad then \quad \begin{cases} pbest_i = F(x_i(k)) \\ x_{pbest_i} = x_i(k) \end{cases} \quad (11)$$

In the third stage, the best particle performance of each individual is evaluated as follows:

$$if \quad F(x_i(k)) < gbest_i \quad then \quad \begin{cases} gbest_i = F(x_i(k)) \\ x_{gbest_i} = x_i(k) \end{cases} \quad (12)$$

In the fourth stage, for each individual, the velocity vector is changed using the following equation:

$$v_i(k) = wv_i(k-1) + r_1 C_1(x_{pbest_i} - x_i(k)) + r_2 C_2(x_{gbest_i} - x_i(k)) \quad (13)$$

where r and C are random parameters. The r_1 and r_2 parameters are in the range of (0, 1) while C_1 and C_2 are positive constant values. Ebtehaj and Bonakdari (2016b) concluded that the best result is achieved when the sum of these two parameters is not more than 4 $(C_1 + C_2 \leq 4)$.

The w parameter is known as the weight parameter in the above equation. Careful selection of these parameters leads to a balance between local and global swarm performance, which reduces the iteration number. The value of the w parameter using the equation proposed by Shi and Eberhart (1998, 1999) is calculated as follows:

$$w = w_{max} - \frac{w_{max} - w_{min}}{iter_{max}} \cdot iter \quad (14)$$

where w_{min} and w_{max} are the initial and final weights, respectively.

Also, itr_{max} is the maximum iteration value and itr is the iteration number.

In the fifth step, every particle transforms to its new location using the following equation:

$$x_i(k) = x_i(k-1) + v_i(k) \quad (15)$$

2.2.3. Differential Optimization (DE)

Differential Evolution (DE) belongs to evolutionary algorithm ancestors presented by Storn and Price (1997). DE is a random population-based algorithm. The difference between this method and other evolutionary algorithms is the use of differential mutation. In a desired solving population with n-dimensional space, fixed vector numbers are created randomly. To understand the different search spaces and reach the minimum objective function, evolution over time is needed.

A mutation function in DE ($F: I^\mu \to I^\mu$) consists of producing a mutated vector (μ) using the following equation:

$$\vec{v}_i = \vec{a}_{r1} + F(\vec{a}_{r2} - \vec{a}_{r3}) \quad i = 1,2,...,\mu \quad (16)$$

where r_1, r_2, $r_3 \in [1, 2,... \mu]$ are selected randomly. These parameters differ from each other and index i. $F \in [0\ 2]$ is a constant parameter affecting the differential variation between two vectors. Larger F or population size (μ) quantities tend to increase the capacity of the global search algorithm because a new area is known to the search space.

The crossover operator in DE ($CR: I^\mu \to I^\mu$) mutates the vectors ($\vec{v}_i = [\vec{v}_{1i}, \vec{v}_{2i},...,\vec{v}_{di}]$) with a target function ($\vec{a}_i = [\vec{a}_{1i}, \vec{a}_{2i},...,\vec{a}_{di}]$) (an answer to the previous population parent) to produce a trial vector combination ($\vec{a}'_i = [\vec{a}'_{1i}, \vec{a}'_{2i},...,\vec{a}'_{di}]$) using the following equation:

$$a'_{ji} = \begin{cases} v_{ji} & if \ (randb(j) \leq CR) \ or \ j = rnbr(i) \quad j = 1,2,....,d \\ a_{ji} & if \ (randb(j) > CR) \ and \ j \neq rnbr(i) \quad i = 1,2,....,\mu \end{cases} \quad (17)$$

where $randb(j) \in [0\ 1]$ is the jth assessment of a uniform random generator and $rnbr(i) \in 1,2,...,d$ is the random selection index. $CR \in [0\ 1]$ is the crossover parameter that increases the variety of individuals in populations. Larger CR values cause increased child vectors (a_i) similar to mutated vectors (v_i). Thus, the algorithm convergence speed is increased. According to Eq. 17, each objective function has a target function rule. If CR is considered zero, vectors of parents and children differ from at least one variable (Eq. 17).

The selection operator in DE ($s: I\mu \to I\mu$) selects the best costly target function (a_i) and associated trial vector (a_i) as a part of the population for the next generation.

$$If \ \Phi(\vec{a}'_i(g)) < \Phi(\vec{a}_i(g)), then \ \vec{a}_i(g+1) = \vec{a}'_i(g)$$
$$else \qquad \qquad \vec{a}_i(g+1) = \vec{a}_i(g+1) = \vec{a}_i(g)) \quad (18)$$

where g is the current generation. The DE flowchart with the main operators is presented in Fig. 5.

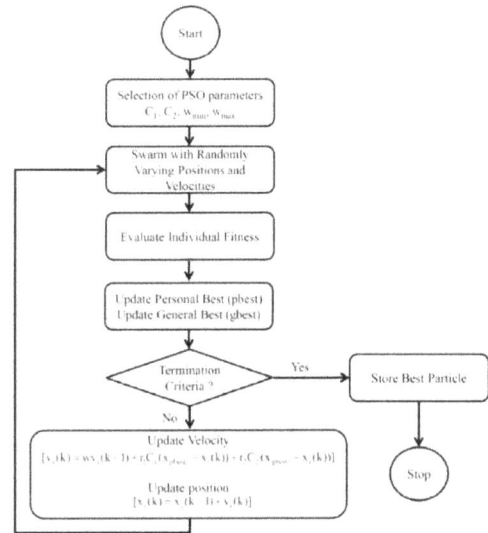

Fig. 4. PSO flowchart.

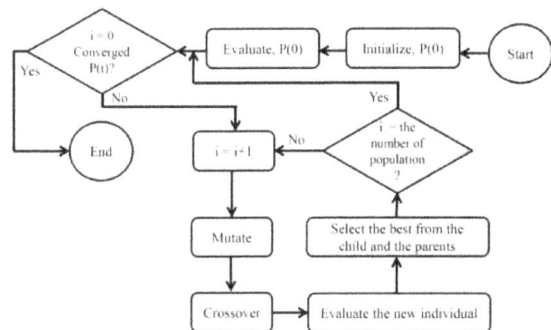

Fig. 5. DE flowchart.

3. Methodology

To determine the minimum velocity required to avoid sediment deposition in open channels (limiting velocity), factors affecting flow velocity should be determined first. Experimental studies in the field of sediment transport with non-deposition condition consider parameters such as flow depth, volumetric sediment concentration, particle size and pipe diameter as effective parameters in determining the limiting velocity. Since each of the parameters includes different dimensions, the dimensionless parameters are used to determine the limiting velocity. Several studies have considered (Nalluri and Ab Ghani 1996; Ab Ghani and Azamathulla 2010; Azamathulla et al. 2012; Ebtehaj et al. 2013) limiting velocity parameters in assessing the functional relationship below:

$$Fr = V/\sqrt{g(s-1)d} = f(C_V, D_{gr}, d/D, d/R, R/D, D^2/A, \lambda_s) \qquad (19)$$

where Fr is the densimetric Froude number, V is the limiting velocity, g is the gravitational acceleration, s is the specific gravity of sediment ($=\rho/\rho_s$), C_V is the volumetric sediment concentration, $D_{gr}(=d(g(s-1)/v^2)^{1/3})$ is the dimensionless particle number, d is the median particle diameter, D is the pipe diameter, R is the hydraulic radius, A is the cross sectional area of the flow, and λ_s is the overall friction factor of sediment.

Ebtehaj and Bonakdari (2014a) classified the dimensionless parameters provided in Eq.19 in 5 different categories: movement (Fr), transport (C_V), sediment (D_{gr}, d/D), transport mode (d/R, D^2/A, R/D) and flow resistance (λ_s). Because there is only one parameter in the transport and flow resistance groups, this parameter is considered fixed. Sediment and transport have two and three different parameters, respectively; therefore, to consider the effect of all groups in Fr parameter estimation related to the movement group, six different models are presented as follows:

Model 1: Fr = f (C_V, D_{gr}, d/R, λ_s)
Model 2: Fr = f (C_V, D_{gr}, D^2/A, λ_s)
Model 3: Fr = f (C_V, D_{gr}, R/D, λ_s)
Model 4: Fr = f (C_V, d/D, d/R, λ_s)
Model 5: Fr = f (C_V, d/D, D^2/A, λ_s
Model 6: Fr = f (C_V, d/D, R/D, λ_s)

In this study, to predict the Fr parameter of the movement group in order to sediment transfer in non-deposition condition, three different data sets, including by Ab Ghani (1993), Ota and Nalluri (1999) and Vongvisessomjai et al. (2010), with a total of 218 different data are used. This data was obtained in different experimental conditions (pipe diameter, sediment and flow characteristics). Details of the experiments were presented in previous studies (i.e., Ebtehaj and Bonakdari 2014a, 2014b, 2016b).

For modeling, the data were divided into two categories in this study: 70 % of data (150) for training and 30% (68) for testing model performance. The control parameters for each evolutionary algorithm employed in this study are GA, DE and PSO (Table 1).

Table 1. Control parameter of evolutionary algorithms

	Parameters	Value
PSO	Number of iterations	5000
	Number of Particles	50
	Initial inertia weight w_{min}	0.9
	Final inertia weight w_{max}	0.3
	Cognitive acceleration C_1	2
	Social acceleration C_2	2
DE	Number of dimensions (D)	4
	Population size (NP)	20
	Mutation constant (F)	0.5
	Crossover constant (CR)	0.9
	parameters; boundaries Vj(U)	12
	parameters; boundaries Vj(L)	-12
GA	Population size	30
	Number of generations	60
	Crossover rate	0.8
	Mutation rate	0.2
	Selection Method	Roulette wheel selection

3.1. Statistical measure

The employed statistical indices to performance evaluation of densimetric Froude number are as follows:

$$R^2 = \left[\frac{\sum_{i=1}^{n}\left(Fr_{Expi} - \overline{Fr_{Exp}}\right)\left(Fr_{Modeli} - \overline{Fr_{Model}}\right)}{\sqrt{\sum_{i=1}^{n}\left(Fr_{Expi} - \overline{Fr_{Exp}}\right)^2 \sum_{i=1}^{n}\left(Fr_{Modeli} - \overline{Fr_{Model}}\right)^2}} \right]^2 \qquad (20)$$

$$RMSE = \sqrt{\frac{1}{n}\sum_{i=1}^{n}\left(Fr_{Expi} - Fr_{Modeli}\right)^2} \qquad (21)$$

$$MARE = \frac{1}{n}\sum_{i=1}^{n}\frac{\left|Fr_{Expi} - Fr_{Modeli}\right|}{Fr_{Expi}} \qquad (22)$$

$$SI = \frac{RMSE}{Fr_{Model}} \qquad (23)$$

$$BIAS = \frac{1}{n}\sum_{i=1}^{n}\left(Fr_{Expi} - Fr_{Modeli}\right) \qquad (24)$$

4. Results and discussion

Fig. 6 shows the performance of the three hybrids, ANFIS-DE, ANFIS-GA and ANFIS-PSO as well as general ANFIS in estimating the limiting velocity, which is expressed as a dimensionless parameter, Fr, for 6 models proposed in the study. For each of the three hybrid models proposed using an evolutionary algorithm, model 1 produced good results. The values estimated by the three methods often had less than 10 % relative error. But clearly, the hybrid algorithm in ANFIS performance training did not do as well as other methods, with a relative error of more than 10% as under and overestimation was used. Results obtained from Table 2 show that the statistical indicators for six different models in both testing and training datasets and using evolutionary algorithms for all indexes leads to better estimation.

Compared to model 1, model 2 showed a completely different situation. Evidently, the performance of all models diminished. Since the only difference between this model and model 1 is using the D^2/A parameter instead of the d/R parameter, it can be concluded that with fixed parameters, including transport (C_V), flow resistance (λ_s) and sediment (D_{gr}), using parameter D^2/A rather than d/R leads to at least 10% increase in relative error with all methods. However, the relative error increase in the ANFIS-DE and ANFIS-GA methods is about 15%. Therefore, the second model in any method presented in this study is uncertain. The performance comparison between evolutionary and hybrid algorithms for the presented input combination in model 2 as obtained from Table 2 indicates the superiority of evolutionary algorithms.

Similar to model 2, model 3 may not perform so well. Nonetheless, the best performance among the parameters related to the transport mode is attained when parameter d/R is used as the group representative. The prediction progress of this model is quite different than the second. In Model 3, most estimates are lower than the experimental data values. Therefore, using the model with the input combinations suggested in model 3 leads to high sediment deposition on the channel bed.

The models' quantitative performance indicates that except for the ANFIS-PSO method, which increased the estimation accuracy less than the ANFIS model, ANFIS-DE and ANFIS-GA performed better than the ANFIS model. Using the DE and GA algorithms in ANFIS training decreased the relative error by about 12% compared to the hybrid algorithm. In comparing models 4 and 2, the parameters related to sediment from D_{gr} to d/D changed and other groups' parameters remained fixed.

Fig. 6 shows that the performance of model 1 is as good as model 4, and all estimated values had less than 10% relative error. Table 2 shows that all methods, namely ANFIS, ANFIS-DE, ANFIS-GA and ANFIS-PSO, performed better than model 1. In fact, selecting the sediment parameter using d/D leads to better results than d/R. But it is noteworthy that there is no significant difference in the index values of both models. Model 5, much like model 1, did not perform well; as a result, using d/D instead of d/R did not significantly change the outcomes, but there was improvement overall. The performance of model 6 compared to model 3 is similar to that of model 5 compared with 2. Using the d/D parameter related to sediment leads to better model performance than D_{gr}.

The comparison between the presented models and results obtained from Table 2 indicate the superior performance of model 4 among all methods. However, there were no significant differences between the hybrid methods, but the ANFIS-PSO Model 4 (R^2=0.976, RMSE=0.26, MARE=0.057, BIAS=-0.004 and SI=0.059) exhibited the best performance among the methods.

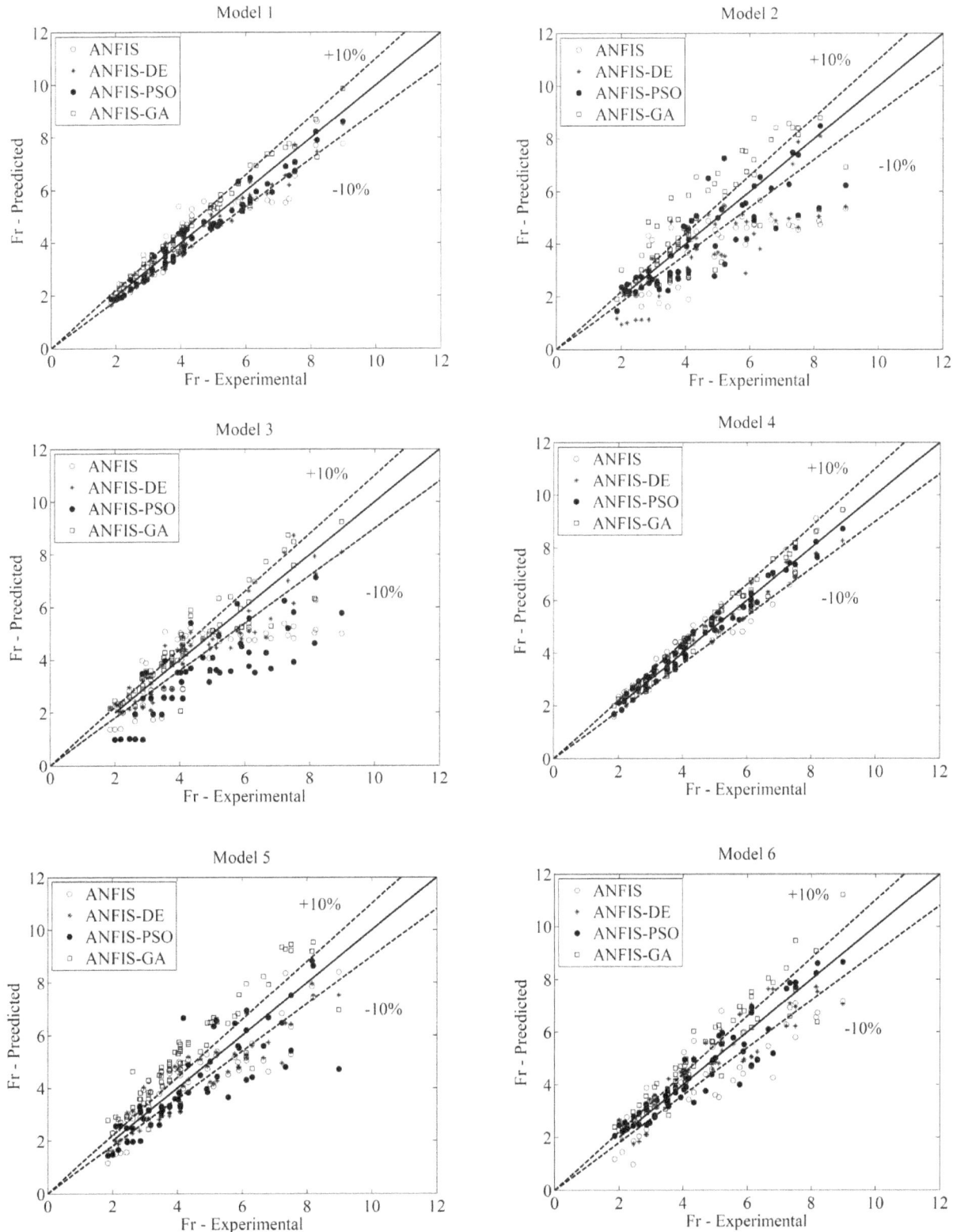

Fig. 6. Scatter plot of Fr prediction (testing).

Table 3 presents the DR values (relative predicted Fr to observed Fr) of different methods and model 4 in this study. In this table, the differences of 0.05, 0.10, 0.15 and 0.20 to the value of the unit is provided for all methods (DR ±0.05, DR ±0.1, DR ±0.15, DR ±0.2). For the values with a difference of 0.05 to the unit (DR ± 5 %), ANFIS and two evolutionary methods, namely ANFIS-DE and ANFIS-GA showed approximately the same results; but it is clear that ANFIS-PSO had superior performance to the other methods.

Immense difference between the gross evolutionary ANFIS and general ANFIS was observed. Hence, the performance of the general ANFIS method was verified with values having difference of 0.1 to the unit (DR ± 10%). It can be concluded from Table 3 that all methods

presented in this study had the maximum difference of 0.15 to DR = 1 (DR ± 15 %), which is 0.2 for general ANFIS.

Table 3. DR values for different ANFIS methods (Model 4).

Method (Model 4)	(DR±5%)	(DR±10%)	(DR±15%)	(DR±20%)
ANFIS	0.27	0.53	0.87	1
ANFIS-DE	0.29	0.82	1	1
ANFIS-PSO	0.4	0.9	1	1
ANFIS-GA	0.25	0.86	1	1

Table 2. Performance evaluation of ANFIS methods in prediction of Fr (All models)

		Models	R^2	RMSE	MARE	BIAS	SI
Train	Model 1	ANFIS	0.933	0.579	0.120	-0.051	0.146
		ANFIS-PSO	0.976	0.342	0.062	0.058	0.089
		ANFIS-DE	0.972	0.370	0.074	0.008	0.095
		ANFIS-GA	0.977	0.399	0.088	-0.201	0.102
	Model 2	ANFIS	0.707	1.136	0.260	-0.031	0.289
		ANFIS-PSO	0.905	0.638	0.099	0.002	0.164
		ANFIS-DE	0.840	0.831	0.167	-0.086	0.213
		ANFIS-GA	0.869	0.997	0.231	-0.527	0.256
	Model 3	ANFIS	0.637	1.255	0.282	-0.099	0.314
		ANFIS-PSO	0.874	0.736	0.160	-0.028	0.187
		ANFIS-DE	0.906	0.637	0.121	0.059	0.163
		ANFIS-GA	0.836	0.996	0.196	-0.457	0.255
	Model 4	ANFIS	0.963	0.406	0.094	-0.027	0.103
		ANFIS-PSO	0.984	0.262	0.060	-0.012	0.067
		ANFIS-DE	0.973	0.341	0.072	0.033	0.087
		ANFIS-GA	0.986	0.318	0.071	-0.185	0.082
	Model 5	ANFIS	0.665	1.198	0.267	-0.027	0.305
		ANFIS-PSO	0.899	0.659	0.102	-0.003	0.169
		ANFIS-DE	0.913	0.632	0.125	0.001	0.162
		ANFIS-GA	0.878	0.887	0.186	-0.424	0.227
	Model 6	ANFIS	0.733	1.080	0.221	0.153	0.288
		ANFIS-PSO	0.962	0.404	0.088	0.000	0.103
		ANFIS-DE	0.916	0.613	0.105	0.041	0.157
		ANFIS-GA	0.927	0.689	0.143	-0.343	0.177
Test	Model 1	ANFIS	0.882	0.590	0.099	0.072	0.137
		ANFIS-PSO	0.966	0.356	0.073	0.173	0.085
		ANFIS-DE	0.963	0.392	0.076	0.216	0.089
		ANFIS-GA	0.965	0.402	0.082	-0.249	0.092
	Model 2	ANFIS	0.611	1.375	0.240	0.895	0.394
		ANFIS-PSO	0.728	1.027	0.173	0.518	0.266
		ANFIS-DE	0.720	1.248	0.232	0.881	0.285
		ANFIS-GA	0.711	1.155	0.205	-0.525	0.263
	Model 3	ANFIS	0.572	1.340	0.254	0.762	0.370
		ANFIS-PSO	0.702	1.564	0.237	1.262	0.501
		ANFIS-DE	0.869	0.663	0.128	0.264	0.151
		ANFIS-GA	0.829	0.762	0.122	-0.235	0.174
	Model 4	ANFIS	0.929	0.452	0.091	-0.077	0.101
		ANFIS-PSO	0.976	0.260	0.057	-0.004	0.059
		ANFIS-DE	0.965	0.323	0.065	0.076	0.074
		ANFIS-GA	0.972	0.346	0.069	-0.175	0.079
	Model 5	ANFIS	0.463	1.223	0.229	0.023	0.280
		ANFIS-PSO	0.722	0.981	0.147	0.376	0.245
		ANFIS-DE	0.868	0.801	0.154	0.507	0.183
		ANFIS-GA	0.841	1.165	0.253	-0.882	0.266
	Model 6	ANFIS	0.538	1.201	0.179	0.348	0.297
		ANFIS-PSO	0.905	0.536	0.088	0.095	0.125
		ANFIS-DE	0.869	0.606	0.112	-0.023	0.138
		ANFIS-GA	0.899	0.785	0.138	-0.477	0.179

5. Conclusions

Sediment transport capacity is reduced by solid deposition in open channel flow. Therefore, an approach of estimating minimum velocity to prevent sediment deposition is required. This study presented three different evolutionary algorithms, i.e. differential evolution (DE), genetic algorithm (GA) and particle swarm optimization (PSO) based on adaptive neuro fuzzy inference systems (ANFIS), as new methods for estimating the limiting velocity (Fr). The new methods are ANFIS-DE, ANFIS-GA and ANFIS-PSO. To estimate Fr, previously conducted studies and dimensional analysis were used, and different

dimensionless parameters were identified in 5 groups: movement, transport, sediment, transport mode and flow resistance. Studies have shown that the d/D parameter in sediment and the d/R parameter in transport mode perform the best in their groups. Therefore, the best model was chosen as Fr = f (C_V, d/R, d/D, λ_s). Comparing the proposed procedure performance with general ANFIS represents the ascending performance of ANFIS when using evolutionary algorithms in hybrid algorithms. Among the proposed methods, ANFIS-PSO (R^2 = 0.976, RMSE = 0.26, MARE = 0.057, BIAS = -0.004 and SI = 0.059) performed the best. Therefore, using evolutionary algorithms as an optimization algorithm method is useful to hybrid performance.

References

Abdi J., Moshiri B., Abdulhai B., Sedigh A.K., Forecasting of short-term traffic-flow based on improved neurofuzzy models via emotional temporal difference learning algorithm, Engineering Applications of Artificial Intelligence 25 (2012) 1022-1042.

Ab Ghani, A., Sediment transport in sewers, Ph.D. Thesis, University of Newcastle Upon Tyne, UK, 1993.

Ab Ghani A, Azamathulla H.M., Gene-expression programming for sediment transport in sewer pipe systems, Journal of Pipeline Systems Engineering and Practice 2 (2010) 102-106.

Afshar A., Shojaei N., Sagharjooghifarahani M., Multiobjective calibration of reservoir water quality modeling using multiobjective particle swarm optimization (MOPSO), Water Resources Management 27 (2013) 1931-1947.

Akbari-Alashti H., Bozorg Haddad O., Fallah-Mehdipour E., Mariño, M.A., Multi-reservoir real-time operation rules: a new genetic programming approach, Proceedings of the Institution of Civil Engineers-Water Management 167 (2014) 561-576.

Akrami S.A., Nourani V., Hakim S.J.S., Development of nonlinear model based on wavelet-ANFIS for rainfall forecasting at Klang Gates Dam, Water Resources Management 28 (2014) 2999-3018.

Amiri N., Ahmadi M., Pirsaheb M., Vasseghian Y., Amiri P., Combination of ozonation with aerobic sequencing batch reactor for soft drink wastewater treatment: experiments and neural network modeling, Journal of Applied Research in Water and Wastewater 2 (2015) 156-163.

Azamathulla H.M., Ab Ghani, A., Fei, S.Y., ANFIS - based approach for predicting sediment transport in clean sewer, Applied Soft Computing 12 (2012) 1227-1230.

Banasiak R., Hydraulic performance of sewer pipes with deposited sediments, Water Science & Technology 57 (2008) 1743–1748.

Behera B.K., Guruprasad, R., Predicting bending rigidity of woven fabrics using adaptive neuro-fuzzy inference system (ANFIS), Journal of The Textile Institute 103 (2012) 1205-1212.

Bilgehan M., Comparison of ANFIS and NN models—With a study in critical buckling load estimation, Applied Soft Computing 11 (2011) 3779-3791.

Bonakdari H., Ebtehaj, I., Study of sediment transport using soft computing technique, In 7th International Conference on Fluvial Hydraulics, RIVER FLOW 2014, Lausanne, Switzerland, 3-5 September, pp. 933-940, 2014a.

Bonakdari H., Ebtehaj I., Verification of equation for non-deposition sediment transport in flood water canals, In 7th International Conference on Fluvial Hydraulics, RIVER FLOW 2014, Lausanne, Switzerland, 3-5 September, 1527-1533, 2014b.

Bosque G., Del Campo I., Echanobe J., Fuzzy systems, neural networks and neuro-fuzzy systems: A vision on their hardware implementation and platforms over two decades, Engineering Applications of Artificial Intelligence 32 (2014) 283-331.

Ciuprina G., Ioan D., Munteanu I., Use of intelligent-particle swarm optimization in electromagnetics, Magnetics IEEE Transactions on 38 (2002) 1037–1040.

Del Valle Y., Venayagamoorthy G.K., Mohagheghi S., Hernandez J.C., Harley R.G., Particle swarm optimization: basic concepts, variants and applications in power systems, Evolutionary Computation IEEE Transactions on 12 (2008) 171-195.

Eberhart R.C., Kennedy J., A new optimizer using particle swarm theory, In Proceedings of the sixth international symposium on micro machine and human science 1 (1995) 39-43.

Eberhart R.C., Hu X., Human tremor analysis using particle swarm optimization. In Evolutionary Computation, 1999. CEC 99, Proceedings of the 1999 Congress on, IEEE 3 (1999) 1927–1930.

Ebtehaj I., Bonakdari H., Evaluation of Sediment Transport in Sewer using Artificial Neural Network, Engineering Applications of Computational Fluid Mechanics 7 (2013) 382–392.

Ebtehaj I., Bonakdari H., Performance Evaluation of Adaptive Neural Fuzzy Inference System for Sediment Transport in Sewers, Water Resources Managemeent 28 (2014a) 4765–4779.

Ebtehaj I., Bonakdari H., Comparison of genetic algorithm and imperialist competitive algorithms in predicting bed load transport in clean pipe, Water Science & Technology 70 (2014b) 1695-1701.

Ebtehaj I., Bonakdari H., Bed load sediment transport estimation in a clean pipe using multilayer perceptron with different training algorithms, KSCE Journal of Civil Engineering 20 (2016a) 581–589

Ebtehaj I., Bonakdari H., Assessment of evolutionary algorithms in predicting non-deposition sediment transport. Urban Water Journal 13 (2016b) 499-510

Ebtehaj I., Bonakdari H., Khoshbin F., Azimi H., Pareto Genetic Design of GMDH-type Neural Network for Predict Discharge Coefficient in Rectangular Side Orifices, Flow Measurement and Instrumentation 41 (2015a) 67-74.

Ebtehaj I., Bonakdari H., Sharif A., Design criteria for sediment transport in sewers based on self-cleansing concept, Journal of Zhejiang University Science A. 15 (2014) 914-924.

Goldberg DE (1989) Genetic Algorithms in Search, Optimization and Machine Learning, Addison-Wesley, 1989.

Güneri A.F., Ertay T., YüCel A., An approach based on ANFIS input selection and modeling for supplier selection problem, Expert Systems with Applications 38 (2011) 14907-14917.

Heo J.S., Lee K.Y., Garduno-Ramirez R., Multiobjective control of power plants using particle swarm optimization techniques, Energy Conversion IEEE Transactions on 21 (2006) 552–561.

Holland J., Genetic algorithms, Scientific American, pp. 66–72, 1992.

Jang J.S.R., Neuro-fuzzy modeling: architecture, analyses and applications. Dissertation, University of California, Berkeley, 1992.

Karasakal O., Guzelkaya M., Eksin I., Yesil E., Kumbasar T., Online tuning of fuzzy PID controllers via rule weighing based on normalized acceleration, Engineering Applications of Artificial Intelligence 26 (2013) 184-197.

Khan M., Azamathulla H.M., Bridge pier scour prediction by gene expression programming, Proceedings of the Institution of Civil Engineers –Water management 165 (2012) 481-493.

Kisi Ö., River suspended sediment concentration modeling using a neural differential evolution approach, Journal of hydrology 389 (2010) 227-235.

Mamdani E.H., Assilan S., An experiment in linguistic synthesis with a fuzzy controller. International journal of man-machine studies 7 (1975) 1–13.

Melanie M., An Introduction to Genetic Algorithms. MIT Press, 1996.

Mendes R., Kennedy J., Neves., J., The fully informed particle swarm: simpler, maybe better, Evolutionary Computation, IEEE Transactions on 8 (2004) 204–210.

Mousa A.A., El-Shorbagy M.A., Abd-El-Wahed W.F., Local search based hybrid particle swarm optimization algorithm for multiobjective optimization, Swarm and Evolutionary Computation 3 (2012) 1-14.

Najafzadeh M., Barani G.A., Azamathulla H.M., Prediction of pipeline scour depth in clear-water and live-bed conditions using group method of data handling, Neural Computing and Applications 24 (2014) 629-635.

Naka S., Genji T., Yura T., Fukuyama Y., Hybrid particle swarm optimization based distribution state estimation using constriction factor approach. In Proceeding International Conference SCIS & ISIS 2, pp. 1083–1088, 2002.

Nalluri C., Ab Ghani A., Design Options for Self-Cleansing Storm Sewers, Water Science & Technology 33 (1996) 215-220.

Ocak H., A medical decision support system based on support vector machines and the genetic algorithm for the evaluation of fetal well-being, Journal Medical Systems 37 (2013) 1-9.

Ota J.J., Nalluri C., Graded sediment transport at limit deposition in clean pipe channel, 28th International Association Hydro-Environmental Engineering Research, Graz, Austria, (1999).

Ota J.J., Nalluri C., Urban storm sewer design: Approach in consideration of sediments, Journal of Hydraulic Engineering 129 (2003) 291-297.

Parsaie A., Haghiabi A.H., Assessment of some famous empirical equation and artificial intelligent model (MLP, ANFIS) to predicting the side weir discharge coefficient, Journal of Applied Research in Water and Wastewater 2 (2014) 75-79.

Pedersen M.E.H., Chipperfield A.J., Simplifying particle swarm optimization, Applied Soft Computing 10 (2010) 618-628.

Premkumar K., Manikandan B.V., Fuzzy PID supervised online ANFIS based speed controller for brushless dc motor, Neurocomputing 157 (2015) 76-90.

Ratnaweera A., Halgamuge S., Watson H., Self-organizing hierarchical particle swarm optimizer with time-varying acceleration coefficients, Evolutionary Computation IEEE Transactions on 8 (2004) 240–255.

Safarzadeh A., Mohajeri, S.H. On the fine sediment deposition patterns in a gravel bed open-channel flow, Journal of Applied Research in Water and Wastewater 3 (2016) 188-192.

Sahay R.R., Srivastava A., Predicting monsoon floods in rivers embedding wavelet transform, genetic algorithm and neural network, Water Resources Management 28 (2014) 301-317.

Shi Y., Eberhart R.C., Parameter selection in particle swarm optimization. In: V.W. Porto, N. Saravanan, D.Waagen and A.E. Eiben, eds. Evolutionary programming VII, Lecture Notes in Computer Science, 1447. Berlin, Heidelberg: Springer, pp. 611–616, 1998.

Shi Y., Eberhart R.C., Empirical study of particle swarm optimization. In: Proceedings of IEEE, International Congress Evolutionary Computation. Piscataway, NJ, USA: IEEE Xplore Digital Library, pp. 1945–1950, 1999.

Singh R., Kainthola A., Singh T.N., Estimation of elastic constant of rock using an ANFIS approach, Applied Soft Computing 12 (2012) 40–45.

Sobhani J., Najimi M., Pourkhorshidi A.R., Parhizkar T., Prediction of the compressive strength of no-slump concrete: a comparative study of regression, neural network and ANFIS models, Construction and Building Materials 24 (2010) 709-718.

Storn R., Price K., Differential evolution – A simple and efficient heuristic for global optimization over continuous spaces, Journal of global optimization 11 (1997) 341–359.

Takagi T., Sugeno M., Fuzzy identification of systems and its applications to modeling and control. Systems, Man and Cybernetics, IEEE Transactions on 15 (1985) 116–132.

Tayfur G., Karimi Y., Singh V.P., Principle component analysis in conjuction with data driven methods for sediment load prediction, Water Resources Management 27 (2013) 2541-2554.

Valizadeh N., El-Shafie A., Forecasting the level of reservoirs using multiple input fuzzification in ANFIS, Water Resources Management 27 (2013) 3319-3331.

Vongvisessomjai N., Tingsanchali T., Babel M.S., Non-deposition design criteria for sewers with part-full flow, Urban Water Journal 7 (2010) 61-77.

Yoshida H., Kawata K., Fukuyama Y., Takayama S., Naknishi Y., A particle swarm optimization for reactive power and voltage control considering voltage security assessment. Power Systems, IEEE Transactions on 15 (2000) 1232–1239.

Yu W., Li X., Fuzzy identification using fuzzy neural networks with stable learning algorithms, Fuzzy Systems, IEEE Transactions on 12 (2004) 411–420.

Zaji A.H., Bonakdari H., Application of artificial neural network and genetic programming models for estimating the longitudinal velocity field in open channel junctions, Flow Measurement and Instrumentation 41 (2015) 81-89.

Zanaganeh M., Mousavi S.J., Shahidi A.F.E., A hybrid genetic algorithm–adaptive network-based fuzzy inference system in prediction of wave parameters, Engineering Applications of Artificial Intelligence 22 (2009) 1194-1202.

Calibration of sluice gate in free and submerged flow using the simulated annealing and ant colony algorithms

Majid Heydari[1,*], Shima Abolfathi[1], Saeid Shabanlou[2]

[1]Department of Water Science and Engineering, Faculty of Agriculture, Bu-Ali Sina Univ., Hamadan, Iran
[2]Department of Water Engineering, Kermanshah Branch, Islamic Azad University, Kermanshah, Iran.

ARTICLE INFO	ABSTRACT
Keywords: Sluice gate Free and Submerged flow Calibration Optimization	There are found numerous methods to measure flow in open channels. The simulation of water flow in channel requires mathematic calibration of the structures in channel so that the water level and the discharge become compatible with demand. Sluice gate is one of the most important structure which can perform in free and submerged flow. In this research, there were experiments on a sluice gate mounted in lab flume of 12.5 m, 0.6 and 0.65 length, width and height, respectively, in the slope of 0.0002. Some equations of measuring the discharge from the sluice gate extracted from Energy equations and Momentum were calibrated using two metaheuristic algorithms of simulated annealing and ant colony. After the sensitivity analysis of algorithm was done, the optimal coefficients of discharge obtained for the Conventional equation of discharge in free and submerged flow was obtained 0.686, and 0.881. Also, in calibration of Energy-Momentum method for submerged flow, the optimal contraction coefficient was 0.533. finally, the methods were assessed and compared for which the statistical indexes show the favorability of results.

1. Introduction

Controlling the water volume released from dams and water level in feeding channels requires the installation of suitable structures on dams and channels so that water level and the discharge are matched with the demand. There are different methods and devices to measure the flow in open channels (Clemmens. 2002) the simulation of water flow in channel needs mathematic description and calibration of structures in channel amongst which the gates are the most important applied on free overflow or inside the water catchment and irrigation channels. (Abbaspour et al. 2001). To achieve the optimal use of these structures, and regarding the recent advances in automatic regulation of flow for spillways and conviyence networks, it is necessary to calculate the discharge coefficients in gates accurately. The operations to get the coefficients of flow equations for the structure through measurement is called calibration. Thus, gate calibration is performed to measure the flow accurately to increase the efficiency in the distribution network, and accuracy in water delivery. Accordingly, researchers have sought to find better methods to calibrate the equations which can be presented in high speed computer programs.

In irrigation networks, the gates are widely used to transmit water or control structures which can act as free or submerged in downstream. (Bijankhan et al. 2012). Sluice gate is widely used in the irrigation channels in many countries, either as checks in the canals or as flow controllers at channel turnouts (Mahmudian Shooshtari 2008). Fig. 1 shows the flow through the gate in free and submerged flow.

In this Fig. Y_1 is the upstream water depth, Y_2 is the minimum water depth after the gate in free flow state (vena contracta) [L], Y is water depth immediately after the gate in submerged flow state[L], Y_3 is water depth at the downstream gate and after turbulence[L], W is opening height[L] and Q is flow discharge [L3/T].

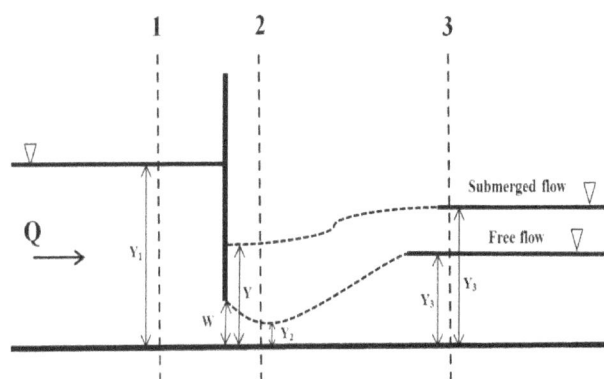

Fig. 1. The schematic figure of the flow through sluice gate in free and submerged flow.

Henry (1950) did an extensive study on sluice gates and presented graphic solutions to get the Discharge coefficient in free and submerged flow. Rajaratnam snd Subramanya (1967) applied the Energy and Momentum equation to find a rating curve equation in free and submerged flow and presented a general equation for discharge through the sluice gates in free and submerged flow.

Clemmens et al. (2003) introduced a method for calibration using Momentum equation for gate downstream and Energy equation for gate upstream, called Energy-Momentum. Lozano et al (2009) investigated some calibration methods of sluice gate using field data and found that Energy-Momentum method will have acceptable accuracy with compactness coefficient. Castro-Orgaz et al. (2010) presented a new method using Energy-Momentum principle and

*Corresponding author E-mail: mheydari_ir@yahoo.com

combining Energy velocity coefficients and Momentum to calibrate the submerged sluice gate and provided an acceptable accuracy for calibration method with field data. Bijankhan and Kouchakzadeh (2010) introduced a equation for transition flow from free to submerged flow derived from simultaneous solution of these two flows and found that the equations of free and submerged flows give the same results for transition flow. They used Ferro method (2001) to find

the curve of flow conditions. Finally, the curve obtained showed remarkable deviation compared with that of the previous curve by Rajaratnam snd Subramanya (1967), Lin et al (2002) and Swamee (1992). Bijankhan et al (2012) introduced a dimensionless Equation for calibration which could be used in the total range of flows. Table (1) shows the equations provided by researchers to obtain Discharge coefficient.

Table 1. Some equations for Discharge coefficient of flow through sluice gate.

Equation	Flow state	Presenter
$C_d = C_c \sqrt{\dfrac{1}{1 + C_c\left(\dfrac{W}{Y_1}\right)}}$	Free	(Rajaratnam 1976)
$C_d = 0.6468 - 0.1641 \sqrt{\dfrac{W}{Y_1}}$	Free	(Garbrecht 1977)
$C_d = 0.62 - 0.15 \sqrt{\dfrac{W}{Y_1}}$	Free	Noutsopoulos and Fanariotis (Spheerli and Hager 1999)
$C_d = 0.6\exp\left(-0.3\dfrac{W}{Y_1}\right)$	Free	Nago(Spheerli and Hager 1999)
$C_d = 0.615\left(1 + 0.3\dfrac{W}{Y_1}\right)^{-1}$	Free	Cozzo(Spheerli and Hager 1999)
$C_d = 0.611\left(\dfrac{Y_1 - W}{Y2 + 15W}\right)^{0.072}$	Free	(Swamee 1992)
$C_d = 0.611\left(\dfrac{Y_1 - W}{Y2 + 15W}\right)^{0.072}\left(0.32\left(\dfrac{0.81Y_3\left(Y_3/W\right)^{0.72} - Y_1}{Y_1 - Y_3}\right) + 1\right)^{-1}$	Submerged	(Swamee 1992)

In this Table C_d is discharge coefficient C_c is the contraction coefficient.

Therefore, selecting a discharge coefficient for the gate must be carried out with specific calibration of gate and supported ideally by validation (Lozano et al. 2009). Hence, to find the optimum coefficient for calibration of discharge equation, there are different methods, each has a specific error. The complexity of some equations and being time-consuming for field methods make it necessary to use intelligent methods as a substituent.

In the recent years, metaheuristic algorithms have been used in complex problems and optimization issue. The developed methods are routed in nature to solve optimization problem. The Simulated Annealing optimization method (SA) is a numerical optimization method with smart random structure which has been simulated based on annealing physical process. Some methods are formed from the study on social insects' behavior such as ant colony method (Jalali 2007). This research aims to review the Energy and Energy-Momentum methods in two algorithms of SA and continuous Ant Colony (ACO$_R$) along with intelligent methods in calibration of sluice gate in free and submerged flow.

2. Methods

In this research, the Energy equation of Rajaratnam snd Subramanya (1967) is used for free and submerged flow through sluice gate as in equations 1 and 2 obtained from the application of Energy equation for downstream and upstream flow and their discharge coefficients were calibrated (Fig. 1)

$$Q = C_d bW\sqrt{2gY_1} \qquad \text{Free flow} \qquad (1)$$

$$Q = C_d bW\sqrt{2g(\Delta Y)} \qquad \text{Submerged flow} \qquad (2)$$

In these equations b is gate width [L], g is gravity acceleration [L/T^2], ΔY is difference between upstream and downstream depth. The value of ΔY was obtained from Eq. (3) (the difference between upstream and downstream depth immediately after the gate (submerged depth) presented by Clemmens et al (2003) and Rajaratnam snd Subramanya (1967). Based on the equation, the

value of discharge coefficient of submerged flow was calibrated using both Eq. (3) and (4) and the accuracy of each was investigated.

$$\Delta Y = Y_1 - Y_3 \qquad (3)$$

$$\Delta Y = Y_1 - Y \qquad (4)$$

Another method to calibrate the gate in submerged flow is the combined method of Energy-Momentum (Clemmens et al. 2003) in which Energy equation between section 1 and 2 is used due to ignoring Energy loss and Momentum equation is used in the range from section 2 to 3 due to hydraulic jump and energy dissipation. The mentioned method was used for calibration as the following. First, the gates of observational discharge Y_1 and Y_2 are inserted into Momentum Eq. (5) and the submersion depth is obtained.

$$\frac{q^2}{gY_2} + \frac{Y^2}{2} = \frac{q^2}{gY_3} + \frac{Y_3^2}{2} \qquad (5)$$

In this equation q is the discharge value in the width unit of gate and the value of Y_2, is calculated from Eq. (6).

$$Y_2 = C_C.W \qquad (6)$$

Inserting the values of the above equations in Eq. (7), we get q value.

$$Y_1 + \frac{q^2}{2gY_1^2} = Y + \frac{q^2}{2gY_2^2} \qquad (7)$$

In these equations, Cc requiring calibration. One of the important bases of optimization and estimation of model parameters is to choose objective function. As gate discharge calibration aims to find experimental coefficients to calculate the discharge, these coefficients must be estimated in a way that there is a negligible difference between calculated discharge from theoretical equation and the observed discharge in practice. Therefore, using square sum of observed and calculated discharges as objective function for minimization can lead us to good solutions (Eq. 8).

$$OF = \sum_{i=1}^{n} \left(Q_{c,i} - Q_{o,i}\right)^2 \qquad (8)$$

$Q_{o,i}$ is the values of observed discharge[L^3/T], $Q_{c,i}$ is the calculated discharge value obtained from Energy or Energy–Momentum methods, i is the counter n is the number of observations.

The decision variables for minimization of this function include flow coefficient for calibration of gate at free flow and submerged flow using Energy method and contraction coefficient for calibration using Energy–Momentum method. Optimization was carried out by simulate annealing and Ant colony optimization which were under sensitivity analysis in Matlab R2009a.

SA algorithm is a Metaheuristic algorithm which uses simulation of simulate annealing to calculate optimal value. The main idea of SA method was introduced by Metropolis et al (1953) without optimization content. Then, this idea was developed by Kirk patrik et al (1983) and Cerny (1985) independently for optimization and SA method became introduced.

Artificial ants introduced in 1991 by Colorni et al, artificial ants search a wide area with simulation of real ants foraging for food. They showed that the b q is the discharge value in the width unit of gate and Cc is the contraction coefficient behavior of food search in real ants can be adapted on optimization problems with small charges on which Ant Colony algorithm was developed. Dorigo (1992) developed the first algorithm to show the ants' behavior for food. With the assumption of continuous variable space, the algorithm is able to move on R space of real numbers. In ACO$_R$ algorithm, the continuity of space in decision variables is carried out in a probability density function. (Socha and dorigo 2008).

In the present research, a rectangular flume of 12.5 m, 0.6 m, and 0.62m in length, width and height with a slope of 0.0002 equipped with a sluice gate and calibrated butterfly valve to measure discharge. The measurements of water level with point gauge of 0.1 mm accuracy were done. The data required for free flow and submerged flow were measured at 35 and 41 test series. These data include observed discharge, gate opening and upstream depth for all flow states in addition to downstream depths for submerged flow.

After the competition of tests and measurements, 70% were selected for calibration (optimization process) and 30% for validation of solutions randomly. In SA method, the quality of solutions is sensitive to the existing parameters and it is important to determine the parameters which produce suitable solutions. (Zegordi et al. 1995). The results of SA applications show that the computation time and the efficiency of this model depend on the setting of its parameters. (Kouvelis and Chiang 1992). Therefore, in this research, the algorithm of related problems underwent sensitivity analysis. Different parameters used to do sensitivity analysis are provided in Table 2.

The ACO$_R$ algorithm is sensitive to the parameter change and to get the best solution, the algorithm must undergo sensitivity analysis using parameter change. The parameters influential which were changed in sensitivity analysis are in Table 3. The intensification factor (q), the positive number inversely related to the importance of good solutions (the inverse of pheromone concentration in that the less the concentration, the greater importance) and parameter ζ which is a positive coefficient influential on probe. In fact, this coefficient acts as the pheromone evaporation ratio. The less the coefficient, the faster the convergence. And it increases the probability of being trapped in local optimum. The great value of this coefficient makes the memory full and decreases the accuracy.

To do the calculations, the algorithms and the equations were written in a computer program along with parameters and objective function. All algorithms underwent the sensitivity analysis. As some parameters were dependent on each other, the sensitivity analysis and parameter change were done simultaneously. Based on sensitivity analysis, SA algorithm had a slower performance to convergence of solution meaning that there is need to more iterations to get the optimum solution, while continuous ant colony algorithm converges to the optimal solution in the initial iterations and algorithm run with great number of iterations increases the time without any effect in quality of solution. Therefore, ACO$_R$ is allocated less time than SA algorithm due to faster convergence. In this research, the value differences of objective function obtained compared to parameter change in ACO$_R$ is less than that in SA algorithm and objective function tolerance in sensitivity analysis was small showing a less sensitivity to parameter change. As seen, the minimum value of objective function and the coefficient of both algorithms are the same in all objective functions. The most optimum solution from SA algorithm and ACO$_R$ for objective function with the most optimum setting case are presented in tables 4 and 5, respectively.

3. Results and discussion

Table 2. SA algorithm parameters and the tested values

Temperature update function	Annealing Function	Initial temperature	Max. iteration	Tolerance function	Reanneal interval
Exponential		100	500	0.000001	100
	Fast	50	400	0.00001	50
Linear		20	300	0.0001	40
	Baltzmann	10	200	0.001	30
Logarithmic		5	100	0.01	20
				0.1	10

Table 3. ACO algorithm parameters and tested values.

Initial population	New population	Max iteration	q	ζ
2	2	5	0.2	0.01
10	5	10	2	0.1
20	10	20	5	1
50		50		

Table 4. The values of objective function and coefficient of SA algorithm for the most optimum algorithm parameters.

Objective function	The best coefficient	The lowest objective function*10³	Temperature update function	Annealing Function	Initial temperature	Max. iteration	Tolerance function	Reanneal interval
Conventional discharge Equation for sluice gate in Free flow	0.686	1.6367	Exponential	Fast	0.000001	300	100	40
Conventional discharge Equation for sluice gate In Submerged flow using Y	0.881	2.4315	Exponential	Fast	0.000001	300	100	50
Conventional discharge Equation for sluice gate In Submerged flow using Y₃	1.179	3.1314	Exponential	Fast	0.000001	200	100	40
Energy- Momentum method for Sluice gate in submerged flow	0.553	0.6915	Exponential	Fast	0.000001	300	100	40

Table 5. The values of objective function and coefficient of ACO_R algorithm for the most optimum algorithm parameters.

Objective function	The best coefficient	The lowest objective function*10^3	Initial population	New population	Max iteration	q	ζ
Conventional discharge equation for sluice gate in free flow	0.686	1.6367	10	5	5	0.2	1
Conventional discharge equation for sluice gate in submerged flow using Y	0.881	2.4315	10	10	5	0.2	1
Conventional discharge equation for sluice gate in submerged flow using Y_3	1.179	3.1314	2	10	5	0.2	1
Energy- Momentum method for sluice gate in submerged flow	0.553	0.6915	2	10	5	0.2	1

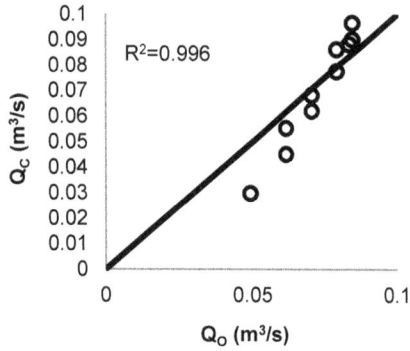

a. Calibration data

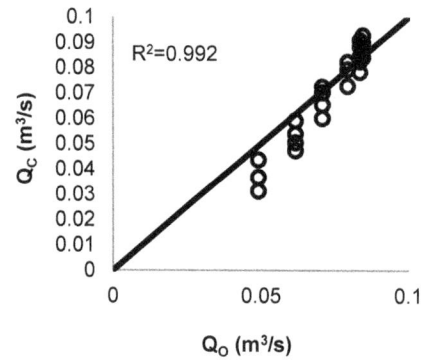

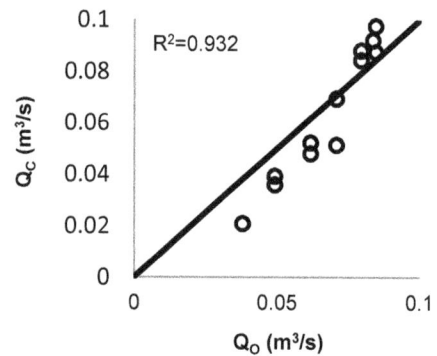

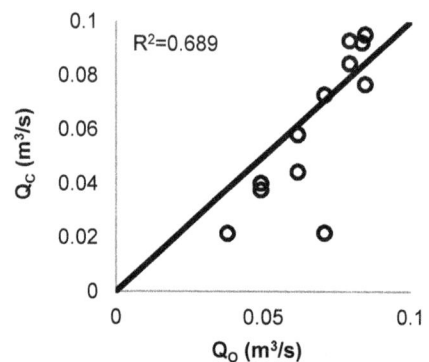

b. Validation data

Fig. 2. Correlation diagram of Conventional discharge equation for sluice gate in free flow.

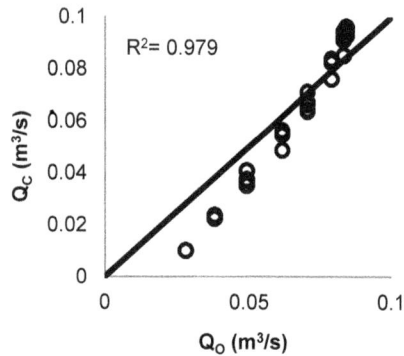

a. Calibration data

b. Validation data

Fig. 3. Correlation diagram of Conventional discharge equation for sluice gate in submerged flow using Y

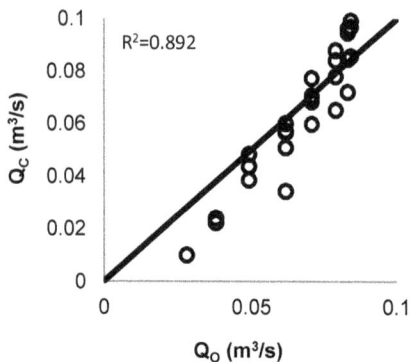

a. Calibration data

b. Validation data

Fig. 4. Correlation diagram of Conventional discharge equation for sluice gate in submerged flow using Y3.

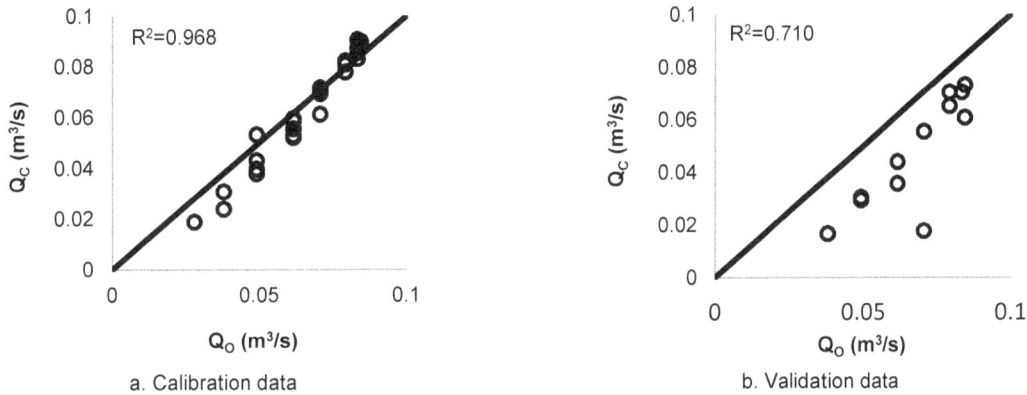

a. Calibration data   b. Validation data

Fig. 5. Correlation diagram of Energy-Momentum method for sluice gate in submerged flow.

Table 6. The equations of calculating statistical indexes to assess results.

Index	equation of calculating index
$R^2 = \dfrac{\left[\sum_{i=1}^{n}\left(Q_{o,i}-\overline{Q_o}\right)\left(Q_{c,i}-\overline{Q_c}\right)\right]^2}{\sum_{i=1}^{n}\left(Q_{o,i}-\overline{Q_o}\right)^2 \sum_{i=1}^{n}\left(Q_{c,i}-\overline{Q_c}\right)^2}$	Coefficient of Determination
$MBE = \dfrac{\sum_{i=1}^{n}\left(Q_{c,i}-Q_{o,i}\right)}{n}$	Mean Bias Error
$RMSE = \sqrt{\dfrac{\sum_{i=1}^{n}\left(Q_{c,i}-Q_{o,i}\right)^2}{n}}$	Root Mean Square Error
$E_r = \left|\dfrac{Q_{o,i}-Q_{c,i}}{Q_{o,i}}\right|*100$	Relative Error

In Figs (2) to (5), the correlation graphs of calibration data and validation of each objective function are presented separately. The value of the observed discharge is the length of points and the value of calculated discharge is the width of points based on the obtained coefficients. The index R^2 is a numerical criterion to make an assessment the closeness of which to one shows the favorability of solution. In these diagrams, it is seen that for validation data, discharge is underestimated. Since calculated and measured data are compared to assess the quality of solution from algorithm. In this research, to assess the results, some statistical indexes are used as in Table 6.

As the results of both algorithms are the same, the values of statistical indexes are simultaneously calculated as seen in Tables 7 and 8. Although all the indexes are located in an acceptable range, using calibrated coefficient for calibration data gives better results as the values of R^2 confirm the issue. Regarding the diagrams and indexes in the conventional discharge equation in submerged flow, using the difference between Y and Y_1 provides more accuracy than that of Y_1 and Y_3. In using the Energy-Momentum equation, it is seen that this method gives better results for calibration data. $\overline{E_r}$ is the mean of relative error percentage.

Table 7. The values of statistical indexes for calibration data.

Objective function	RMSE	MBE	$\overline{E_r}$
Conventional discharge equation for sluice gate in Free flow	0.0081	-0.0019	9.46
Conventional discharge equation for sluice gate In submerged flow using Y	0.0093	-0.0032	15.04
Conventional discharge equation for sluice gate In submerged flow using Y_3	0.0106	-0.0032	31.61
Energy- Momentum method for sluice gate in submerged flow	0.0063	-0.0026	9.49

Table 8. The values of statistical indexes for validation data.

Objective function	RMSE	MBE	$\overline{E_r}$
Conventional discharge equation for sluice gate in Free flow	0.0101	-0.0026	13.05
Conventional discharge equation for sluice gate In submerged flow using Y	0.0115	-0.0042	17.19
Conventional discharge equation for sluice gate In submerged flow using Y_3	0.0175	-0.0063	20.67
Energy- Momentum method for sluice gate in submerged flow	0.0232	-0.0204	32.58

In Figs. 6-9, the diagrams of relative error percentage (E_r) versus relative gate opening (W/Y_1) and that against calculated discharge have been shown for different methods of research, using the whole data. In these diagrams, to measure E_r, the equation in Table 6 has been used without absolute value. As seen in diagrams, the calibration of free flow with relative opening increase (W/Y_1), the relative error percentage decreases moving towards overestimation. While for most data, underestimation in discharge calculation has been observed. In the range medium discharges, there is seen to be less error. In addition, increasing discharge makes the relative error percentage proceed from underestimation for small discharges to

overestimation for larger discharges. The submerged flow has the same procedure, except for the lack of specific correlation between relative opening and relative error percentage. In the methods used for submerged flow, most of the points have underestimation and the overestimation points in less relative opening are seen more. Point distribution and points with large relative error percentage on the relative opening diagram versus relative error percentage are greater than that of free flow, which is more visible for Energy-Momentum method diagram and the relative error percentage to observed discharge.

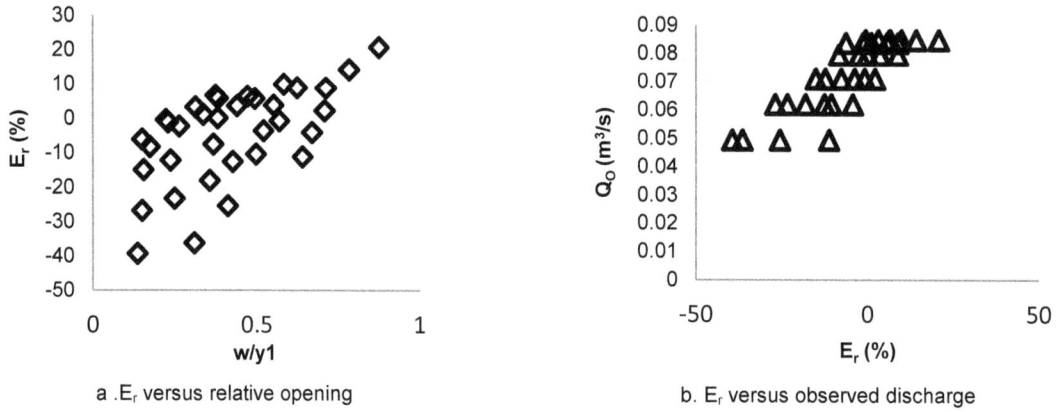

a. E_r versus relative opening   b. E_r versus observed discharge

Fig. 6. Conventional discharge equation for sluice gate in free flow.

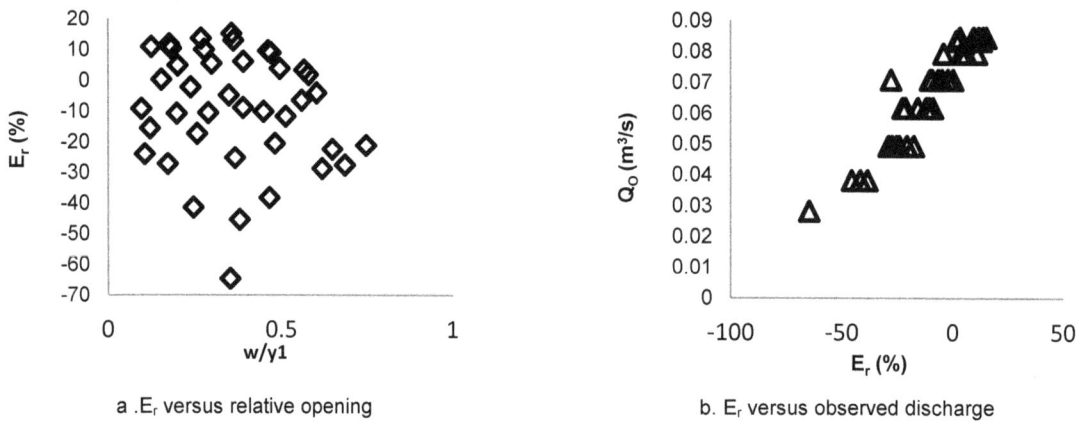

a. E_r versus relative opening   b. E_r versus observed discharge

Fig. 7. Conventional discharge equation for sluice gate in submerged flow using Y.

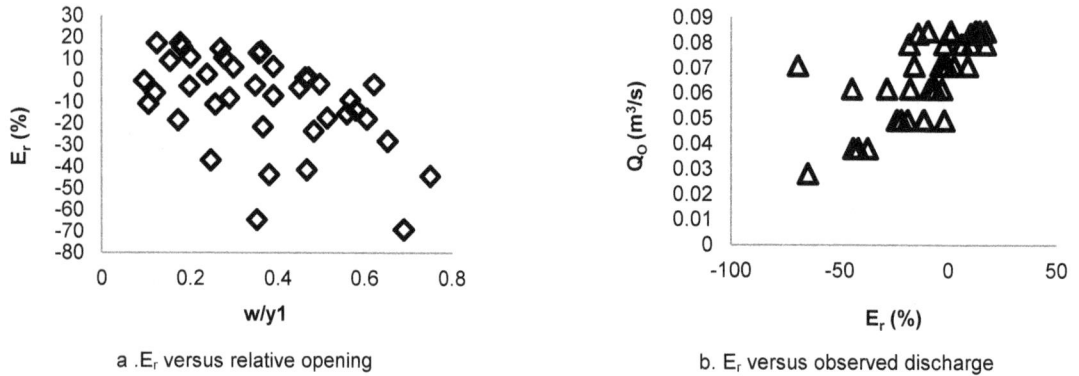

a. E_r versus relative opening   b. E_r versus observed discharge

Fig. 8. Conventional discharge equation for sluice gate in submerged flow using Y3.

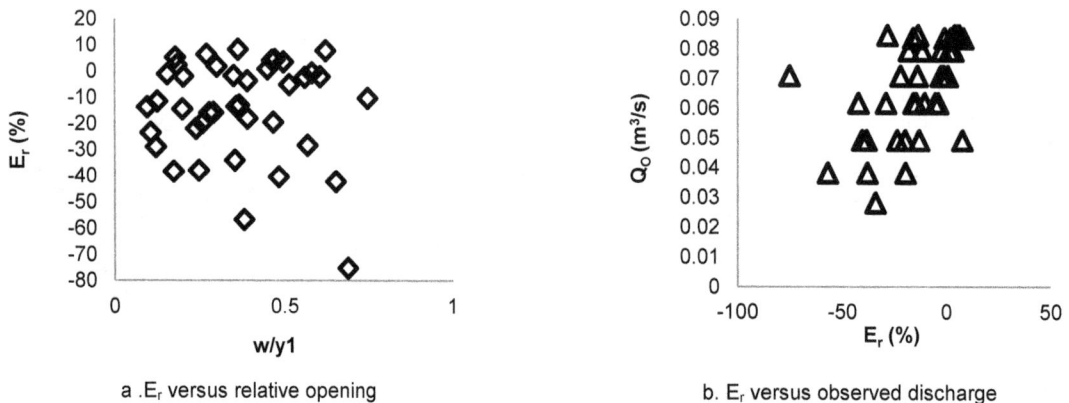

a. E_r versus relative opening   b. E_r versus observed discharge

Fig. 9. Energy- Momentum method for sluice gate in submerged flow.

To compare the results of this research with those of previous researches, the data used in this research were placed in Table 1 and the coefficients were used to calculate discharge. Finally, two parameters R^2 and E_r were calculated for these equations and the methods presented in this research using all validation and calibration data. The results of these parameters are presented in Table 9. Regarding Table 9 about the free flow, the equation presented by Swamee has provided weak results for the present data while Noutsopoulos and Fanariotis's equation and Nago and Cozzo's have good R^2 index. Considering both R^2 index and $\overline{E_r}$ the present method

has provided good results. For submerged flow, Swamee's equation has weak results and the conventional discharge equation for Y at downstream depth with high values of R^2 and low values of mean relative error percentage is used as the best method to calculate Discharge and calibration of sluice gate in submerged flow. Although the $\overline{E_r}$ is less in Energy-Momentum method showing the accuracy, the weak statistical indexes for validation data and low R^2 decrease the suitability of the method.

Table 9. Comparison of results of the previous equations with those of the present research for use in all data.

Presenter	Flow state	R^2	% Er_{mean}
(Rajaratnam. 1976)	free	0.81	22.19
(Garbrecht. 1977)	free	0.81	91.5
Noutsopoulos and Fanariotis(Spheerli and Hager 1999)	free	0.98	27.41
Nago(Spheerli and Hager 1999)	free	0.98	26.95
Cozzo(Spheerli and Hager 1999)	free	0.98	24.43
(Swamee. 1992)	free	0.11	82.63
Conventional discharge equation (This research)	free	0.93	10.49
(Swamee. 1992)	submerged	0.37	92.29
Conventional discharge equation using Y(This research)	submerged	0.96	15.68
Conventional discharge equation using Y_3(This research)	submerged	0.82	16.53
Energy- Momentum method	submerged	0.76	14.42

4. Conclusions

In the present research, there were attentions to calibration of sluice gate in free and submerged flow using two metaheuristic algorithms of SA and continuous and ant colony. The target functions were defined using equations that researchers introduced before and the optimization was carried out with MATLAB R2009. The algorithms underwent sensitivity analysis to get most optimum solution. Finally, optimum coefficients for theoretical equations were obtained.

After completion of algorithms and sensitivity analysis, the optimum discharge coefficient of for conventional discharge equation in sluice gate for free flow is 0.686, in submerged flow with upstream depth difference (Y_1) and the depth immediately after gate (Y), is

0.881 and for upstream depth difference (Y_1) and downstream depth after(Y_3) perturbations was 1.179. Also, the contraction coefficient in Energy-Momentum method for submerged sluice gate was 0.533.

Regarding the analysis and calculation of statistical indexes, it can be seen that using Metaheuristic algorithms of SA and ACO_R can be suitable to calibrate the equations of sluice gate. Also, in calibration of submerged flow of sluice gate, using conventional discharge equation the better results can be obtained with upstream depth difference (Y_1) and depth after gate (Y). In addition, it was concluded that Energy-Momentum method can provide good results for data with it is calibrated, while the results of other data are weak yet acceptable.

References

Abbaspour K.C., Schulin R., van-Genuchten M.T., Estimating unsaturated soil hydraulic parameters using ant colony optimization, Advances in Water Resources 24 (2001) 827-841.

Bijankhan M., Kouchakzadeh S., Discussion of 'benchmark of discharge calibration methods for submerged sluice gates' by Carlos Sepu´ Iveda, Manuel Go´ mez, and Jose´ Rodellar, Journal of Irrigation and Drainage Engineering 137 (2010) 56-57.

Bijankhan M., Darvishi E., Kouchakzadeh S., Discussion of 'Energy and Momentum velocity coefficients for calibrating submerged sluice gates in irrigation canals' by Oscar Castro-Orgaz, David Lozano, and Luciano Mateos, Journal of Irrigation and Drainage Engineering, 138 (2012) 852-854.

Bijankhan M., Ferro V., Kouchakzadeh S., New stage–discharge relationships for free and submerged sluice gates, Flow Measurement and Instrumentation 28 (2012) 50-56.

Castro-Orgaz O., Lozano D., Mateos L., Energy and Momentum velocity coefficients for calibrating submerged sluice gates in irrigation canals, Journal of Irrigation and Drainage Engineering 136 (2010) 610-616.

Cerny V., A thermodynamical approach to traveling salesman problem: Aefficient simulation algorithm, Journal of Optimization Theory and Application 45 (1985) 41-51.

Clemmens A.J., New calibration method for submerged radial gates, Paper presented at the USCID/EWRI Conference on Energy, Climate, Environment and Water, San Luis Obispo, USA, (2002) 399-408.

Clemmens A.J., Strelkoff T.S., Replogle J.A., Calibration of submerged radial gates, Journal of Hydraulic Engineering129 (2003) 680-687.

Colorni A., Dorgio M., Maniezzo V., Distributed optimization by ant-colonies. 1st Europian Conference on Artificial life (ECAL'91), Paris, France, Elsevier Publishing (1991) 134–142.

Dorigo M., Optimization, learning and natural algorithm. PhD thesis, Politecnico di Milano, Italy (1992).

Ferro V., Closure to 'simultaneous flow over and under a gate' by V. Ferro., Journal of Irrigation and Drainage Engineering 127 (2001) 326-328.

Garbrecht G., Discussion of discharge computation at river control structures, Journal of Hydraulic Division 103 (1977) 1481-1484.

Henry H., Discussion: diffusion of submerged jet, Transaction Proceeding ASCE 115 (1950) 687-697.

Jalali M.R., Afshar A., Marino, M.A. Multi-Colony ant algorithm for continuous multi-reservoir operation optimization problem, Water Resources Management 21 (2007) 1429-1447.

Kirk patrik S., Gellat, C.D.Jr., Vecchi M.P., Optimization by Simulated Anneling, Science 220 (1983) 671-680.

Kouvelis P., Chiang W., A simulated annealing procedure for single row lay out problems inflexible manufacturing systems, International Journal of production Research 30 (1992) 717-732.

Lin C., Yen J., Tsai C., Influence of sluice gate contraction coefficient on distinguishing condition, Journal of Irrigation and Drainage Engineering 128 (2002) 249-252.

Lozano D., Mateos L., Merkley G.P., Clemmens A.J., Field Calibration of Submerged Sluice Gates in Irrigation Canals. Joutnal of Irrigation and Drainage Engineering 6 (2009) 763-772.

Mahmudian Shooshtari, M., flow Principles in open channels.Volume 2, Shahid Chamran University Press, Iran.(In Persian), (2008).

Metropolis N., Rosenbluth A., Teller A., Teller E., Equation of state calculations by computing machines, Journal of Chemistry Physics 21 (1953) 1087-1092.

Rajaratnam N., Subramanya K., Flow equation for the sluice gate, Journal of Irrigation and Drainage Engineering 93 (1967) 57-77.

Socha K., Dorigo M., Ant Colony optimization for continuous domains, European Journal of Operational Research 185 (2008) 1155-1173.

Spheerli J., Hager W.H., Discussion of irrotationa flow and real fluid effects under planer gates, by J. S. montes, Journal of Hydraulic Engineering 125 (1999) 208-210.

Swamee P., Sluice-gate discharge equations. Journal of Irrigation and Drainage Engineering, 118 (1992) 56-60.

Zegordi S.H., Itoh K., Enkawa T., Aknowledgeable simulated annealing scheme for the early/tardy flow shop scheduling problem, International Journal of Production Research 33 (1995) 1449-1466.

Permissions

All chapters in this book were first published in JARWW, by Razi University; hereby published with permission under the Creative Commons Attribution License or equivalent. Every chapter published in this book has been scrutinized by our experts. Their significance has been extensively debated. The topics covered herein carry significant findings which will fuel the growth of the discipline. They may even be implemented as practical applications or may be referred to as a beginning point for another development.

The contributors of this book come from diverse backgrounds, making this book a truly international effort. This book will bring forth new frontiers with its revolutionizing research information and detailed analysis of the nascent developments around the world.

We would like to thank all the contributing authors for lending their expertise to make the book truly unique. They have played a crucial role in the development of this book. Without their invaluable contributions this book wouldn't have been possible. They have made vital efforts to compile up to date information on the varied aspects of this subject to make this book a valuable addition to the collection of many professionals and students.

This book was conceptualized with the vision of imparting up-to-date information and advanced data in this field. To ensure the same, a matchless editorial board was set up. Every individual on the board went through rigorous rounds of assessment to prove their worth. After which they invested a large part of their time researching and compiling the most relevant data for our readers.

The editorial board has been involved in producing this book since its inception. They have spent rigorous hours researching and exploring the diverse topics which have resulted in the successful publishing of this book. They have passed on their knowledge of decades through this book. To expedite this challenging task, the publisher supported the team at every step. A small team of assistant editors was also appointed to further simplify the editing procedure and attain best results for the readers.

Apart from the editorial board, the designing team has also invested a significant amount of their time in understanding the subject and creating the most relevant covers. They scrutinized every image to scout for the most suitable representation of the subject and create an appropriate cover for the book.

The publishing team has been an ardent support to the editorial, designing and production team. Their endless efforts to recruit the best for this project, has resulted in the accomplishment of this book. They are a veteran in the field of academics and their pool of knowledge is as vast as their experience in printing. Their expertise and guidance has proved useful at every step. Their uncompromising quality standards have made this book an exceptional effort. Their encouragement from time to time has been an inspiration for everyone.

The publisher and the editorial board hope that this book will prove to be a valuable piece of knowledge for researchers, students, practitioners and scholars across the globe.

List of Contributors

Hadi Ghaebi
Department of Mechanical Engineering, Faculty of Engineering, University of Mohaghegh Ardabili

Mehdi Bahadorinejad and Mohammad Hassan Saidi
Center of Excellence in Energy Conversion (CEEC), School of Mechanical Engineering, Sharif University of Technology, Tehran, Iran

Abbas Parsaie and Amir Hamzeh Haghiabi
Department of water Engineering, Lorestan University, Khorramabad, Iran

Matthieu Dufresne and José Vazquez
National School for Water and Environmental Engineering of Strasbourg (ENGEES), Strasbourg cedex France
ICube (University of Strasbourg, CNRS, INSA of Strasbourg, ENGEES), Mechanics Department, Fluid Mechanics Team

Abbas Parsaie and Amir Hamzeh Haghiabi
Water Engineering Department, Lorestan University Khoram Abad, Lorestan Province, Iran

Sajad Shahabi and Masoud Reza Hessami Kermani
Department of Civil Engineering, Bahonar University, Kerman, Iran

Matthieu Dufresne and José Vazquez
National School for Water and Environmental Engineering of Strasbourg (ENGEES) – ICube (University of Strasbourg, CNRS, INSA of Strasbourg, ENGEES), Mechanics Department, Fluid Mechanics Team ENGEES, Strasbourg cedex France

Javad Ahmadi, Davood Kahforoushan, Esmaeil Fatehifar, Khaled Zoroufchi Benis and Manouchehr Nadjafi
Environmental Engineering Research Center, Faculty of Chemical Engineering, Sahand University of Technology, Iran

Amin Hajiahmadi
Department of Civil Engineering, University of Sistan and Baluchestan, Zahedan, Iran

Mojtaba Saneie
Soil Conservation and Watershed Management Research Institute (SCWMRI), Tehran, Iran

Mehdi Azhdari Moghadam
Department of Civil Engineering, University of Sistan and Baluchestan, Zahedan, Iran

Akbar Safarzadeh
Faculty of Civil Engineering, University of Mohaghegh Ardabili, Ardabil, Iran

Seyed Hossein Mohajeri
Department of Civil Engineering, Science and Research Branch, Islamic Azad University, Tehran

Reza Karimi and Ali Akbar Akhtari
Department of Civil Engineering, Razi University, Kermanshah, Iran

Omid Seyedashraf
Department of Civil Engineering, Kermanshah University of Technology, Kermanshah, Iran

Reza Jalilzadeh Yengejeh
Department of Environmental Engineering, Science and Research Branch, Islamic Azad University, Khouzestan, Iran

Jafar Morshedi
Department of Geoghraphy, Ahvaz Branch, Islamic Azad University, Ahvaz, Iran

Razieh Yazdizadeh
Department of Environmental Sciences, Science and Research Branch, Islamic Azad University, Khouzestan, Iran

Azadeh Gholami, Hossein Bonakdari and Ali Akbar Akhtari
Department of Civil Engineering, Razi University, Kermanshah, Iran
Water and Wastewater Research Center, Razi University, Kermanshah, Iran

Ehsan Fadaei Kermani, Gholam Abbas Barani and Mohamad Javad Khanjani
Department of Civil Engineering, Shahid Bahonar University, Kerman, Iran

AbdollahTaheri Tizro and Maryam Ghashghaie
Department of Water Engineering, College of Agriculture, Bu-Ali Sina University, Iran

Pantazis Georgiou
Deptartment of Hydraulics, Soil Science & Agriculture Engineering, School of Agriculture, Faculty of Agriculture, Forestry and Natural Environment, Aristotle University of Thessaloniki, Thessaloniki, Greece

Konstantinos Voudouris
Laboratory of Engineering Geology & Hydrogeology, Department of Geology, Aristotle University, Egnatia St., Thessaloniki, Greece

Akram Fatemi
Soil Science Department, Razi University, Kermanshah, Iran

Hossein Bonakdari
Department of Civil Engineering, Razi University, Kermanshah, Iran

Gislain Lipeme-Kouyi
Department of Civil and Environmental Engineering, University of Lyon, Villeurbanne, France

Girdhari Lal Asawa
Department of Civil Engineering, GLA University, Mathura, India

Ali Mahdavi
Department of Civil Engineering, Arak University, Arak, Iran

Majid Heydari and Milad Faridnia
Department of Water Science and Engineering, Faculty of Agriculture, Bu-Ali Sina University, Hamadan, Iran

Jalal Sadeghian
Department of Civil Engineering, Faculty of Technical and Engineering, Bu-Ali Sina University, Hamadan, Iran

Saeid Shabanlou
Department of Water Engineering, Kermanshah Branch, Islamic Azad University, Kermanshah, Iran

Akbar Safarzadeh
Department of Engineering, University of Mohaghegh Ardabili, Iran

Babak Khaiatrostami
Research Department, Ardabil Regional Water Co., Iran

Hamid Najaf Zadeh and Karim Hosseinzadeh Dalir
Geography and Urban Planning Department, Marand Branch, Islamic Azad University, Marand, Iran

Mohammad Reza Pourmohammadi
Geography and Urban Planning Department, Faculty of Planning and Environment Sciences, University of Tabriz, Iran

Mehdi Nezhad Naderi and Omid Zolfaghari
Department of Civil Engineering, Tonekabon Branch, Islamic Azad University, Tonekabon, Iran

Azam Akhbari and Mohsen Vafaeifard
Department of Civil Engineering, Faculty of Engineering, University of Malaya, Kuala Lumpur, Malaysia

Amir Hossein Zaji and Hamed Azimi
Young Researches Club, Kermanshah Islamic Azad University

Hamed Azimi
Young Researchers Club, Kermanshah Branch, Islamic Azad University, Kermanshah, Iran

Majeid Heydari
Department of Science and Water Engineering, Faculty of Agriculture, Bu-Ali Sina University, Hamedan, Iran

Saeid Shabanlou
Department of Water Engineering, Kermanshah Branch, Islamic Azad University, Kermanshah, Iran

Mohammad Hossein Karimi Pashaki, Amir Khosrojerdi and Hossein Sedghi
Department of Water Science and Engineering, Islamic Azad University, Science and Research Branch, Tehran, Iran

Hossein Mirzaei-Takhtgahi and Houshang Ghamarnia
Department of Water Engineering, Campus of Agriculture and Natural Resources, Razi University, Kermanshah, Iran

Meghdad Pirsaheb
Department of Environmental Health Engineering, School of Public Health, Kermanshah University of Medical Sciences, Kermanshah, Iran

Arash Azari
Department of Water Engineering, Razi University, Iran

Milad Asadi
Department of Hydrology and Water Resources, Shahid Chamran University, Iran

Bahram Rezaei and Alireza Safarzade
Department of Civil Engineering, Bu-Ali Sina University, Hamedan, Iran

Hamidreza Babaali
Department of Civil Engineering, Khorramabad Branch, Islamic Azad University, Khorramabad, Iran

Zohreh Ramak
Department of Civil Engineering, Science and Research of Branch, Islamic Azad University, Tehran, Iran

Reza Sepahvand
Faculty of Civil Engineering, Isfahan University of Technology, Isfahan, Iran

Nafiseh Aghababaei
Department of Chemical Engineering, Tafresh University, Tafresh, Iran

Sultan Noman Qasem
Computer Science Department, College of Computer and Information Sciences, Al Imam Mohammad Ibn Saud, Islamic University (IMSIU), Riyadh, Saudi Arabia

Isa Ebtehaj
Young Researchers and Elite Club, Kermanshah Branch, Islamic Azad University, Kermanshah, Iran

Hossien Riahi Madavar
Department of water engineering, Vali-e-Asr University of Rafsanjan, Rafsanjan, Iran

Majid Heydari and Shima Abolfathi
Department of Water Science and Engineering, Faculty of Agriculture, Bu-Ali Sina Univ., Hamadan, Iran

Saeid Shabanlou
Department of Water Engineering, Kermanshah Branch, Islamic Azad University, Kermanshah, Iran

Index